SURFACE WAVES
IN PLASMAS AND SOLIDS

ICSW - 85

SURFACE WAVES
IN PLASMAS AND SOLIDS

Edited by
S. VUKOVIC
Institute of Advanced Studies
The Australian National University
Canberra

World Scientific

Published by

World Scientific Publishing Co Pte Ltd.
P. O. Box 128, Farrer Road, Singapore 9128

Library of Congress Cataloging-in-Publication data is available.

SURFACE WAVES IN PLASMAS AND SOLIDS

ISBN 9971-50-139-2

Printed in Singapore by Fong and Sons Printers Pte. Ltd.

ICSW - 85

SECOND INTERNATIONAL CONFERENCE ON SURFACE WAVES IN PLASMAS AND SOLIDS

Ohrid, Yugoslavia, September 5-11, 1985

organized by: Institute of Physics, Belgrade

sponsored by: European Physical Society
Union of Societies of Mathematicians, Physicists and Astronomers of Yugoslavia

INTERNATIONAL ADVISORY COMMITTEE:

Yu. M. Aliev	USSR	M. Moisan	Canada
A. D. Boardman	UK	Yu. A. Romanov	USSR
C. M. Ferreira	Portugal	L. Stenflo	Sweden
Yu. V. Gulyaev	USSR	A. W. Trivelpiece	USA
A. Hasegawa	USA	M. Y. Yu	FRG
A. A. Maradudin	USA	I. Zhelyazkov	Bulgaria

ORGANIZING COMMITTEE:

S. Vukovic, V. Cadez, N. Aleksic, S. Vujicic, O. Zivkovic and Z. Dragojlovic

PREFACE

This book contains invited lectures, progress reports and contributed papers presented at the second International Conference on Surface Waves in Plasmas and Solids (ICSW - 85), which was held in Ohrid, Yugoslavia, September 5 - 11, 1985. The Conference helped to stimulate discussions between solid state and plasma physicists, on the following topics: general theory, SW in fusion plasmas, astrophysical phenomena, SW produced plasmas and applications, solid state plasmas, SW on elastic and piezoelectric media, superlatticies, magnetostatic waves, thin films etc. I belive that the present proceedings will be of great interest to the specialists as well as to the colleagues entering the field in both plasma and solid state physics.

The International Advisory Committee has decided that the third ICSW will take place in Portugal in 1989 and will be organized by Professor C. M. Ferreira.

The Editor *

* Present address: Department of Engineering Physics, The Australian National University, P. O. Box 4, Canberra ACT 2601, Australia.

CONTENTS

Part II: Contributed Papers

Part I: Invited Lectures & Progress Reports

Nonlinear Electromagnetic Surface and Guided Waves: Theory

A. D. Boardman and P. Egan
Department of Physics
University of Salford
Salford, M5 4WT, UK

Abstract

The current developments in the theory of optically nonlinear surface and guided waves are given within a general framework. The nonlinearity is introduced in the form of an intensity-dependent refractive index, and the assumptions necessary to develop the theory of both TE and TM waves are carefully examined. Several surface guiding systems are considered that include p- and s-polarized waves at (1) a single interface between nonlinear dielectrics, (2) a linear waveguide bounded by nonlinear media, and (3) a nonlinear guiding layer. The paper includes a discussion of the kind of dispersion equations that arise and a full analysis of the power flow characteristics, together with some detailed methods of solution.

1. Introduction

Recently there has been a rapid development in the theory of surface and guided waves in media with a Kerr-type nonlinearity.[1-46] It is easy to see that this type of nonlinearity arises from the third-order term of an expansion[24] of the electric polarization vector in powers of the electric field. This procedure generally produces coupled intractable equations that must be reduced through judicious modeling of the dielectric tensor if substantial analytical progress is to be made.

Broadly speaking, since only the third-order nonlinearity is used, then a bulk electromagnetic wave will experience an intensity-dependent refractive index that, in turn, leads to self-focusing or self-defocusing. Although such effects occur for bulk waves, it is well known that in an unbounded nonlinear medium, propagation distances are diffraction-limited.[24] One of the attractions of using guided waves is the absence of such a limitation. Another is the ability of a guiding structure to produce the sort of high optical power densities that are necessary for some device applications and to introduce beneficial waveguide dispersion.[27]

The significance of some early work on surface waves went unappreciated until the appearance of independent investigations of (TM) p-polarized and (TE) s-polarized nonlinear surface polaritons at a single interface.[1,2,36] In the p-polarized case[1] it was shown that the wave number of the usual surface polariton depends on the intensity of the wave at the interface. This is an entirely expected feature of a nonlinear wave that has a linear limit, but the existence of nonlinear s-polarized polaritons[2,36] without such a limit was to some extent an unexpected new feature involving a power threshold. The existence of such a threshold in a nonlinear medium is, as will be discussed later, connected with the establishment of a self-focused channel[2,36,47] in the vicinity of the interface.

The use of the third-order term in the polarization vector implies that if only one frequency ω is present, then neglecting a possible 3ω process, the nonlinearity is experienced at the input frequency ω. This type of process is of great importance in integrated optics because through the intensity-dependent refractive index a power-dependent guiding wave vector occurs. This feature has led to a growing number of investigations of thin film structures. As we have already stated, the input and output frequency of the system remains the same, but it is the movement of the field maximum across the interfaces that turns out to be of principal interest because this phenomenon can lead to two possible wave number values for a given power level. This, in turn, implies the existence of spatial switching and other applications.

The investigation of these effects requires an exact treatment of nonlinear field equations. This can now be achieved, up to a point, in an analytical fashion but beyond that the solution assumes the form of a computational algorithm. The latter may, in some circumstances, require exhaustive iterative techniques. The class of problems for which a perturbation approach, involving a small, nonlinearly-induced, deviation from the linear state are, of course, important but at such a limit the field distribution is essentially undisturbed.

In this review we concentrate upon rather higher power level phenomena at which the field distribution is substantially modified. The major current developments are presented in great detail, and the extent to which the approximations, currently used, are valid are examined. No published work exists on the stability of the nonlinear solutions of the guiding problems studied to date, so this area will be ignored for the moment.

After discussing the various forms of the dielectric tensor and the optimum form of the field equations, the paper then deals with the (TE) s-polarized and (TM) p-polarized waves in separate sections. The infra-structure of the TE wave calculations is much more highly developed because the assumptions on which the theory is based are more secure than in the TM case. As a consequence, the TM section will be briefly developed. It will, nevertheless, review, in sufficient detail, the progress that has been made with the understanding that the results are much more model-dependent than in the TE case.

2. The Dielectric Tensor and the Nonlinear Field Equations

The polarization vector of an optically nonlinear medium can be expanded as a power series in the electric field components if, at best, the amplitudes of these components vary only slowly compared to the rate at which the nonlinearity is established. If this is the case then, in general, the expansion takes the form

$$P_i = \chi^{(1)}_{ij} E_j + \chi^{(2)}_{ijk} E_j E_k + \chi^{(3)}_{ijkl} E_j E_k E_l + \dots , \tag{2.1}$$

where the variables are the time Fourier field components and a nonlinear mixing process is being considered. The susceptibility tensors $\chi^{(n)}_{ijkl\dots}$ must transform according to the point group symmetry of the medium so that, for highly symmetric materials, many of the elements of these tensors will be zero. If, for example, the material has a center of symmetry, then **all** second-order nonlinear effects are identically zero. It is only by **breaking** the symmetry that they re-assert themselves. The third-order nonlinear effects, on the other hand, are never identically zero under any point group symmetry operation, and it is this type of nonlinearity that is the focus of our attention here. As is the case with other orders of nonlinearity, the third-order term can produce combination frequencies. For example, from a single laser beam with a frequency ω a third harmonic 3ω can be produced through the nonlinear polarization

$$P_i(3\omega) = \chi^{(3)}_{ijkl} (-3\omega;\omega,\omega,\omega)\ E_j(\omega)E_k(\omega)E_l(\omega) , \tag{2.2}$$

but it is also possible to produce a polarisation at one and the same frequency ω through the terms

$$P_i(\omega) = \chi^{(3)}_{ijkl} (-\omega;\omega,-\omega,\omega)\ E_i(\omega)E^*_k(\omega)E_l(\omega) . \tag{2.3}$$

It is this latter possibility that has attracted a great deal of attention recently. The 3ω generation shown in equation (2.2) requires a phase-matching that can be achieved but it is not likely to occur at the same time as large intensity dependent effects.

Obviously, since $\chi^{(3)}_{ijkl}$ has 81 elements, it is a very difficult term to handle and considerable work remains to be done before it is modeled correctly for guided waves. Nevertheless, the simplest system of any practical importance is completely isotropic and, for such a material, symmetry demands that $\mathbf{E}(\omega)$ and $\mathbf{P}(\omega)$ be either parallel or antiparallel. The products of $\mathbf{E}(\omega)$, $\mathbf{E}^*(\omega)$ and $\mathbf{E}(\omega)$ that satisfy this criterion are well-known[51] so that for an isotropic material the polarization vector is, rigorously, of the form[51,52]

$$\mathbf{P} = \beta_1(\mathbf{E}\cdot\mathbf{E}^*)\mathbf{E} + \beta_2(\mathbf{E}\cdot\mathbf{E})\mathbf{E}^* , \qquad (2.4)$$

where β_1 and β_2 are, in principle, to be evaluated as functions of frequency, but will turn out, in the theory given here, to be constants. Note that the practice of writing $\mathbf{E}$ as dependent upon ω is now suppressed, for convenience.

Maxwell's equations reduce to the wave equation, in terms of the nonlinear polarization $\mathbf{P}$,

$$\left[\varepsilon - \frac{c^2k^2}{\omega^2}\right] \mathbf{E} + \frac{c^2}{\omega^2}\, \mathbf{k}(\mathbf{k}\cdot\mathbf{E}) + c^2\mu_0\mathbf{P} = 0, \qquad (2.5)$$

where $\mathbf{k}$ is the wave number, ω is the angular frequency of the wave, c is the velocity of light in vacuo and ε is the linear part of the dielectric function.

The consequences of Eq. (2.5) are quickly derived for an infinite medium. If, for example, a linearly polarized wave is propagating along the x-axis with wave number k_x and a single electric field component E_y, then

$$n^2 = \frac{c^2k^2}{\omega^2} = \varepsilon + c^2\mu_0(\beta_1 + \beta_2)|E_y|^2 \, . \tag{2.6}$$

From this equation we collect the proof that this type type of third-order optical process really does manifest itself as an intensity-dependent refractive index. If a circularly polarized wave is now considered, for which $E_z/E_y = \pm i$, then, from Eq. (2.5)

$$\left[\left[\varepsilon - \frac{c^2k^2}{\omega^2}\right] + c^2\mu_0\beta_1(|\mathbf{E}|^2)\right] (E_z + iE_y) = 0 \tag{2.7}$$

since $E_z^2 + E_y^2 = 0$. This result shows that even for something as straightforward as bulk, unguided, plane waves the polarization of the wave is important because, it has introduced $(\beta_1 + \beta_2)$ for linearly polarized waves but only β_1 for circularly polarized waves. If the field is not circularly polarized, as would be the case for guided waves, $(\beta_1 + \beta_2)$ is replaced by a guiding structure-dependent factor.

The general wave equation (2.5) shows that all the field components are coupled to each other. The uncoupling achieved by **assuming** a state of polarization needs to be justified. It is precisely this problem that is encountered for guided waves, even for an isotropic material. If the coupling is broken by asserting that the waves are s-polarized (TE) or p-polarized (TM), then they will remain uncoupled unless there are off-diagonal elements in the dielectric tensor. It is an interesting question for the future as to whether the nonlinear systems examined below have mixed s-p stationary states. The answer will have a strong bearing upon the excitation mechanisms of the nonlinear waves considered here.

For s-polarized waves propagating down a guiding x-axis, say, only the E_y field component appears in the field equations. For p-polarized waves both E_x and E_z appear, however. It is, precisely, the phase relationship of E_x to E_z to each other that creates an additional serious difficulty in the exact theory of nonlinear p-polarized waves.

For an isotropic material the polarization vector is[51]

$$P_i = c\varepsilon_0^2\varepsilon n_{2I}\left[\frac{2}{3}E_iE_jE_j^* + \frac{1}{3}E_i^*E_jE_j\right]$$

$$= 2\varepsilon_0\sqrt{\varepsilon}\, n_{2E}\left[\frac{2}{3}E_iE_jE_j^* + \frac{1}{3}E_i^*E_jE_j\right] . \qquad (2.8)$$

Here n_{2I} is the intensity-related nonlinear coefficient which is defined through $n = \sqrt{\varepsilon} + n_{2I}I$. This definition arises as follows. Given an infinite nonlinear medium supporting plane waves of intensity I, then it appears to have an intensity-dependent refractive index $n = \sqrt{\varepsilon} + n_{2I}I$, where ε is the linear part of the dielectric function and n_{2I} is a convenient way to characterize the material. Now $n^2 \simeq \varepsilon + 2\sqrt{\varepsilon}\, n_{2I}I$ and I is the time-averaged Poynting energy flow $\sqrt{\varepsilon}/2\mu_0c\ |E|^2$, where **E** is the field vector of the wave. Hence another coefficient n_{2E} can also be introduced through $n^2 \simeq \varepsilon + 2\sqrt{\varepsilon}\, n_{2E}\ |E|^2$ so that either n_{2I} or $n_{2E} = \sqrt{\varepsilon}/2\mu_0c\ n_{2I}$ may be used.

For TE waves propagating along the x-axis, the electric field is $(0, E_y, 0)$ so that the only nonzero component of polarization is

$$P_y = 2\varepsilon_0\sqrt{\varepsilon}\, n_{2E}\left[\frac{2}{3}E_y|E_y|^2 + \frac{1}{3}E_y^*E_y^2\right] = 2\varepsilon_0\sqrt{\varepsilon}\, n_{2E}|E_y|^2E_y . \qquad (2.9)$$

On the other hand, for TM waves, the electric field is $(E_x, 0, E_z)$ so that both P_x and P_z are finite where

$$P_x = 2\varepsilon_0\sqrt{\varepsilon}\ n_{2E} \left[\frac{2}{3} E_x(|E_x|^2 + |E_z|^2) + \frac{1}{3} E_x^*(E_x^2 + E_z^2)\right]$$

$$= 2\varepsilon_0\sqrt{\varepsilon}\ n_{2E} \left[(|E_x|^2 + \frac{1}{3} |E_z|^2) E_x + \frac{E_z}{3} (E_z^* E_x + E_x^* E_z)\right] \tag{2.10}$$

$$P_z = 2\varepsilon_0\sqrt{\varepsilon}\ n_{2E} \left[(|E_z|^2 + \frac{1}{3} |E_x|^2) E_z + \frac{E_x}{3} (E_z E_x^* + E_z^* E_x)\right] . \tag{2.11}$$

It is at this point that the problem of the phase relationship between E_x and E_z arises if $P_{x,z}$ are to be reduced to more tractable forms. Suppose that $E_x = r_1 e^{i\vartheta}$ and $E_z = r_2 e^{i(\vartheta+\varphi)}$, then

$$\frac{1}{3} E_z^2 E_x^* = \left[\frac{1}{3} |E_z|^2 E_x\right] \left[\frac{E_x^* E_z}{E_z^* E_x}\right] = \frac{e^{2i\varphi}}{3} |E_z|^2 E_x \tag{2.12}$$

so that P_x becomes

$$P_x = 2\varepsilon_0\sqrt{\varepsilon}\ n_{2E} \left[|E_x|^2 + \left[\frac{2}{3} + \frac{e^{2i\varphi}}{3}\right] |E_z|^2\right] E_x \tag{2.13}$$

For stationary states of a waveguide structure φ is always $\pi/2$. In fact, it will be seen from Maxwell's equations, later on, that if $\varphi=\pi/2$ anywhere in the system then it will be $\pi/2$ everywhere. For an excitation process involving, say, an incident laser beam on a nonlinear guide, it is clear that $\varphi\neq\pi/2$. Hence, for stationary states of a nonlinear guided TM wave,

$$P_x = 2\varepsilon_0\sqrt{\varepsilon}\ n_{2E} \left[|E_x|^2 + \frac{1}{3} |E_z|^2\right] E_x \tag{2.14}$$

$$P_z = 2\varepsilon_0\sqrt{\varepsilon}\ n_{2E} \left[|E_z|^2 + \frac{1}{3} |E_x|^2\right] E_z \tag{2.15}$$

Since, for TM waves, it is the polarizations P_x and P_z that arise in an isotropic medium only the ε_{xx} and ε_{zz} components of the nonlinear dielectric tensor enter into Maxwell's equations. The propagation of TM waves can, therefore, be analyzed in terms of the 2x2 sub-tensor

$$\varepsilon^{TM} = \begin{bmatrix} \varepsilon + \alpha_1|E_x|^2 + \alpha_2|E_z|^2 & 0 \\ 0 & \varepsilon + \alpha_2|E_x|^2 + \alpha_1|E_z|^2 \end{bmatrix} \qquad (2.16)$$

where $\alpha_1 = 2\sqrt{\varepsilon}\, n_{2E}$, $\alpha_2 = 2\sqrt{\varepsilon}\, n_{2E}[\frac{2}{3} + \exp(2i\varphi)/3]$.

If more generality is required, then it has been shown for a uniaxial material, with perpendicular and parallel components of the dielectric constant being written as $\varepsilon_\perp$ and $\varepsilon_\parallel$, that a reasonably complete 3x3 nonlinear tensor is[3]

$$\varepsilon = \begin{bmatrix} \varepsilon_\parallel + a|E_x|^2 + b|E_y|^2 + c|E_z|^2 & 0 & 0 \\ 0 & \varepsilon_\parallel + b|E_x|^2 + a|E_y|^2 + c|E_z|^2 & 0 \\ 0 & 0 & \varepsilon_\perp + d|E_x|^2 + d|E_y|^2 + e|E_z|^2 \end{bmatrix} \qquad (2.17)$$

Although this tensor already has some approximations built into it, it is not worthwhile to go into them here. It represents the best that can be achieved from the point group symmetry operations and some neglect of unimportant terms. Further **arbitrary** truncation is needed if analytical progress toward a dispersion relationship is required. For a completely isotropic material a lot of theoretical progress has been made in recent years because the dielectric tensor seems to be disarmingly simple.

The field equation for TE waves in an isotropic nonlinear medium will now be derived. For such waves the electromagnetic field vectors will be defined here as

$$\mathbf{E} = (0, E_y, 0), \qquad \mathbf{H} = (H_x, 0, H_z)$$

Hence, for this coordinate system, only the ε_{yy} component of the nonlinear dielectric tensor enters into Maxwell's equations and this will be written as

$$\varepsilon_{NL} \equiv \varepsilon_{yy} = \varepsilon + \alpha|E_y|^2 \tag{2.18}$$

where $\alpha \equiv \alpha_1$. Suppose now that the E_y field components have the form $E_y(z)$ $\exp[i(k_x - \omega t)]$ where $E_y(z)$ is now only an amplitude function and the nonlinear wave travels down the x-axis with a wave number k_x and angular frequency ω. The relevant components of Maxwell's equations are

$$\frac{dE_y}{dz} = -i\omega\mu_0 H_x \tag{2.19}$$

$$k_x E_y = \omega\mu_0 H_z, \quad \frac{dH_x}{dz} - ik_x H_z = i\omega\varepsilon_0\varepsilon_{NL} E_y . \tag{2.20}$$

Equations (2.19) and (2.20) give the field equation

$$\frac{d^2E_y}{dz^2} - \left[k_x^2 - \frac{\omega^2}{c^2}\varepsilon\right] E_y + \frac{\omega^2}{c^2}\alpha|E_y|^2 E_y = 0 . \tag{2.21}$$

At this stage it is clear that a complication has arisen through the term $|E_y|^2E_y$. It is common practice to assume, usually without proof, that the field is real so that $|E_y|^2E_y$ simply becomes E_y^3. This action makes Eq. (2.21) easy to solve analytically.

Suppose[11], however, that the field component E_y is not, initially, assumed to be real but is taken, in general, to be of the form $E_y(z) = E(z)\, e^{i\varphi(z)}$. This solution, on substitution into Eq. (2.21) yields, after equating real and imaginary parts to zero,

$$\frac{d^2E(z)}{dz^2} - E(z)\left[\frac{d\varphi}{dz}\right]^2 - \left[k^2 - \frac{\omega^2}{c^2}\,\varepsilon\right]E(z) + \frac{\omega^2}{c^2}\alpha E(z)^3 = 0 \tag{2.22}$$

$$\frac{d}{dz}\left[E(z)^2\,\frac{d\varphi}{dz}\right] = 0. \tag{2.23}$$

The last equation expresses the conservation of energy flux[11] in the nonlinear medium and integrates to a constant K. Therefore, the complete equation for real E(z) is

$$\frac{d^2E(z)}{dz^2} - \left[k^2 - \frac{\omega^2}{c^2}\,\varepsilon\right]E(z) - \frac{K^2}{E(z)^3} + \frac{\omega^2}{c^2}\,\alpha E(z)^3 = 0. \tag{2.24}$$

The presence of the term involving K is very interesting and should be considered very carefully. If $k_x^2 > \omega^2/c^2\,\varepsilon$, then the amplitude E(z) in both a nonlinear medium and a linear medium decays away to zero at infinity. In a linear medium this is just a straightforward exponential decay. In a nonlinear medium it is a more complex decay, yet to be discussed. In either case since $E(z)^2 d\varphi/dz$ is conserved, the decay is characterized by K = 0 and the real field assumption is a satisfactory one. If $k_x^2 < \omega^2/c^2\,\varepsilon$, or we have a more general guiding situation, then K = 0 corresponds to **only one possibility** i.e. to the stationary guided wave state. The consequences for guiding problems involving an excitation process for which for $K \neq 0$ are not generally known. In the radiative **linear case**, it is easy to see that E(z) = constant and $\varphi = \beta z$ where $\beta = \sqrt{\omega^2\varepsilon/c^2 - k_x^2}$.

The situation for TM waves is even more complicated since the dielectric tensor is difficult, anyway, to reduce to any tractable form. If **both** E_x and E_z are retained then, for guided waves, no loss of generality occurs if E_x and E_z are set to be $\pi/2$ out of phase. If an argument can be invoked that allows the neglect of either E_x or E_z, then much more analytical progress can be made since the remaining single field component is real, just as in the TE case.

The first p-polarized case to be investigated[1] was based upon a now popular, but arbitrary, uniaxial assumption that the part of the dielectric tensor that couples E_x to E_z is

$$\varepsilon^{TM} = \begin{bmatrix} \varepsilon + \alpha E_x^2 & 0 \\ 0 & \varepsilon \end{bmatrix} \tag{2.25}$$

where α=a and E_x is the x component of the total field (E_x, 0, E_z). Since only E_x is involved, a field equation can be obtained that has almost the same form as the field equation for E_y in the TE case i.e.

$$\frac{d^2E_x}{dz^2} - \frac{\varepsilon_\perp}{\varepsilon_\parallel}\left[k_x^2 - \frac{\omega^2}{c^2}\varepsilon_\parallel\right]E_x - \frac{\alpha}{\varepsilon_\parallel}\left[k_x^2 - \frac{\omega^2}{c^2}\varepsilon_\parallel\right]E_x^3 = 0. \tag{2.26}$$

An important feature has now appeared, however, and it is that in Eqs. (2.26) and (2.21), the nonlinear terms for $\alpha > 0$, i.e., a self-focusing medium, have opposite signs. This awkward fact will receive further attention in a later section on p-polarized waves.

The advantage of dealing with only one field component is now abundantly clear, but it is legitimate to question[21] whether E_x is the correct field component in the TM case, since strong arguments can be evoked to say that it is E_z that should be calculated. This point will also be returned to later. It is sufficient, for the moment, to say that if applications for nonlinear guided waves are being sought then the region of possible optical device interest and, indeed, the optically accessible region of the surface polariton curve, is very close to and fairly well down the light-line. In this region the field in the nonlinear part of the system is reasonably transverse and thus E_z and <u>not</u> E_x is the dominant field component. If this is the case, then the tensor that couples E_x and E_z should be[21,23,25], from Eq.(2.16),

$$\varepsilon^{TM} = \begin{bmatrix} \varepsilon & 0 \\ 0 & \varepsilon + \alpha E_z^2 \end{bmatrix} \tag{2.27}$$

where in Eq. (2.16) we have set $\alpha_1=\alpha$ and have neglected $\alpha_1|E_x|^2, \alpha_2|E_z|^2$ and $\alpha_2|E_x|^2$.

If Eq. (2.27) is valid, then the field equation involves only the magnetic field component H_y, i.e.,

$$\frac{d^2H_y}{dz^2} - \left[k^2 - \frac{\omega^2}{c^2}\varepsilon\right] H_y + \frac{\alpha k_x^4}{\omega^2 \varepsilon_0^2 \varepsilon^3} H_y^3 = 0 , \tag{2.28}$$

where the approximation $E_z^2 \simeq k^2/\omega^2\varepsilon_0^2\varepsilon\ H_y^2$ has been used.

Nevertheless, there have been some recent attempts to proceed to more arbitrary forms of dielectric tensors and hence to a more general form of field equation.[32,42,43] Suppose that the nonlinear dielectric tensor is of the form

$$\varepsilon = \begin{bmatrix} \varepsilon_{xx} & 0 & 0 \\ 0 & \varepsilon_{yy} & 0 \\ 0 & 0 & \varepsilon_{zz} \end{bmatrix} \tag{2.29}$$

and that TM wave solutions $(E_x, 0, E_z)$ $(0, B_y, 0)$ are being sought. The basic equations are then

$$\frac{\partial B_y}{\partial z} = \frac{i\omega}{c^2} \varepsilon_{xx} E_x \tag{2.30}$$

$$i\omega B_y = \frac{\partial E_x}{\partial z} - ik_x E_z \tag{2.31}$$

$$k_x B_y = -\frac{\omega}{c^2} \varepsilon_{zz} E_z \tag{2.32}$$

from these equations it is easy to obtain[42]

$$\frac{\partial}{\partial z}\left[\frac{\varepsilon_{zz}}{(k_x^2 - \frac{\omega^2}{c^2}\varepsilon_{zz})}\frac{\partial E_x}{\partial z}\right] = \varepsilon_{xx}E_x \tag{2.33}$$

Alternatively, in terms of B_y, it is equally clear that[32,38,39,48]

$$\frac{\partial}{\partial z}\left[\frac{1}{\varepsilon_{xx}}\frac{\partial B_y}{\partial z}\right] = (k_x^2 - \frac{\omega^2}{c^2}\varepsilon_{xx})\frac{B_y}{\varepsilon_{zz}} \tag{2.34}$$

The equation in terms of E_x expands to

$$\frac{\partial A}{\partial z}\frac{\partial E_x}{\partial z} + A\frac{\partial^2 E_x}{\partial z^2} = \varepsilon_{xx} E_x \tag{2.35}$$

where $A = \varepsilon_{zz}/(k^2 - \frac{\omega^2}{c^2}\varepsilon_{zz})$. Hence if Eq. (2.35) is multiplied through by A dE_x/dz, it becomes

$$\frac{1}{2}\frac{\partial}{\partial z}\left[A^2 (\frac{\partial E_x}{\partial z})^2\right] = \varepsilon_{xx} A E_x \frac{\partial E_x}{\partial z} \tag{2.36}$$

This equation, upon integration, yields[42]

$$A^2 (\frac{\partial E_x}{\partial z})^2 = \int \varepsilon_{xx} A \, d(E_x^2) + C \tag{2.37}$$

where, provided that the tensor elements are functions of E_x^2 alone, the quadrature can, in principle, be readily effected. This generalization obviously lifts the uniaxial assumption but restricts the calculation to a type of nonlinearity that involves only one real field component that is parallel to the direction of propagation. Such a model will apply more to nonlinear plasmas than to solid state integrated optics devices.

It is essential, therefore, to include E_z and recent attempts[32,38,39,48] to do this begin with Eq. (2.34) for B_y and introduce the subsidiary condition

$$|E|^2 = |E_x|^2 + |E_z|^2 = \left|\frac{c^2}{\omega\varepsilon_{xx}} \frac{\partial B_y}{\partial z}\right|^2 + \left|\frac{c^2 k_x}{\omega\varepsilon_{zz}} B_y\right|^2 \tag{2.38}$$

In Eq. (2.38) E_x and E_z are given **equal** weight[32,38,39] yet, as was shown earlier, this is not the case for guided waves in an isotropic medium. This does simplify the problem considerably, especially since we can, quite generally, write $E_x=A_x$ and $E_z=iA_z$ where A_x and A_z are real to obtain

$$E^2 = E_x^2 + E_z^2 = \frac{c^4}{\omega^2}\left[\frac{1}{\varepsilon_{xx}^2}\left(\frac{\partial B_y}{\partial z}\right)^2 + \frac{k_x^2}{\varepsilon_{zz}^2} B_y^2\right] \tag{2.39}$$

where ε_{xx} and ε_{zz} involve only A_x^2, A_z^2 and B_y is real. Once it is realised that E_x and E_z are $\pi/2$ out of phase and they are given equal weight then quadrature of the type shown in Eq. (2.37), or of a more general type based on Eqs. (2.39) and (2.34), is **always** possible.

Finally, since the nonlinearity for some purposes can be regarded as very <weak, it is still of interest to develop a perturbative[41] approach to the development of the field equations. Suppose that the dielectric tensor is assumed to be of the form

$$\varepsilon = \begin{bmatrix} \varepsilon+\delta\varepsilon_{xx} & 0 & 0 \\ 0 & \varepsilon+\delta\varepsilon_{yy} & 0 \\ 0 & 0 & \varepsilon+\delta\varepsilon_{zz} \end{bmatrix} \tag{2.40}$$

where $\delta\varepsilon_{ij}$ is, for the moment an arbitrary a function of **all three** field components (E_x, E_y, E_z). This type of tensor quickly leads to the set of equations

$$\frac{dE_x}{dz} + i\frac{\varkappa^2}{\omega\varepsilon_0\varepsilon}E_z = -i\frac{k_x}{\varepsilon\varepsilon_0}\,\delta\varepsilon_{zz}\,E_z \tag{2.41}$$

$$\frac{dH_y}{dz} - i\omega\varepsilon_0\varepsilon E_x = i\omega\varepsilon_0\delta\varepsilon_{xx}E_x \tag{2.42}$$

$$\frac{dE_y}{dz} + i\omega\mu_0 H_x = 0 \tag{2.43}$$

$$\frac{dH_x}{dz} - i\frac{\varkappa^2}{\omega\mu_0}\,E_y = -i\omega\varepsilon_0\delta\varepsilon_{yy}E_y \tag{2.44}$$

where $\varkappa^2 = k_x^2 - \varepsilon\omega^2/c^2$.

If Eqs. (2.41) to (2.44) are solved perturbatively, then a valid scheme of calculation is to introduce solutions in the form

$$\mathbf{E} = \mathbf{E}^{(0)} + \mathbf{E}^{(1)} + \mathbf{E}^{(2)} + \ldots \tag{2.45}$$

$$\mathbf{H} = \mathbf{H}^{(0)} + \mathbf{H}^{(1)} + \mathbf{H}^{(2)} + \ldots \tag{2.46}$$

where the boundary condition is

$$E_x^{(0)}(0) = E\ , \quad E_x^{(1)}(0) = 0, \quad E_x^{(2)}(0) = 0, \ \ldots \tag{2.47}$$

For such a perturbative scheme, typical zeroth-order and first-order equations are

$$\text{TM:}\quad \frac{dE_x^{(0)}}{dz} + i\frac{\varkappa^2}{\omega\varepsilon_0\varepsilon}\,H_y^{(0)} = 0\ ; \qquad \text{TE:}\quad \frac{dE_y^{(0)}}{dz} + i\omega\mu_0 H_x^{(0)} = 0 \tag{2.48}$$

$$\text{TM:}\quad \frac{dE_x^{(1)}}{dz} + i\frac{\varkappa^2}{\omega\varepsilon_0\varepsilon} H_y^{(1)} = -i\frac{k}{\varepsilon\varepsilon_0}\,\delta\varepsilon_{zz}E_z^{(0)} \tag{2.49a}$$

$$\text{TE:}\quad \frac{dE_y^{(1)}}{dz} + i\omega\mu_0 H_x^{(1)} = 0 \tag{2.49b}$$

It has been shown earlier that the effective tensor that TM waves couple into is

$$\varepsilon^{TM} = \begin{bmatrix} \varepsilon + a|E_x|^2 + b|E_z|^2 & 0 \\ 0 & \varepsilon + a|E_z|^2 + b|E_x|^2 \end{bmatrix} \tag{2.50}$$

where a is not, in general, equal to b. For such a TM wave we obtain in this perturbative sheme

$$\frac{d^2E_x^{(0)}}{dz^2} - \varkappa^2 E_x^{(1)} = -\frac{E^3}{\varepsilon}\, e^{-3\varkappa z}\left[3k_x^2\left(\frac{ak_x^2}{\varkappa^2} + b\right) - \varkappa^2\left(a + \frac{k_x^2}{\varkappa^2}b\right)\right] \tag{2.51}$$

This equation is of the form

$$\frac{d^2E_x^{(1)}}{dz^2} - \varkappa^2 E_x^{(1)} = Ge^{-3\varkappa z}\,, \tag{2.52}$$

and its solution is

$$E_x^{(1)} = \frac{G}{8\varkappa^2}\left[e^{-3\varkappa z} - e^{-\varkappa z}\right]\,. \tag{2.53}$$

$H_y^{(1)}$ can be determined in a similar manner. The consequences of using a perturbative approach will be returned to in the later section on p-polarized waves. It can be said at this stage, however, that at this level of approximation it is going to be good enough to take the **linear dispersion equation** of any guiding structure and then replace one, or all, of the dielectric tensor elements that appear in it with its nonlinear form.

3. s-Polarized Waves

A. Single Interface

A brief, initial, study of nonlinear s-polarized waves at a single interface was performed some time ago.[47] The topic has been given an important impetus, however, by some new studies[2,36] that emphasize both their relationship to the traditional surface polaritons and to the possibility of establishing self-focused channels close to the interface.

In this paper the interface between two semi-infinite nonlinear dielectrics is considered for generality and the starting point is the basic equations for the y-component of the electric field in each region. If the field component is written as E_i where i denotes the dielectric medium, then, if E_i are real

$$\frac{d^2E_1}{dz^2} - (k_1^2 - 2\Lambda_1 E_1^2)E_1 = 0 \quad , \qquad z<0 \tag{3.1}$$

$$\frac{d^2E_2}{dz^2} - (k_2^2 - 2\Lambda_2 E_2^2)E_2 = 0 \quad , \qquad z>0 \tag{3.2}$$

where $k_i^2 = k_x^2 - \varepsilon_i\omega^2/c^2$, and the ε_i are the linear parts of the dielectric functions. The nonlinear coefficients are $\Lambda_i = \omega^2\alpha_i/2c^2$, and the quantity α_i is defined in Eq. (2.18).

In general the fields E_i are not real but of the form $E_i(z)\exp[i\varphi_i(z)]$ where $E_i^2(z)\, d\varphi_i/dz = K_i$ and K_i is a constant. This means that

$$\frac{d^2E_i(z)}{dz^2} - [k_i^2 - 2\Lambda_i\, E_i^2(z)]E_i(z) - \frac{K_i^2}{E_i^3(z)} = 0. \tag{3.3}$$

Now $\frac{d}{dz}$ $[dE_i(z)/dz]^2 = 2[dE_i(z)/dz]x[d^2E_i(z)/dz^2]$ so that Eq. (3.3) can be written as

$$\frac{d}{dz}\left[\frac{dE_i(z)}{dz}\right]^2 - 2\left[k_i^2 + \frac{K_i}{E_i^4(z)} - 2\Lambda_i E_i^2(z)\right] E_i(z)\,\frac{dE_i(z)}{dz} = 0. \qquad (3.4)$$

This equation integrates to

$$\left[\frac{dE_i(z)}{dz}\right]^2 - k_i^2 E_i(z) + \frac{K_i}{E_i(z)^2} + \Lambda_i\, E_i(z)^4 = C_{2i}, \qquad (3.5)$$

where the C_{2i} are more integration constants. This interesting expression that can be put in the form[43]

$$\frac{1}{2}\left[\frac{dE_i(z)}{dz}\right]^2 + V_i[E_i(z)] = C_{2i}, \qquad (3.6)$$

where $V_i[E_i(z)]$ is the potential function

$$V_i[E_i(z)] = \frac{K_i}{2E_i(z)^2} - \frac{1}{2}k_i^2 E_i(z)^2 + \frac{\Lambda_i\, E_i(z)^4}{2}\,. \qquad (3.7)$$

For a single interface, the nonlinear waves are bound to the neighborhood of the surface with the fields on either side decaying to zero as infinity is approached. Thus, in the case to be considered here, $K_i = 0$ and $C_{2i} = 0$ so that

$$V_i[E_i(z)] = -\frac{1}{2}k_i^2 E_i(z)^2 + \frac{1}{2}\Lambda_i E_i(z)^4\,. \qquad (3.8)$$

This representation has some value in discussing this type of problem[43] in terms of the potential wells $V_i[E_i(z)]$. If, for example, the nonlinear waves are propagating along an interface between a linear and a nonlinear medium, then

$$z>0: \quad V_2[E_2(z)] = -\frac{1}{2}k_2^2E_2(z)^2 + \frac{\Lambda_2}{2}E_2(z)^4 \; ; \qquad z<0: \quad V_1[E_1(z)] = -\frac{1}{2}k_1^2\, E_1(z)^2 \tag{3.9}$$

For this situation, since $E_i(z)$ and $dE_i(z)/dz = 0$ at $z = \pm\infty$, it is possible to think in terms of a mechanical analogy in which a particle, starting at rest with $E_1(z) = 0$ at $z = -\infty$, rolls **down** the potential $V_1[E_1(z)]$ and rolls back **up** the potential $V_2[E_2(z)]$ to end up at $E_2(z) = 0$ at $z = +\infty$. Such a particle would obviously, have to cross from $V_1[E_1(z)]$ to $V_2[E_2(z)]$ at $E_1(0) = E_2(0)$. It seems that this is as far as it is necessary to take this analogy for the $K = 0$, bounded case, but it should be emphasised that in any excitation of guided waves the $K \neq 0$ cases could be important because of the interpretation of Eq.(2.23) as the conservation of energy flux.

For the problem under discussion now, the first integrations of Eqs. (3.1) and (3.2) are

$$\left[\frac{dE_1(z)}{dz}\right]^2 - [k_1^2 - \Lambda_1E_1(z)^2]E_1(z)^2 = 0 \,, \tag{3.10}$$

$$\left[\frac{dE_2(z)}{dz}\right]^2 - [k_2^2 - \Lambda_2E_2(z)^2]E_2(z)^2 = 0 \,, \tag{3.11}$$

where $E_i(z)$, $dE_i(z)/dz$ are required to vanish as $z \to \pm\infty$.

The form of Eqs. (3.10) or (3.11), after setting $E \equiv E_i(z)$, $\Lambda = \Lambda_i$, $\kappa = k_i$ factorizes to

$$\left[\frac{dE}{dz} - \sqrt{(\kappa^2 - \Lambda E^2)}E\right]\left[\frac{dE}{dz} + \sqrt{(k^2 - \Lambda E^2)}E\right] = 0. \tag{3.12}$$

Therefore the solutions are of the form

$$\Lambda>0: \quad \int \frac{dE}{E\sqrt{\frac{\varkappa^2}{\Lambda} - E^2}} = -\sqrt{\frac{\Lambda}{\varkappa^2}} \cosh^{-1}\left|\left[\frac{\varkappa^2}{\Lambda}\right]^{1/2} E\right| = \pm\Lambda^{1/2}z + A , \qquad (3.13)$$

$$\Lambda<0: \quad \int \frac{dE}{E\sqrt{\frac{\varkappa^2}{\Lambda} - E^2}} = -\sqrt{\frac{|\Lambda|}{\varkappa^2}} \sinh^{-1}\left|\left[\frac{\varkappa^2}{|\Lambda|}\right]^{1/2} E\right| = \pm\Lambda^{1/2}z + A , \qquad (3.14)$$

where A is an integration constant of the form $\Lambda^{1/2}z_0$. The functional form of the field E is, in compact form,

$$\Lambda_i > 0: \quad E_i(z) = \frac{c}{\omega}\sqrt{\frac{2}{\alpha_i}} \frac{k_i}{\cosh[k_i(z \pm z_0)]} , \qquad (3.15)$$

$$\Lambda_i < 0: \quad E_i(z) = \pm \frac{c}{\omega}\sqrt{\frac{2}{|\alpha_i|}} \frac{k_i}{\sinh[k_i(z \pm z_0)]} . \qquad (3.16)$$

In Eq. (3.15) it does not matter which sign is taken, the application of the boundary conditions will assure the appropriate choice.

For waves propagating along the interface between two nonlinear media with Λ_i(or α_i) > 0

$$z>0: \quad E_2(z) = \frac{c}{\omega}\sqrt{\frac{2}{\alpha_2}} \frac{k_2}{\cosh[k_2(z - z_{02})]} , \qquad (3.17)$$

$$z<0: \quad E_1(z) = \frac{c}{\omega}\sqrt{\frac{2}{\alpha_1}} \frac{k_1}{\cosh[k_1(z + z_{01})]} . \qquad (3.18)$$

The boundary conditions are that $E_i(z)$ and $dE_i(z)/dz$ are continuous and their imposition immediately gives

$$k_1\tanh(k_1 z_{01}) + k_2\tanh(k_2 z_{02}) = 0 \ . \tag{3.19}$$

If $z_{02} > 0$, $z_{01} > 0$, a maximum is required at $z = z_{02}$ and $z = -z_{01}$ but, as will be seen below, what will happen is that either $z_{02} > 0$ and $z_{01} < 0$ corresponding to a maximum in the upper medium but none in the lower medium, or vice versa. Obviously for $\alpha_i < 0$ the tanh functions in (3.19) are replaced by cotanh functions. The field shapes in the nonlinear media for a $\cosh^{-1}$ function can therefore peak to $z_{01,2}$ or not. This means that it is possible for the field to have a positive or negative slope at the boundary. If it has a positive slope, then the nonlinearity will prevent the field from growing indefinitely through a self-focusing increase in the refractive index.

A knowledge of the field shapes is **not necessary**, however, in order to determine the dispersion relationships of the problem. These dispersion relationships may arise from a straight application of the boundary conditions, as was done in Eq. (3.19), or they may emerge from a calculation of the power flow down the guide.

Suppose that the electric field at the boundary is E_0, then

$$E_1(0) = E_2(0) = E_0 \ ; \qquad \frac{dE_1(z)}{dz}\Big|_{z=0} = \frac{dE_2(z)}{dz}\Big|_{z=0} \ , \tag{3.20}$$

so that Eqs. (3.10) and (3.11) yield[5]

$$E_0^2 = \frac{2(\varepsilon_1 - \varepsilon_2)}{\alpha_2 - \alpha_1} \tag{3.21}$$

Also, if one medium is linear, we recover the now well-known result[2,47] that

$$E_0^2 = \frac{2(\varepsilon_2-\varepsilon_1)}{\alpha_1} \quad ; \qquad \alpha_2 = 0 \ . \tag{3.22}$$

Equation (3.19) does, in fact, give the same results. This can be shown from the relationship

$$\tanh^2(k_i z_{0i}) = 1 - \frac{\omega^2 \alpha_i E_0^2}{2c^2 k_i^2} \ . \tag{3.23}$$

If the limit $\alpha_2 \to 0$, $\alpha_1 \to 0$ is taken in Eq. (3.21) then $E_0 \to \infty$. This is one way of showing that s-polarized waves do not exist in the linear limit for a single interface. Also since E_0 is **real** the following restrictions also occur

$$\varepsilon_1-\varepsilon_2 < 0 \iff \alpha_2-\alpha_1 < 0 \ ; \qquad \varepsilon_1-\varepsilon_2 > 0 \iff \alpha_2-\alpha_1 > 0 \ . \tag{3.24}$$

Equations (3.21) and (3.22) are of some immediate interest because if ε_1 and ε_2 are constants, as they would be for an interface between, say, glass and a nonlinear medium like CS_2, then the field amplitude is <u>fixed</u> at the boundary.

The conditions (3.21) and (3.22) can also be expressed in terms of an optical intensity at the boundary[25] because the time-averaged Poynting vector along the interface is

$$S_x = \langle S_x \rangle = \frac{c^2 \varepsilon_0 k}{2\omega} E_0^2 \ .$$

Hence,

$$S_x = \frac{ck}{\omega} \frac{\varepsilon_1 - \varepsilon_2}{\varepsilon_2 n_{2l}} \quad \text{(Linear/Nonlinear)} \ ,$$

$$S_x = \frac{ck}{\omega} \frac{\varepsilon_1 - \varepsilon_2}{[\varepsilon_2 n_{2lc} - \varepsilon_1 n_{2ls}]} \quad \text{(Nonlinear/Nonlinear)} \ , \tag{3.25}$$

where the nonlinear coefficient is $\alpha = c\varepsilon_0\varepsilon n_{2l}$, n_{2l} appears through the intensity-dependent change $n_2 I$ in the refractive index, and the subscripts c and s mean cladding (medium 2) and substrate (medium 1) respectively.[25]

For $\alpha < 0$, i.e., a self-defocusing medium, the inverse sinh functions in Eq. (3.16) diverge whenever $(z \pm z_0) = 0$ and therefore physically acceptable solutions exist only if the singularity is virtual.

Note that Eq. (3.19) and hence Eqs. (3.21) or (3.22) contain no wave number dependence and, for constant ε_i, do not even have any frequency dependence. It is for this reason that they cannot be regarded as dispersion equations. Perhaps, since the equations arise from the application of the boundary conditions it would be better to call them eigenvalue equations. If a linear medium such as a metal, say, with a negative frequency-dependent $\varepsilon_1(\omega)$, abutts a nonlinear medium, then Eq. (3.22) serves to make the frequency depend upon field amplitude at the boundary. It will still be **without** a wave number dependence. We will return to the question of how to generate wave dispersion equations later.

The power flow parallel to the interface is given by

$$P = \int_{-\infty}^{\infty} \langle S\rangle\cdot\bar{x}\,dz = \frac{c^2\varepsilon_0 k_x}{2\omega}\int_{-\infty}^{\infty} E_i(z)^2\,dz\ , \tag{3.26}$$

where $\bar{\mathbf{x}}$ is a unit vector in the x-direction and $\langle S\rangle$ is the time-averaged Poynting vector

$$\langle S\rangle = \frac{1}{2}\,\mathrm{Real}(\mathbf{E}\times\mathbf{H}) = \frac{c^2\varepsilon_0}{2\omega}\,\mathrm{Real}\left(k_x E_y(z)^2,\ 0,\ -iE_y(z)\frac{dE_y(z)}{dz}\right)\ . \tag{3.27}$$

Here Real[ ] denotes the real part of the expression and **H** is the associated magnetic field. If equations (3.10) and (3.11) are now differentiated with respect to z

$$\frac{d}{dz}\left[\frac{\frac{d}{dz}E_i(z)}{E_i(z)}\right] = -\Lambda_i E_i(z)^2 \ . \qquad (3.28)$$

This remarkably simple result means that the integral in the power formula (3.26) can be immediately evaluated **without actually performing the integration or requiring any knowledge of the field distribution.** In fact, the total power transported by TE surface waves guided by the interface between two semi-infinite isotropic nonlinear dielectrics can be written down immediately as

$$P_{(+)}^{(N,N)} = \frac{c^2\varepsilon_0 k_x}{2\omega}\left[\frac{k_1-\gamma}{\Lambda_1} + \frac{k_2+\gamma}{\Lambda_2}\right] , \qquad (3.29)$$

$$P_{(-)}^{(N,N)} = \frac{c^2\varepsilon_0 k_x}{2\omega}\left[\frac{k_1+\gamma}{\Lambda_1} + \frac{k_2-\gamma}{\Lambda_2}\right] , \qquad (3.30)$$

where

$$\gamma = \sqrt{k_x^2 - \frac{\omega^2}{c^2}\left[\frac{\alpha_2\varepsilon_1-\alpha_1\varepsilon_2}{\alpha_2-\alpha_1}\right]} \ . \qquad (3.31)$$

Note that equations (3.29) and (3.30) are identical to the result presented many years ago[47] without derivation, substantial comment or numerical investigation.

The terms in the power formulae (3.29) and (3.30) can, if Eq.(3.28) is not used, be obtained by direct integration over the whole structure. For example, in the upper cladding medium the formula that leads to the $(k_2-\gamma)/\Lambda_2$ term is

$$P_c = \frac{ck_x}{\omega}\,\frac{ck_2}{\omega}\,\frac{c^2\varepsilon_0}{\omega\alpha_2}\,[1 - \tanh(k_2 z_{02})] \ . \qquad (3.32)$$

Equations (3.23) and (3.21) show that the term 1-tanh $(k_2-\gamma)/k_2$ and (3.29) is once again obtained.

The subscript (+) denotes a positive field gradient at the boundary and hence a field maximum or bulge in the upper medium. Similarly the subscript (-) refers to a maximum in the lower medium. A further interesting[36], but not vital, point is that if the field maximum occurs, for example, in the upper nonlinear medium, then

$$\int_0^{z(E_{max})} E_2(z)^2\,dz = \frac{\gamma}{\Lambda_2}\,, \qquad \int_{z(E_{max})}^{\infty} E_2(z)^2\,dz = \frac{k_2}{\Lambda_2}\,. \tag{3.33}$$

The power flow therefore consists of three contributions, one for the region $z < 0$, a second for $0 < z(E_{max})$ and a thirc for $z(E_{max}) < z < \infty$ as shown in Fig. 1. This feature may be of use in any future development of the theory.

It is often the case that one medium is optically linear. The linear limit of the upper medium, for example, is

$$\lim_{\Lambda_2\to 0}\left[\frac{k_2-\gamma_2}{\Lambda_2}\right] = \lim_{\Lambda_2\to 0}\left[\frac{[k_2 - (k_2^2 - \Lambda_2 E_0^2)^{1/2}]}{\Lambda_2}\right] = \frac{\varepsilon_2-\varepsilon_1}{\alpha_1 k_2} = \frac{E_0^2}{2k_2}\,. \tag{3.34}$$

Hence the power flow between a linear and a nonlinear medium is

$$P^{(N,L)} = \frac{c^2\varepsilon_0 n_x}{\omega\alpha_1}\left[n_1 + n_2 + \frac{\varepsilon_2 - \varepsilon_1}{2n_2}\right], \tag{3.35}$$

where $n_x = ck_x/\omega$, $n_i = ck_i/w$. Now as $n_2\to 0$ (i.e. $n_x \to \sqrt{\varepsilon_2}$) it can be seen that large powers are needed to maintain the surface wave and that most of it is carried within the linear medium.

We return now to the question of dispersion equations. It was commented earlier that Eqs. (3.21) on (3.22) cannot be regarded as a dispersion equations for s-polarised waves, although it is tempting to regard the absence of k dependence in (3.21) or (3.22) as simply the lack of a linear limit. Indeed this is true, but in this type of nonlinear problem, however, it is also correct to view the power formula as a dispersion equation so that (3.29) and (3.30) are really the frequency (ω) wave number (k) relationships for a given power level. A simple example is plotted in Fig. 2 which shows that a certain critical power level must be reached before a nonlinear surface wave and hence a self-focused chanel is created.

B **Linear Guide with Nonlinear Boundaries**

The type of guiding structure that will be considered here is shown in Fig. 3. In the linear layer the first integration of the field equation is

$$\left[\frac{dE_2(z)}{dz}\right]^2 + k_2^2 E_2(z) = C_2 \ , \qquad (3.36)$$

where $k_2^2 = \varepsilon_2\omega^2/c^2 - k_x^2$ and the fields in regions 1, 2, and 3 are now, respectively, E_1, E_2 and E_3. Note that E_1 and E_3 satisfy equations (3.10) or (3.11) with $k_i^2 = k_x^2 - \varepsilon_i\omega^2/c^2$, i = 1, 3. The fields would therefore decay in the usual exponential fashion if it were not for the nonlinearity. Also, since the guide is asymmetric, all modes are cut-off in the linear regime, whereas, for a symmetric guide, the lowest mode is not cut-off. The first point that should be made is that, because the guide is now nonlinear, there is a distinction between $k_x^2 < \varepsilon_2\omega^2/c^2$ that corresponds to guided waves and $k_x^2 > \varepsilon_2\omega^2/c^2$ that corresponds to surface waves. In the latter case the fields in the dielectric are hyperbolic and are in fact, localised, or peaked, close to the boundaries. This corresponds to the self-focused s-polarised surface waves encountered earlier for a single interface. The difference now is that guided waves are also possible and a linear limit exists. The type of field solutions that are possible are shown in Fig. 4 where it can be seen that not all the guided modes have field maxima or <u>bulges</u> within the guide. Also a guided mode need not have a bulge outside the guide. Surface waves, however, must have a bulge in either the upper or lower nonlinear medium, or both.

A number of further points must now be noted. Firstly, the bulge arises when a field solution sets out at the boundary as a growing solution. As the field grows in intensity the nonlinearity of the medium causes a self-focusing action. A bulge appears and the field decays safely away towards infinity. Secondly, it is convenient to label the waves according to how many times times the field crosses the axis inside the guide. This borrowing from linear waveguide terminology permits the introduction of the notation TE_0, TE_1, ...TE_n waves. An immediate point of interest is that only TE_0 and TE_1, waves can match onto surface waves solutions. This is because for surface waves the fields inside the guide must be purely hyperbolic and may only cross the z-axis once.

The fields at the boundaries of the film will be defined as $E_0=E_2(0)$, $E_b=E_2(d)$ but, because both $E_2(z)$ and $dE_2(z)/dz$ are continuous at z=0 and z=d, they are related to each other through the following equation for a conic section[7,8,9]

$$\frac{\alpha_1{}^2\alpha_3}{\alpha_3\eta_1{}^2 - \alpha_1\eta_3{}^2}(E_0{}^2 - \frac{\eta_1}{\alpha_1})^2 - \frac{\alpha_1\alpha_3{}^2}{\alpha_3\eta_1{}^2 - \alpha_1\eta_3{}^2}(E_b{}^2 - \frac{\eta_3}{\alpha_3})^2$$

$$= S_1(E_0{}^2 - \frac{\eta_1}{\alpha_1})^2 - S_2(E_b{}^2 - \frac{\eta_3}{\alpha_3})^2 = 1 \ , \qquad (3.37)$$

where $\eta_i = \varepsilon_2 - \varepsilon_i$ with i=1, 3. If S_1 and S_2 are both positive, a hyperbola is produced. But if S_1 and S_2 have opposite signs, an ellipse is produced. Eq. (3.37) is a useful device especially since the assumption of real fields allows interest to center only upon the first quadrant. Fig. 5 shows a typical hyperbola for an asymmetric guide that has both upper and lower branches. In any problem for which they arise, they need to be dealt with separately with each one giving their own dispersion curves and power flow contours. If a purely symmetric guide is considered in which $\varepsilon_1 = \varepsilon_3$ then the conic section reduces to straight lines, as also shown in Fig. 5. These lines are perpendicular to each other and the interesting point is that one of the two lines corresponds to $E_0^2 = E_b^2$ i.e. symmetric and antisymmetric modes as expected for a symmetric guide, but the other corresponds to $E_0^2 \neq E_b^2$. This is unexpected since it corresponds to nonlinearly induced asymmetric modes.

For the asymmetric linear guide there are really four eigenvalue equations, three of which lead to non-physical solutions in the sense that they are associated with waves that grow with distance away from the boundaries.

If nonlinearity is included these four equations are now all needed because the nonlinearity, as we have already noted previously, contours the field away from divergence and in so doing creates a field maximum or bulge. The four equations for guided waves, with $k_{2g}^2 = \frac{\varepsilon_2 \omega^2}{c^2} - k_x^2$, are[8]:

EVEN PARITY: $E_b > 0$, $E_0 > 0$ or $E_b < 0$, $E_0 < 0$

$$\cos(k_{2g}d) = (k_{2g}^2 + \gamma_1\gamma_3)/M \quad \text{[single bulge in one nonlinear medium]}\ , \tag{3.38}$$

$$\cos(k_{2g}d) = (k_{2g}^2 - \gamma_1\gamma_3)/M \quad \text{[bulges or decay in both nonlinear medium]} \tag{3.39}$$

ODD PARITY: $E_0 > 0$, $E_b < 0$ or $E_0 < 0, E_b > 0$

$$\cos(k_{2g}d) = (\gamma_1\gamma_3 + k_{2g}^2)/M \quad \text{[single bulge]}\ , \tag{3.40}$$

$$\cos(k_{2g}d) = (\gamma_1\gamma_3 - k_{2g}^2)/M \quad \text{[double bulge or double decay]}\ , \tag{3.41}$$

where $M = (\gamma_1^2 + k_{2g}^2)^{1/2}\,(\gamma_3^2 + k_{2g}^2)^{1/2}$.

For a surface wave, the field in the guide is hyperbolic, a feature that owes its existence entirely due to the optical nonlinearity. This occurs when $k_x^2 > \varepsilon_2\omega^2/c^2$ so that k_{2g}^2 is replaced by $k_{2s}^2 = k_x^2 - \varepsilon_2\omega^2/c^2$ but the γ_i actually remain exactly the same. For example, the even parity eigenvalue equation for surface modes is

$$\cosh(k_{2s}d) = \frac{\pm\gamma_1\gamma_3 - k_{2s}^2}{(\gamma_1^2 - k_{2s}^2)^{1/2}\,(\gamma_3^2 - k_{2s}^2)^{1/2}} \,. \tag{3.42}$$

It is not now difficult to show that surface modes exist for both $\alpha_1 E_0^2/2 < \varepsilon_2-\varepsilon_1$ and $\alpha_1 E_0/2 > \varepsilon_2-\varepsilon_1$, whereas guided modes exist only for $\alpha_1 E_0^2/2 < \varepsilon_2-\varepsilon_1$.

The numerical analysis of the nonlinear waves begins with the determination of how the wave number k_x or the effective refractive index $n_x = ck_x/\omega$ depends upon $\alpha_1 E_0^2/2$. The data that will be used to illustrate the method corresponds to a general hyperbolic (E_0^2, E_b^2) relationship that has both an upper and lower branch and whose centre is in the first quadrant. The lower E_b curve rises from zero at $\alpha_1 E_0^2/2$ to a maximum, and then decreases to zero at $\alpha_1 E_0^2/2 = \varepsilon_2-\varepsilon_1$. This lower branch is quite distinct from the upper E_b curve that begins at a finite E_b value, falls to a minimum and then rises indefinitely into the surface wave region $\alpha_1 E_0^2/2 > \varepsilon_2-\varepsilon_1$.

As an example consider the even parity guided mode equation fully expressed in the form

$$\cos(n_{2g}D) = \frac{n_{2g}^2 - \left(\varepsilon_2-\varepsilon_1-\frac{\alpha_1 E_0^2}{2}-n_{2g}^2\right)^{1/2}\left(\varepsilon_2-\varepsilon_3-\frac{\alpha_3 E_b^2}{2}-n_{2g}^2\right)^{1/2}}{\left(\varepsilon_2-\varepsilon_3-\frac{\alpha_1 E_b^2}{2}\right)^{1/2}\left(\varepsilon_2-\varepsilon_3-\frac{\alpha_3 E_b^2}{2}\right)^{1/2}} \tag{3.43}$$

where $n_{2g} = ck_{2g}/\omega$ and $D = \omega d/c$. The linear limit of Eq. (3.42) is

$$\cos(n_{2g}D) = \frac{n_{2g}^2 - (\varepsilon_2-\varepsilon_1-n_{2g}^2)^{1/2}(\varepsilon_2-\varepsilon_3-n_{2g}^2)^{1/2}}{(\varepsilon_2-\varepsilon_1)^{1/2}(\varepsilon_2-\varepsilon_3)^{1/2}} = L . \quad (3.44)$$

Of the group of Eqs. (3.38) to (3.43) only (3.43) has a linear limit. Note also that even the lowest (TE_0) wave is cut-off when $n_{2g}^2 = \varepsilon_2-\varepsilon_1$ (i.e. $n_x = \sqrt{\varepsilon_1}$) because the guide is asymmetric. Also in order to be specific it is assumed here that $\varepsilon_1<\varepsilon_3$ and $\varepsilon_2>\varepsilon_1,\varepsilon_3$ so that cut-off will first occur when n_x reaches $\sqrt{\varepsilon_1}$. All the linear modes of the system can be found from the intersection of the function L, where

$$-1 < L < \sqrt{\frac{\varepsilon_2-\varepsilon_1}{\varepsilon_2-\varepsilon_3}} , \quad (3.45$$

and the $\cos(n_{2g}D)$ curves for various values of D. The same method will now be adopted for the nonlinear case, and also the TE_n terminology used to label the nonlinear waves will be retained. From Eq. (3.43) there appears to be a nonlinear cut-off curve where $n_{2g}^2 = \varepsilon_2-\varepsilon_1-\alpha_1E_0^2/2$ -dealing only for moment with the lower E_b curve. The curves corresponding to the RHS of Eq. (3.42) all begin at -1 where $n_{2g} = 0$, and end up on this apparent nonlinear cut-off curve as $\alpha_1E_0^2/2$ is varied. Acceptable values of n_x correspond to real intersections of $\cos(n_{2g}D)$, as shown in Fig. 6, with this family of curves and, as D increases, the cosine curve crosses the n_{2g} axis much closer to $n_{2g}= 0$ so that higher TE_n waves then become possible.

All the waves derived from Eq. (3.43) correspond to solutions that have field maxima inside the guide and decay in the usual exponential manner into the substrate and the cladding. This is not enough, however, because on the locus $n_{2g}^2 = \varepsilon_2-\varepsilon_1-\alpha_1E_0^2/2$, the field maximum in the guide is on the point of moving out into the nonlinear medium, even though the wave remains, for the time being, as a guided wave. In fact this is where the other equations in the set (3.38) to (3.41) enter into the problem and it is Eq. (3.38) that stands to be satisfied as the maximum moves across the guide boundary. Indeed, at this point γ_1 changes sign and would, in the linear limit, correspond to an exponentially diverging field.

As D decreases, either ω decreases or the guide thickness decreases so that the $\cos(n_{2g}D)$ curves cross the n_{2g} axis at successively higher values. At the point $n_x = \sqrt{\varepsilon_1}$, a critical thickness is reached that is

$$D_{c1}(L)) = \frac{1}{\sqrt{\varepsilon_2-\varepsilon_1}} \cos^{-1}\left[\frac{\varepsilon_2-\varepsilon_1}{\varepsilon_2-\varepsilon_3}\right]^{1/2} . \tag{3.46}$$

For $D < D_{c1}(L)$, further solutions are allowed because of the nonlinearity, until the smallest permissible thickness $D_{c2}(L)$ is reached where

$$D_{c2}(L) = \left| \frac{(\varepsilon_2-\varepsilon_1)^{1/2} - (\varepsilon_2-\varepsilon_3)^{1/2}}{(\varepsilon_2-\varepsilon_1)^{1/2}\ (\varepsilon_2-\varepsilon_3)^{1/2}} \right| . \tag{3.47}$$

For a symmetrically loaded slab $D_{c1}(L) = 0$, $D_{c2}(L) = 0$ showing that it will always support a wave no matter what the guide thickness is.

The upper E_b curve has quite different properties because, for this branch of the boundary fields relationship, the changeover locus is $n_{2g}{}^2 = \varepsilon_2-\varepsilon_1-\alpha_1E_0{}^2/2$ or $n_{2g}{}^2 = \varepsilon_2-\varepsilon_3-\alpha_3E_b{}^2/2$, whichever happens to be satisfied first. There are critical values $D_{ci}(U)$ associated with the upper E_b curve and these are different from $D_{ci}(L)$ in so far as they now depend upon the nonlinear cofficients α_i.

The above analysis enables the dependence of n_x upon $\alpha_1E_0{}^2/2$ to be determined as a prelude to determining the dependence of the power flow down the guide upon n_x or $\alpha_1E_0{}^2/2$ (cf. Fig. 7 for example). Since n_x (i.e. the effective index) depends upon $E_0{}^2$, the linear guide with nonlinear substrate and/or superstrate is already quite different, physically, from the single interface. If the lower surface of the guide is held in place and the upper surface is taken to infinity, then $\alpha_1 E_0^2/2 \rightarrow \varepsilon_2-\varepsilon_1$, $\alpha_3 E_b^2/2 \rightarrow 0$ and n_x becomes **independent** of $\alpha_1 E_0^2/2$ as was shown previously for the single interface. Similarly, for the upper E_b curve, if the upper surface remains fixed in place while the lower surface is removed to infinity, then $\alpha_1 E_0^2/2 \rightarrow 0$, $\alpha_3 E_b^2/2 \rightarrow \varepsilon_2-\varepsilon_3$ with n_x ceasing to depend upon $\alpha_3 E_b^2/2$, as expected.

Odd parity modes can be analysed in exactly the same way. The small n_x solutions for the odd parity cases are modes that in both nonlinear media are double bulge nonlinear guided or surface waves. As D decreases, it is only the double bulge waves that are encountered, in sharp contrast to the even parity modes. Odd parity surface modes must now have bulges in both nonlinear media. Also for both upper and lower E_b curves $n_x \rightarrow \infty$ at $\alpha_1 E_0^2/2 = \varepsilon_2-\varepsilon_1$ and $\alpha_1 E_0^2/2 = 0$ so that odd parity modes do not exist beyond $\alpha_1 E_0^2/2 = \varepsilon_2-\varepsilon_1$. The complete situation, for this type of problem, is summarised in table 1.

From Eq. (3.28) the power integration[7,8,9] for medium i reduces to

$$\int_a^b E_i(z)^2\, dz = -\frac{1}{\Lambda_i}\left[\frac{1}{E_i(z)}\,\frac{dE_i(z)}{dz}\right]_a^b . \tag{3.48}$$

As $z \rightarrow -\infty$, the ratio $[dE_1(z)/dz]/E_1(z) \rightarrow k_1$ and, as $z \rightarrow \infty$, the ratio $[dE_3(z)/dz]/E_3(z) \rightarrow -k_3$. Also $[dE_i(z)/dz]/E_i(z) \rightarrow \pm\gamma_i$ on the boundaries. Thus, in medium 1, the contribution to the power flow is

$$P_1 = \frac{c^2\varepsilon_0 k_x}{\omega}\,\frac{k_1 \pm \gamma_1}{\Lambda_1}, \tag{3.49}$$

and in medium 3 it is

$$P_3 = \frac{c^2\varepsilon_0 k_x}{\omega} \frac{k_3 \pm \gamma_3}{\Lambda_3} . \qquad (3.50)$$

For the guided wave region 2

$$\frac{dE_2(z)}{dz} = \pm\sqrt{C_2 - k_{2g}{}^2 E_2(z)^2} , \qquad (3.51)$$

where

$$C_2 = (k_{2g}{}^2 + \gamma_1{}^2)E_0{}^2 = (k_{2g}{}^2 + \gamma_3{}^2)E_b{}^2 . \qquad (3.52)$$

Hence

$$\left[E_2(z)\frac{dE_2(z)}{dz}\right]^2 + k_{2g}{}^2 E_2(z)^4 - C_2 E_2(z)^2 = 0 \qquad (3.53)$$

which, upon differentiation with respect to z, gives

$$2E_2(z)\frac{dE_2(z)}{dz} \frac{d}{dz}\left[E_2(z)\frac{dE_2(z)}{dz}\right] + 4k_{2g}{}^2 E_2(z)^3\frac{dE_2(z)}{dz} - 2C_2E_2(z)\frac{dE_2(z)}{dz} = 0 , \qquad (3.54)$$

so that

$$E_2(z)^2 = \frac{1}{2k_{2g}{}^2}\left[C_2 - \frac{d}{dz}\left[E_2(z)\frac{dE_2(z)}{dz}\right]\right] , \qquad (3.55)$$

and the power integral for the linear layer is

$$\int_0^d E_2(z)^2\, dz = \frac{1}{2k_2{}^2}\left[C_2 d - \left[E_2(z)\frac{dE_2(z)}{dz}\right]_d + \left[E_2(z)\frac{dE_2(z)}{dz}\right]_o\right]$$

$$= \frac{1}{2k_2^2}\left[(k_2^2 + \gamma_1^2)E_0^2 d \mp \gamma_3 E_b^2 \pm \gamma_1 E_0^2\right] , \tag{3.56}$$

where the upper signs indicate

$$\left[\frac{dE_2(z)/dz}{E_2(z)}\right]_{z=0} > 0, \qquad \left[\frac{dE_2(z)/dz}{E_2(z)}\right]_{z=d} > 0$$

and vice-versa. If $\beta = \alpha_3/\alpha_1$ and $P_0 = c^2\varepsilon_0/\omega^2\alpha_3$, then the total power flow is rationalized to

$$\begin{aligned} P = P_0 n_x[\ \beta[\sqrt{n_x^2-\varepsilon_1} \pm \sqrt{n_x^2-\varepsilon_1-\alpha_1 E_0^2/2}\,] + \sqrt{n_x^2-\varepsilon_3} \\ \pm \sqrt{n_x^2-\varepsilon_3-\alpha_3 E_b^2/2} \\ + \frac{1}{2(\varepsilon_2-n_x^2)}\,[(\varepsilon_2-\varepsilon_1-\alpha_1 E_0^2/2)\beta D\alpha_1 E_0^2/2 \mp \sqrt{n_x^2-\varepsilon_1-\alpha_1\ E_0^2/2}\beta\ \alpha_1 E_0^2/2 \\ \mp (n_x^2-\varepsilon_3-\alpha_3 E_b^2/2)\alpha_3 E_b^2/2]\ ]\ . \end{aligned} \tag{3.57}$$

In the linear limit, P_1 and P_3 become $P_1 = E_0^2/2k_1$ and $P_3 = E_b^2/2k_3$ so that it is quite easy to project out from (3.57) the limits of a linear guide on a linear substrate that is bounded by a nonlinear superstrate, or vice-versa. It is also easy to project out the symmetric case that consists of a linear guide bounded by two identical media i.e. setting $\alpha_1 = \alpha_3$ and $\varepsilon_1 = \varepsilon_3$ gives

$$P = P_0 k_x\left[\frac{2(k_1 - \gamma_1)}{\Lambda_1} + \frac{E_0^2}{2k_2^2}[(k_2^2 + \gamma_1^2)d + 2\gamma_1]\right] . \tag{3.58}$$

In this symmetric case the boundary field relationship[6] is simply

$$(E_0^2 - E_b^2)\left[\varepsilon_2 - \varepsilon_1 - (E_0^2 + E_b^2)\frac{\alpha_1}{2}\right] = 0 . \tag{3.59}$$

This implies that there exists symmetric and antisymmetric modes as would be expected for such a linearly symmetric guide i.e.

$$E_0^2 = E_b^2 , \tag{3.60}$$

The second option is

$$E_b^2 = \frac{2(\varepsilon_2 - \varepsilon_1)}{\alpha_1} - E_0^2 , \tag{3.61}$$

which is an asymmetric wave. In the linear limit this cannot exist but it can exist in the nonlinear structure because the nonlinearity drives it into an asymmetric state. Hence, for the symmetric or antisymmetric modes with $E_0^2 = E_b^2$, the power formula just involves E_0^2 that, in turn, is obtained from

$$\gamma_1^2 = k_2 \tan\left(\frac{k_2 d}{2}\right); \quad E_0^2 = \frac{\left[k_1^2 - k_2^2\tan^2\left(\frac{k_2 d}{2}\right)\right]}{\Lambda_1} . \tag{3.62}$$

This means that E_0^2 can be eliminated entirely in favor of a function of $(k_2 d)$. In fact, this can only be done, analytically, for the symmetric guide. For other asymmetric guides, the full (E_b^2, E_0^2) relationship must be used. The algorithm then is to determine the (E_b^2, E_0^2) conic section from the physical input parameters of the system, then determine the (n_x, E_0^2) curves and follow this with the calculation of the power flow in terms of n_x.

In the linear limit the power flow is given by

$$P_L = \frac{c^2\varepsilon_0 E_0^2}{4\omega^2}\left\{\frac{1}{\sqrt{n_x^2-\varepsilon_1}} + \left(\frac{\varepsilon_2-\varepsilon_1}{\varepsilon_2-\varepsilon_3}\right)\frac{1}{n_x^2-\varepsilon_3}\right.$$

$$\left. + \frac{1}{\varepsilon_2-n_x^2}\left[(\varepsilon_2-\varepsilon_1)D + (n_x^2-\varepsilon_1) + \left(\frac{\varepsilon_2-\varepsilon_1}{\varepsilon_2-\varepsilon_3}\right)(n_x^2-\varepsilon_3)\right]\right\}, \tag{3.63}$$

where it is interesting to see that, in spite of the linear assumption, E_0^2 is still in evidence. This is not unexpected but the power does not depend upon n_x through E_0^2. Instead it rises vertically from the linear value of n_x as E_0^2 is increased. Of course, for a fixed E_0^2, the power varies quite strongly with n_x. It is important to understand this property of the power flow in a linear guide because it is the feature of varying E_0^2, as n_x varies, that will make the nonlinear power curves look so different and so interesting.

The power-n_x variation is illustrated here both for a guide with only one nonlinear boundary[16] (Fig. 8) and an asymmetric guide[7,8] (Fig. 9). For the symmetric guide there are, as discussed above, symmetric, antisymmetric and asymmetric waves. But what is immediately apparent is that, as E_0^2 is varied, the n_x variation of the power curve can exhibit a maximum for certain values of D. This peak occurs for larger D but, in view of the behaviour of a single surface, it is, again, not an unexpected feature for the following reasons. Surface modes correspond to wave indices to the **right** of $n_x = \sqrt{\varepsilon_2}$ so as $D \to \infty$, all that is left in the possibility of self-focussed surface waves with all wave numbers being beyond $n_x = \sqrt{\varepsilon_2}$. In this limit the power curve drops from infinity at $\sqrt{\varepsilon_2}$ and passes through a minimum. This means that a power **threshold** is needed to generate such a surface wave. At finite D the system can also support **guided** waves at low powers and for wave indices below $\sqrt{\varepsilon_2}$. It is this interplay between the ability of TE waves to propagate both as surface and guided waves that leads to a maximum in the power as D increases. Also as D decreases below the linear cut-off value, nonlinearly induced values are still possible that do not now have a linear limit. The complex structure of the asymmetric power curves is on this basis completely understandable since two non-degenerate surfaces exist. Each produces its own minima as D increases and the system also accesses the sum of such mimima.

Quite apart from the fact that the lower and upper E_b curves give quite different results, the question of whether the dispersion curves of the system should be labelled with $\alpha_1 E_0^2/2$ or the power flow down the guide must be addressed.

The procedure developed up to now provides the $(n_x, \alpha_1 E_0^2/2)$ variation and the $(P/P_0, n_x)$ variation. From these the dispersion curves (n_x, D) labelled with constant power values or labelled with constant $(\alpha_1 E_0^2/2)$ values can be determined. These families of curves are actually quite different from each other and they are shown in Figs. 10 and 11, on which the linear light lines $n_x = \sqrt{\varepsilon_1}$ and $n_x = \sqrt{\varepsilon_2}$ are also drawn. Once again it can be argued that the power labelled curves are the true dispersion curves since only they remain as $D \to \infty$. In this limit the semi-infinite cases are recovered, but also the $\alpha_1 E_0^2/2$ labelled curves disappear.

C. The Nonlinear Film

The nonlinear film presents a different set of mathematical problems.[9,33] The fundamental nonlinear equation is, after integrating once, of the form shown in Eq. (3.5). If the substrate of the film has a dielectric constant ε_1 and the superstrate has dielectric constant ε_3, then, for a film with a linear dielectric constant ε_2 and a nonlinear coefficient α_2, the K=0 real field solutions are

$$\left[\frac{dE_1(z)}{dz}\right]^2_{z=0} = k_1^2 E_0^2 \,, \qquad \left[\frac{dE_3(z)}{dz}\right]^2_{z=d} = k_3^2 E_b^2 \,, \tag{3.64}$$

$$\left[\frac{dE_2(z)}{dz}\right]^2 = k_2^2 E_2(z)^2 - \Lambda_2 E_2(z)^4 + C_2 \,, \tag{3.65}$$

where C_2 is an integration constant that, because, $E_i(z)$ and $dE_i(z)/dz$ are continuous at the boundaries, is given by

$$C_2 = E_0^2 \frac{\omega^2}{c^2} (\varepsilon_2-\varepsilon_1+\frac{\alpha_2}{2}E_0^2) = E_b^2 \frac{\omega^2}{c^2} (\varepsilon_2-\varepsilon_3+\frac{\alpha_2}{2}E_b^2) \tag{3.66}$$

In this case the (E_0^2, E_b^2) relationship is

$$\frac{\alpha_2^2}{(\varepsilon_3-\varepsilon_1)(\varepsilon_1+\varepsilon_3-2\varepsilon_2)} \left[E_b^2 - \frac{\varepsilon_3-\varepsilon_2}{\alpha_2}\right]^2 - \frac{\alpha_2^2}{(\varepsilon_3-\varepsilon_1)(\varepsilon_1+\varepsilon_3-2\varepsilon_2)} \left[E_0^2 - \frac{\varepsilon_1-\varepsilon_2}{\alpha_2}\right]^2 = 1 \ . \tag{3.67}$$

$E_2(z)$, the field in the thin film will be obtained from

$$\left[\frac{dE_2(z)}{dz}\right]^2 = \Lambda_2 \left[\frac{\sqrt{k_2^4 + 4\Lambda_2 C_2} + k_2^2}{2\Lambda_2} - E_2(z)^2\right] \left[\frac{\sqrt{k_2^4 + 4\Lambda_2 C_2}}{2\Lambda_2} + E_2(z)^2\right] , \tag{3.68}$$

which since each factor is greater than zero leads to

$$\int \frac{dE_2(z)}{\sqrt{b^2 - E_2(z)^2}\sqrt{a^2 + E_2(z)^2}} = \Lambda_2^{1/2}(z+z_0) \ , \tag{3.69}$$

where

$$b^2 = \frac{\sqrt{k_2^4 + 4\Lambda_2 C_2} + k_2^2}{2\Lambda_2} \ , \quad a^2 = \frac{\sqrt{k_2^4 + 4\Lambda_2 C_2} - k_2^2}{2\Lambda_2} \ , \tag{3.70}$$

and z_0 is another constant of integration. Now the integration in Eq. (3.69) is a standard form expressible in terms of the Jacobian elliptic function cn $(x|m)$ where x is the argument of the function and m is called the modulus. For economy, however, these will be refered to as Jacobi functions and, more often than not, the modulus will be omitted by simply writing cn(x). The field[9] inside the film is , formally of the form

$$E_2(z) = \sqrt{\frac{q^2 + k_2^2}{2\Lambda_2}}\ \mathrm{cn}\left[\frac{q(z+z_0)|(q^2+k_2^2)}{2q^2}\right] = p\ \mathrm{cn}[q(z+z_0)|m] , \tag{3.71}$$

where $q = (k_2^4 + 4\Lambda_2 C_2)^{1/4}$. At the lower boundary of the film $E_2=E_0$ so that

$$z_0 = \frac{1}{q}\ \mathrm{cn}^{-1}\left(\frac{E_0}{p}\right) . \tag{3.72}$$

Hence, since $E_2= E_b$ at the upper boundary

$$qd = \mathrm{cn}^{-1}\left(\frac{E_b}{p}\right) - \mathrm{cn}^{-1}\left(\frac{E_0}{p}\right) \tag{3.73}$$

which is formally the eigenvalue equation. Since the inverse Jacobi functions are incomplete elliptic integrals, Eq. (3.73) is not as simple, or convenient, as it appears to be. It would be very useful to regain the advantage of using Jacobi functions since they are fairly easy to compute if a Landen transformation is used. This can be done by making qd the argument of another Jacobi function. Exactly which one is quite arbitary so cn is chosen once again to give

$$\begin{aligned}\mathrm{cn}[qd] &= \mathrm{cn}[qz_0 - q(d + z_0)] \\ &= \frac{\mathrm{cn}[qz_0]\mathrm{cn}[q(d+z_0)]+\mathrm{sn}[qz_0]\mathrm{dn}[qz_0]\mathrm{sn}[q(d+z_0)]\mathrm{dn}[q(d+z_0)]}{1-m\mathrm{sn}^2[qz_0]\ \mathrm{sn}^2[q(d+z_0)]}\end{aligned} \tag{3.74}$$

In the substrate $E_1(z) = E_0 e^{k_1 z}$, and in the superstrate $E_3(z) = E_b e^{k_3(d-z)}$ where $k_{1,3}^2 = k_x^2 - \varepsilon_{1,3}\omega^2/c^2$. Therefore, from the boundary conditions

$$\mathrm{cn}[qd] = \frac{2E_0E_b(q^2 - k_1k_3)}{[k_1^2E_0^2 + k_3^2E_b^2 + q^2(E_0^2 + E_b^2) + \Lambda_2(E_0^2 - E_b^2)^2]} \ . \tag{3.75}$$

This is a compact expression that depends explicitly upon the physical input parameters and the boundary field relationship. It can, therefore, be used in a very direct manner to obtain the variation (Fig. 12) of n_x with $\alpha_2E_0^2/2$ as a prelude to determining the dependence of the power flow on n_x. Naturally Eq. (3.75) is only valid for $C_2 > 0$ and $k_2^2 > 0$ where $C_2 > 0$ implies that $\alpha_2E_0^2/2 > \varepsilon_1-\varepsilon_2$. If, for any other data, the parameter of the Jacobi function becomes negative, or greater than unity, then Jacobi´s real transformation must be used to pass to another Jacobi function that will have a different q and m.

In the belief that E_0 is a better physical parameter to work with than the integration constant z_0 or the Jacobi function modulus, $E_2(z)$ is transformed to

$$E_2(z) = qE_0 \left[\frac{q\,\mathrm{cn}[qz] + k_1\mathrm{sn}[qz]\,\mathrm{dn}[qz]}{q^2\mathrm{dn}^2[qz] + \Lambda\ E_0^2\mathrm{sn}^2[qz]}\right] , \tag{3.76}$$

which can also be easily transformed again when m gets out of the range $0 < m < 1$.

The power flow down the guide is $(P_1+P_2+P_3)$ where[9]

$$P_1 = \frac{c^2k_x\varepsilon_0}{\omega\alpha_2}\ \frac{\alpha_2E_0^2}{2k_1} \ ; \qquad P_3 = \frac{c^2k_x\varepsilon_0}{\omega\alpha_2}\ \frac{\alpha_2E_b^2}{2k_3} \ , \tag{3.77}$$

and

$$P_2 = \frac{2c^2k_x\varepsilon_0}{\omega\alpha_2}\ \frac{\alpha_2E_0^2}{2}\ \frac{c}{\omega}\int_0^D Q^2\left[\frac{Q\mathrm{cn}[Qz] + n_1\mathrm{sn}[QZ]\mathrm{dn}[QZ]}{Q^2\mathrm{dn}^2[QZ] + \frac{\alpha_2E_0^2}{2}\mathrm{sn}^2[QZ]}\right]dZ \ , \tag{3.78}$$

where $Z = z\omega/c$, $D = d\omega/c$, $Q = cq/\omega$ and $n_1 = ck_1/\omega$.

The variation of n_x with $\alpha_2 E_0^2/2$ can be found from Eq. (3.75) where, in contrast to the linear layer with nonlinear boundaries, it is the Jacobi function that varies strongly with the optical power density $\alpha_2 E_0^2/2$. The right-hand side of the equation is somewhat insensitive to any change in $\alpha_2 E_0^2/2$. The principal results shown in the figures include the field profiles across the film and the power-wave number relationship.

The power flow down the guide decreases as n_x increases from $\sqrt{\varepsilon_1}$, the cut-off point. It then goes through a minimum and rises quite rapidly as n_x grows indefinitely. For the data chosen this feature occurs for all TE waves so all orders possess a power threshold. For other data only the highest order modes have thresholds.

If the power is split into P_1 flowing in the substrate and P_2 in the nonlinear film, as n_x decreases from the right of the minimum, P_1 increases and P_2 decreases, as shown in Fig. 13. As this occurs the field amplitude in the layer naturally falls. After passing through the minimum in the power, the power in the substrate continues to increase because, as n_x falls towards the cut-off value $\sqrt{\varepsilon_1}$, the gradient of the field amplitude at the boundary approaches zero. This signals the onset of radiative modes at which point an infinite power is required to drive a guided mode. The variation of the field amplitude and the power thresholds for all modes are illustrated in Figs. 14 and 15.

4. P-Polarised Waves

A. Single Interface

The initial impetus to the study of p-polarised waves came from a study[1] based upon the tensor given by Eq. (2.25) that will be called[25] the $\varepsilon_{xx}(E_x^2)$ uniaxial approximation. Eq. (2.25) shows that the essential assumption is that only E_x^2 appears as the nonlinear term. The material is also assumed to be uniaxial in that $\varepsilon_1 \neq \varepsilon_3$. The latter step is not essential and, in view of the fact that even to get to Eq. (2.17) involves some arbitrariness, it could be considered to be an unnecessary complication. For this reason $\varepsilon_1 = \varepsilon_3$ is often used[25] so then it is the nonlinearity that makes the material uniaxial.

The basic equations are (2.30) to (2.32), and, if the diagonal tensor

$$\varepsilon = \begin{bmatrix} \varepsilon_{xx} & 0 & 0 \\ 0 & \varepsilon_{yy} & 0 \\ 0 & 0 & \varepsilon_{zz} \end{bmatrix} \tag{4.1}$$

is assumed, Maxwell's equations reduce to

$$\frac{d^2E_x}{dz^2} - (k_x^2 - \frac{\omega^2}{c^2}\varepsilon_{zz}) \frac{\varepsilon_{xx}}{\varepsilon_{zz}} E_x = 0 \tag{4.2}$$

where $\varepsilon_{xx}=\varepsilon_{\perp} + \alpha(E_x)^2$, and $\varepsilon_{zz}=\varepsilon_{\parallel}$. Eq.(4.2), after making the $\varepsilon_{xx}(E_x^2)$ uniaxial assumption, and after integrating once gives for medium 1 ($z<0$) and medium 2 ($z>0$)

$$z<0: \quad \left[\frac{dE_1(z)}{dz}\right]^2 - \frac{k_1^2}{\varepsilon_{\parallel}^{(1)}} (\varepsilon_{\perp}^{(1)} + \alpha_1 \frac{E_1(z)}{2})E_1(z)^2 = 0 \ , \tag{4.3}$$

$$z>0: \quad \left[\frac{dE_2(z)}{dz}\right]^2 - \frac{k_2^2}{\varepsilon_{\parallel}^{(2)}} (\varepsilon_{\perp}^{(2)} + \alpha_2 \frac{E_2(z)}{2})E_2(z)^2 = 0 \ , \tag{4.4}$$

where $k_i^2 = k_x^2 - \varepsilon_{\parallel}^{(i)}\omega^2/c^2$. The boundary conditions at $z = 0$ are

$$E_1(0) = E_2(0) = E_0 \ , \tag{4.5}$$

$$\frac{\varepsilon_{\parallel}^{(1)}}{k_1^2} \frac{dE_1(z)}{dz} = \frac{\varepsilon_{\parallel}^{(2)}}{k_2^2} \frac{dE_2(z)}{dz} \ . \tag{4.6}$$

These upon application give

$$\frac{\varepsilon_{\parallel}(1)}{k_1^2}\left[\varepsilon_{\perp}(1) + \frac{\alpha_1 E_0^2}{2}\right] = \frac{\varepsilon_{\parallel}(2)}{k_2^2}\left[\varepsilon_{\perp}(2) + \frac{\alpha_2 E_0^2}{2}\right] \qquad (4.7)$$

Eq. (4.7) can be readily re-arranged to yield[1,3]

$$\frac{c^2 k_x^2}{\omega^2} = \frac{\varepsilon_{\parallel}(1)\varepsilon_{\parallel}(2)\left[\varepsilon_{\perp}(1) - \varepsilon_{\perp}(2) + \frac{\alpha_1-\alpha_2}{2}E_0^2\right]}{\left[\varepsilon_{\parallel}(1)\varepsilon_{\perp}(1) - \varepsilon_{\parallel}(2)\varepsilon_{\perp}(2) + \frac{[\alpha_1\varepsilon_{\parallel}(1) - \alpha_2\varepsilon_{\parallel}(2)]}{2}E_0^2\right]} \,. \qquad (4.8)$$

If $\varepsilon_{\parallel}(1) = \varepsilon_{\perp}(1) = \varepsilon_1$, $\varepsilon_{\parallel}(2) = \varepsilon_{\perp}(2) = \varepsilon_2$ and $\alpha_2 = 0$, $\alpha_1 = \alpha$, then Eq. (4.8) simplifies to

$$\frac{c^2 k_x^2}{\omega^2} = \frac{\varepsilon_1\varepsilon_2\left[\varepsilon_1 - \varepsilon_2 + \frac{\alpha E_0^2}{2}\right]}{\left[\varepsilon_1^2 - \varepsilon_2^2 + \frac{\alpha\varepsilon_1 E_0^2}{2}\right]} \,. \qquad (4.9)$$

These equations show that the boundary value relationship, as opposed to the situation encountered for s-polarised waves, does give a dispersion equation in which ω depends upon k and the field amplitude at the boundary. Furthermore, when $\alpha = 0$, Eq. (4.9) reduces to the well-known linear dispersion relationship for p-polarised surface polaritons.

The solutions appear to be extremely satisfactory until it is realized that in Eq. (4.3) and (4.4) the sign of the nonlinear term is **opposite** to that in the TE wave case.[21] This means that the only bounded self-focussing solutions of the form given by Eqs. (3.17) and (3.18) are for **negative** ($\alpha<0$) nonlinear coefficients.

This is curious to say the least. But there is away out because, as was discussed in section 2, there are excellent arguments[21] for believing that it is the E_z

component of the p-polarised polariton field that dominates over E_x in the nonlinear medium.

Some insight into this can be obtained from a consideration of the low power linear limit[49] of p-polarised surface polariton propagation for which it is well-known that, for the constituent medium with dielectric function ε

$$\left|\frac{E_x}{E_z}\right|^2 = \left|\frac{\frac{c^2k_x^2}{\omega^2} - \varepsilon}{\frac{c^2k_x^2}{\omega^2}}\right| \qquad (4.10)$$

Now the linear dispersion relationship is

$$k_x^2 = \frac{\omega^2}{c^2}\,\frac{\varepsilon_1\varepsilon_2}{\varepsilon_1 + \varepsilon_2}\,, \qquad (4.11)$$

where ε_2 is the dielectric function of a metal, for example, and ε_1 is the dielectric function of a cladding or bounding dielectric that is going to become nonlinear. For all applications to metals, and hence surface plasmons, ε_2 is simply taken as large and negative. From Eqs. (4.10) and (4.11) it is clear that, in the linear case,

$$\left|\frac{E_x}{E_z}\right|^2 = \left|\frac{\varepsilon_1}{\varepsilon_2}\right| \quad \text{(cladding)}; \qquad \left|\frac{E_x}{E_z}\right|^2 = \left|\frac{\varepsilon_2}{\varepsilon_1}\right| \quad \text{(metal)}\,. \qquad (4.12)$$

In a metal, as we have said before, $|\varepsilon_2| \gg 1$ so that E_z dominates in the cladding while E_x dominates in the metal. The metal will remain linear even at high input powers so it is the cladding that becomes nonlinear in any application. Therefore, it appears to be correct to use a dominant E_z formulation. An additional argument in favor of this is as follows. If the ck_x/ω in Eq. (4.10) is taken to be <u>approximately</u> the nonlinear refractive index n_{NL} then[21]

$$\left|\frac{E_x}{E_z}\right|^2 \simeq \frac{n_{NL}^2 - \varepsilon_1}{n_{NL}^2} \simeq \frac{2(n_{NL} - \sqrt{\varepsilon_1})}{\sqrt{\varepsilon_1}} , \tag{4.13}$$

where $(n_{NL} - \sqrt{\varepsilon_1})$ is the change in refractive index due to the nonlinearity. Since this usually saturates at about 0.03 , E_z must be the dominant field component in any practical application.

If E_z is used then the whole problem can be formulated in terms of H_y, as was discussed earlier. In fact, the first integration of Eq. (2.28) for H_y is

$$\left[\frac{dH_y}{dz}\right]^2 - (k_x^2 - \frac{\omega^2}{c^2}\varepsilon)H_y^2 + \frac{\alpha k_x^4}{\omega^2\varepsilon_0^2\varepsilon^3} H_y^4 = 0 . \tag{4.14}$$

The field components in media 1 (lower) and 2 (upper) are H_1 and H_2 respectively where

$$\left[\frac{dH_1}{dz}\right]^2 - k_1^2H_1^2 + \frac{\alpha_1 k_x^4}{\omega^2\varepsilon_0^2\varepsilon_1^3} H_1^4 = 0 , \tag{4.15}$$

$$\left[\frac{dH_2}{dz}\right]^2 - k_2^2H_2^2 + \frac{\alpha_2 k_x^4}{\omega^2\varepsilon_0^2\varepsilon_2^3} H_2^4 = 0 , \tag{4.16}$$

and $k_i^2 = k_x^2 - \varepsilon_i\omega^2/c^2$. The boundary conditions to use at $z = 0$ now are

$$\frac{1}{\varepsilon_1^2}\left[\frac{dH_1}{dz}\right]^2 = \frac{1}{\varepsilon_2^2}\left[\frac{dH_2}{dz}\right]^2 \; ; \quad H_1 = H_2 = H_0 . \tag{4.17}$$

Their application , for $\alpha_1 = 0$, leads to

$$\frac{\alpha_2\varepsilon_1 H_0{}^2}{\omega^2\varepsilon_0{}^2\varepsilon_2{}^3}k_x{}^4 + (\varepsilon_2{}^2 - \varepsilon_1{}^2)k_x{}^2 + \frac{\omega^2\varepsilon_2\varepsilon_1}{c^2}(\varepsilon_1 - \varepsilon_2) = 0 \ , \tag{4.18}$$

which is completely different from Eq. (4.9).

The Poynting vector for the p-polarised case is $(-E_z H_y/2)$ and can be expressed as

$$S = \frac{-E_z H_y}{2} = \frac{k_x H_y{}^2}{2\omega\varepsilon_{zz}\varepsilon_0} \ . \tag{4.19}$$

This is a quantity that is discontinuous and ´jumps´ in value as the boundary is crossed.

It is also useful for the $\varepsilon_{xx}(E_x{}^2)$ formulation to note that, designating the upper medium as the cladding (c) and the lower medium as the substrate (s), we have

$$E_x = \sqrt{\frac{2\varepsilon_c}{|\alpha_c|}}\ \frac{1}{\cosh[k_c(z-z_c)]} \ , \tag{4.20}$$

where $k_c{}^2 = k_x{}^2 - \varepsilon_c\omega^2/c^2$. Therefore,

$$H_y = i\frac{\omega\varepsilon_c\varepsilon_0}{k_c}E_x\sqrt{1 + \frac{\alpha_c}{2\varepsilon_c}E_x{}^2} \ . \tag{4.21}$$

Hence, the intensity at the surface is obtainable in terme of $H_y = H_0$ in the $\varepsilon_{zz}(E_z{}^2)$ approximation, or in terms of $E_x = E_0$ in the $\varepsilon_{xx}(E_x^2)$ formulation.

For a linear/nonlinear interface, the surface intensity is (in the $\varepsilon_{zz}(E_z{}^2)$ model)[25]

$$S = \frac{\beta(n_s^2 - n_c^2)\,[\beta^2(n_s^2 + n_c^2) - n_s^2 n_c^2]}{n_{2c} n_c^2 n_s^4}\,, \tag{4.22}$$

where $\beta = ck_x/\omega$ and n_{2c} is the nonlinear coefficient defined earlier on. It is clear now that $n_s > n_c$ is needed for valid solutions to exist and since $\beta \geq n_s$, a minimum value S_{min} must exist where[25]

$$S_{min} = \frac{n_s(n_s^2 - n_c^2)}{n_c^2 n_{2c}}\,. \tag{4.23}$$

If $\alpha < 0$, then physically acceptable solutions do not exist for the $\varepsilon_{zz}(E_z^2)$ approximation. They do exist in the $\varepsilon_{xx}(\varepsilon_x^2)$ case but, because large changes in refractive index are implied it is not probable that they are of any practical value.

For two nonlinear media, we have in the $\varepsilon_{zz}(E_z^2)$ model

$$S = \frac{\beta[\beta^2(n_s^2 + n_c^2) - n_c^2 n_s^2][n_s^2 - n_c^2]}{n_c^2[n_s^4 n_{2c} - n_c^4 n_{2s}]}\,. \tag{4.24}$$

For p-polarised waves in the $\varepsilon_{zz}(E_z^2)$ approximation, a power threshold exists for all dielectric media when at least one of the semi-infinite media is self-focussing and the linear parts of the dielectric functions are positive. Achievable differences, however, in refractive indices at low powers and the nonlinearity saturation value must be traded off against each other.

The perturbative scheme[41] that uses a more general form of dielectric tensor can be applied to the single interface problem provided that the nonlinearity is sufficiently weak. The final result of using the scheme described in detail earlier on, is

$$\frac{\varepsilon_1}{k_1} + \frac{\varepsilon_2}{k_2} + \frac{E_0^2}{4k_2}\left[a(1 + \frac{k_x^4}{k_2^4}) + \frac{2bk_x^2}{k_2^2}\right] = 0 \tag{4.25}$$

where E_0 is the electric field amplitude at the boundary, a and b can be assigned and no uniaxial assumption has been made. The lifting of this assumption is only possible analytically, under restrictive conditions and, unfortunately, begs the general question because the nonlinearity has been taken to be so **weak.** Nevertheless, it is a useful approach if we really do want to know how the weak regime is likely to **evolve.** This is the sort of information that is necessary to the study of envelope solitons.[53,54] The setting of a = b and a further minor rearrangement of Eq. (4.25) brings it into line with a previous ad hoc procedure.[40,45] This was to take the **linear** dispersion equation and then allow part of the guiding structure to have a nonlinear refractive index that depends weakly on the electric field amplitude at the boundary.

B. **Linear Guide with Nonlinear Boundaries**

Although the $\varepsilon_{xx}(E_x^2)$ approximation has been criticised here, it has been used in a number of publications.[1,6,28,29,34,35,37,31,3] The essential formulae are best derived in a general way for three nonlinear media in which a nonlinear film is bounded by two nonlinear media. The integrated field equations are

$$\left[\frac{dE_i}{dz}\right]^2 - \frac{k_i^2}{\varepsilon^{(i)}}\left(\varepsilon^{(i)} + \frac{\alpha_i E_i^2}{2}\right)E_i^2 = C_i \;, \qquad (4.26)$$

where the media are i=1 ($z \le 0$), i=2 ($0 \le z \le d$) and i=3 ($z \ge d$). The integration constants are $C_1 = 0$, $C_3 = 0$ and $C_2 \neq 0$. The boundary conditions are, as usual,

$$[E_1]_{z=0} = [E_2]_{z=0} \;; \qquad [E_2]_{z=d} = [E_2]_{z=d} \;, \qquad (4.27)$$

and

$$\left[\frac{\varepsilon_{\parallel}^{(1)}}{k_1^2}\frac{dE_1}{dz}\right]_{z=0} = \left[\frac{\varepsilon_{\parallel}^{(2)}}{k_2^2}\frac{dE_2}{dz}\right]_{z=0} \;, \qquad (4.28)$$

$$\left[\frac{\varepsilon_{\parallel}(2)}{k_2^2}\frac{dE_2}{dz}\right]_{z=d} = \left[\frac{\varepsilon_{\parallel}(3)}{k_3^2}\frac{dE_3}{dz}\right]_{z=d} , \tag{4.29}$$

These, immediately, give

$$C_2 = \frac{k_2^2 E_0^2}{\varepsilon_{\parallel}(2)}\left[\frac{\varepsilon_{\parallel}(1)}{\varepsilon_{\parallel}(2)}\frac{k_2^2}{k_1^2}\left[\varepsilon_{\perp}(1) + \frac{\alpha_1 E_0^2}{2}\right] - \left[\varepsilon_{\perp}(2) + \frac{\alpha_2 E_0^2}{2}\right]\right] , \tag{4.30}$$

and the field in the film for $C_2 > 0$, $k_2\varepsilon_{\perp}(2)/\varepsilon_{\parallel}(2) > 0$ and $k_2^2\alpha_2/2\varepsilon_{\parallel}(2) > 0$ is

$$E_2 = p\ \mathrm{sc}[q(z+z_0)|m] , \tag{4.31}$$

where

$$q = \left[\frac{k_2^2\varepsilon_{\perp}(2)}{\varepsilon_{\parallel}(2)} + [k_2^4\left[\frac{\varepsilon_{\perp}(2)}{\varepsilon_{\parallel}(2)}\right]^2 - \frac{2k_2^2\alpha_2 C_2}{\varepsilon_{\parallel}(2)}]^{1/2}\right]^{1/2} , \tag{4.32}$$

$$p = \left[\frac{2(k_2^2\varepsilon_{\perp}(2) - q\varepsilon_{\parallel}(2))}{\alpha_2 k_2^2}\right]^{1/2} \tag{4.33}$$

$$m = 2 - \frac{k_2^2\varepsilon_{\perp}(2)}{q^2\varepsilon_{\parallel}(2)} . \tag{4.34}$$

Following the same procedure as we did for TE waves on a nonlinear film, the following eigenvalue equation is obtained

$$\mathrm{sn}[qd] = \frac{\pm E_0 E_b[\varphi_1(p^2 + E_0^2) \pm \varphi_3(p^2 + E_b^2)]}{q\sqrt{p^2 + E_0^2}\ \sqrt{p^2 + E_b^2}\ [(p^2 + E_0^2)(p^2 + E_b^2) - mE_0^2E_b^2]} , \tag{4.35}$$

where

$$\varphi_1 = k_1\sqrt{\frac{\varepsilon_\perp^{(1)} + \alpha_1 E_0^2/2}{\varepsilon_\parallel^{(1)}}}\ , \qquad \varphi_3 = k_3\sqrt{\frac{\varepsilon_\perp^{(3)} + \alpha_3 E_0^2/2}{\varepsilon_\parallel^{(3)}}}\ . \tag{4.36}$$

From this expression all the limiting TM cases that exist in the literature can be obtained.

The classic application[17,23] of the p-polarised theory is to the case of a thin metal film bounded on one or both sides by a nonlinear medium. This case is of special interest because of the existence of long-range plasmons in the linear limit. The basic field equation, assuming that E_z is the dominant field component is Eq. (4.14). The solutions for $\alpha > 0$ are as follows.

In the upper (cladding) medium,[23] as in the formally equivalent TE theory,

$$H_y = \frac{c}{\omega}\sqrt{\frac{2}{\alpha_c'}}\ \frac{k_c}{\cosh[k_c(z-z_c)]}\ , \tag{4.37}$$

where $k_c^2 = k_x^2 - \varepsilon_c\omega^2/c^2$ and $\alpha_c' = c^2k_x^4\alpha_c/\omega^4\varepsilon_0^2\varepsilon_c$. In the film it is possible to write the H_y field as

$$H_y = \frac{c}{\omega}\sqrt{\frac{2}{\alpha_c}}\ \frac{k_c}{\cosh[k_c z_c]}\left[\cosh(k_m z) + \frac{\bar{k}_c}{k_m}\tanh(k_c z_c)\ \sinh(k_m z)\right]\ , \tag{4.38}$$

where $\bar{k}_c = k_c\varepsilon_m/\varepsilon_c$, $k_m = k_x^2 - \varepsilon_m\frac{\omega^2}{c^2}$ and E_m is the dielectric function of the metal. If the film is of thickness d and if the substrate is also nonlinear, it will have a field

$$H_y = \frac{c}{\omega}\sqrt{\frac{2}{\alpha_s}}\ \frac{k_s}{\cosh[k_s(z-d+z_s)]}\ . \tag{4.39}$$

Matching the boundary conditions gives,[23]

$$\tanh(k_m d) = \frac{k_m\left[\bar{k}_c\tanh(k_c z_c) + \bar{k}_s\tanh(k_s z_s)\right]}{-k_m{}^2 - \bar{k}_c\bar{k}_s\tanh(k_c z_c)\ \tanh(k_s z_s)} \tag{4.40}$$

which is the eigenvalue equation. The power and attenuation can also be calculated now. This theory is illustrated numerically in the paper by Stegeman and Seaton so only the conclusions will be summarised here.

The numerical study of nonlinear surface polaritions associated with thin metal films has been performed[23] for data that is typical of InSb in a self-focussing or a self-defocussing state. Attenuation has been "patched" into the problem and it is found, for example, that long-range plasmons cannot be excited at power levels above 10mw/mm if self-focussing is assumed.

REFERENCES

1. V. M. Agranovich, V. S. Babichencko, V. Ya Chernyak, Sov. Phys. JEPT. Lett., **32**, 512 (1981).
2. A. A. Maradudin, Z. Phys. B. **41**, 341 (1981).
3. A. A. Maradudin, in Optical and Acoustic Waves in Solids-Modern Topics, M. Borissov, ed., (World Scientific Publ., Singapore) (1983).
4. N. N. Akmediev, Sov. Phys. JETP, **56**, 299 (1982).
5. A. D. Boardman, P. Egan and A. Shivarova, App. Sci. Res., **41**, 345 (1984).
6. A. D. Boardman and P. Egan, Journ. de Physique, Coll. C., **45**, 291 (1984).
7. A. D. Boardman and P. Egan, Phil. Trans. Roy. Soc. Lond. A., 313, 173 (1984).
8. A. D. Boardman and P. Egan, IEEE J. Quantum Elec., (October) (1985).
9. A. D. Boardman and P. Egan, IEEE J. Quantum Elec., (February) (1986).
10. A. E. Kaplan, Sov. Phys. JEPT, **45**, 896 (1977).
11. A. E. Kaplan, IEEE J. Quantum Elec. **QE-17**, 336 (1981).
12. P. W. Smith, W. Tomlinson, P. J. Maloney, and J. P. Hermann, IEEE J. Quantum Elec. **QE-17**, 340 (1981).
13. W. J. Tomlinson, J. P. Gordon, P. W. Smith, and A. E. Kaplan, Appl. Optics **21**, 2041 (1982).
14. H. Vach, C. T. Seaton, G. I. Stegeman, and I. C. Khoo, Opt. Lett., **9**, 238 (1984).
15. N. N. Akmediev, Sov. Phys. JETP, **57**, 111 (1983).
16. G. I. Stegeman, C. T. Seaton, J. Chilwell, and S. W. Smith, App. Phys. Lett., **44**, 830 (1984).
17. G. I. Stegeman and C. T. Seaton, Opt. Lett., 9, 235 (1984).
18. G. I. Stegeman, J. D. Valera, C. T. Seaton, J. Sipe, and A. A. Maradudin, Sol. St. Comm., 52, 293 (1984).
19. C. T. Seaton, J. D. Valera, R. L. Shoemaker, and G. I. Stegeman, App. Phys. Lett., **45**, 1162 (1984).
20. G. I. Stegeman, C. T. Seaton and H. G. Winful, Phil. Tran. R. Soc. Lond. A.

313, 321 (1984).

21. C. T. Seaton, J. D. Valera, B. Svensson and G. I. Stegeman, Opt. Lett., **10**, 149 (1985).
22. G. I. Stegeman and C. T. Seaton, Opt. Lett., **9**, 235 (1984).
23. J. Ariyasu, C. T. Seaton, G. I. Stegeman, A. A. Maradudin, and R. F. Wallis, (in press) (1985).
24. C. T. Seaton, G. I. Stegeman, W. M. Hetherington III, and H. G. Winful Proc. 3rd European Integrated Optics Conf. "Series in Optical Sciences" H. P. Nolting and R. Ulrich, Ed. pp. 178 Springer (1985).
25. G. I. Stegeman, C. T. Seaton, J. Ariyasu, R. F. Wallis, and A. A. Maradudin, (preprint) (1985).
26. C. Liao and G. I. Stegeman, Appl. Phys. Lett., **44**, 164 (1984).
27. G. I. Stegeman, IEEE J. Quantum Electron, **QE-18**, 1619 (1982).
28. V. K. Fedyanin and D. Mihalache, Zeit. Fur Physik B, **47**, 167 (1982).
29. A. I. Lomtev, Opt. Soectr., **55**, 656 (1984).
30. U. Langbein, F. Lederer and H. E. Ponath, Opt. Comm., **46**, 167 (1983).
31. D. Mihalache and H. Totia, Rev. Roum. Phys., **29**, 365 (1984).
32. K. Leung, (preprint) (1985).
33. N. N. Admediev, K. O. Boltar, and V. M. Eleonskii, Opt. Spect., **53**, 654 (1982).
34. F. Lederer, U. Langbein, and H. E. Ponath, App. Phys. B., **31**, 69 (1983).
35. F. Lederer, U. Langbein, and H. E. Ponath, App. Phys. B., **31**, 187 (1983).
36. W. J. Tomlinson, Opt. Lett., **5**, 323 (1980).
37. D. Mihalache and V. K. Fedyanin, Theor. and Math Physics, **54**, 289 (1983).
38. A. G. Boev, Sov. Phys. JETP, **50**, 47 (1979).
39. A. G. Boev, sov. Phys. JETP, **52**, 67 (1980).
40. G. M. Carter, Y. J. Chen, and S. U. Tripathy, App. Phys. Lett. **43**, 891 (1983).
41. V. M. Agranovich and V. Ya Chernyak, Sol. St. Comm. **8**, 1309 (1982).
42. M. Y, Yu, Phys. Rev. A, **28**, 1855 (1983).
43. K. M. Leung, Phys. Rev. A, **31**, 1189 (1985).

44. C. Liao, G. I. Stegeman, C. T. Seaton, R. L. Shoemaker, J. D. Valera, and H. G. Winful (preprint) (1985).

45. Y. J. Chen and G. M. Canter, App. Phys. Lett., **44**, 164 (1984).

46. D. J. Robbins, Opt. Comm., **47**, 309 (1983).

47. A. G. Litvak and V. A. Mironov, ISV. Vysch. Uch. Zav.-radiofiskika, **11**, 1911 (1968). (One of the authors (ADB) is extremely grateful to Professor Litvak for pointing out the existence of this paper to him and for valuable private discussions at the First International Conference on Surface Waves in Solid State and Gaseous Plasmas at Sofia University, Bulgaria 1981).

48. Y. R. Alanakyan, Sov. Phys. Tech. Phys., **12**, 587 (1967).

49. A. D. Boardman, Ed. **Electromagnetic Surface Modes**, New York, Wiley, 1982.

50. A. D. Boardman, D. E. O´Connor, and P. A. Young, **Symmetry and Its Applications in Science**, London, McGraw-Hill, (1973).

51. P. D. Maker and R. Terhune, Phys. Rev. **137**, A801 (1964)

52. V. I. Karpman, **Nonlinear Waves in Dispersive Media** Pergamon (1975).

53. A. D. Boardman,G.S.Cooper, A. A. Maradudin and J. P. Shen, Phys. Rev. (to be published) (1985).

54. A.D.Boardman,G.S.Cooper and P.Egan, Journ. de Physique, Coll.C., **45**, 197 (1984)

	$\alpha_1 E_0^2/2 < \varepsilon_2-\varepsilon_1$			
	Even parity		Odd parity	
	Single bulge	Double bulge decay	Single bulge	Double bulge decay
SURFACE $\cosh(k_{2s}d) =$	$\frac{\gamma_1\gamma_3-k_{2s}^2}{(\gamma_1^2-k_{2s}^2)^{\frac{1}{2}}(\gamma_3^2-k_{2s}^2)^{\frac{1}{2}}}$	NO EIGENVALUE EQUATIONS		$\frac{\gamma_1\gamma_3+k_{2s}^2}{(\gamma_1^2-k_{2s}^2)^{\frac{1}{2}}(\gamma_3^2-k_{2s}^2)^{\frac{1}{2}}}$
GUIDED $\cos(k_{2g}d) =$	$\frac{k_{2g}^2+\gamma_1\gamma_3}{(\gamma_1^2+k_{2g}^2)^{\frac{1}{2}}(\gamma_3^2+k_{2g}^2)^{\frac{1}{2}}}$	$\frac{k_{2g}^2-\gamma_1\gamma_3}{(\gamma_1^2+k_{2g}^2)^{\frac{1}{2}}(\gamma_3^2+k_{2g}^2)^{\frac{1}{2}}}$	$\frac{-k_{2g}^2-\gamma_1\gamma_3}{(\gamma_1^2+k_{2g}^2)^{\frac{1}{2}}(\gamma_3^2+k_{2g}^2)^{\frac{1}{2}}}$	$\frac{-k_{2g}^2+\gamma_1\gamma_3}{(\gamma_1^2+k_{2g}^2)^{\frac{1}{2}}(\gamma_3^2+k_{2g}^2)^{\frac{1}{2}}}$
	$\alpha_1 E_0^2/2 > \varepsilon_2-\varepsilon_1$			
SURFACE $\cosh(k_{2s}d) =$	$\frac{k_{2s}^2-\gamma_1\gamma_3}{(k_{2s}^2-\gamma_1^2)^{\frac{1}{2}}(k_{2s}^2-\gamma_3^2)^{\frac{1}{2}}}$	$\frac{k_{2s}^2+\gamma_1\gamma_3}{(k_{2s}^2-\gamma_1^2)^{\frac{1}{2}}(k_{2s}^2-\gamma_3^2)^{\frac{1}{2}}}$	NO EIGENVALUE EQUATIONS	
GUIDED	NO EIGENVALUE EQUATIONS		NO EIGENVALUE EQUATIONS	

TABLE 1

FIGURE CAPTIONS

Figure 1. Field profiles for (a) field amplitude maximum in the upper layer (medium 2) and (b) a field amplitude maximum in the lower layer (medium 1). the shading divides the total field profile into the three terms of the power flow equation. The medium sustaining the field amplitude maximum must have a positive nonlinear coefficient.

Figure 2. P/P_0 versus n_x for the case when the upper medium (medium 2) is linear.

Figure 3. The nonlinear optical waveguide consisting of a thin linear dielectric film, with dielectric constant ε_2, sandwiched between two optically non-linear semi-infinite dielectrics, with dielectric functions ε_{NL1} and ε_{NL2}. In each case the nonlinear mechanism is the optical Kerr effect and operates at the fundamental frequency ω.

Figure 4. Field pattern distributions across the waveguide. $\varepsilon_1 = 2.45$, $\varepsilon_2 = 2.6$, $\varepsilon_3 = 2.3$, $\beta = \alpha_3/\alpha_1 = 1.05$

Figure 5. The effect of increasing the symmetry of the waveguide on the boundary field amplitude relation. In the limit of a symmetrically loaded film the hyperbolae degenerate to straight lines. Note that each set of hyperbolae has a different vertical scale. The unusual asymmetric waves in a completely symmetric structure are clearly shown here.

Figure 6. (a) Graphical TE_0 guided wave solutions of the second equation in Table 1 for the lower E_b curve, and a typical D. The shaded part of the cosine curve denotes the physically acceptable solutions. $n_{2g}^2 = \varepsilon_2-\varepsilon_1-\alpha_1 E_0^2/2$, $n_x = \sqrt{\varepsilon_1}$, at B, the linear cut-off point. $n_x = \sqrt{\varepsilon_2}$ at $n_{2g} = 0$ and $\alpha_1 E_0^2/2 = \varepsilon_2-\varepsilon_1$. At point A, a bulge in the field is on the point of entering the nonlinear medium.

(b) Graphical TE_0 guided wave solutions for the lower E_b curve of both the first and second equations in Table 1. The shaded part of the cosine curve shows acceptable "bulge" guided wave solutions.

Figure 7. Lower E_b curve, TE_0 solutions. The dotted line is drawn at $n_x = \sqrt{\varepsilon_2}$ that is one of the light lines of the linear system. The guided wave region lies in the range $\sqrt{\varepsilon_1} \leq n_x \leq \sqrt{\varepsilon_2}$ and the surface wave region is $n_x \geq \sqrt{\varepsilon_2}$. The data is the same as in Fig. 6 and all the curves are labelled with $\delta = D/2\pi = \omega d/2\pi c$.

Figure 8. The effective refractive index $\beta(=n_x)$ versus the guided wave power[16] for TE_0(A) and TE_1(B) waves guided by a film of thickness 2 μm.

Figure 9. Variation of energy flow[7,8] down the guide with effective refractive index, n_x, for both the lower and the upper E_b curves. The surface wave region begins at $n_x = \sqrt{\varepsilon_2}$ and TE_0, TE_1 and TE_2 solutions are shown for the typical data set of Fig. 4, section 4. The curves are labelled with $\delta = D/2\pi$.

Figure 10. Dispersion curves labelled with optical power density, for lower E_b curve. Data as for Fig. 4.

Figure 11. Dispersion curves labelled with power (i.e. energy) flow P down the guide, for lower E_b curve. Data as for Fig. 4.

Figure 12. Dependence of wave index of TE_0 waves on the optical power density at the lower boundary of a nonlinear film. The curves are labelled with values of $\omega d/2\pi c$ and they do not exist below the cut-off line $n_x = \sqrt{\varepsilon_1}$.

Figure 13. Separation of the total power flow of a TE_0 wave into constituent power flows in the substrate (P_1) and in a nonlinear film (P_2) sitting on that substrate. $\varepsilon_1 = 2.45$, $\varepsilon_2 = 2.3$, $\varepsilon_3 = 1$.

Figure 14. Electric field profiles of a TE_0 wave inside a nonlinear film as the power flow charges with wave index. Note that as $n_x \rightarrow E_1$ the gradient of the field amplitude at the boundary approaches zero.

Figure 15. Illustration of the minimum power levels for TE_0, TE_1, and TE_2 waves. Further increases in film thickness will not significantly lower the minima.

Figure 16. Variation in the real ($\beta = ck_x/\omega$) and imaginary (β_I) part of the effective refractive index as a function of guided wave power for surface plasmons guided by the interface between a metal ($\varepsilon_m = -1000 -160i$) and a nonlinear dielectric ($\varepsilon_c = 16 -0.0096i$) with $n_{2c} = 10^{-7} m^2/W$ (dashed line) and $n_{2c} = -10^{-7} m^2/W$ based upon the $E_{zz}(E_z^2)$ approximation. For β_I, the lower curve corresponds to a lossless dielectric, and the upper curve includes losses in both the metal and the nonlinear dielectric. The dash-dot line corresponds to the $\varepsilon_{xx}(E_x^2)$ approximation.

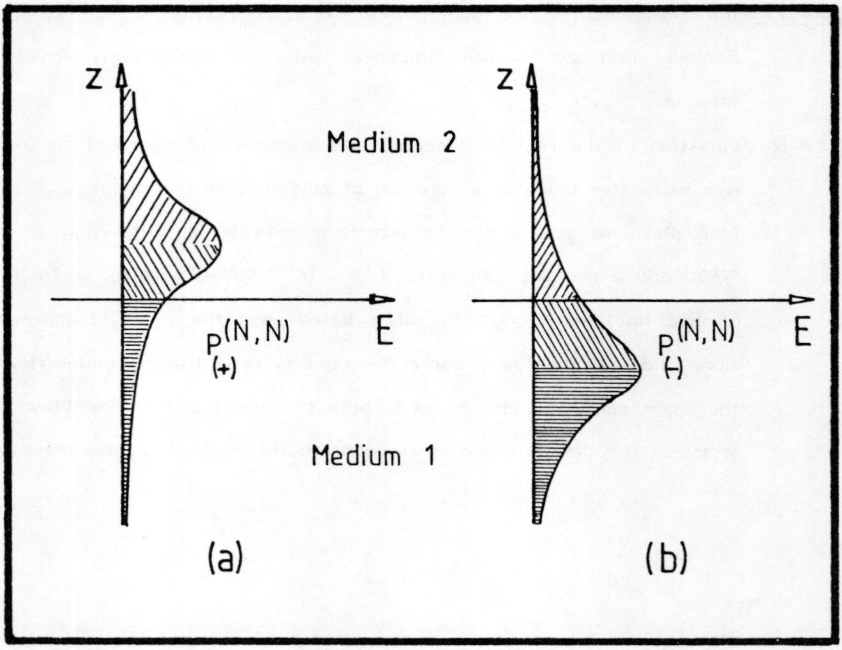

Fig. 1

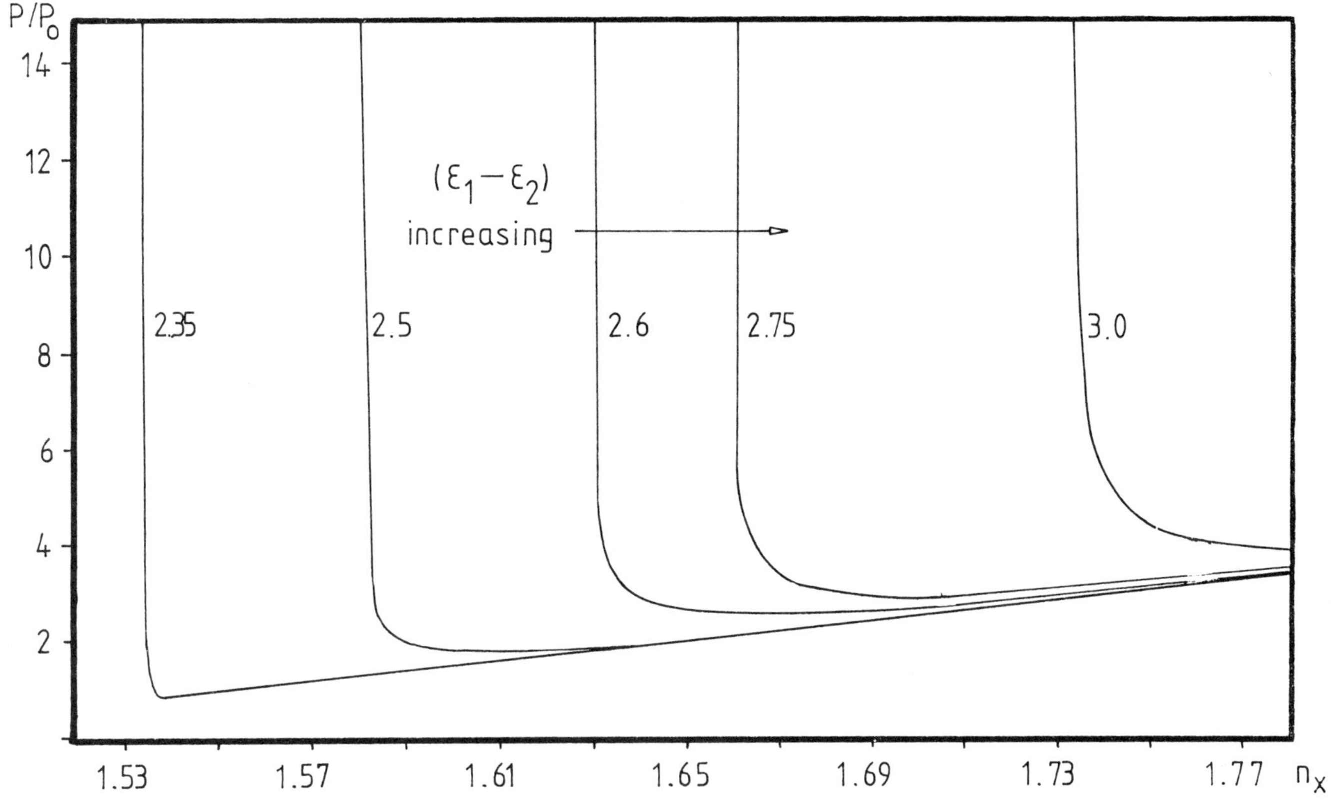
P/P_o
14
12
10
8
6
4
2
1.53
1.57
1.61
1.65
1.69
1.73
1.77
n_x
2.35
2.5
2.6
2.75
3.0
(ε_1 − ε_2)
increasing

Fig. 2

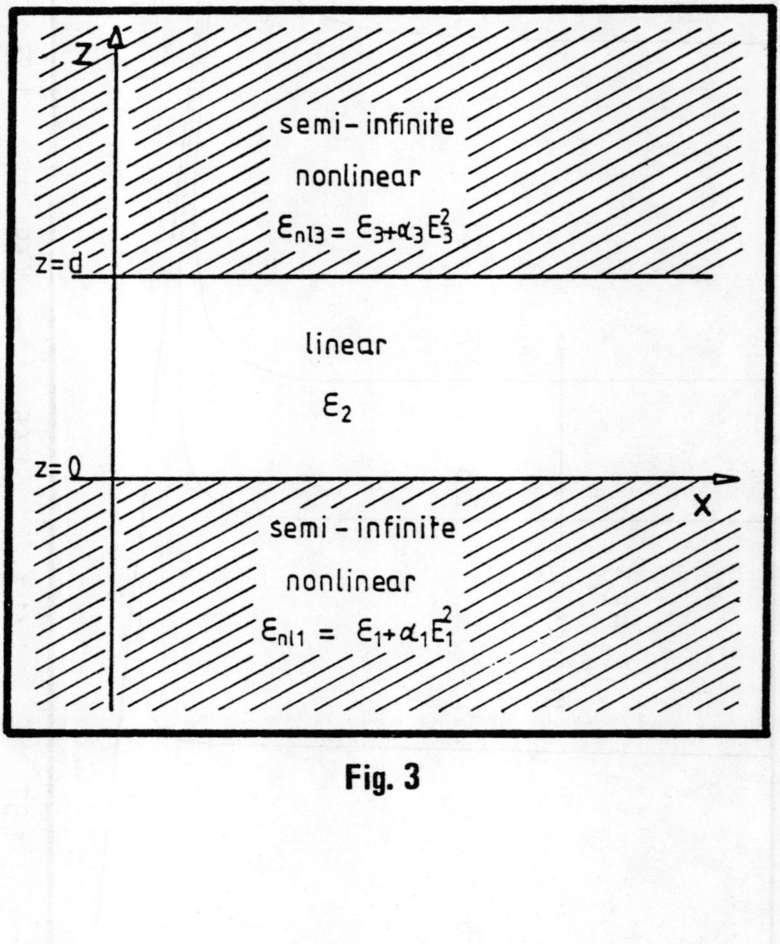
z
semi-infinite
nonlinear
$\varepsilon_{nl3} = \varepsilon_3 + \alpha_3 E_3^2$
z=d
linear
ε_2
z=0
x
semi-infinite
nonlinear
$\varepsilon_{nl1} = \varepsilon_1 + \alpha_1 E_1^2$

Fig. 3

Fig. 4

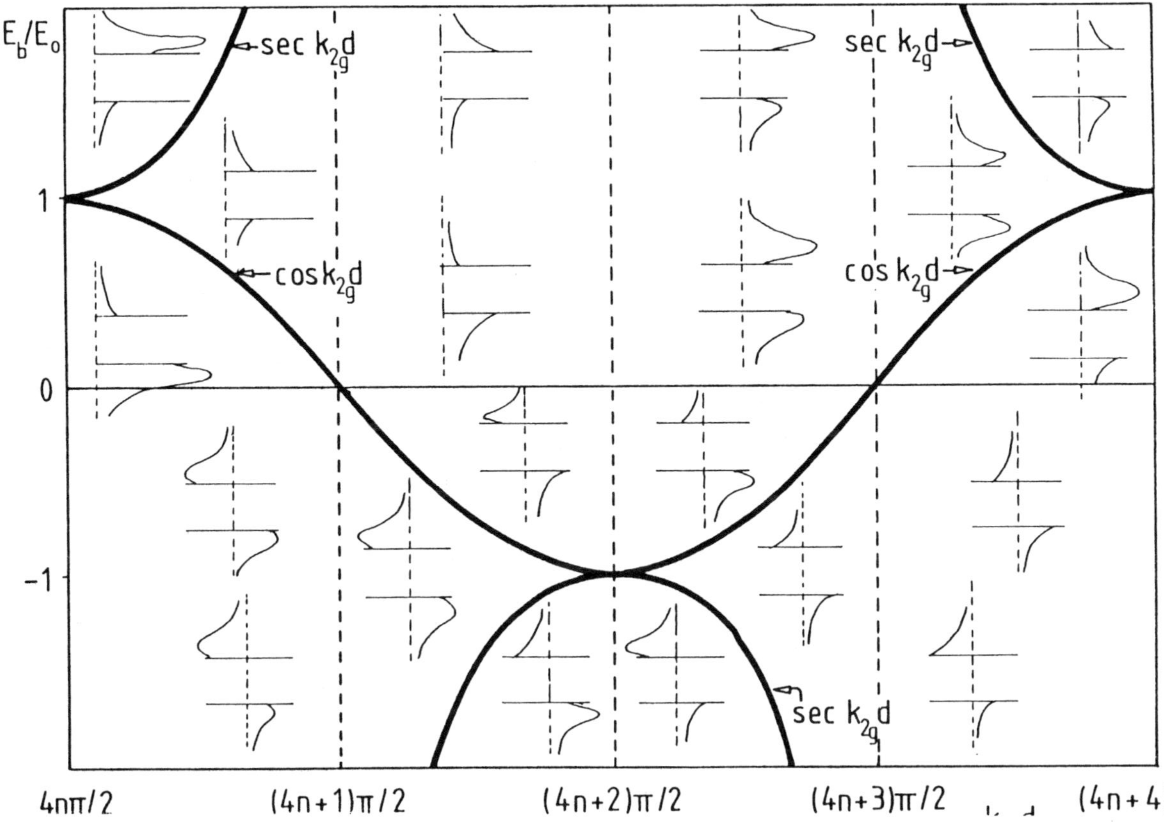

E_b/E_o
1
0
−1
sec k_2g d
cos k_2g d
sec k_2g d
cos k_2g d
sec k_2g d
4nπ/2
(4n+1)π/2
(4n+2)π/2
(4n+3)π/2
(4n+4)π/2

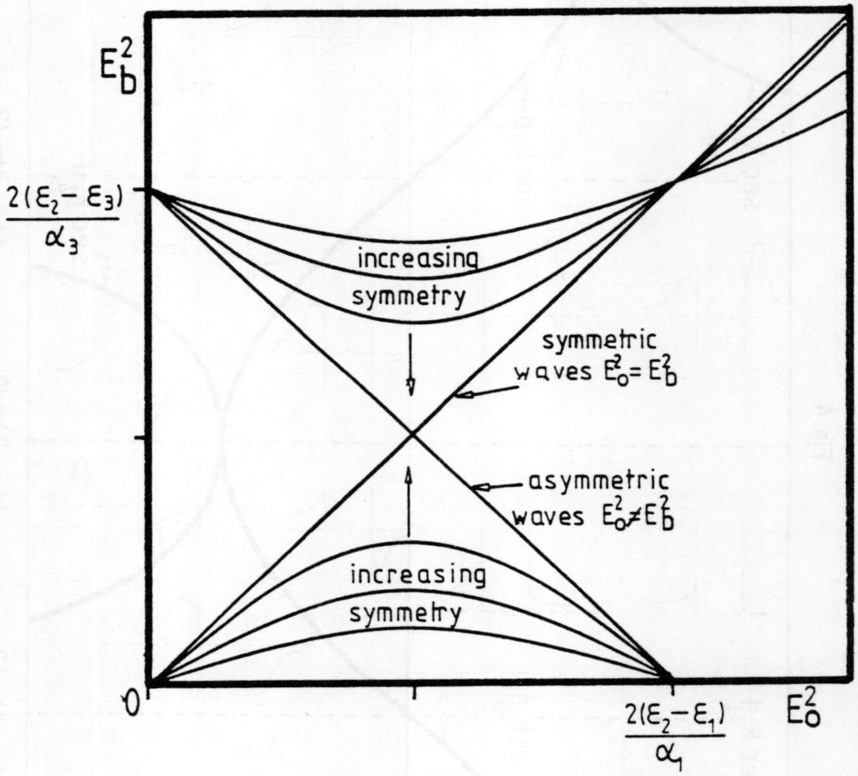

E_b^2
$\frac{2(\varepsilon_2-\varepsilon_3)}{\alpha_3}$
increasing
symmetry
symmetric
waves $E_o^2=E_b^2$
asymmetric
waves $E_o^2\neq E_b^2$
increasing
symmetry
0
$\frac{2(\varepsilon_2-\varepsilon_1)}{\alpha_1}$
E_o^2

Fig. 5

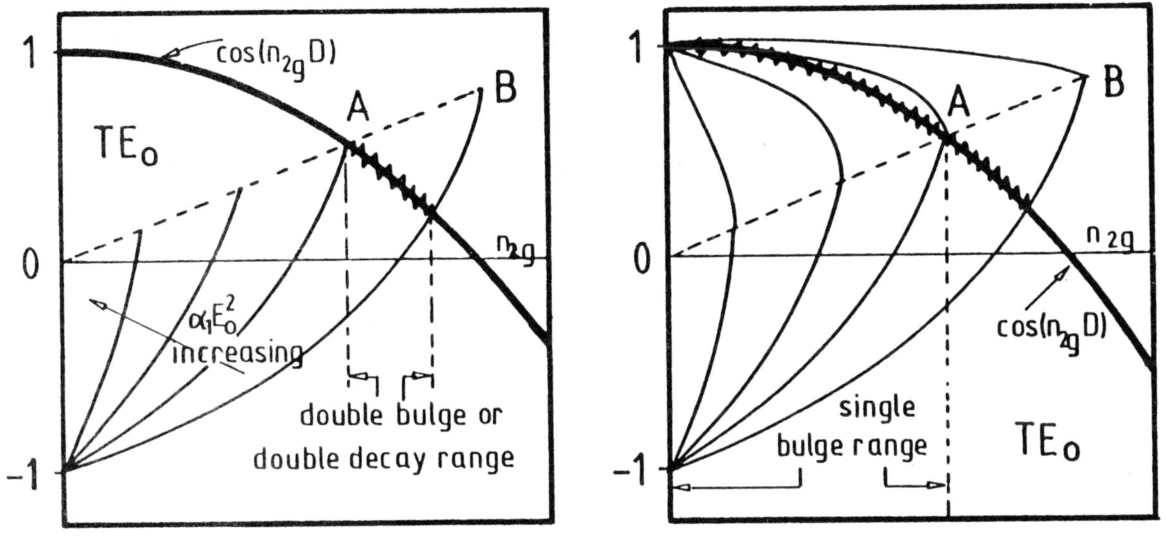
1
0
-1
TE0
cos(n2gD)
A
B
n2g
α1E0²
increasing
double bulge or
double decay range
cos(n2gD)
single
bulge range

(a) (b)

Fig. 6

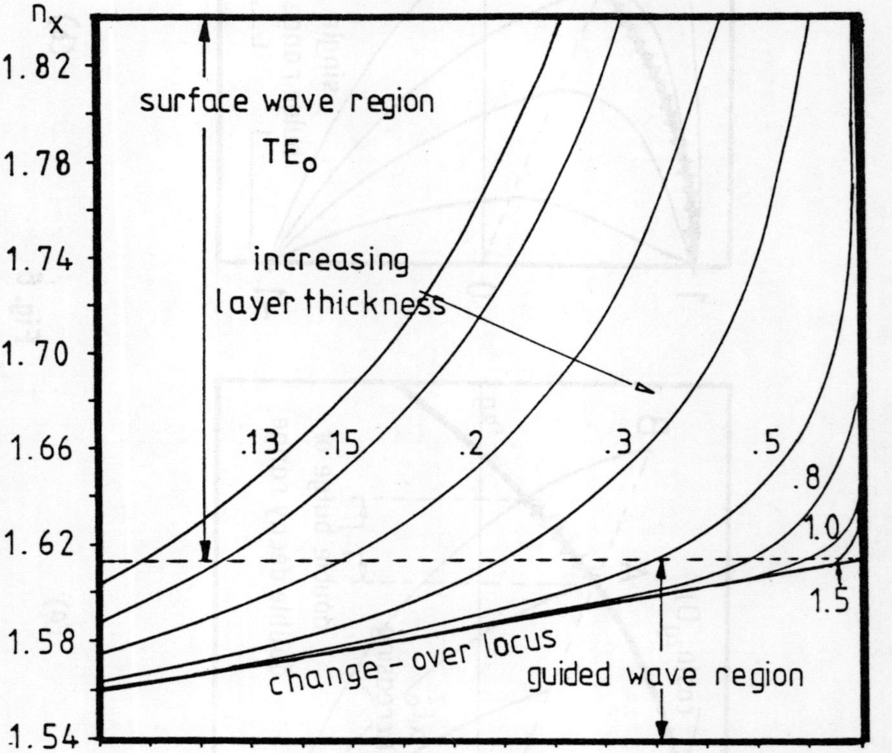

n_x
1.82
1.78
1.74
1.70
1.66
1.62
1.58
1.54
surface wave region
TE_o
increasing
layer thickness
.13
.15
.2
.3
.5
.8
1.0
1.5
change - over locus
guided wave region
0.05
0.10
0.15
$\alpha_1 E_o^2/2$

Fig. 7

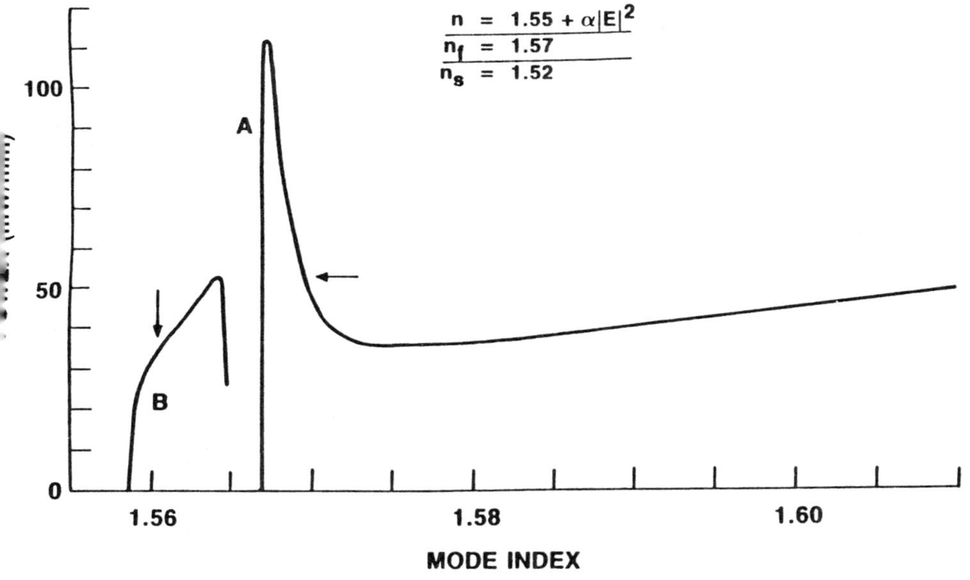

Fig. 8

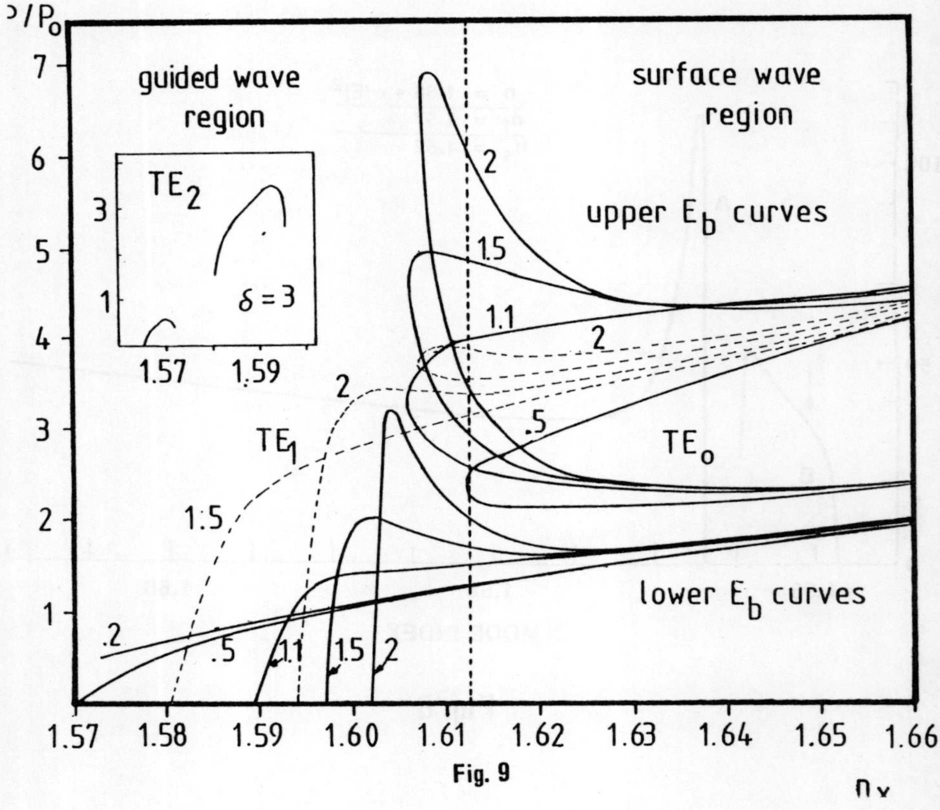
P/Po
guided wave region
surface wave region
TE2
δ = 3
upper Eb curves
lower Eb curves
TE1
TE0
nx

Fig. 9

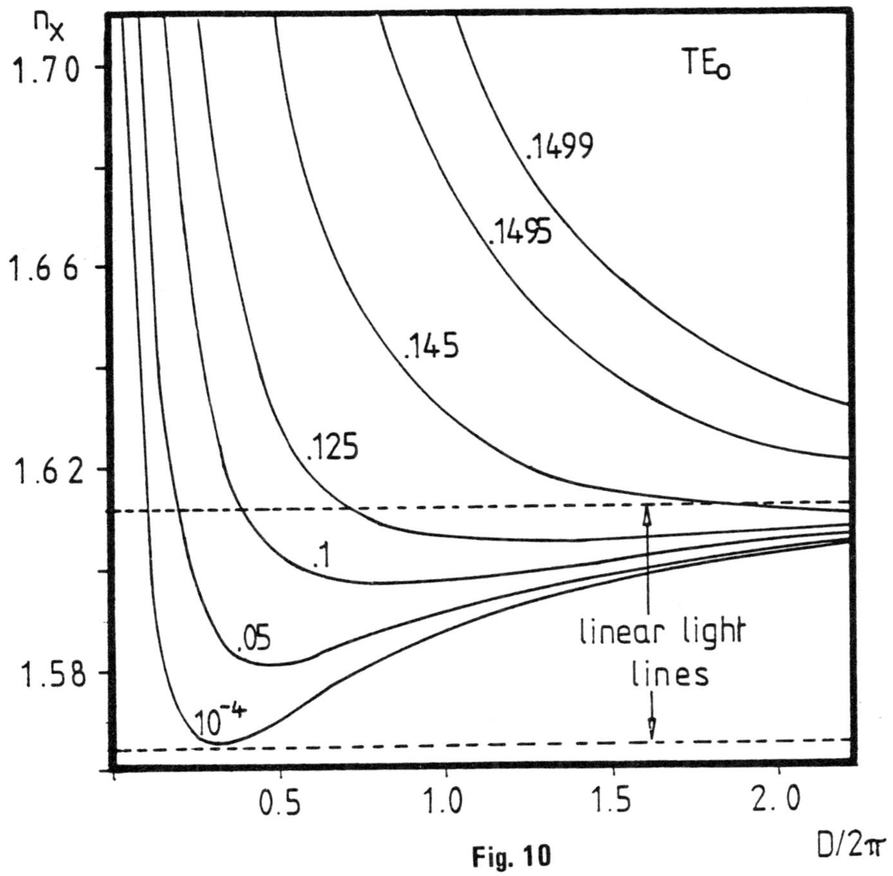

n_x
1.70
1.66
1.62
1.58
TE$_0$
.1499
.1495
.145
.125
.1
.05
10^{-4}
linear light lines
0.5
1.0
1.5
2.0
D/2π

Fig. 10

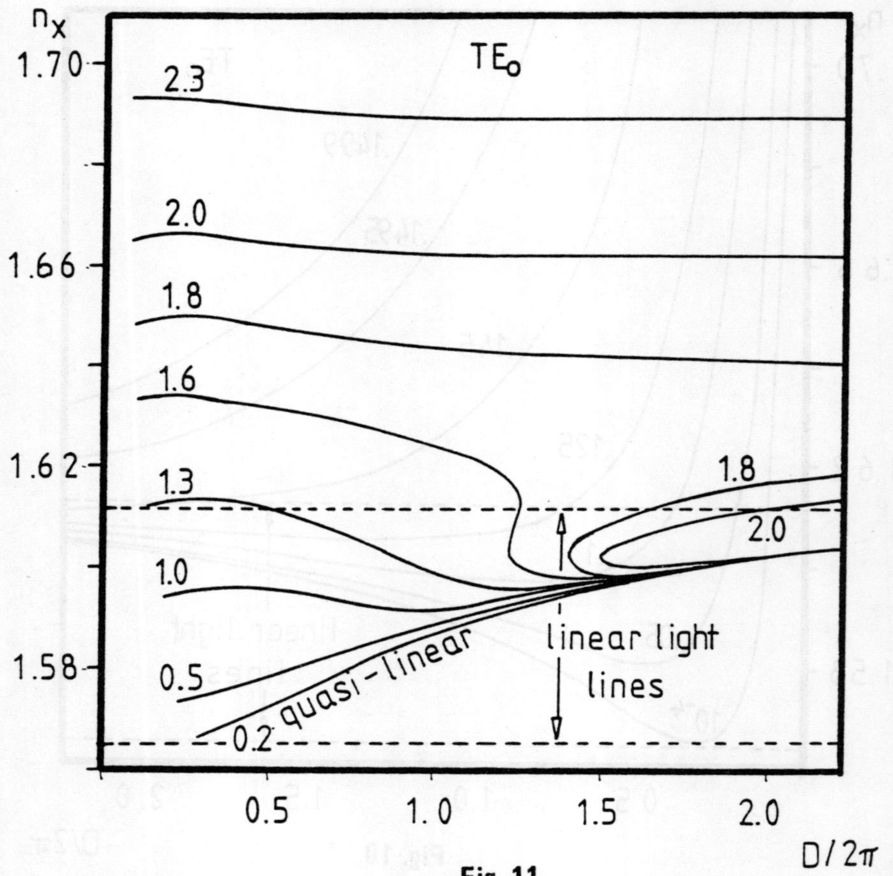

n_x
1.70
1.66
1.62
1.58
TE_0
2.3
2.0
1.8
1.6
1.3
1.0
0.5
0.2
1.8
2.0
quasi-linear
linear light
lines
0.5
1.0
1.5
2.0
D/2π

Fig. 11

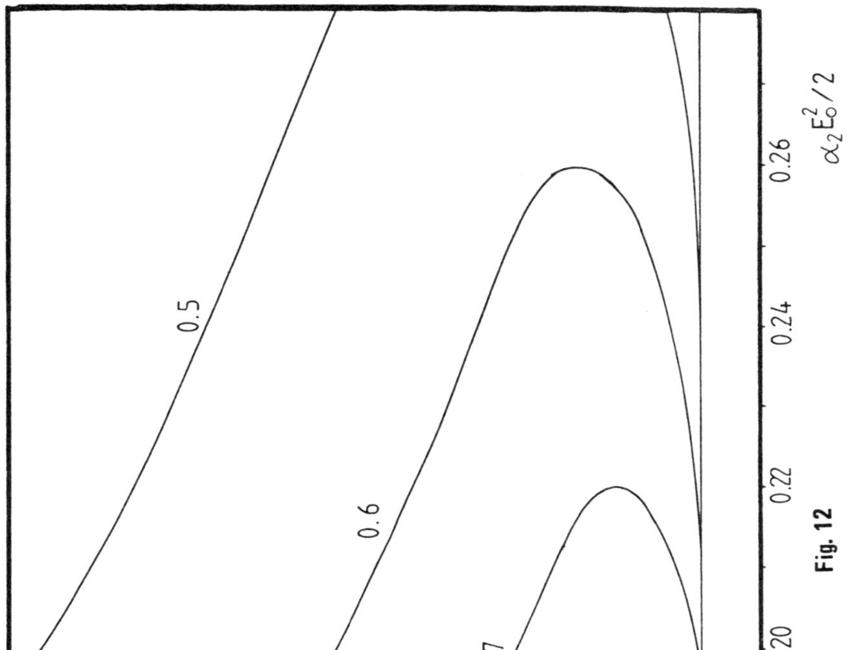

n_x
2.3
2.1
1.9
1.7
1.5
0.16
0.18
0.20
0.22
0.24
0.26
$\alpha_2 E_o^2/2$
0.5
0.6
0.7
0.8
1.0
1.2
1.5

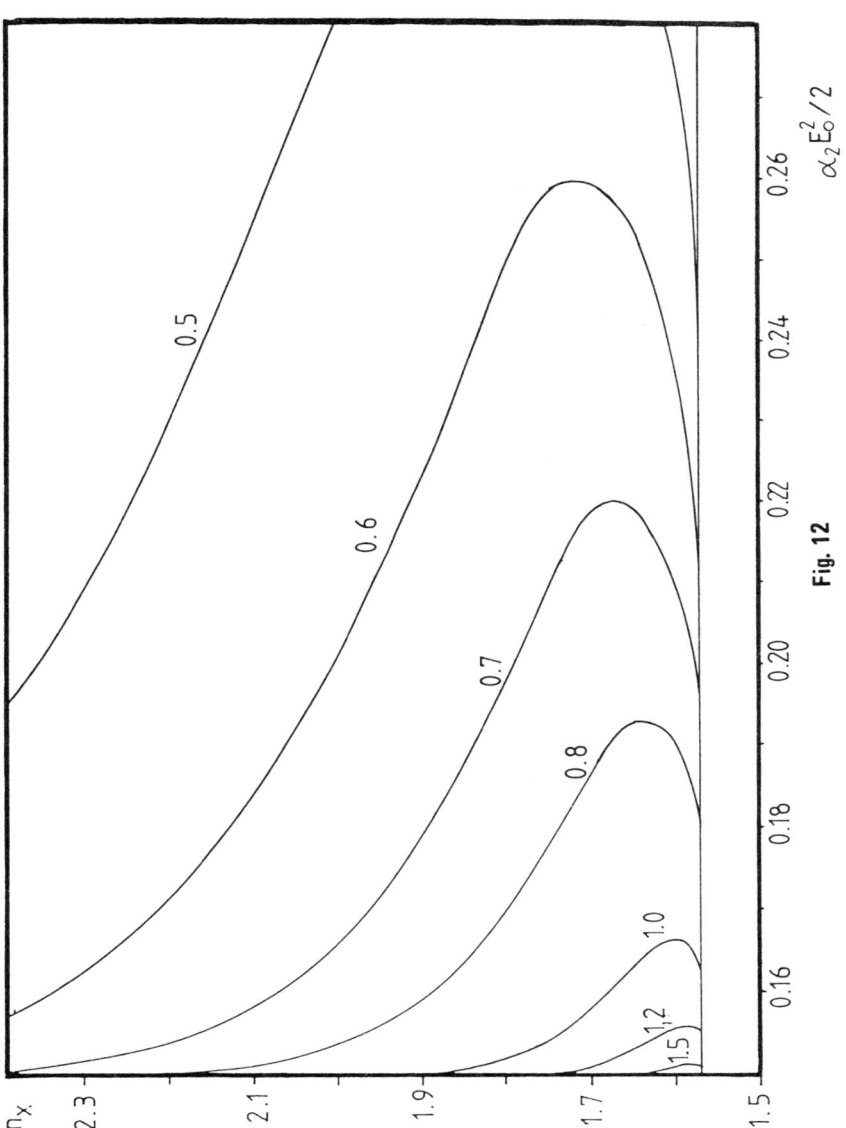

Fig. 12

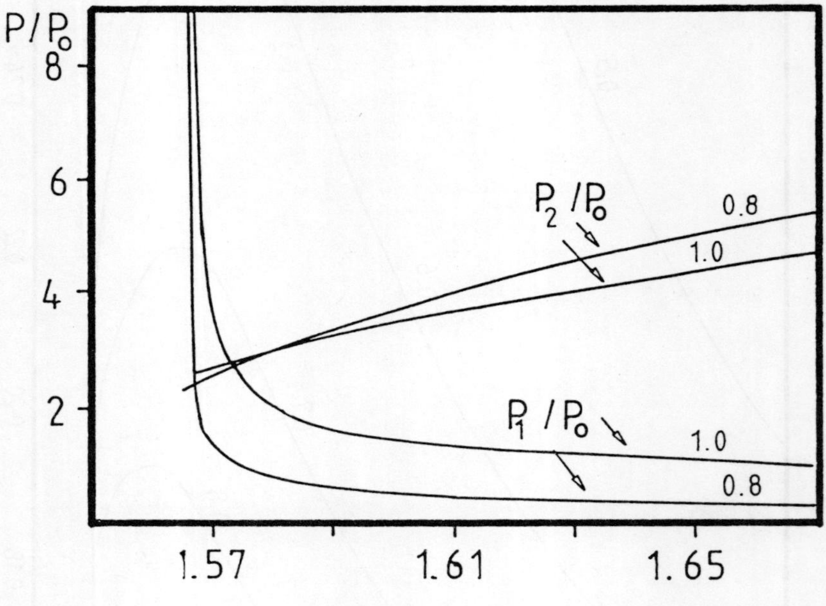

Fig. 13

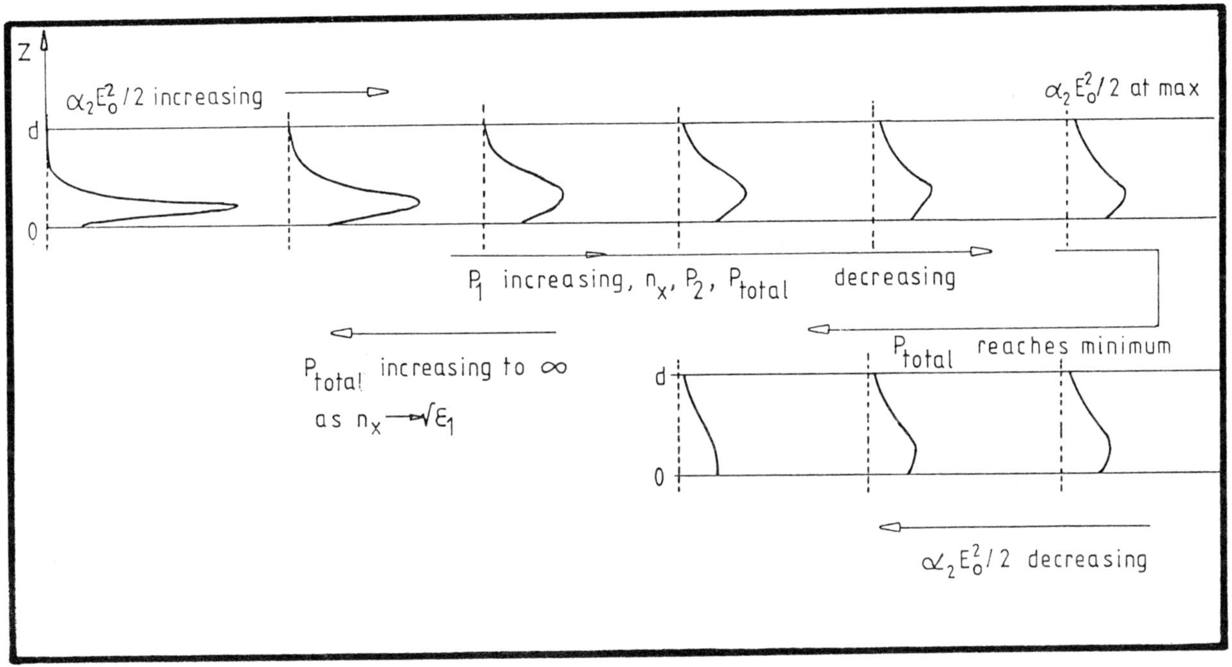

Z
$\alpha_2 E_0^2/2$ increasing
$\alpha_2 E_0^2/2$ at max
d
0
P_1 increasing, n_x, P_2, P_{total} decreasing
P_{total} increasing to ∞
as $n_x \to \sqrt{\varepsilon_1}$
P_{total} reaches minimum
$\alpha_2 E_0^2/2$ decreasing

Fig. 14

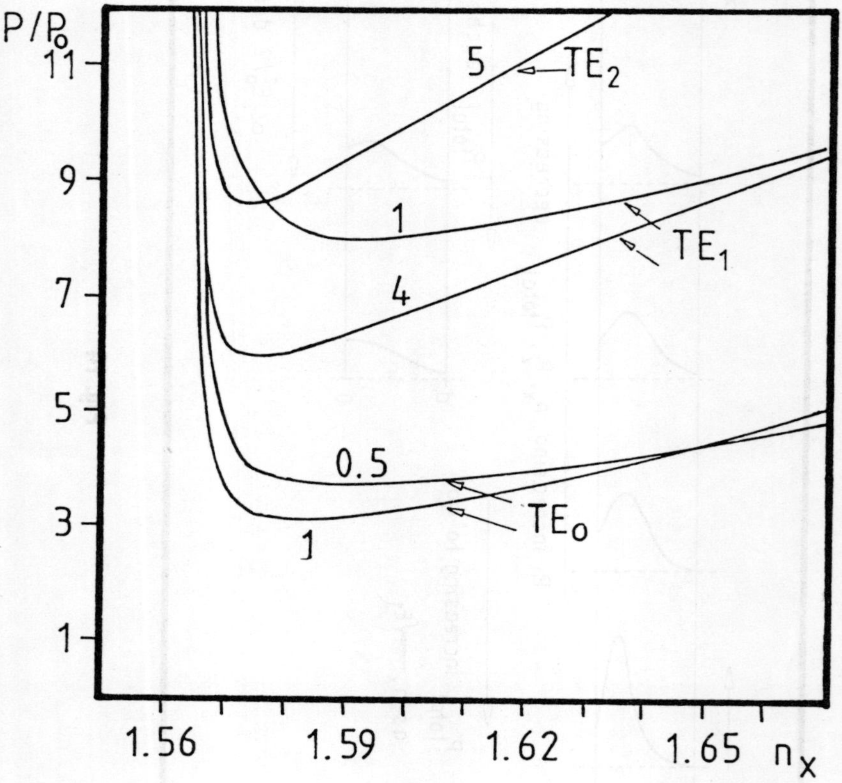

Fig. 15

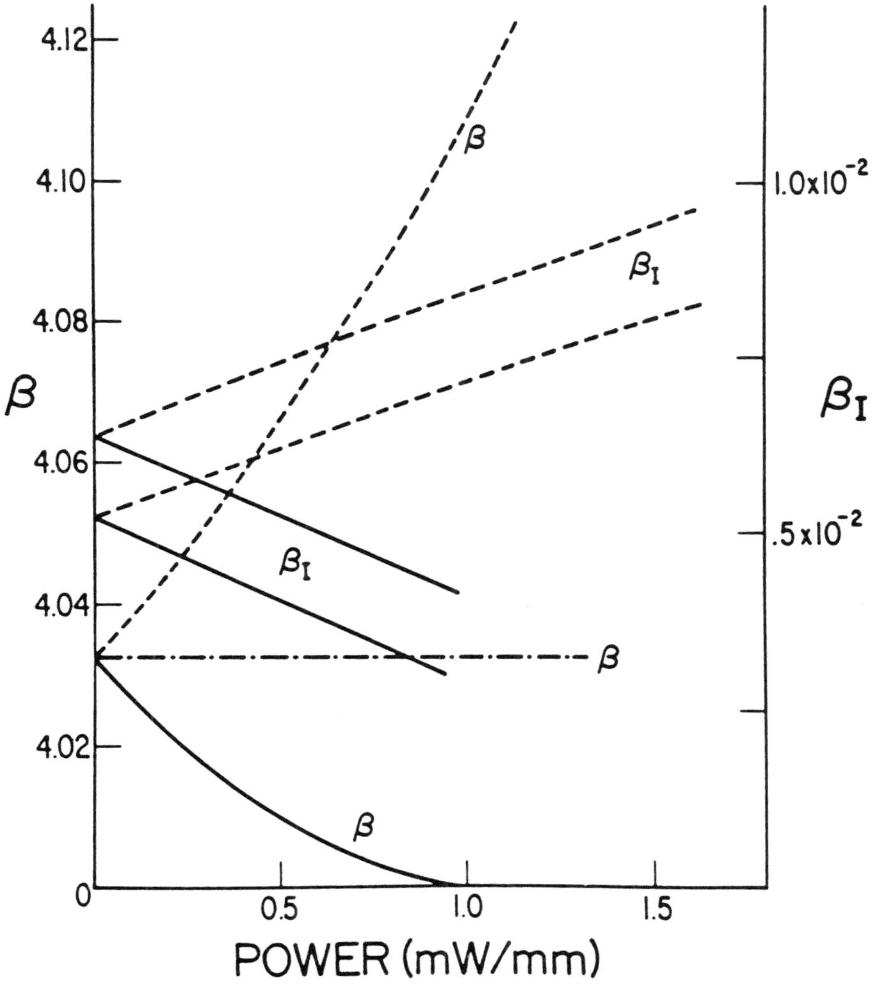

Fig. 16

GUIDED WAVES IN SOLAR MAGNETIC STRUCTURES

P. M. Edwin

The Open University in Scotland, Edinburgh,

and

B. Roberts

Department of Applied Mathematics, University of St. Andrews, North Haugh, St. Andrews, KY16 9SS, Fife, Scotland.

1. Introduction. The solar atmosphere is usually described in terms of three main regions, which, with increasing radius from the solar limb are the photosphere, chromosphere and magnetically dominated corona. In the past the atmosphere has been modelled as a planar one stratified by gravity and much attention was concentrated (and is even now) on producing a model atmosphere which correctly represents the temperature, pressure and density variation in this horizontally-averaged, radially-varying structure. However, as a result of information gained over the past decade or so, particularly from Skylab, a picture of the solar atmosphere has evolved[1,2,3] that more or less represents current thinking as to the atmosphere's description (Figure 1). Each region has its own typical structures capable, it appears, of supporting magnetohydrodynamic (mhd) waves.

In the photosphere one of the most obvious features is the sunspot, typified by a cool dark region, with a strong (3kG) magnetic field, some 10^4km in diameter with temperatures of about 4000K (some 2000K cooler than the spot's surroundings). Sunspots are not homogeneous, but generally exhibit an irregular pattern of bright points, called umbral dots (see reviews[4,5]), some 150km in diameter with brightness similar to that of the surrounding 6000K photosphere and with lifetimes of 1500s. Magnetic fields are suspected to be smaller in these dots than in the surrounding umbra. Beckers[6] pointed out that there are no detectable intensity oscillations at periods of 1 to 300s as might associate mhd waves with these dots, though periods of 180s (the 'chromospheric' oscillation) and 300s (the 'photospheric' oscillation) are associated with sunspot umbrae in general. Observations of sunspot oscillations have been summarised (e.g.[7,8]) as comprising vertical oscillations within the umbrae (with periods from 130-185s), and horizontal penumbral waves running outwards at speeds of 10-20kms^{-1} and periods in the range 210-300s.

Sunspots apart, the rest of the photosphere is far from homogeneous. It is now generally accepted that the magnetic field over most of the solar surface is concentrated into regions of magnetic flux (e.g.[9,10,11]) ranging from small ($\sim$100km diameter,1500G) flux tubes, through knots (1.5kG,500km diameter) and pores ($\sim$2kG,10^{3}km diameter) to sunspots.

Isolated flux tubes are at the limit of telescopic resolution so it is extremely difficult to detect oscillatory behaviour within them. Nonetheless Giovanelli and Brown[12] and Giovanelli et al.[13] have observed vertical velocity fluctuations in photospheric flux tubes but further observations, particularly from space (such as from the planned Solar Optical Telescope) are awaited with interest.

It is now generally agreed (e.g.[6,10,14,15,16]) that the magnetic flux concentrations of the photosphere (usually at supergranule convection cell boundaries) appear to be cospatial with features higher in the solar atmosphere, in the chromosphere. Briefly, the principal features of the chromosphere may be classfied into horizontal ($<$300km above limb) fibrils which appear to outline magnetic flux tubes joining areas of opposite polarity[17] and spicules, radial jets of gas which shoot upwards with average velocities of 20kms^{-1} from the chromospheric network along magnetic field lines. Attempts have been made to explain spicular behaviour by the buffeting of granule or supergranule cells on magnetic flux tubes, thus causing waves[18,19,20,21]. Giovanelli[22] has described in detail the propagation of waves observed in chromospheric fibrils. In general observations of chromospheric oscillations are limited[23-27].

On the other hand our knowledge of the solar corona has benefited from the EUV and soft X-ray observations of the Skylab mission, the overall picture being of a region of coronal loops, emitting structures which seem to act as tracers of coronal magnetic field. Loops vary in their characteristics. The tiny X-ray bright points which have time scales of the order of 1/3 day are thought to be loop structures in one extreme whilst the huge solar prominences, extending some 5×10^4km above the limb and which last from days to months, are thought to represent the other extreme. Some of the loops are 'closed' being anchored by magnetically opposite footpoints in the denser chromosphere and photosphere, extending no higher than half a solar radius (R_o = 6.96×10^5km) whereas others are 'open' fading high in the corona and coinciding with the source of the high-speed solar wind. Because of the large variation in size and type of loop their properties vary considerably and therefore it is difficult to describe a typical loop. Webb[28] quotes field strengths of 300G, Wentzel[29] 100G for small compact loops and gives a typical Alfvén speed of 3×10^3kms^{-1}. Suggestions have been made that some loops have cool cores[30-33] and that the particle density is greater by as much as a factor of 10 on the axis of the loop than at its outer surface[32,34,35]. The values usually refer to loops in active regions, being the easier ones to observe, especially those appearing shortly after a solar flare.

Information concerning the oscillatory behaviour of the plasma following a flare comes from radio wave data[36] and more recently from the hard X-ray data of the Solar Maximum Mission Satellite (e.g.[37,38]). There is also some coronagraph evidence of oscillations (e.g.[39,40]). Most oscillatory behaviour appears to be impulsively generated though not necessarily associated with flares[41,42]. There is a diversity of periodicities, from seconds to minutes, but interest lies in explaining the shorter (~ 1s) periods observed in the corona.

Here we describe the types of mhd waves that flux tubes, fibrils, spicules and coronal loops may be expected to support. The general problem is a complicated one but various realistic simplifications may be made to assist the analysis. For example, in the magnetically dominated corona the zero beta (or 'cold plasma')approximation involving the neglect of the sound speed compared with the Alfvén speed, is an acceptable simplification, as is the neglect of gravity. One property common to the variety of solar phenomena mentioned above which can often be exploited mathematically is that transverse dimensions are much smaller than longitudinal ones and as a result the <u>slender flux tube</u> approximation[43,44] can be applied.

The fact that these solar features have lateral boundaries means that the usual magnetoacoustic waves of an infinite medium are modified by the structuring. Indeed, the mathematical analogies between this ducting of mhd waves in the solar context and the guided waves in other fields (such as seismology, oceanography, fibre optics and laboratory plasmas) has proved useful in examining the solar problem.

2. Mhd waves in structured media. In order to examine structuring in the plasma (one fluid mhd approximation*) it will be assumed that the structure has infinite extent in the z-direction of a Cartesian coordinate xyz system. The effects of stratification due to gravity will be ignored initially. In the equilibrium state it is supposed that the plasma pressure p, density ρ and magnetic induction (field) $B\hat{\underline{z}}$ are x-dependent. In equilibrium the total pressure, made up of gas pressure p(x) and magnetic pressure $B^2(x)/2\mu_0$ must be constant:

$$\frac{d}{dx}(p + B^2/2\mu_0) = 0. \tag{1}$$

Linear, isentropic perturbations (velocity $\underline{v} = (v_x, v_y, v_z)$ and magnetic field $\underline{b} = (b_x, b_y, b_z)$) about the basic state (1) may then be Fourier analysed by writing the variables in the form

$$v_x = \hat{v}_x(x)e^{i(\omega t+\ell y+kz)}, \text{ etc.,}$$

where the x-dependent amplitude of the velocity component, v_x, satisfies

$$\frac{d}{dx}\left\{\frac{\rho(x)(k^2v_A^2(x)-\omega^2)}{m^2(x)+\ell^2}\frac{d\hat{v}_x}{dx}\right\} - \rho(x)[k^2v_A^2(x)-\omega^2]\hat{v}_x = 0. \tag{2}$$

Here

$$m^2(x) = \frac{(k^2c_s^2(x)-\omega^2)(k^2v_A^2(x)-\omega^2)}{(c_s^2(x)+v_A^2(x))(k^2c_T^2(x)-\omega^2)}, \tag{3}$$

where $c_s(x) = (\gamma p(x)/\rho(x))^{\frac{1}{2}}$ and $v_A(x) = B(x)/(\mu_0\rho(x))^{\frac{1}{2}}$ are the sound and Alfvén speeds in the unperturbed plasma of adiabatic index γ. The speed c_T, given by

$$c_T(x) = c_s(x)v_A(x)/(c_s^2(x) + v_A^2(x))^{\frac{1}{2}},$$

*See Appendix

is the slow wave cusp speed; it is both sub-sonic and sub-Alfvénic.

Equation (2) is the Cartesian equivalent of the Hain-Lüst equation[45] arising in a cylindrical geometry. It has been obtained by, for example, Goedbloed[46], Chen and Hasegawa[47], Wentzel[48] and Roberts[49].

Discussion of (2) in its general form is a difficult task, complicated by the presence of two singularities[50] - one at the Alfvén frequency ($\omega^2 = k^2 v_A^2(x)$) and the other at the cusp frequency ($\omega^2 = k^2 c_T^2(x)$). These singularities are associated with the presence of <u>continuous spectra</u> (the Alfvén and cusp continua)[51-55]. When the medium is discretely structured, so as to be comprised of slabs or cylinders of piecewise uniform magnetic field and gas density distributions, Eq.(2) may be solved in a straightforward manner.

Consider, first, a <u>slab</u> of magnetic field with equilibrium given by

$$p(x),\rho(x),B(x),c_s(x) = \begin{cases} p_0,\rho_0,B_0,c_0, & |x| < x_0, \\ \\ p_e,\rho_e,B_e,c_e, & |x| > x_0 . \end{cases} \tag{4}$$

The structure is taken as infinite in the y-direction and, for simplicity, we set $v_y = 0$ and $\ell = 0$. We consider oscillations that are confined to the inhomogeneity $|x| \le x_0$. Oscillations of this type would seem to be in accordance with observations (see, for example, Giovanelli's comments[22] on the observed properties of chromospheric fibril waves). Thus the inhomogeneity acts as a wave guide, with disturbances in the structure's exterior $|x| > x_0$ being laterally evanescent; $\hat{v}_x \to 0$ as $|x| \to \infty$. The solutions of (2) are then described by

$$\hat{v}_x(x) = \begin{cases} \alpha_e e^{-m_e(x-x_0)}, & x > x_0 \\ \alpha_0 \cosh m_0 x + \beta_0 \sinh m_0 x, & |x| < x_0 \\ \beta_e e^{m_e(x+x_0)}, & x < -x_0 \end{cases} \tag{5}$$

where $\alpha_0,\beta_0,\alpha_e,\beta_e$ are arbitrary constants, m_0 and m_e are the values of m in $|x| < x_0$ and $|x| > x_0$, respectively, and it is supposed that $m_e > 0$.

For an incompressible fluid, corresponding to $\gamma \to \infty$, both m_0 and m_e tend to $|k|$ and the dispersion relation resulting from solving (2) in $|x| < x_0$ and $|x| > x_0$ and matching velocity and total pressure across $x = \pm x_0$ takes the explicit form:

$$\frac{\omega^2}{k^2} = \frac{\rho_0 v_A^2 + \rho_e v_{A_e}^2 \begin{Bmatrix}\tanh\\ \coth\end{Bmatrix} |k| x_0}{\rho_0 + \rho_e \begin{Bmatrix}\tanh\\ \coth\end{Bmatrix} |k| x_0}, \tag{6}$$

where $v_{Ae} = B_e/(\mu_0\rho_e)^{\frac{1}{2}}$ is the Alfvén speed in the environment of the slab. This relation was first obtained by Kruskal and Schwarzschild[56]. The 'tanh' function corresponds to symmetric oscillations of the slab (sausage modes), the 'coth' function to asymmetric modes (the kink waves).

The solutions of (6) are Alfvénic surface waves which owe their existence to the presence of the boundary and which have wave amplitudes which take maximum values on the slab boundaries and decline on either side of $x = \pm x_0$ (see Equations (5) with $m_0 = m_e = |k|$).

For long wavelength disturbances in a slender structure so that $|k|x_0 << 1$, Eq.(6) reduces to

$$\omega^2 \simeq k^2 v_A^2 \left\{1 + \frac{\rho_e}{\rho_0}\left(\frac{v_{Ae}^2}{v_A^2} - 1\right)|k|x_0\right\}, \tag{7a}$$

for the sausage wave, and to

$$\omega^2 \simeq k^2 v_{Ae}^2 \left\{1 + \frac{\rho_0}{\rho_e}\left(\frac{v_A^2}{v_{Ae}^2} - 1\right)|k|x_0\right\} \tag{7b}$$

for the kink wave. In a wide structure ($|k|x_0 >> 1$), both modes have phase speed given approximately by $\omega^2 = k^2 c_k^2$, where

$$c_k^2 = \frac{\rho_e v_{Ae}^2 + \rho_0 v_A^2}{\rho_e + \rho_0}.$$

This is coincident with the result for a single magnetic interface[56].

Returning to the compressible problem, the four constants in (5) may be eliminated by invoking the continuity of normal component of velocity and total pressure across $x = \pm x_0$. This gives the dispersion relation

$$\rho_e(k^2 v_{Ae}^2 - \omega^2) m_0 \begin{Bmatrix}\tanh\\ \coth\end{Bmatrix} m_0 x_0 + \rho_0(k^2 v_A^2 - \omega^2) m_e = 0. \tag{8}$$

Equation (8), being a transcendental equation, possesses a rich spectrum of solutions. It has been derived in slightly more general form by several authors (e.g.[57,58]) who then subjected it to partial scrutiny as they pursued the particular problem in hand, and in limited form by others (e.g.[48,49,59,60,61]). Certainly the complete problem is a complicated one but by modelling the particular solar features of interest (for example, a photospheric flux tube as a magnetically dominant structure within a field-free region) the mathematical and thence physical description of possible mhd waves becomes more amenable.

It is perhaps appropriate to mention at this stage the terminology used here for the various solutions and modes of (8) since descriptions vary in the literature. Following Roberts[49] solutions of (8) with $m_0^2 > 0$ will be referred to as surface waves and those with $m_0^2 < 0$ as body waves (see Figure 2). Thus the distinction pertains only to their spatial behaviour within the structure; both these categories are termed surface waves by, for example, Wentzel[48] and Moisan <u>et al.</u>[62] in keeping with the fact that both sets of modes are confined to the neighbourhood of the slab.

In the unstructured situation i.e. the case of an infinite atmosphere, one expects three groups of waves, the slow and fast sets of magnetoacoustic gravity waves and the set of transverse Alfvén waves. Our analysis has uncoupled the latter and so these waves will not be considered further but we must note that the slow and fast waves of the infinite medium are altered in a complicated way depending on the relative magnitudes of the sound and Alfvén speeds inside and outside the structure[63,64]. In some cases, for example, that of a cold plasma, in which the Alfvén speed exterior to the structure is greater than that within, only two sets of body waves (slow and fast) occur whereas in a photospheric situation, where the Alfvén speed inside the structure is dominant, the slow mode can always propagate (either as a body or surface wave) but the fast mode may only propagate in a structure which is cooler than its surroundings (i.e. $c_e > c_0$)[61]. Solutions to Eq.(8) appropriate to photospheric circumstances (i.e. a strong field surrounded by almost field-free plasma) are summarized in Figure 3.

<u>Coronal structures</u>. The behaviour of mhd waves in structures in the magnetically dominated corona, some 3×10^3km above the photosphere and virtually gravity-free, is more easily seen by applying the low beta or cold plasma ($c_e = c_0 = 0$) approximation of (8) to a coronal loop, considered as an infinitely long slab. The analysis may be applied quite generally provided $v_A, v_{Ae} \gg c_e, c_0$[63,64]. In the cold plasma case the fast

body waves only remain. Writing $n_0^2 = -m_0^2$, Eq.(8) with $c_e = c_0 = 0$ reduces to

$$\tan(n_0 x_0) = -\frac{n_0}{m_e} \tag{9}$$

for the fast sausage (body) waves, and

$$\tan(n_0 x_0) = \frac{m_e}{n_0} \tag{10}$$

for the fast kink (body) waves. In the cold plasma limit, n_0 and m_e are given by

$$n_0^2 = (\omega^2 - k^2 v_A^2)/v_A^2, \qquad m_e^2 = (k^2 v_{Ae}^2 - \omega^2)/v_{Ae}^2. \tag{11}$$

It is interesting to note that equations equivalent to (9) and (10) have arisen in a number of other, physically dissimilar, situations. Indeed, Equation (9) arises in seismological and oceanographical studies; it was studied in detail by Pekeris[65] (see also Ewing et al.[66]). Equation (10) is Love's equation; it was studied earlier this century by Love[67] (see also[66,68]) in his investigation of waves in the elastic layers of the Earth's crust. Furthermore, equations similar to (9) and (10) have arisen in other areas. For example, waves ducted in the tropospheric layer in the lee of mountains[69,70] and in the shallower water of the continental shelf[71] obey similar equations. Also, an analogous situation arises in optical fibre communications. Indeed, the analogous equation to the cold plasma version of (2) is well known in the optical fibre literature (e.g.[72-78]).

It is of interest also to consider the case of a cylindrical inhomogeneity. For simplicity, we will suppose that $c_e = c_0 = 0$; the more general case of a low beta plasma (viz, $v_A, v_{Ae} \gg c_e, c_0$) gives similar results[64]. For a cylinder of radius a and gas density ρ_0 surrounded by gas of density ρ_e, all embedded in a uniform field $B\underline{\hat{z}}$, waves of the form

$$v_r = \hat{v}_r(r)e^{i(\omega t+n\vartheta+kz)}$$

satisfy the dispersion relation

$$\frac{J_n'(n_0a)}{J_n(n_0a)}\,\frac{K_n(m_ea)}{K_n'(m_ea)} = -\frac{n_0}{m_e}, \tag{12}$$

and J_n and K_n are Bessel functions of order n, possessing derivatives J_n' and K_n'. The expressions for n_0 and m_e (>0) are as before (Eq.11).

Equation (12) is the cylindrical version of (9) and (10) with n = 0,1. In particular, sausage modes (n = 0) satisfy

$$-n_0\,\frac{J_0(n_0a)}{J_1(n_0a)} = m_e\,\frac{K_0(m_ea)}{K_1(m_ea)}. \tag{13}$$

Equation (13) is identical to the dispersion relation for the lowest order transverse electric and magnetic modes of a cylindrical core surrounded by an infinite cladding (see, for example, Okamoto and Okoshi[79]).

Of course, the above dispersion relations apply to a cylinder with a discontinuity in density at the radius r = a. In reality, we expect a continuous distribution of density to arise in coronal loops. Similarly, in glass fibres the boundary between core and cladding can become smooth, due to diffusion of dielectric constituents, and so <u>graded index</u> glass rods have been investigated (e.g.[72,73,76,78,80-84]). Analogous investigations may be made for coronal loops[85] though they will not be pursued here other than to remark that smooth changes in density exhibit similar features to the simpler discontinuous profiles assumed in deriving (12) and (13).

The behaviour of the phase speed ω/k as a function of ka is sketched in Figure 4, allowing for the effects of non-zero sound speeds (but maintaining the low β ordering, $v_A, v_{Ae} >> c_0, c_e$). These modes are the ducted waves of a region of low Alfvén speed (i.e. $v_A < v_{Ae}$), corresponding to a region of high density ($\rho_0 > \rho_e$). Regions of high Alfvén speed ($v_A > v_{Ae}$) give rise to radial propagation in the region outside the cylinder (e.g.[86-88]). Ducting, then, arises whenever $v_A < v_{Ae}$.

There are no surface waves in a coronal loop, i.e. there are no solutions with $n_0^2 < 0$. As is evident from Figure 4, there are in fact two classes of body waves, widely separated in phase-speed (in a low β plasma). The slower of the two sets of body waves are sound waves constrained to propagate one dimensionally along the almost rigid field lines. These waves are only mildly dispersive; in a slender structure (ka << 1)

$$\omega \sim kc_T \sim kc_0 \quad (\text{for } v_A >> c_0).$$

On the other hand, the fast oscillations of a high density, low Alfvén velocity structure are highly dispersive, with the exception of the fundamental kink mode which has a phase speed (if ka is not too large) $\omega/k \sim c_k$.

In a closed coronal loop with its footpoints anchored in the high density chromosphere/photosphere standing modes may occur. For a loop of length L and $k = j\pi/L$ (integer j) the time scales are

$$\tau_s \sim \frac{2L}{c_0}, \quad \text{and} \quad \tau_f \sim \frac{2L}{c_k}. \tag{14}$$

For typical coronal sound and Alfvén speeds of $\sim 200\text{kms}^{-1}$ and $\sim 2000\text{kms}^{-1}$, respectively, and loop lengths of $2 \times 10^4 - 10^5$km Eq.(14) gives periods of 200 - 1000s for slow modes and ~70s for the fast modes. Such time scales have been observed[39,89,90]. However, these slow and fundamental fast kink modes do not explain the observed second and sub-second time scales.

To attempt an explanation of periodicities of the order of a second we must turn to the higher order (sausage and kink) fast modes of Figure 4, and consider their impulsive generation. The impulsive generation may be the result of a solar flare, or some other increase in solar activity which may act as a single or multiple source of the disturbance. So the situation under consideration is that of an impulsively generated fast mode (described by Equation (9), (10) or (12)) being ducted along a dense coronal structure. This is analogous to the situation considered by Pekeris[65] in his oceanographic studies. A typical group velocity plot for such a mode is shown in Figure 5. The significant feature is the occurrence of a minimum. Figure 6 shows the qualitative appearance of a wave packet some distance $z = h >> a$ along the structure from the impulsive source ($x = z = 0$). The initial disturbance (of frequency ω_c) will reach the point $z = h$ at a time $t = h/v_{Ae}$ after the impulse (at $t = 0$), as the group velocity c_g has a maximum value of v_{Ae}. The frequency of this initial signal will slowly increase reaching the value ω_x (Figure 5) at time $t = h/v_A$. The amplitude of these initial oscillations will be small as only a small part of bandwidth of the entire signal is contained between ω_c and ω_x. At time $t = h/v_A$ the high frequency part of the signal will arrive and the two frequencies contained in the signal approach each other until, at time $t = h/c_g^{min}$, they both equal ω^{min}, after which the signal will rapidly decay in the

Airy phase. The three phases of temporal behaviour are labelled the periodic phase, the quasi-periodic phase and the decay phase. The greatest part of the signal bandwidth (from $\omega = \omega_x$ to $\omega \to \infty$) will arrive during the quasi-periodic phase; the wave packet amplitude will be much larger during this phase than during the earlier periodic phase, which may indeed never be observed but lost in 'noise'. The exact appearance of the wave packet will, of course, depend on the frequency content and spatial extent and structure of the impulsive source. But it is clear that the vast majority of wave power will arrive in a well defined packet between times $t = h/v_A$ and $t = h/c_g^{min}$, and during this time much of the power will be centred near the frequency ω^{min}. This resembles the pulsations recorded in the radio and X-ray data[41,91-96]. Hence the suggestion was put forward in Roberts *et al.*[97] that short period (~ 1s) pulsations may be produced by impulsively generated fast magnetoacoustic waves propagating in a dense region of the corona. We note, too, that similar ideas have recently been put forward as an explanation of Pi2 oscillations in the magnetosphere[98].

In order to draw comparisons with theory the particular frequencies and time scales associated with Figure 6 must be identified. Consider the lowest order fast sausage mode of Figure 4. At cut-off $\omega/k \sim v_{Ae} = \omega_c/k_c$ so that for a cylindrical structure in a cold plasma,

$$\tau_c = \frac{2\pi a}{j_1^{(0)} v_A}\left(1 - \rho_e/\rho_0\right)^{\frac{1}{2}},$$

where $j_1^{(0)}$ is the first zero of the Bessel function J_0. For a dense inhomogeneity ($\rho_0 >> \rho_e$)

$$\tau_c \simeq 2.6\left(\frac{a}{v_A}\right),$$

and so for typical coronal parameters ($2a \sim 2 \times 10^3$km) τ_c is the order of 1s. Notice that ω_c is the smallest frequency of the disturbance so τ_c is the largest periodicity; $\tau^{min} = 2\pi/\omega^{min}$ is much smaller and so more rapid oscillations occur, as indeed observational data reveal.

The duration time

$$\tau_{dur} = h[1/c_g^{min} - 1/v_A]$$

of the quasi-periodic phase is theoretically distinctive so comparison with estimates of its duration and observed signals of similar form should be possible[97]. It may also be possible to use the theory and observed data to help determine physical conditions in the corona. For example, observations will usually provide the onset time $\tau_{onset} = h/v_A$ and the pulse periodicity τ^{min}. By plotting a/h (in units of τ^{min}/τ_{onset}) against τ_{dur} (in units of τ_{onset}) the scale of the inhomogeneity (a/h) can be determined, and thence the width of the inhomogeneity 2a and the associated Alfvén speeds[99].

Photospheric structures. It has already been mentioned that structuring is manifest in the photosphere in the form of flux tubes, from the readily visible sunspots to the sub-telescopic intense tubes. The lower solar atmosphere is divided into strong-field media (the intense tubes) or field-free media (the tubes' surroundings) though within sunspots themselves there is fine structure in the form of umbral dots. Thus some progress in understanding these features can be made[49,61,85] using Equation (8) in the extremes of $v_{Ae} = 0$, and in turn $v_A = 0$. However, the photosphere is gravitationally stratified and in order to understand the behaviour of waves in both a structured and stratified medium it is instructive to consider the slender flux tube approximation, in which the discussion is restricted to waves of long (relative to the width of the inhomogeneity) wavelength.

The equations describing the longitudinal, isentropic motion $v(z,t)$ of a gas of density $\rho(z,t)$ and pressure $p(z,t)$ confined within an elastic tube of cross-sectional area $A(z,t)$ may be written[43,44]:

$$\left.\begin{aligned}
&\frac{\partial(\rho A)}{\partial t} + \frac{\partial}{\partial z}(\rho v A) = 0, && (15a)\\
&\rho\left(\frac{\partial v}{\partial t} + v\frac{\partial v}{\partial z}\right) = -\frac{\partial p}{\partial z} - \rho g, && (15b)\\
&p + \frac{B^2}{2\mu_0} = p_e, && (15c)\\
&\frac{\partial p}{\partial t} + v\frac{\partial p}{\partial z} = \frac{\gamma p}{\rho}\left[\frac{\partial \rho}{\partial t} + v\frac{\partial \rho}{\partial z}\right], && (15d)\\
&BA = \text{constant}. && (15e)
\end{aligned}\right\}$$

Here gravity is assumed to be aligned with the z-axis of the tube ($\underline{g} = -g\underline{z}$) and p_e is the gas pressure in the tube's exterior. Equations 15(a)-(c) apply to any elastic tube in which motions are predominantly along the tube (see Lighthill[100] for the gravity-free case) but here both magnetic and gravitational effects are included and Equations (15) are non-linear.

The simplest way forward in investigating Equations (15) is to linearize them about the equilibrium

$$\left.\begin{aligned}
&\rho_0(z) = \frac{\rho_0(0)A_0(0)e^{-n_0(z)}}{A_0(z)}, && p_0(z) = p_0(0)e^{-n_0(z)},\\
&A_0(z) = A_0(0)e^{\frac{1}{2}n_0(z)}, && B_0(z) = B_0(0)e^{-\frac{1}{2}n_0(z)},
\end{aligned}\right\} \quad (16)$$

where $n_0(z) = \int_0^z \frac{dz'}{\Lambda_0(z')}$ and $\Lambda_0(z) = \frac{p_0(z)}{\rho_0(z)g} \equiv \frac{k_B T_0(z)}{mg}$ is the pressure scale height (for an ideal gas) of the tube's atmosphere. (The temperature $T_0(z)$ is the same inside and outside the tube.)

Assuming that p_e remains equal to its equilibrium value, it may be shown[101,102] that longitudinal motions

$$v(z,t) = [\rho_0(z)A_0(z)c_T^2(z)]^{-\frac{1}{2}} \quad \varphi(z,t)$$

are governed by an equation of the Klein-Gordon type:

$$\frac{\partial^2\varphi}{\partial t^2} - c_T^2(z)\frac{\partial^2\varphi}{\partial z^2} + \omega_v^2(z)\varphi = 0. \tag{17}$$

The general expression for $\omega_v^2(v)$ is rather involved[103]. But for an <u>isothermal</u> atmosphere (Λ_0 = constant) it reduces to the form

$$\omega_v = \left(\frac{9}{16} - \frac{1}{2\gamma} + \left(\frac{\gamma-1}{\gamma}\right)\frac{c_0^2}{v_A^2}\right)^{\frac{1}{2}} \frac{c_T}{\Lambda_0} . \tag{18}$$

Thus the geometry and elasticity of a magnetic flux tube give rise to a cut-off frequency ω_v, which is analogous to the cut-off

$$\omega_a = \frac{c_0}{2\Lambda_0}$$

for vertically propagating acoustic-gravity waves in a field-free medium[104]. The presence of these cut-offs implies that only frequencies above the cut-off propagate under adiabatic conditions; frequencies below are evanescent. Roberts[103] suggests that the outwardly propagating disturbances observed by Giovanelli <u>et al.</u>[13] are the non-adiabatic versions of the solutions to Equations (15), acoustic gravity waves propagating along slender magnetic flux tubes. The tube both guides, and through its elasticity, slows down the waves, which may be further modulated by acoustic gravity waves propagating in the field-free environment of the tube.

Rae and Roberts[102] show that an impulsively generated disturbance results in a wave-front propagating at the sub-sonic and sub-Alfvénic cusp or tube speed c_T. The wave-front trails a wake oscillating at the tube frequency ω_v, a behaviour typical of solutions of the Klein-Gordon equation. Hollweg[20] considers the non-linear evolution of a quasi-impulsive source propagating upwards along a rigid tube. The rebound shocks which develop in the chromosphere impinge on the transition region and thrust the underlying chromosphere upwards, the rebound shock train being identifiable with the elongated cool, dense jets of gas known as spicules.

If the rigid tube assumption is not employed then the non-linear equations (15) may still be examined analytically. For the gravity-free case, a flux tube is seen to support solitary waves, the dispersive effects of the tube's environment balancing the gas dynamic non-linearities[21]. Here again the analogies with other branches of physics become apparent, the structured mhd situation being similar to the waveguide formed by a density inhomogeneity within a fluid[105]. In fact, equations (15) together with the gas dynamic equations describing the (magnetic) environment of the tube may be combined to give the equation[21]:

$$\frac{\partial v}{\partial t} + c_T \frac{\partial v}{\partial z} + \beta v \frac{\partial v}{\partial z} + \alpha \frac{\partial^2}{\partial z^2} \mathcal{H}[v(z,t)] = 0, \tag{19}$$

where

$$\alpha = \frac{\rho_e (c_0^2 - c_T^2)(c_T^2 - v_{Ae}^2)^{\frac{1}{2}} (c_e^2 + v_{Ae}^2)^{\frac{1}{2}} (c_T^2 - c_{Te}^2)^{\frac{1}{2}}}{\rho_0 (c_e^2 - c_T^2)^{\frac{1}{2}} (c_0^2 + v_A^2) c_T}$$

is the coefficient of the dispersive term and $\beta = \frac{1}{2} v_A^2 \dfrac{[3c_0^2 + (\gamma+1) v_A^2]}{(c_0^2 + v_A^2)^2}$ is the coefficient of the non-linear term. $\mathcal{H}$ is the Hilbert transform. Equation (19) is the Benjamin-Ono equation[106,107]. The single soliton solution of (19) is[106]

$$v(z,t) = \frac{N}{1 + \left(\frac{z-st}{\ell}\right)^2}, \tag{20}$$

for velocity amplitude N, speed s and scale ℓ related by

$$s = c_T + \tfrac{1}{4}\beta N, \qquad \ell = \frac{4\alpha}{N\beta}. \tag{21}$$

If the effect of weak dissipation, in the form of thermal conduction or radiative decay, is included in the Equation (15d) then (19) generalizes to the Benjamin-Ono-Burgers Equation[85,108]

$$\frac{\partial v}{\partial t} + c_T\frac{\partial v}{\partial z} + \beta v\frac{\partial v}{\partial z} - \mu\frac{\partial^2 v}{\partial z^2} + \mu' v + \alpha\frac{\partial^2}{\partial z^2}\mathcal{H}[v(z,t)] = 0, \tag{22}$$

where

$$\mu = \frac{c_T(\gamma-1)\,Q}{2c_0^2\rho_0 c_p x_0} \quad \text{and} \quad \mu' = \frac{c_T(\gamma-1)x_0}{2c_0^2\gamma\tau_R} \quad \text{and}$$

c_p is the specific heat at constant pressure, Q is the (constant) heat transfer coefficient and τ_R is the radiative relaxation time, known for the solar atmosphere from tables[109-111]. For such a weakly dissipative medium, an analysis similar to that of Ott and Sudan[112] may be employed to show that the soliton amplitude N of Equation (20) satisfies

$$N = N_0\left[1 + \beta^2\frac{N_0^2\mu c_T t}{x_0 8\alpha^2}\right]^{-\frac{1}{2}} \qquad (\mu' = 0) \tag{23}$$

for the thermally conducting case ($\mu' = 0$), and

$$N = N_0 e^{-\frac{2\mu' c_T t}{x_0}} \qquad (\mu = 0) \tag{24}$$

for the case of radiative damping. (The other factors of (20) are also slowly varying functions of time, of course.)

It was suggested[21] that solitons may be manifest as spicules. For a magnetic flux tube in the upper photosphere with equal sound and Alfvén speeds, say, $c_0 = v_A = 7.5\text{kms}^{-1}$, and a typical radiative decay time of 350s[109], the soliton amplitude would decay by a factor of e in a time of $5\tau_R$ seconds = 1750s). The initial soliton speed is given by (21) ($s = 5.3 + 0.17\ N_0\ \text{kms}^{-1}$) and since velocity amplitudes at photospheric levels are unlikely to exceed those generated by granules ($N_0 \sim 1\text{-}3\text{kms}^{-1}$), the initial soliton speed is at most 6kms^{-1}. So such a solitary wave could travel through the whole of the solar photosphere and chromosphere without suffering significant attenuation from radiative losses[85].

Chromospheric features. Solitons are seen as being the chromospheric manifestations of non-linear wave behaviour within intense flux tubes in the photosphere. Another observational feature of the chromosphere are the fibril waves. Wentzel[113] has suggested that fibril waves are likely to be flux tube modes, the tube having an internal Alfvén speed that is lower than that in its surroundings. Spruit[3], too, in his discussion of transversal (kink) tube waves comments that the expected wave amplitudes and periods agree well with observations of wave motions in H_α fibrils.

In the terminology used here the fibril wave would be an example of the fast kink mode of Figure 4, a magnetic Love wave[64] propagating in a density enhancement for which $v_{Ae} > v_A$ ($> c_e, c_0$). For example, Giovanelli[22] estimated a magnetic field of 10G and used a density, higher inside the fibril, of 10" particles cm^{-3}. Using these values together with an observed period of 170s in Equation (14) would imply a wavelength of 8.3×10^3 km, which is entirely consistent with the observed fibril lengths of 10-15 arc sec.

To summarize, we have suggested that there is a diversity of waves that a structured (and stratified) medium such as the solar atmosphere can support; the Pekeris and Love magnetic waves of the coronal, higher atmosphere; Benjamin-Ono solitons and the flux tube modes of the chromosphere which may describe spicular and fibril behaviour, respectively, and Klein-Gordon solutions and their associated wakes which may offer some insight into photospheric, outwardly-propagating waves.

APPENDIX

Following Goedbloed[52] the equations of mhd for a hypothetical medium, 'a plasma', are used to describe the interaction of a perfectly conducting fluid with a magnetic field:

the induction equation,

$$\frac{\partial \underline{B}}{\partial t} = \underline{\text{curl}}(\underset{\sim}{v} \times \underset{\sim}{\bar{B}});$$

the 'frozen-in' condition,

$$\int_S \underset{\sim}{B}.d\underset{\sim}{S} = \text{constant};$$

the continuity equation,

$$\frac{d\rho}{dt} + \rho \text{ div } \underset{\sim}{v} = 0,$$

where $\frac{d}{dt} \equiv \frac{\partial}{\partial t} + \underset{\sim}{v}.\underline{\text{grad}}$;

the equation of motion

$$\rho \frac{d\underline{v}}{dt} = -\underline{\nabla}p + \frac{1}{\mu_0}(\underline{\nabla} \times \underline{B}) \times \underline{B} + \rho\underline{g};$$

the thermodynamic equation,

$$\rho\, c_p \frac{dT}{dt} - \frac{dp}{dt} = Q\nabla^2 T - \mathcal{R},$$

where Q is a heat transfer coefficient (joule $m^{-1}s^{-1}K^{-1}$) and $\mathcal{R}$ is a radiative loss term;

and the perfect gas equation,

$$p = R\rho T.$$

References

1. Parker, E.R., Cosmical Magnetic Fields Their Origin and Their Activity, Oxford (1979).
2. Noyes, R.W., The Sun, Our Star, Harvard Univ. Press, Cambridge, Mass. (1982).
3. Spruit, H.C., Theory of Photospheric Mag. Fields, Proc. of I.A.U. Symp. 102, 'Solar and Stellar Magnetic Fields', Zurich (1983).
4. Moore, R.L., Space Science Reviews, 28, 387 (1981).
5. Moore, R.L., Dynamic Phenomena in Sunspots in 'The Physics of Sunspots', Proc. of Sac. Pk. Obs. Workshop, Sunspot, New Mexico (1981).
6. Beckers, J.M., Dynamics of the Solar Photosphere in The Sun as a Star (ed. Jordan, S.) NASA SP-450 (1981).
7. Athay, R.G., Observations of Mass Motions in Active Regions in Solar Active Regions (ed. Orrall, F.Q.) Colorado Assoc. Univ. Press (1981).
8. Lites, B.W., White, O.R. & Packman, D., Astrophys. J., 253, 386 (1982).
9. Frazier, E.N., Phil. Trans. R. Soc. Lond. A 281, 295 (1976).
10. Stenflo, J.O., IAU Colloq. 36, 143 (1976).
11. Zwaan, C., Solar Magnetic Structure and the Solar Activity Cycle Review of Observational Data in The Sun as a Star (ed. Jordan, S.) NASA SP-450 (1981).
12. Giovanelli, R.G. & Brown, N., Solar Phys. 52, 27 (1977).
13. Giovanelli, R.G., Livingston, W.C. & Harvey, J.W., Solar Phys. 59, 49 (1978).
14. Beckers, J.M., Proc. of Int. Symp. on Solar-Terrestrial Physics (ed. Williams, D.J.) Am. Geo. Union, p.89 (1976).

15. Harvey, J., Highlights of Astronomy (ed. E.A. Müller) 4 (II), 223 (1977).

16. Sheeley, N.R. Jnr., The Overall Structure and Evolution of Active Regions in Solar Active Regions (ed. Orrall, F.Q.) Colorado Univ. Press (1981).

17. Athay, R.G., The Chromosphere and Transition Region in The Sun as a Star (ed. Jordan, S.) NASA SP-450 (1981).

18. Parker, E. N., Astrophys. J., 190, 429 (1974).

19. Roberts, B., Solar Phys., 61, 23 (1979).

20. Hollweg, J.V., J. Geophys. Res., 87 (A10), 8065 (1982).

21. Roberts, B. & Mangeney, A., Mon. Not. R. Ast. Soc. 198, 7P (1982).

22. Giovanelli, R.G., Solar Phys. 44, 299 (1975).

23. Liu,S-Y, Astrophys J., 189, 359 (1974).

24. Beckers, J.M. & Artzner, G., Solar Phys., 37, 309 (1974).

25. White, O.R. & Athay, R.G., Astrophys. J. Supp., 39, 317 (1979).

26. Athay, R.G. & White, O.R., Astrophys. J. Supp., 39, 333 (1979).

27. Giovanelli, R.G. & Beckers, J., I.A.U. Symp. 102, 407 (1983).

28. Webb, D.F., Active Region Structures in the Transition Region and Corona in Solar Active Regions (ed. Orrall, F.Q.) Colorado Assoc. Univ. Press (1981).

29. Wentzel, D.G., Coronal Heating in The Sun as a Star (ed. Jordan, S.) NASA SP-450 (1981).

30. Foukal, P.V., Solar Phys. 43, 327 (1975).

31. Foukal, P.V., Astrophys. J., 210, 575 (1976).

32. Jordan, C., I.A.U. Symp. 68, 109 (1975).

33. Levine, R.H., Future Active Region Observations in Solar Active Regions (ed. Orrall, F.Q.) Colorado Assoc. Univ. Press (1981).

34. Pick, M., Trottet, G. & MacQueen, R.M., Solar Phys. 63, 369 (1979).

35. Spruit, H.C., Magnetic Flux Tubes in The Sun as a Star (ed. Jordan, S.) NASA SP-450 (1981).

36. Krüger, A., Introduction to Solar Radio Astronomy and Radio Physics, Reidel (1979).

37. Kiplinger, A.L., Dennis, B.R., Emslie, A.G., Frost, K.J. & Orwig, L.E., Astrophys. J., 265, L99 (1983).

38. Kiplinger, A.L., Dennis, B.R., Frost, K.J. & Orwig, L.E., Astrophys. J., 273, 783 (1983).

39. Koutchmy, S., Žugžda, Y.D. & Ločans, V., Astron. & Astrophys. 120, 185 (1983).

40. Pasachoff, J.M. & Landman , D.A., Solar Phys. 90, 325 (1984).

41. Tapping, K.F., Solar Phys., 59, 145 (1978).

42. Gaizauskas, V., & Tapping, K.F., Astrophys. J., 241,804 (1980).

43. Defouw, R.J., Astrophys. J., 209, 266 (1976).

44. Roberts, B. & Webb, A.R., Solar Phys., 56, 5 (1978).

45. Hain, V.K. & Lüst, R., Zeits. für Natur. 13a, 936 (1958).

46. Goedbloed, J.P., Physica 53, 412 (1971).

47. Chen, L. & Hasegawa, A., J. Geophys. Res. 79, 1033 (1974).

48. Wentzel, D.G., Astrophys. J., 227, 319 (1979).

49. Roberts, B., Solar Physics, 69, 27 (1981).

50. Appert, K., Gruber, R. & Vaclavik, J., Phys. Fluids, 17, 1471 (1974).

51. Goedbloed, J.P., Phys. Fluids, 18, 10 (1975).

52. Goedbloed, J.P., Lecture Notes on Ideal Magnetohydrodynamics, Rijnhuizen Report 83-145, Nieuwegein, The Netherlands (1983).

53. Adam, J.A., Astrophysics & Sp. Sci., 78, 293 (1981).

54. Adam, J.A., Phys. Rep., 86, 217 (1982).

55. Rae, I.C. & Roberts, B., Mon. Not. R. Astr. Soc., 201, 1171 (1982).

56. Kruskal, M. & Schwarschild, M., Proc. Roy. Soc. Lond. A223, 348 (1954).

57. Chakraborty, B.B., Prog. Theo. Phys. 40, 210 (1968).

58. McKenzie, J.F., J. Geophys. Res., 75, 5331 (1970).

59. Cram, L.E. & Wilson, P.R., Solar Phys. 41, 313 (1975).

60. Wilson, P.R., Astrophys. J., 221, 672 (1978).

61. Roberts, B., Solar Phys. 69, 39 (1981).

62. Moisan, M., Shivarova, A. & Trivelpiece, A.W., Plasma Phys., 24, 1331 (1982).

63. Edwin, P.M. & Roberts, B., Solar Phys., 76, 239 (1982).

64. Edwin, P.M. & Roberts, B., Solar Phys., 88, 179 (1983).

65. Pekeris, C.L., Theory of Propagation of Explosive Sound in Shallow Water, Geol. Soc. America Memoir 27 (1948).

66. Ewing, W.M., Jardetzky, W.S. & Press, F., Elastic Waves in Layered Media, McGraw Hill (1957).

67. Love, A.E.H., Some Problems of Geodynamics, C.U.P. (1911).

68. Kennett, B.L.N., Seismic Wave Propagation in Stratified Media, C.U.P. (1983).

69. Scorer, R.S., Quart. J. Roy. Met. Soc., 75, 41 (1949).

70. Yih, C-S., Dynamics of Nonhomogeneous Fluids, Macmillan (1965).

71. Summerfield, W., Phil. Trans. Roy. Soc. A272, 361 (1972).

72. Kirchhoff, H., Arch. Elektron. & Übertrangungstech, 26, 537 (1972).

73. Kirchhoff, H., Arch. Elektron. & Übertrangungstech, 27, 13 (1973).

74. Marcuse, D., Light Transmission Optics, Van Nostrand Reinhold (1972).

75. Marcuse, D., Theory of Dielectric Optical Waveguides, Academic Press (1974).

76. Dil, J.G. & Blok, H., Opto-electronics, 5, 415 (1973).

77. Olshansky, R., Rev. Mod. Phys., 51, 341 (1979).

78. Keck, D.B., Optical Fiber Waveguides in Fundamentals of Optical Fiber Communications (ed. Barnoski, M.K.), Academic Press (1981).

79. Okamota, K. & Okoshi, T., IEEE Trans MTT 24, 416 (1976).

80. Conwell, E.M., Appl. Phys. Lett., 23, 328 (1973).

81. Okoshi, T. & Okamoto, K., IEEE Trans MTT 22, 938 (1974).

82. Gloge, D., IEEE Trans MTT 23, 106 (1975).

83. Arnaud, J.A., Beam & Fiber Optics, Academic Press (1976).

84. Unger, H.G., Planar Optical Waveguides & Fibres, Oxford (1977).

85. Edwin, P.M., Ph.D. Thesis, St Andrews University, Scotland (1984).

86. Meerson, B.I., Sasorov, P.V. & Stepanov, A.V., Solar Phys. 58, 165 (1978).

87. Spruit, H.C., Solar Phys., 75, 3 (1982).

88. Cally, P.S., Astron. Astrophys. 136, 121 (1984).

89. Tsubaki, T., Solar Phys., 51, 121 (1977).

90. Trottet, G., Pick, M. & Heyvaerts, J., Astron. Astrophys., 79, 164 (1979).

91. Rosenberg, H., Astron. & Astrophys., 9, 159 (1970).

92. McLean, D.J. & Sheridan, K.V., Solar Phys., 32, 485 (1973).

93. Pick, M. & Trottet, G., Solar Phys., 60, 353 (1978).

94. Orwig, L.E., Frost, K.J. & Dennis, B.R., Astrophys. J. Lett., 244, L163 (1981).

95. Kane, S.R., Kai, K., Kosugi, T., Enome, S., Landecker, P.B. & McKenzie, D.L., Astrophys. J., 271, 376 (1983).

96. Zodi, A.M., Kaufmann, P. & Zirin, H., Solar Phys., 92, 283 (1984).

97. Roberts, B., Edwin, P.M. & Benz, A.O., Nature, 305, 688 (1983).

98. Edwin, P.M., Roberts, B. & Hughes, W.J., J. Geophy. Res. (submitted).

99. Roberts, B., Edwin, P.M. & Benz, A.O., Astrophys. J., 279, 857 (1984).

100. Lighthill, M.J., Waves in Fluids, C.U.P. (1978).

101. Webb, A.R. & Roberts, B., Solar Phys., 59, 249 (1978).

102. Rae, I.C. & Roberts, B., Astrophys. J., 256, 761 (1982).

103. Roberts, B., Solar Phys., 87, 77 (1983).

104. Lamb, H., Hydrodynamics, C.U.P. (1932).

105. Davis, R.E. & Acrivos, A., J. Fluid Mech. 29, 593 (1967).

106. Benjamin, T.B., J. Fluid Mech., 29, 559 (1967).

107. Ono, H., J. Phys. Soc. Japan, 39, 1082 (1975).

108. Edwin, P.M. & Roberts, B., Wave Propagation (submitted).

109. Bray, R.J. & Loughhead, R.E., The Solar Chromosphere, Chapman & Hall (1974).

110. Spruit, H.C., Solar Phys., 34, 277 (1974).

111. Giovanelli, R.G., Solar Phys., 59, 293 (1978).

112. Ott, E. & Sudan, R.N., Phys. Fluids 13, 1432 (1970).

113. Wentzel, D.G., Astron. Astrophys., 76, 20 (1979).

Figure Captions

Figure 1 A schematic picture, varying logarithmically with height h (km) showing solar atmospheric features. Compiled from similar diagrams in, for example, references 1, 2, 3 and 10.

Figure 2 The waves are classified as (a) surface waves ($m_0^2 > 0$) and (b) body waves ($m_0^2 > 0$) according to their spatial behaviour within the structure.

Figure 3 The phase-speed ω/k as a function of kx_0 ($k > 0$) for photospheric conditions ($v_A > c_e > c_0 > v_{Ae}$). Only two of the infinitely many slow body waves are shown. Hatching indicates regions where free modes (real ω and k) do not occur. ——: sausage mode; ---: kink mode.

Figure 4 The phase-speed ω/k as a function of ka ($k > 0$) under coronal conditions ($v_{Ae} > v_A >> c_0, c_e$). Only two of the infinitely many slow body waves are shown. Hatching denotes regions in which there are no free (real ω and k) modes. ——: sausage mode; ---: kink mode.

Figure 5 The group velocity c_g as a function of frequency ω, for the lowest order fast sausage wave of Figure 4, in the low β(cold plasma) limit. The occurrence of a minimum in c_g at ω^{min} is shown.

Figure 6 A sketch of the evolution of the fast sausage wave in the low β extreme ($v_{Ae}, v_A >> c_e, c_0$) showing the various phases in the disturbance as recorded at an observation level $z = h$ away from an impulsive source at $z = 0$. (A similar sketch has been given by Pekeris[65] in his discussion of waves in an ocean layer.)

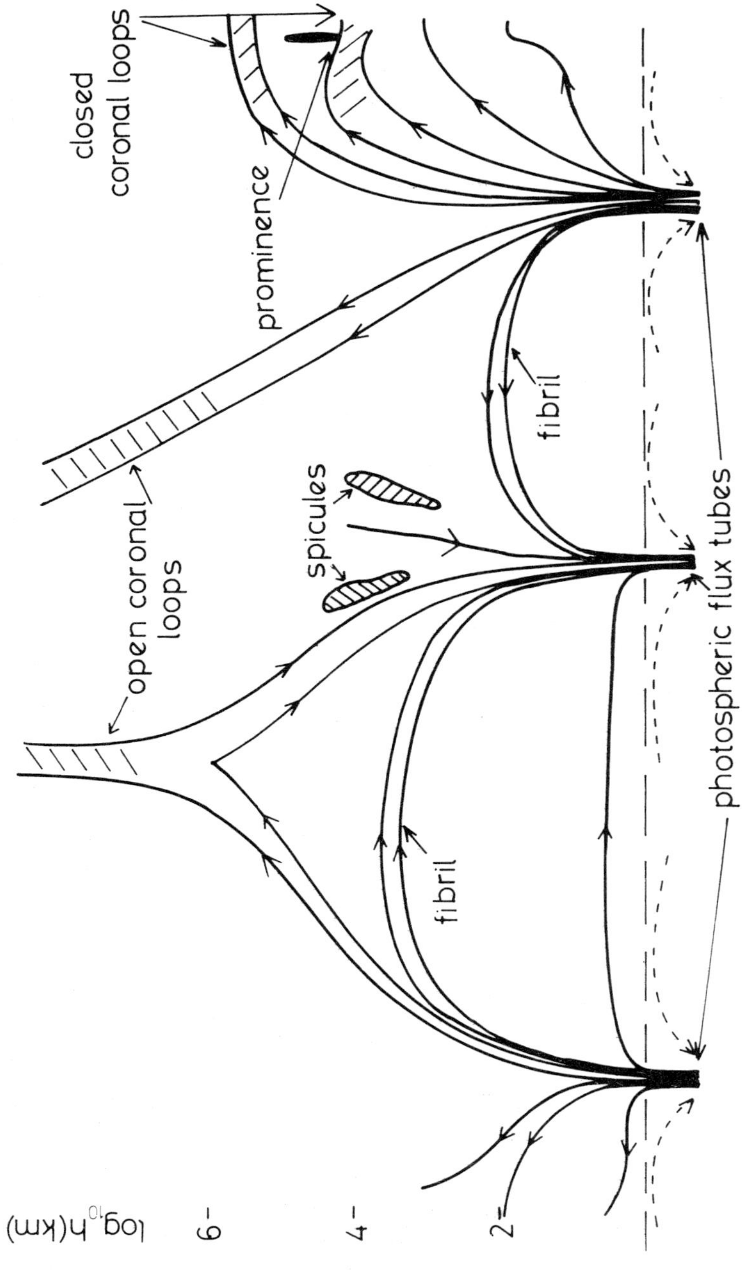
closed coronal loops
prominence
open coronal loops
spicules
fibril
fibril
photospheric flux tubes
6
4
2
log10 h(km)

Fig. 1

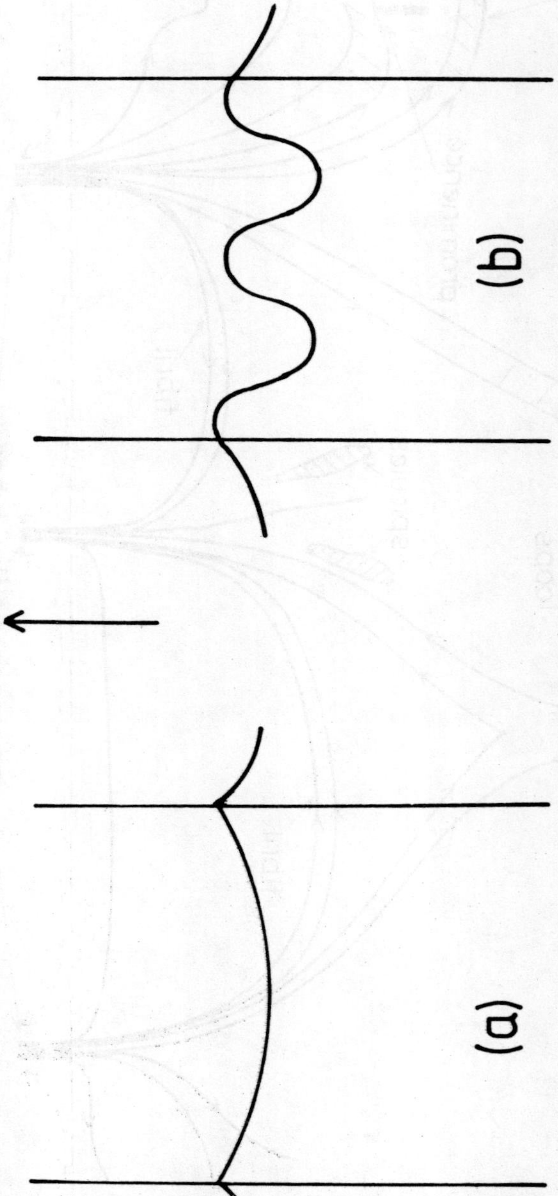

Fig. 2

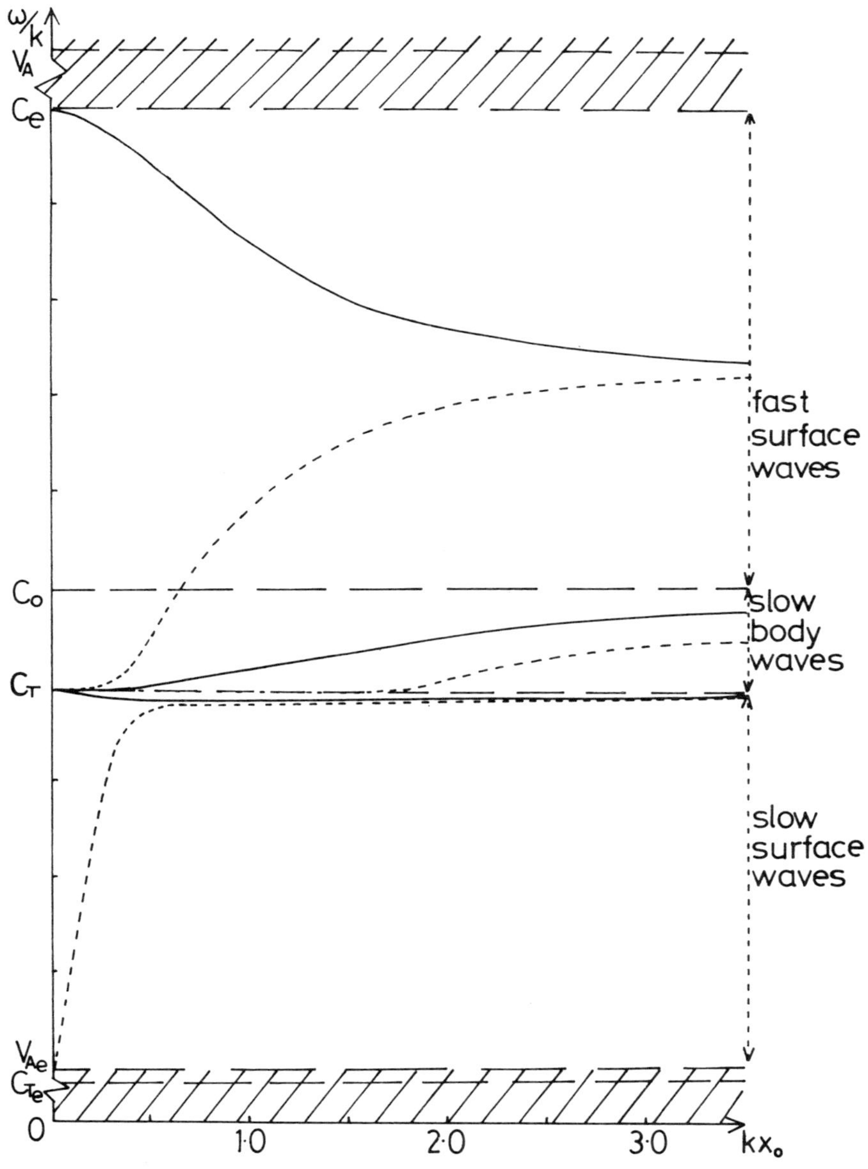
ω/k
V_A
C_e
fast
surface
waves
C_o
slow
body
waves
C_T
slow
surface
waves
V_{Ae}
C_{Te}
0
1·0
2·0
3·0
kx_o

Fig. 3

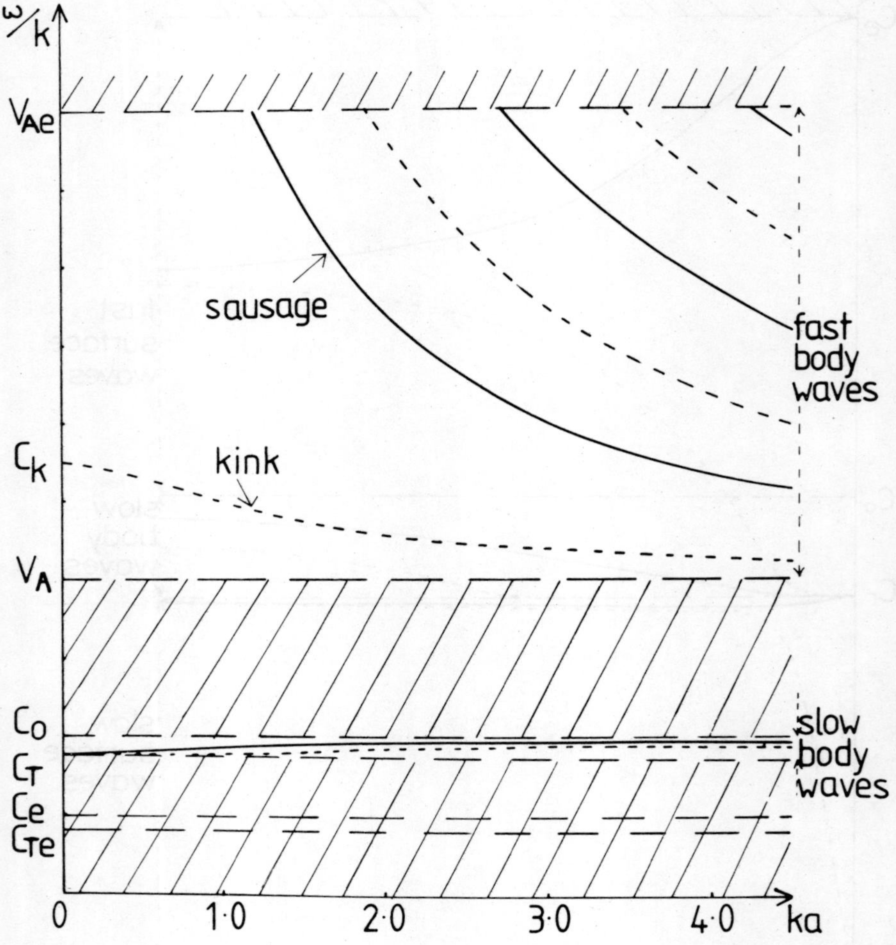
ω/k
V_Ae
sausage
fast
body
waves
C_k
kink
V_A
C_o
C_T
C_e
C_Te
slow
body
waves
0
1·0
2·0
3·0
4·0
ka

Fig. 4

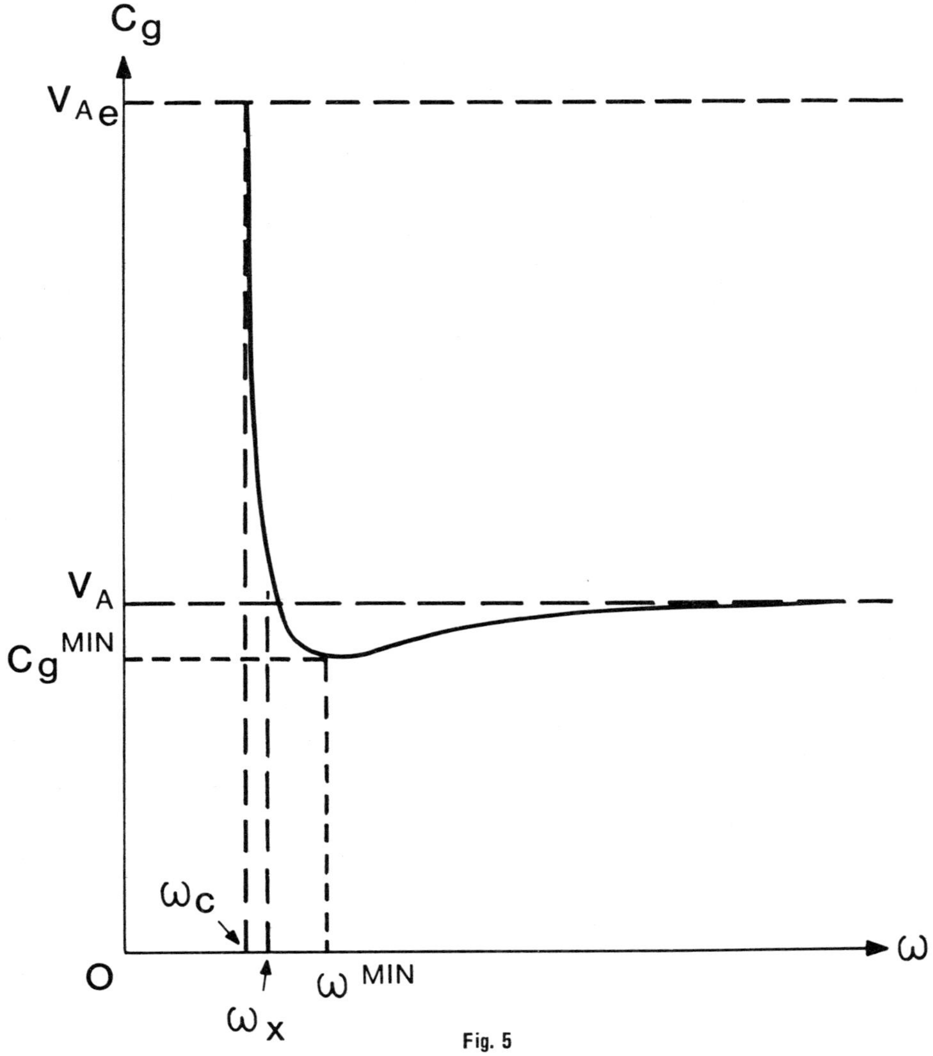

Fig. 5

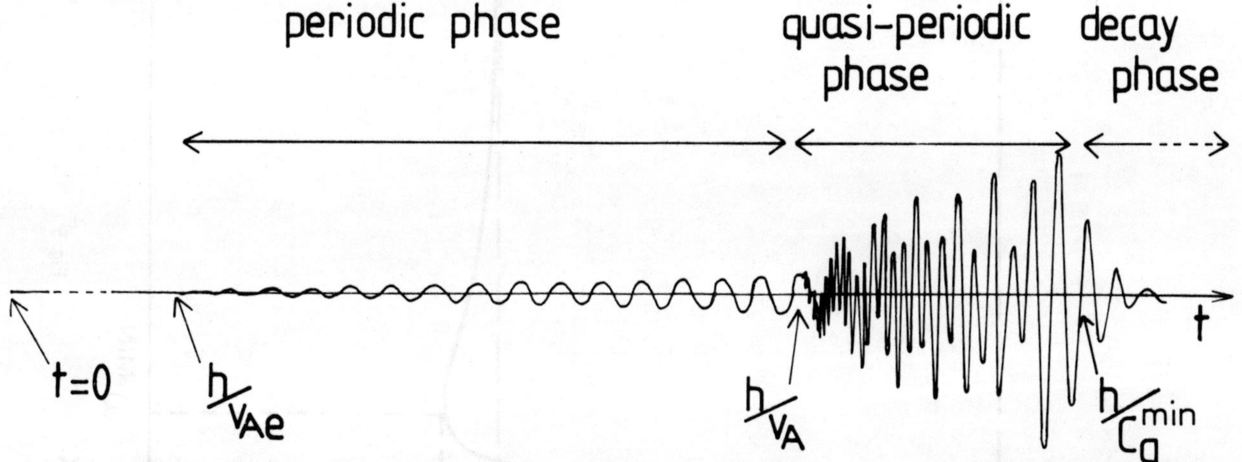
periodic phase
quasi-periodic phase
decay phase
t
t=0
h/vAe
h/vA
h/Cg min

Fig. 6

THE CONTRIBUTION OF SURFACE WAVES TO THE MODELLING OF RF AND MICROWAVE DISCHARGES

C.M. Ferreira[a] and M. Moisan[b]

a) Centro de Electrodinâmica da Universidade Técnica de Lisboa, Instituto Superior Técnico, 1096 Lisboa Codex, Portugal
b) Département de Physique, Université de Montréal, Montréal, Québec H3C 3J7, Canada

I. Introduction. The aim of this presentation is to show that some distinctive features of surface wave produced plasmas are such that they offer a unique opportunity of making progress in the modelling of RF and microwave plasmas in general. The general but simple model for both RF and microwave plasmas arrived at by Moisan and Zakrzewski[1] is an illustration of this potential.

As a plasma source, the surface wave produced plasma has numerous qualities: it can be produced efficiently, it is stable, quiescent and perfectly reproducible under a large variety of gas pressures and tube diameters. All these features are valuable ones, in particular for applications, but they are not distinctive of surface wave plasmas as compared to some other HF plasma sources (we shall use the term "high frequency (HF) plasmas" to designate both RF and microwave plasmas). The reasons for the superiority of the surface wave discharge with regard to the modelling of HF driven plasmas can be considered to result from the following three features:

1) The discharge is sustained by a travelling wave running along the plasma column.

   This means that the surface wave is transferring its energy to the plasma in a continuous way as it propagates. One way of writing this down is to consider P(z), the total power flow across the plane transverse to the direction of propagation at z (cylindrical coordinates are going to be used throughout). The fact that P(z) decreases to P(z) - dP(z) at z + dz implies that dP(z) represents the amount of power absorbed by the plasma column over the corresponding length

interval z, z + dz. Since the value of the power absorbed over any given column length interval z, z + dz, as well as the corresponding number of electrons produced are experimentally easily accessible quantities, the amount of HF power absorbed on the average per electron can be determined over a large range of wave power flow and electron density values.

2) The electric field intensity of the wave in the plasma is radially inhomogeneous and the shape of its distribution can be substantially varied by changing the operating conditions.

This property has been used to bring out two points. It leads to the dependence of the radial density distribution of excited atoms on the radial variation of the electric field intensity, E(r). It also allows to observe that, under ambipolar diffusion conditions, the radially averaged intensity value of the electric field in the plasma remains essentially constant along the plasma column, even though the radial distribution of the field varies substantially as a function of z. In fact, the magnitude of the field required to sustain the plasma against the losses (i.e., the maintenance field) is essentially independent of the distribution of the field inside and outside the plasma column and of the amount of power flowing along the system.

3) The surface wave produced plasma can be operated at frequencies ranging from a few MHz to many GHz, thus covering both the RF and the microwave domain.

The possibility of experimentally varying the wave frequency over such a large domain has led to observe that the maintenance field is predominantly determined by the particle loss mechanisms as in the classical positive column plasma of a DC discharge. In fact, as it will be shown below, one can define an **effective** maintenance field E_e such that the ratio of E_e/N (N being the gas density) corresponds to the value required to sustain the HF discharge for a given gas density and tube radius, for all frequencies (including DC discharges).

The unique frequency flexibility of surface wave discharges also permits to study the influence of the ν/ω ratio (ν and ω are the electron-neutral collision frequency and the wave angular frequency, respectively) on the electron energy distribution function (EEDF). This should prove useful in explaining why it is observed that, with certain gases, the population density of the highly lying levels of excited atoms, including those of ionized particles, is larger with microwaves as compared to RF, though the power density deposited in the plasma is kept constant.

In the sections that now follow, we are going to elaborate on the results concerning the modelling of HF driven plasmas that have been inferred from the investigations of surface wave discharges, as a result of the distinctive features just mentioned. Our presentation will not follow the chronological order of the publications connected with this subject but it will rather start by summarizing a very recently proposed general but simple model for HF driven plasmas. The paper is thus organized as follows. Section II briefly recalls the general model presented by Moisan and Zakrzewski[1] for HF plasmas, introducing at the same time some relevant parameters for the study of HF discharges. Section III deals with the results that can be inferred from the fact that the electric field intensity is radially inhomogeneous. This concerns the radial density distribution of excited atoms as well as some general relations and remarks on the so-called skin effect in overdense HF plasmas. Section IV concerns the influence of the parameter ν/ω on the EEDF and section V, while summarizing the modelling contribution from the surface wave discharges, points out the various steps that marked its development.

II. A general simple model for HF discharges. This model has been recently proposed on the occasion of a review paper by Moisan and Zakrzewski[1]. It results from the experimental observations and the modelling efforts of many authors[2-17] on surface wave discharges. Though this model is believed to be original, some of its elements may be found, sometimes expressed differently, in the vast literature that concerns the various types of RF and microwave produced plasmas.

II.1 - Summary of the model and power balance

An interesting point of this model is that it clearly brings out what are, under given operating conditions, the quantities that remain constant, whatever the way that the HF plasma is produced and, on the other hand, what are the quantities that are specific to each individual setup. In fact, the model is based on the assumption that the formal treatment of the physical processes occurring in HF discharges can be divided into two separate parts: one, concerning the maintenance processes within the plasma and, the other, the electromagnetic behavior of the HF circuit containing the plasma. The other assumptions are:

- The HF electromagnetic field applicator (e.g. coil, resonant cavity, wave launcher) is located outside the plasma tube: there are no electrodes in contact with the gas, as in the well known RF capacitive discharge.

- The applied HF frequency is high enough so the ions, because of their mass, can be considered immobile over one HF period.

- The mean free path is smaller than the plasma dimensions and the electrons oscillate in the gas volume, usually with an amplitude that is small in comparison with the plasma dimensions.

Under these conditions, the model assumes that the power balance equation can be written quite generally for HF discharges as:

$$P_I\, A(\bar{n}) = \Theta(\bar{n})\, V_p\, \bar{n}, \qquad (1)$$

where P_I is the HF power incident on the plasma, $A(\bar{n})$ is the fraction of the incident power that is absorbed within the total plasma volume V_p, whereas $\Theta(\bar{n})$ is the power lost per electron* in collisions of all kinds

*The total power loss, suffered by an average electron in collisions of all kinds, can be expressed as
$\Theta \equiv \frac{2m}{M} \langle \nu_m \varepsilon \rangle + \sum_k \langle h_k \nu_m \rangle eV_k$,
where m and M are the electron and ion mass, ε is the electron energy and ν_m, the collision frequency for momentum transfer, and h_k and V_k are, respectively, the efficiency and the threshold potential for excitation by collisions leading to the k-th level. The brackets $\langle\ \rangle$ denote the averaging over the electron energy distribution.

with neutral atoms, and $\bar{n}$ is the average electron density over V_p. The right hand side of Eq. (1) represents the power loss into the plasma and it is assumed independent of the way that the electric field is imposed on the plasma. As we shall see further, the quantity Θ emerges as a fundamental parameter in the study of HF discharges. In the model, its value is assumed, to a first approximation, to be independent of the applied electromagnetic field configuration surrounding the plasma volume i.e., it should depend only on the plasma tube diameter, the nature of the gas and its pressure, the electron density and the ν/ω ratio (except in the limiting cases of $\nu \gg \omega$ and $\nu \ll \omega$). The left hand side of Eq. (1) represents the electrodynamic behavior of the applicator as loaded with the plasma; phenomenologically, the latter is simply represented as a dielectric material whose permittivity, for a given frequency, is solely determined by the electron density and the collision frequency.

It is important to note that $A(\bar{n})$ is a function that is specific to each applicator and setup, to such an extent that it even depends, for example, on the position of the plasma column relatively to the applicator. For plasmas produced within a resonant cavity, $A(\bar{n})$ has the form of a resonance curve and its amplitude depends on the coupling of the feeder with the cavity. As concerns surface wave produced plasmas and, in general, travelling wave discharges, $A(\bar{n})$ can be resolved into a spatially distributed function, as shown in the next paragraph.

II.2 Application of the model to a travelling wave sustained discharge

The amount of power $dP(z)$ lost by the wave when propagating over an elementary distance dz is connected with the total power flow $P(z)$ through the relation defining $\alpha(\bar{n})$, the attenuation coefficient of the wave:

$$-dP(z) = 2\alpha(\bar{n})\, P(z)\, dz. \qquad (2)$$

Looking at Fig. 1, the power absorbed in the plasma slab defined by z, z + dz, can be equivalently written as:

$$P(z)\, dA(\bar{n}) = 2\,\alpha(\bar{n})\, P(z)\, dz, \tag{3}$$

noting that the integration of $dA(\bar{n})$ over the whole plasma column length leads to $A(\bar{n})$, which now appears as a total or global absorption coefficient. As Θ is assumed to be independent of the applied electromagnetic field configuration and also independent of the level of wave power flowing outside the plasma column, the power balance relation in any plasma slab is thus

$$P(z)\, dA(\bar{n}) = \Theta(\bar{n})\, \bar{n}\, \pi\, a^2\, dz, \tag{4}$$

where a is the plasma radius (In Eq. (4), $\bar{n}$ is the electron density averaged over the corresponding slab). Thus, in a travelling wave discharge, the power balance relation amounts to

$$2\alpha(\bar{n})\, P(z) = \Theta(\bar{n})\, \bar{n}\, \pi a^2, \tag{5}$$

This shows that $\Theta(\bar{n})$ is easy to obtain experimentally.

II.3 The Θ/p versus pa relation is a similarity law

This paragraph will show that the quantity Θ is a fundamental variable in the study of surface wave discharges. Its introduction, as a semi-empirical variable (in fact as Θ^{-1}) by Glaude et al.[3], marked the beginning of the successful modelling of surface wave discharges. It was thereafter theoretically characterized by Ferreira[11] who showed that in a diffusion controlled surface wave plasma under low attenuation conditions ($\nu \ll \omega$), Θ/p has the unique functional dependence

$$\frac{\Theta}{p} = f\left(\frac{\bar{\omega}_{pe}\, a}{c},\ pa \right), \tag{6}$$

where $\bar{\omega}_{pe}$ is the plasma angular frequency corresponding to the average electron density, c is the speed of light in vacuum, and p is the reduced gas pressure. This similarity law applies provided the value of $\bar{\omega}_{pe}$ is not too close to the propagation cut-off value (i.e., excepting the plasma region near the end of the column). Ferreira further showed that the dependence of Θ/p on $\bar{\omega}_{pe}a/c$ is usually small, so we have

$$\frac{\Theta}{p} \simeq f(pa), \tag{7}$$

i.e. Θ can be reasonably well assumed independent of the plasma density. In the limit of $\omega \gg \nu$, Θ is independent of the frequency. However, in situations where $\omega \sim \nu$, one should expect some dependence of Θ on the ν/ω ratio[15], coming from the dependence of the electron energy distribution function (EEDF) on this parameter[12].

The data shown in Fig. 2, obtained with an argon plasma created by an azimuthally symmetric surface wave, show Θ/p values measured by various authors under different experimental conditions. We may conclude that:

1) Though the experimental points correspond to very different discharge conditions (ω, p, a), they all appear to fall on the same straight line which is that calculated by Ferreira. Thus, it appears that the similarity law of Θ/p vs. pa is well satisfied in the operating range of these experiments.

2) The fact that the similarity law is verified over a considerable electron density range ($\sim 10^8 - 10^{12}$ cm^{-3}) indicates that, in the present case, Θ is practically independent of the electron density.

3) The fact that the radial distribution of the electric field intensity considerably varies with the wave frequency and the average electron density (see Sect. III.1) does not appear to significantly reflect on the values of Θ obtained, for example, at 210 and 2450 MHz.

This fact justifies our assumption that in HF plasmas, in general, Θ does not depend, to a first approximation, on the electric field configuration imposed on the plasma column.

Some of these conclusions should be regarded, nevertheless, with some caution. In fact, one would have expected, on theoretical grounds, that Θ had some explicit dependence on $\bar{n}$, (resulting, e.g., from stepwise ionisation processes) and on the ν/ω ratio, when $\nu \sim \omega$ (as, e.g., in the case of the lower frequencies of Fig. 2). The fact that the same similarity law appears to be experimentally verified over such a large range of electron density and field frequency, even in cases where the basic assumptions of Ferreira's theory are no longer strictly valid, still needs a fully reliable theoretical justification.

In any case, we note in conclusion that the study of surface wave discharges has led to identify Θ as a meaningful parameter for the description of HF discharges in general. The similarity law of Θ/p vs. pa is of a great practical value: one full set of Θ measurements obtained with a given plasma column can be used to predict the value of Θ in any other plasma column operated with the same gas.

II.4 The plasma maintenance field

The quantity Θ defined above is intimately related to the concept of the effective maintenance field necessary for the steady-state operation of the discharge. In fact, the functional dependence of Θ/p on pa results, via the electron power balance equation, from the discharge characteristics expressing the maintenance field as a function of the operating conditions. As it is well known, the discharge characteristics for a diffusion controlled discharge plasma are the result of a balance between collisional ionisation of the gas and the loss of electrons and ions by ambipolar diffusion to the wall. The fact that the rate of electron production must compensate exactly for the rate of electron loss (the latter being determined by the collisions of ions and atoms and by the space charge fields) uniquely determines the magnitude of the maintenance field inside the plasma for given operating conditions, irrespectively of the particular device used to apply the field.

To illustrate these ideas, let us consider first a model gas in which ν is independent of the electron velocity (helium can be described, to a first approximation, by such a model). The number of independent parameters in the theory can then be reduced by introducing an effective field[18,19]

$$E_e = E \frac{\nu}{(\nu^2 + \omega^2)^{\frac{1}{2}}} , \tag{8}$$

where E denotes the r.m.s. value of the applied field. The discharge characteristics resulting from the ionisation-loss balance can then be expressed in the form of E_e/p vs. $p\Lambda$ for all frequencies, Λ denoting the characteristic diffusion length for the discharge vessel.

The power absorbed by the electrons from the HF fields, per unit volume, can be expressed as

$$\frac{dP_{abs}}{dV} = R_e(\sigma)E^2 , \tag{9}$$

where

$$R_e(\sigma) = \frac{e^2 n}{m} \frac{\nu}{\nu^2 + \omega^2} \tag{10}$$

is the real part of the plasma complex conductivity σ.

Combining Eqs (8)-(10), we obtain

$$\frac{dP_{abs}}{dV} = \frac{e^2 n}{m} \frac{E_e^2}{\nu} .$$

The total absorbed power in the discharge volume is

$$P_{abs} \equiv P_I \, A(\bar{n}) = \frac{e^2}{m\nu} \int_{V_p} n \, E_e^2 \, dV \qquad (12)$$

and it must compensate for the total power loss by electrons through collisions, which we have written as $\Theta(\bar{n}) \, V_p \, \bar{n}$ (see Eq. 1). It then follows that

$$\Theta = \frac{e^2}{m\nu} \frac{1}{V_p} \int_{V_p} \frac{n}{\bar{n}} E_e^2 \, dV. \qquad (13)$$

In cylindrical or in planar geometry, the reduced field at the axis E_{eo}/p constitutes an eigenvalue for the diffusion problem[27] and may be determined as a function of $p\Lambda$ from the ionisation-loss balance. We may then rewrite Eq. (13) in the form

$$\frac{\Theta}{p} = \frac{e^2}{m(\nu/p)} \left(\frac{E_{eo}}{p}\right)^2 \frac{1}{V_p} \int_{V_p} \frac{n}{\bar{n}} \mathcal{E}^2 \, dV, \qquad (14)$$

where $\mathcal{E} = E_e/E_{eo}$. This equation, taking into account the dependence of E_{eo}/p on $p\Lambda$, explains why the ratio of Θ/p is an explicit function of the $p\Lambda$ (or pa) product. We also note that a dependence of Θ/p on the average electron density $\bar{n}$ arises from the normalized integral in Eq. (14), which actually introduces the effect of the spatial distributions of the electron density and field intensity, and also from the reduced maintenance field E_{eo}/p (for example, in the case of surface wave plasmas, E_{eo}/p depends somewhat on the parameter $\bar{\omega}_{pe}$ a/c[5,11], as shown in Fig. 3).

Alternatively, one could define a weighted average field intensity using the definition

$$\bar{E}_e^2 \equiv \frac{1}{V_p} \int_{V_p} \frac{n}{\bar{n}} E_e^2 \, dV, \qquad (15)$$

and then express Θ/p in the form

$$\frac{\Theta}{p} \equiv \frac{e^2}{m(\nu/p)} \left(\frac{\bar{E}_e}{p}\right)^2 . \tag{16}$$

Since ν/p is a constant, Eq. (16) clearly shows that Θ/p and $(E_e/p)^2$ are equivalent parameters. Furthermore, one notes that the dependence on the detailed spatial distributions of n and E_e, and on the absolute value of the electron density appears only through an averaging process and may be expected, therefore, to be rather smooth.

When ν is a function of the electron velocity, the theory is, in general, more complicated, as the concept of effective field cannot be fully applied. Then, the additional parameter p/ω (or the dimensionless parameter ν_e/ω, ν_e denoting some representative value of the collision frequency for the gas used) comes into play. In argon, however, one can define[15] an effective field for ionisation, such that the electron rate coefficient for ionisation depends only on E_e/p, and not on p/ω for all frequencies. This effective field concept is a valuable one with regard to the calculation of the discharge characteristics, in the form of E_e/p vs. $p\Lambda$. Fig. 3 shows such calculated characteristics for an argon plasma produced by a surface wave for two values of $\bar{\omega}_{pe}$ a/c. Also shown for comparison are such characteristics obtained assuming a spatially homogeneous field distribution.

III. Dependence of plasma parameters on the radial variation of the electric field intensity in an overdense HF plasma.

III.1 Radial density distribution of electrons

In a surface wave produced plasma, the radial variation of the electric field intensity depends on the radial density distribution of electrons that, in turn, depends on the radial variation of the electric field intensity. This is a self-consistent problem. It has been solved by Ferreira[5]. Figure 4 shows the electron density radial distribution n(r) obtained for various average electron density values, while Fig. 5 presents the corresponding radial profiles of the electric field intensity, E(r). One notes that the electron density distribution is comparatively weakly affected when the average plasma density is varied, whereas the radial distribution of the electric field intensity drastically changes.

The radial behavior of the electron density and electric field intensity has not been measured directly, since using probe methods would perturb the plasma too much. However, the results observed for the radial density distributions of excited atoms qualitatively support the results from Ferreira calculations (Note in passing that the surface wave discharge is one of the few HF discharges for which the electric field intensity has been self-consistently calculated).

III.2 Radial density distributions of excited atoms

The radial density distribution of excited atoms in surface wave discharges has been examined experimentally by Ricard and collaborators[4,8,9,14,20-22], using an end-on measurement method on short length plasma columns (10-20 cm) terminated at both ends by observation windows. Typical results of these measurements are shown in Fig. 6 for argon atoms in a radiative state and, in Fig. 7, for argon atoms in a metastable state. Note that, for given gas pressures, the distribution are slightly flatter for the atoms in the metastable state. Figure 8 shows the calculated[5] radial density distribution for a radiative level, for conditions corresponding approximately to those of Fig. 6. Though the calculated curves do not quantitatively match the experimental observations, the evolution of the theoretical curves with the gas pressure closely follows the behavior of the measured density distributions (Note that the calculations do not consider step-wise excitation).

The radial density distribution of excited atoms in a radiative state, $n_j(r)$, can be expressed, in surface wave discharges, in terms of the electric field intensity and the electron density radial distributions as

$$n_j(r) = C_1 \; E^{\alpha}(r) \; n(r), \qquad (17)$$

where C_1 is a constant independent of the radial position, and α is a value determined from fitting a power law to the electron rate coefficient calculated by Ferreira and Loureiro[12] from the Boltzmann equation. For argon, it is 1.7 for the $^3P_2$ and $^3P_0$ metastable levels, 2.0 for the $^3P_1$ and $^1P_1$ resonant levels, 2.9 for the group of higher lying levels and, finally, 4.1 for ionization (A dependence similar to that of Eq. (17) had been suggested many years ago by Allis et al.[23] in connection with an HF plasma produced between two parallel metallic plates).

Many authors[23-26] have reported observations, in various types of HF produced plasmas, of radial distributions of emitted line (or multi-line) intensity similar to that presented in Fig. 6: the emitted intensity is maximum near the wall and it decreases toward the axis. In many instances, the explanation suggested for these observations has been limited to saying that it is due to the skin effect, referring by this to the physical mechanism that makes the HF electric field intensity decrease (exponentially) as it penetrates into an absorbing metallic medium. The next section is concerned with a deeper analysis of this phenomenon.

III.3 The radial variation of the electric field intensity in overdense HF plasmas

Though it followed from the study of surface wave discharges, Eq. (17) is obviously applicable to any type of HF plasmas, whatever the way that the electric field is imposed on the plasma. Thus, for those plasmas for which the emitted line intensity is observed to decrease toward the axis, it means that the electric field intensity is also decreasing toward the axis. However, to what extent can we say, when such a behavior is observed, that it is the result of the classical skin effect already mentioned: a low collision ($\nu \ll \omega$) plasma, as far an electromagnetic field is concerned, is more a dielectric than a metal-like material.

The intensity of a plane wave incident on a homogeneous low collision plasma medium decreases as it penetrates the plasma if it is overdense ($\omega < \omega_{pe}$). This is because, under these conditions, the wave number in the plasma is imaginary. The evanescence of the wave field in such a case is similar to that observed in a waveguide at cut-off. This damping increases with the ratio ω_{pe}/ω. Such a damping mechanism remains qualitatively the same, whatever the wave field configuration. However,it is only partially responsible for the evanescence observed. A second mechanism can be invoked for the decreasing of the field intensity: it is connected with the inhomogeneity of the plasma medium, as we now see using Allis et al.[23] parallel plate plasma model.

In a HF discharge taking place between two parallel plate electrodes separated from the gas by a dielectric barrier, it is clear that the total current density J (including both electron current and displacement current) must remain constant along the direction (x) perpendicular to the two plates, i.e.,

$$J = \sigma(x)E(x) = \text{cst.}, \qquad (18)$$

where σ is the plasma conductivity.

Assuming ν/p = cst., this equation can be written

$$J = \frac{ne^2}{m}\left(\frac{1}{\nu + j\omega}\right) + j\omega\varepsilon_o\, E, \qquad (19)$$

The magnitude of the resistivity being therefore

$$\frac{E}{J} = \frac{1}{\omega\varepsilon_0} \frac{(\delta^2 + 1)^{\frac{1}{2}}}{[(y-1)^2 + \delta^2]^{\frac{1}{2}}}, \qquad (20)$$

where we have defined

$$y \equiv (\frac{\omega_{pe}}{\omega})^2 = \frac{ne^2}{\varepsilon_0 m\omega^2}, \qquad (21)$$

$$\delta \equiv \frac{\nu}{\omega}. \qquad (22)$$

It follows that the magnitude of the field varies across the discharge according to the law

$$\frac{E}{E_0} = [\frac{(y_0 - 1)^2 + \delta^2}{(y-1)^2 + \delta^2}], \qquad (23)$$

where the subscript zero denotes the values at the axis. As the plasma density decreases from the axis towards the wall, we see from Eq. (23) that E is minimum at the axis and increases towards the wall. It reaches a maximum value at the wall, or close to it[23,27] in the case where the plasma density at the sheath boundary is lower than the resonance value (y = 1). Figure 9 shows some results obtained by Allis et al.[23] in a particular experimental situation illustrating the spatial distributions of the plasma density, the magnitude of the electric field, and the ionisation rate (both per electron and per unit volume). The distribution of the light intensity emitted by the discharge is similar to that of the ionisation rate per unit volume (both involve the product of the electron density by some power of the field intensity as in Eq. (17)); it closely resembles the distributions obtained in surface wave produced plasmas as shown in Figs 6 and 8.

Ferreira[27] recently showed that the results of the Allis et al[23] model could, in fact, be qualitatively recovered in a cylindrical surface wave produced plasma, i.e., in a three dimensional situation. The reason for this lies in the fact that the physical processes responsible for the maintenance of the plasma (absorption of HF power from the

E-field, collisional ionisation of the gas, loss by diffusion to the wall) are basically the same in both situations and may be described by closely resembling models. In the case of parallel plane geometry, we only need the hydrodynamic-type equations that describe the balance and the transport of the charged particles, the electron Boltzmann equation, and the equation $\vec{\nabla}.\vec{J} = 0$ (i.e., Eq. (18)) for self-consistency. In the case of a cylindrical plasma sustained by, say, a TM surface wave, the description of the electromagnetic part of the problem is somewhat more complicated: in fact, in addition to the equation $\vec{\nabla}.\vec{J} = 0$, we now need the dispersion relation and the (general) equation that relates the transverse component of $\vec{E}$ to the transverse gradient of the axial component for any TM travelling wave, i.e.,

$$\vec{E}_T = - \frac{jk}{\frac{\omega^2}{c^2} \varepsilon_p - k^2} \vec{\nabla}_T E_z \tag{24}$$

(here, k is the wave number and ε_p is the plasma relative permittivity). We point out that the set of equations $\vec{\nabla}.\vec{J} = 0$ and (24) implies that the field intensity in an overdense diffusion controlled plasma necessarily decreases from the wall towards the axis. In a capacitive discharge between parallel plates, this same behavior is implied by the equation $\vec{\nabla}.\vec{J} = 0$ alone as we have seen.

It is, therefore, clear that spatial density and field distributions exhibiting similar features should be expected as well, in any other situations where HF fields sustain overdense plasmas.

One may note in passing that the radial distribution of the electric field intensity obtained within a cylindrical tube containing any HF overdense plasma always seems to be quite close to the field configuration associated with the propagation of a surface wave. This is probably the reason why so many people have achieved surface wave produced plasmas (as a rule without knowing it), just by applying enough HF power to a cylindrical discharge tube, using almost any kind of applicator.

IV. Influence of the ν/ω ratio on the Electron Energy Distribution Function (EEDF).

IV.1 Physical mechanism of this influence and theoretical predictions

In most gases (e.g., argon), the electron collision frequency for momentum transfer depends on the electron energy. As the energy taken from the electromagnetic field by the electrons through collisions is the largest for the condition $\nu \simeq \omega$, one can forsee that the EEDF will be affected as the wave frequency is varied. Ferreira and Loureiro[12,15] have solved this problem and showed that, in argon, the shape of the EEDF could substantially change with ω: their calculations predict that the EEDF is not maxwellian for $\nu/\omega > 1$ (RF or DC discharges), whereas it resembles more closely a maxwellian distribution as ω becomes large ($\nu/\omega \ll 1$) (and becomes in fact Maxwellized when the electron density starts to exceed some 10^{12} cm^{-3}).

Fig. 10 shows the EEDF in argon calculated from the Boltzmann equation for various values of $0 \leqslant \nu_e/\omega < \infty$ ($\nu_e/N = 2 \times 10^{-7}$ cm^3 s^{-1}) and for the same value of the mean electron energy in all cases (3.5 eV). One consequence of this influence of ν/ω on the EEDF is that, at values of $\nu/\omega \ll 1$, the distribution contains comparatively more high energy electron capable of exciting and ionizing the gas than in the opposite limit (DC and low frequencies). It then turns out, as shown by Ferreira and Loureiro[15], that the power input per electron required to sustain a discharge for a fixed (pa) value is lower when $\nu/\omega \ll 1$ than when $\nu/\omega \gg 1$. For the same density of power deposited in the plasma, the case $\nu/\omega \ll 1$ yields therefore a higher plasma concentration and, as a consequence, a higher concentration of excited states. These predictions seem to agree very well with experimental observations[28]. They also provide an explanation for the apparent paradox that the lines emitted from highly lying levels are more intense in microwave discharges, as compared with RF and DC discharges. Due to the wide range of wave frequencies and of gas pressures in surface wave discharges, this dependence on ν/ω can be used as a convenient means of achieving an external control of the EEDF, to increase the efficiency and the selectivity of chemical reactions in plasma[24,28,29]. Further analysis of the dependence of the EEDF on ν/ω can be found in the work by Ferreira and Loureiro.

These theoretical results on the EEDF were obtained as part of the modelling work done on surface wave discharges at pressures up to about 1 Torr but they clearly apply to any HF produced plasma complying with the assumptions already mentioned.

IV.2 Experimental results

As mentioned earlier, one certainly cannot fully rely on Langmuir probes within surface wave discharges. The few acceptable experimental results available that concern the EEDF are thus indirect results. They come from the study of the gain of a helium-neon laser pumped by a surface wave plasma[30]. Moutoulas et al.[31] have observed that the laser gain, over the interval 100-1000 MHz (i.e. $\nu/\omega \simeq 2$ to 0.2), increases with the wave frequency, for given electron density values. Using a model yielding the population density for the two levels of the lasing transition, Moutoulas et al. show that the variation of the EEDF with ν/ω is responsible for the gain increase with frequency. One, however, notes that the variation of the EEDF with ν/ω is still far from being well documented experimentally. In that respect, it is clear that the surface wave discharge remains the best choice, for future work, as it is the only HF driven plasma that can be operated over the full range of RF and microwave frequencies.

V. Conclusion. It has been shown that the study of surface wave discharges has led to a substantial contribution in the modelling of HF discharges.

The starting point of this successful modelling seems to have been the travelling wave character of the surface wave that spatially spreads the plasma and the wave parameters. This has naturally conducted to the definition of Θ as a meaningful parameter. Then, the second contributing factor has certainly been the ability to vary the operating parameters of the surface wave discharge over an extremely broad range: it has permitted to observe many regularities in the behavior of Θ, namely that it is independent of the power flow and of the electron density (in argon) and, further, that it obeyed a similarity law. From these observations, it was possible to infer a general but simple model for HF discharges. The third contributing factor came from the possibility, with surface wave discharges, to obtain large variations in the radial gradient of the electric field intensity in the plasma by varying the electron density, the wave frequency and the plasma tube diameter. This has led to observe that one could define a maintenance field that is practically independent of the radial distribution of the electric field intensity. At the same time, it was instrumental in understanding the radial density distributions of excited atoms as well as the so-called skin effect observed in overdense HF produced plasmas.

References

[1] M. Moisan and Z. Zakrzewski, in "Radiation processes in discharge plasmas", J.M. Proud and L.H. Luessen, eds., NATO ASI series, Plenum Publishing, New York (1986).

[2] Z. Zakrzewski, M. Moisan, V.M.M. Glaude, C. Beaudry, P. Leprince, Plasma Phys. 19, 77 (1977).

[3] V.M.M. Glaude, M. Moisan, R. Pantel, P. Leprince, and J. Marec, J. Appl. Phys. 51, 5693 (1980).

[4] M. Moisan, R. Pantel, A. Ricard, V.M.M. Glaude, P. Leprince, and W.P. Allis, Rev. Physique Appl. 15, 1383 (1980).

[5] C.M. Ferreira, J. Phys. D: Appl. Phys. 14, 1811 (1981).

[6] M. Chaker, P. Nghiem, E. Bloyet, Ph. Leprince, and J. Marec, J. Physique-Lettres 43, L-71 (1982).

[7] P. Nghiem, M. Chaker, E. Bloyet, P. Leprince, and J. Marec, J. Appl. Phys. 53, 2920 (1982).

[8] M. Moisan, R. Pantel, and A. Ricard, Can. J. Phys. 60, 379 (1982).

[9] M. Moisan, C.M. Ferreira, Y. Hajlaoui, D. Henry, J. Hubert, R. Pantel, A. Ricard, and Z. Zakrzewski, Rev. Physique Appl. 17, 707 (1982).

[10] Z. Zakrzewski, J. Phys. D: Appl. Phys., 16, 171 (1983).

[11] C.M. Ferreira, J. Phys. D: Appl. Phys. 16, 1673 (1983).

[12] C.M. Ferreira and J. Loureiro, J. Phys. D: Appl. Phys. 16, 2471 (1983).

[13] E. Mateev, I. Zhelyazkov, and V. Atanassov, J. Appl. Phys. 54, 3049 (1983).

[14] R. Pantel, A. Ricard, and M.Moisan, Beitr. Plasmaphys. 23, 561 (1983).

[15] C.M. Ferreira and J. Loureiro,J. Phys. D: Appl. Phys. 17, 1175 (1984).

[16] Yu. M. Aliev, A.G. Boev and A.P. Shivarova, J. Phys. D: Appl. Phys. 17, 2233 (1984).

[17] M. Chaker and M. Moisan, J. Appl. Phys. 57, 91 (1985).

[18] D.J. Rose and S.C. Brown, Phys. Rev. 98, 310 (1955).

[19] S.C. Brown, Handb. of Physics 22, 531 (1956).

[20] M.Moisan and A. Ricard, Can. J. Phys. 55, 1010 (1977).

[21] A. Ricard, D. Collobert, and M. Moisan, J. Phys. B: At. Mol. Phys. 16, 1657 (1983).

[22] A. Ricard, J. Hubert, and M. Moisan, Proceedings 17th Int. Conf. on Phenomena in Ionized Gases, Budapest (1985), Contrib. papers, p. 741-3.

[23] W.P. Allis, S.C. Brown, and E. Everhart, Phys. Rev. 84, 579 (1951).

[24] A.T. Bell, Ind. Engng. Chem. Fundam. 9, 160 (1970).

[25] A. Petelin and M. Ury, "Plasma Sources for deep UV Litography", VLSI Electronics Micro-Structure Science, Vol. 8, N.G. Einspruch, Editor, Academic Press, New York (1984).

[26] M. Moisan, Plasma Phys. 16, 1 (1974).

[27] C.M. Ferreira, in "Radiation processes in discharge plasmas", J.M. Proud and L.H. Luessen eds., NATO ASI series, Plenum Publishing, New York (1986).

[28] Yu. A. Lebedev and L.S. Polak, High Energy Chemistry 13, 331 (1979).

[29] M. Wertheimer and M.Moisan, J. Vac. Sci. Tech., to appear (1985).

[30] C. Moutoulas, M. Moisan, L. Bertrand, J. Hubert, J.L. Lachambre, and A. Ricard, Appl. Phys. Lett. 46, 323 (1985).

[31] C. Moutoulas, M. Moisan, and Z. Zakrzewski, CLEO Conf. (Baltimore, 1985).

Figure Captions

Fig. 1 - Flow chart of the power through an elementary plasma slab along a plasma column sustained by a travelling wave (from [1]).

Fig. 2 - Theoretical and experimental values of the ratio of power loss per electron to the gas pressure in an argon discharge as a function of the product "pressure times plasma radius", for various wave frequencies (f) and plasma radii (a)(from [6,17]).

Fig. 3 - Calculated reduced sustaining field at the axis for a surface wave produced plasma vs. Na for $\bar{\omega}_{pe}$ a/c = 0.75 (B) and $\bar{\omega}_{pe}$ a/c = 1.5 (C), assuming a Maxwellian EEDF. Curve A - homogeneous field and Maxwellian EEDF. Curve D - homogeneous field, EEDF calculated from the Boltzmann equation. N is the gas density, a is the plasma radius, and $\bar{\omega}_{pe}$ is the angular electron plasma frequency corresponding to the cross-section averaged electron density.

Fig. 4 - Calculated radial distribution of electron density in an argon surface wave produced plasma column at 0.1 Torr and for various values of the cross-section averaged electron density $\bar{n}$ (from [5]).

Fig. 5 - Calculated radial variation of the axial and radial components of the surface wave electric field in an argon surface wave produced plasma column and for various values of the average electron density $\bar{n}$ (from [5]).

Fig. 6 - Observed radial dependence of the emission intensity of the ArI 549.6 nm thin line on gas pressure (from [9]).

Fig. 7 - Observed radial distribution of the population density of $^3P_2$ argon atoms in a metastable state (from [4]).

Fig. 8 - Calculated radial density distribution of excited argon atoms in the $3p^5$ 6d configuration in a surface wave produced plasma column with 2a = 25 mm and an absorbed power per unit length of 30 W/m (from [5]).

Fig. 9 - Variation of electron density, electric field, and ionization rate across a plane parallel HF discharge, $y = (\omega_{pe}(r)/\omega)^2$ and ν_i is the ionization frequency (from [23]).

Fig. 10 - Calculated electron energy distribution functions in argon with the same mean energy of 3.5 eV, for the following values of ν_e/ω: ∞, (A); 2.0 (B); 1.25 (C); $\ll 1$(D).

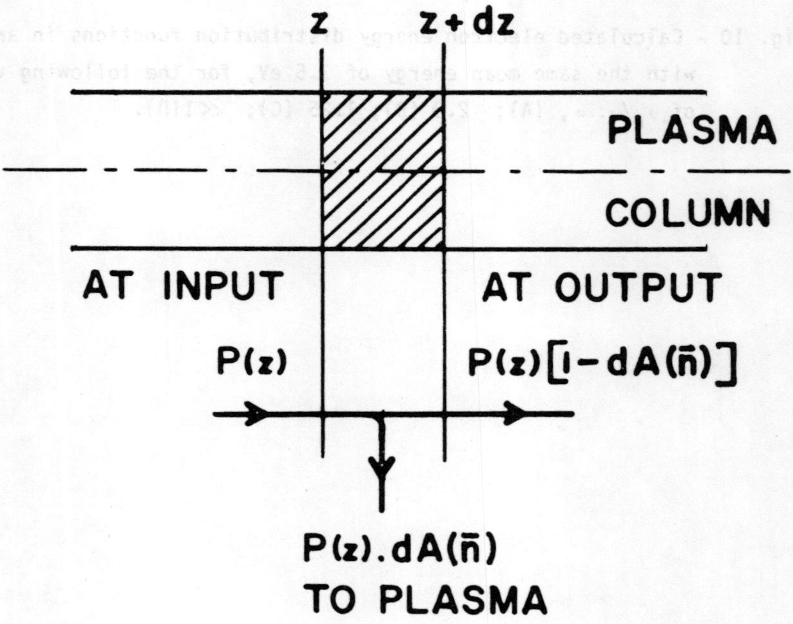
z
z + dz
PLASMA
COLUMN
AT INPUT
AT OUTPUT
P(z)
P(z)[1 - dA(n̄)]
P(z).dA(n̄)
TO PLASMA

Fig. 1

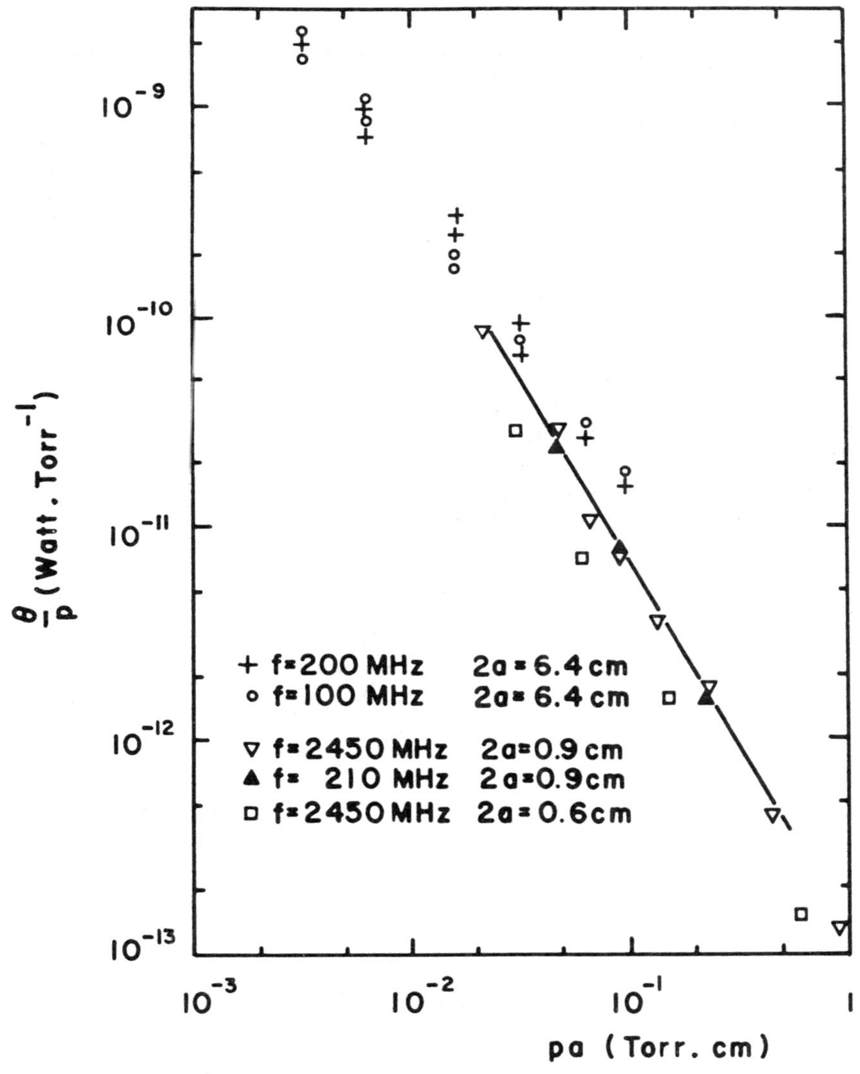
10^{-9}
10^{-10}
10^{-11}
10^{-12}
10^{-13}
$\frac{\theta}{p}$ (Watt . Torr^{-1})
+ f=200 MHz 2a=6.4 cm
o f=100 MHz 2a=6.4 cm
▽ f=2450 MHz 2a=0.9 cm
▲ f= 210 MHz 2a=0.9 cm
□ f=2450 MHz 2a=0.6 cm
10^{-3}
10^{-2}
10^{-1}
1
pa (Torr. cm)

Fig. 2

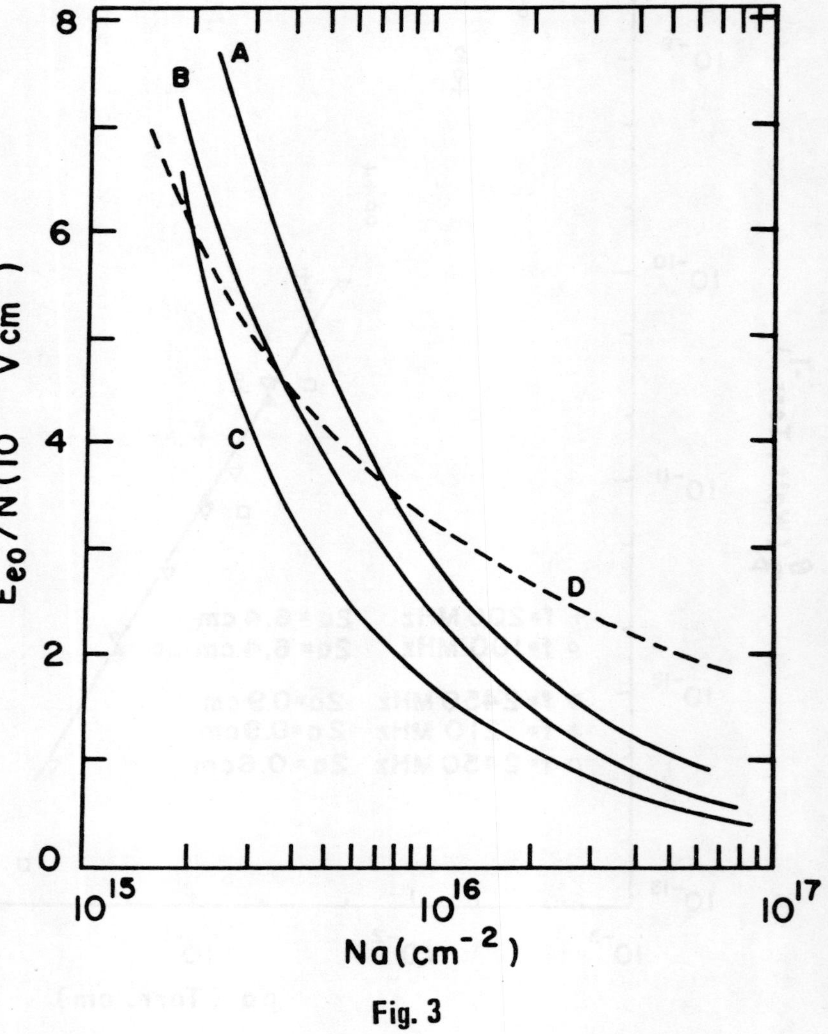

Fig. 3

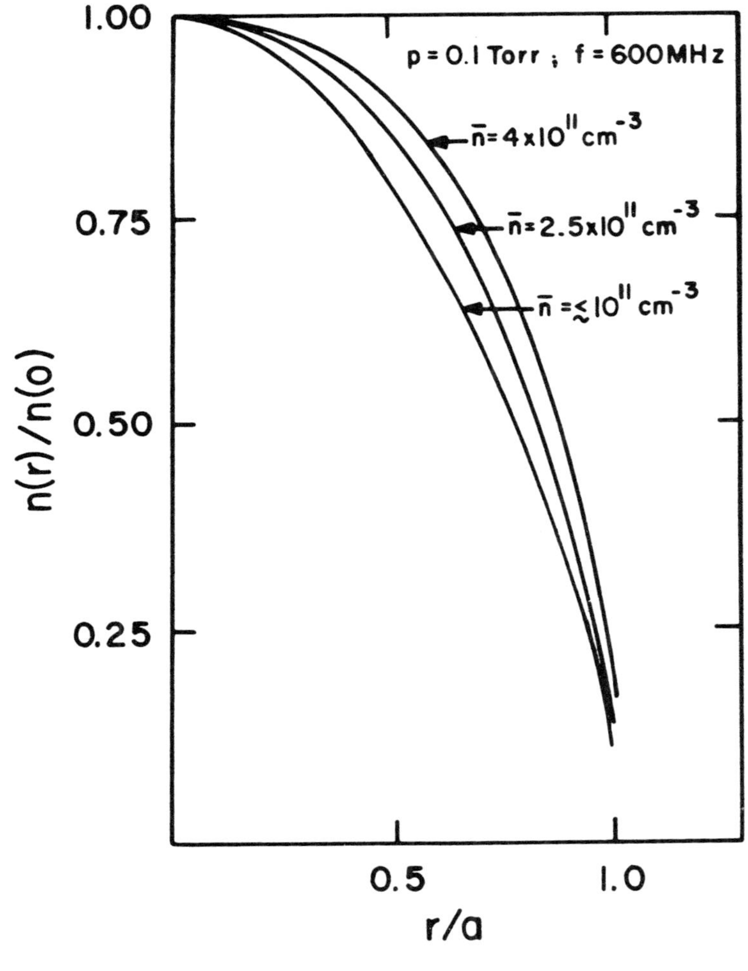

p = 0.1 Torr ; f = 600MHz
n̄ = 4 x 10¹¹ cm⁻³
n̄ = 2.5 x 10¹¹ cm⁻³
n̄ = ≲ 10¹¹ cm⁻³
1.00
0.75
0.50
0.25
n(r)/n(o)
0.5
1.0
r/a

Fig. 4

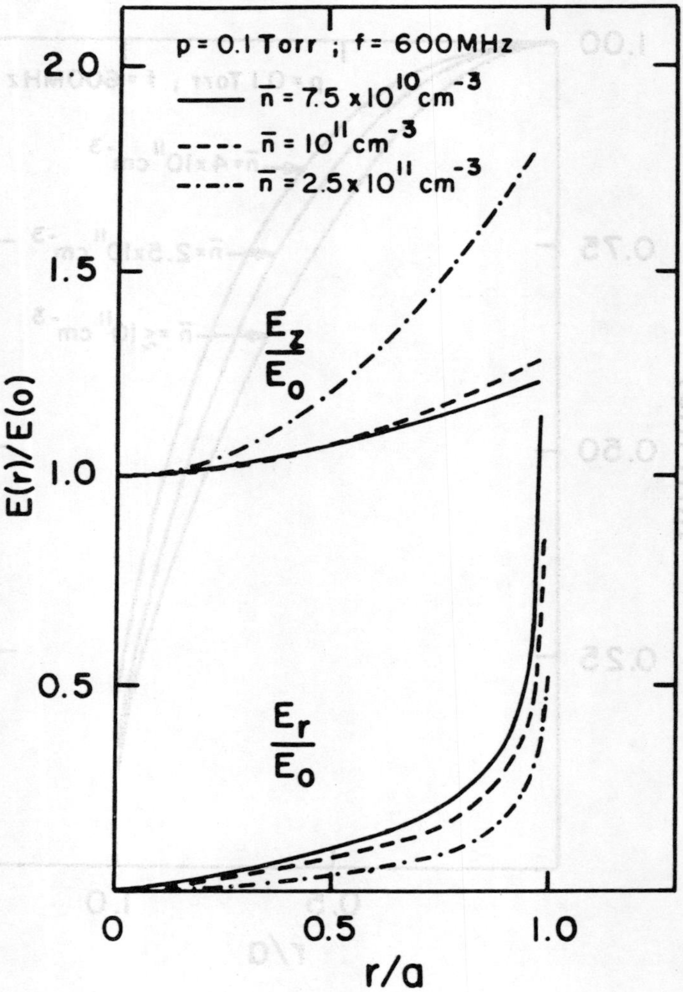
p = 0.1 Torr ; f = 600MHz
n̄ = 7.5 x 10^10 cm^-3
n̄ = 10^11 cm^-3
n̄ = 2.5 x 10^11 cm^-3
E_z / E_o
E_r / E_o
E(r)/E(o)
2.0
1.5
1.0
0.5
0
0.5
1.0
r/a

Fig. 5

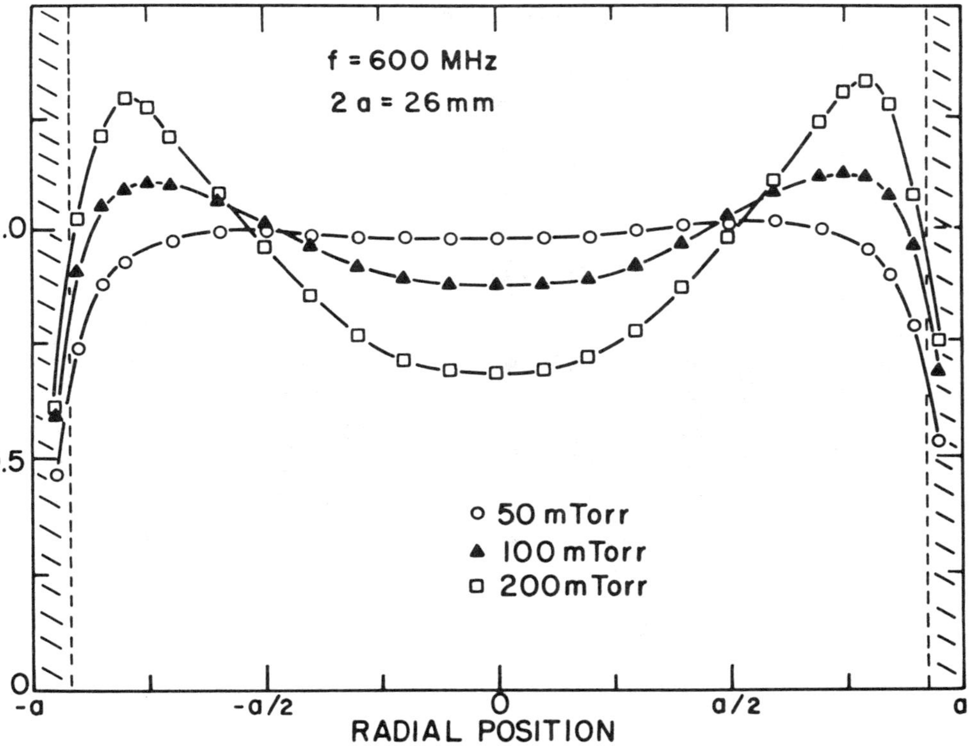
f = 600 MHz
2 a = 26 mm
1.0
0.5
0
-a
-a/2
0
a/2
a
RADIAL POSITION
50 mTorr
100 mTorr
200 mTorr

Fig. 6

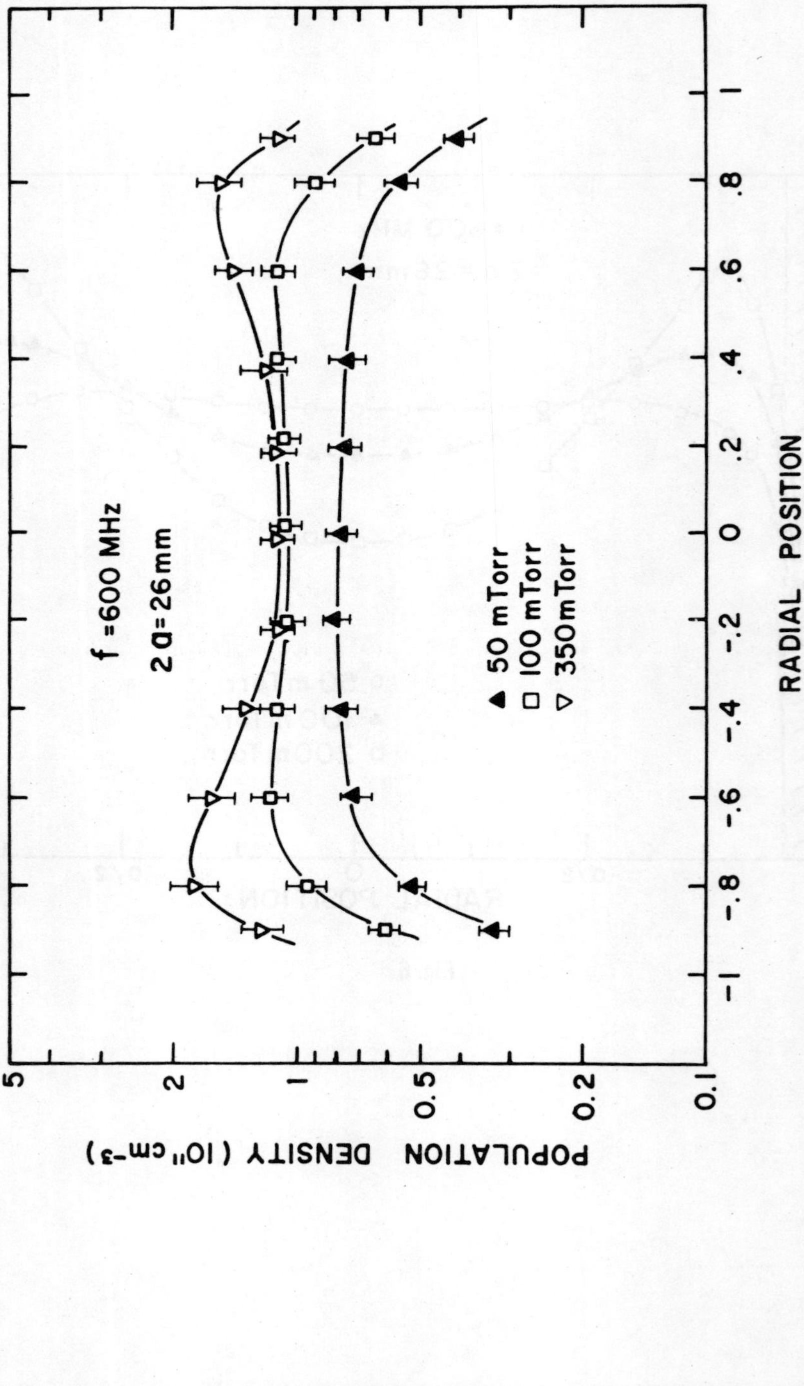
f =600 MHz
2a=26mm
▲ 50 mTorr
□ 100 mTorr
▽ 350mTorr
RADIAL POSITION
POPULATION DENSITY (10¹¹cm⁻³)
-1
-.8
-.6
-.4
-.2
0
.2
.4
.6
.8
1
0.1
0.2
0.5
1
2
5

Fig. 7

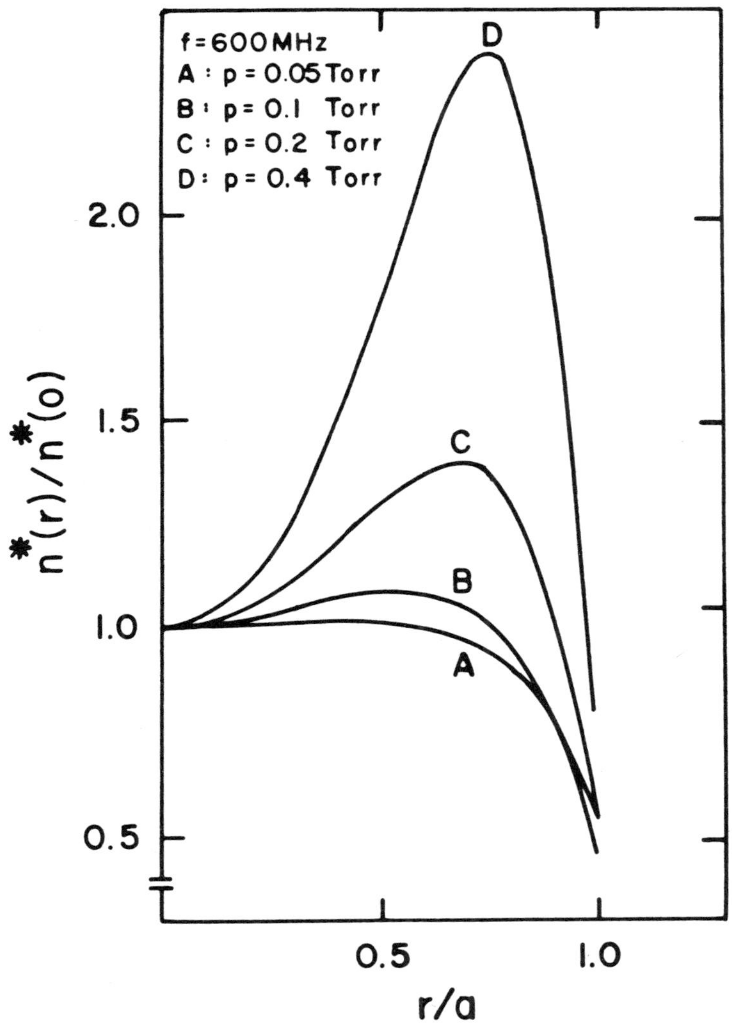
f=600MHz
A : p = 0.05 Torr
B : p = 0.1 Torr
C : p = 0.2 Torr
D : p = 0.4 Torr
D
C
B
A
2.0
1.5
1.0
0.5
n*(r)/n*(o)
0.5
1.0
r/a

Fig. 8

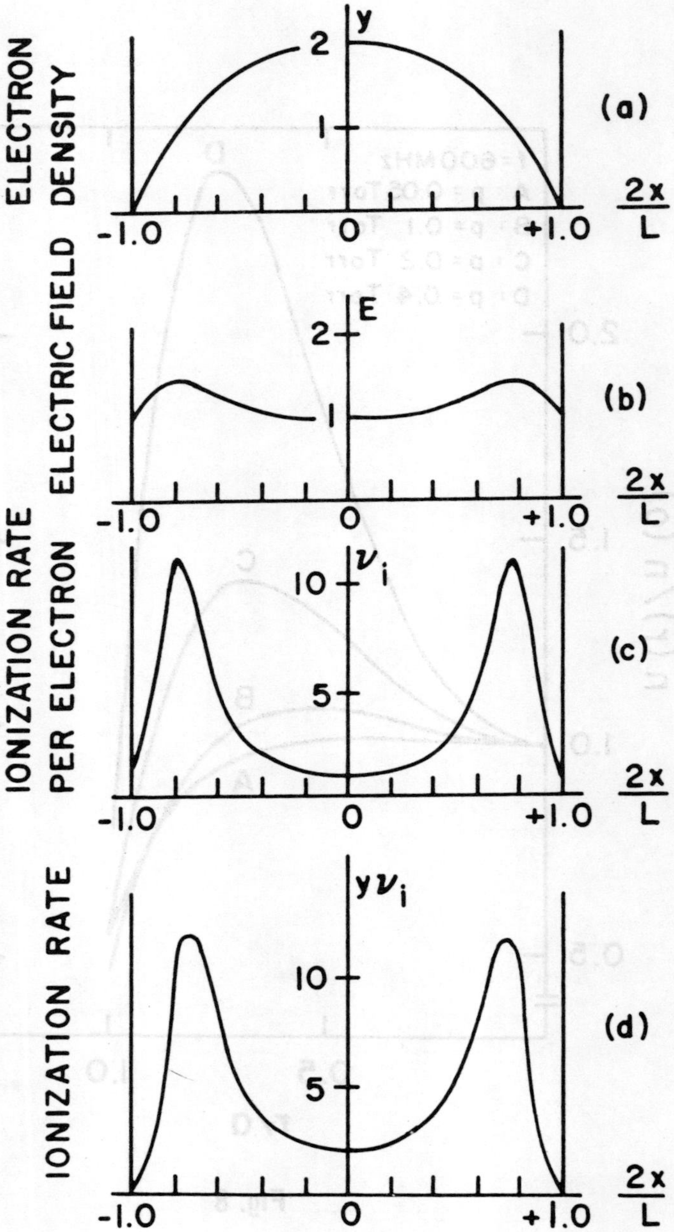
ELECTRON DENSITY
y
2
1
(a)
-1.0
0
+1.0
2x/L
ELECTRIC FIELD
E
2
1
(b)
IONIZATION RATE PER ELECTRON
ν_i
10
5
(c)
IONIZATION RATE
$y\nu_i$
(d)

Fig. 9

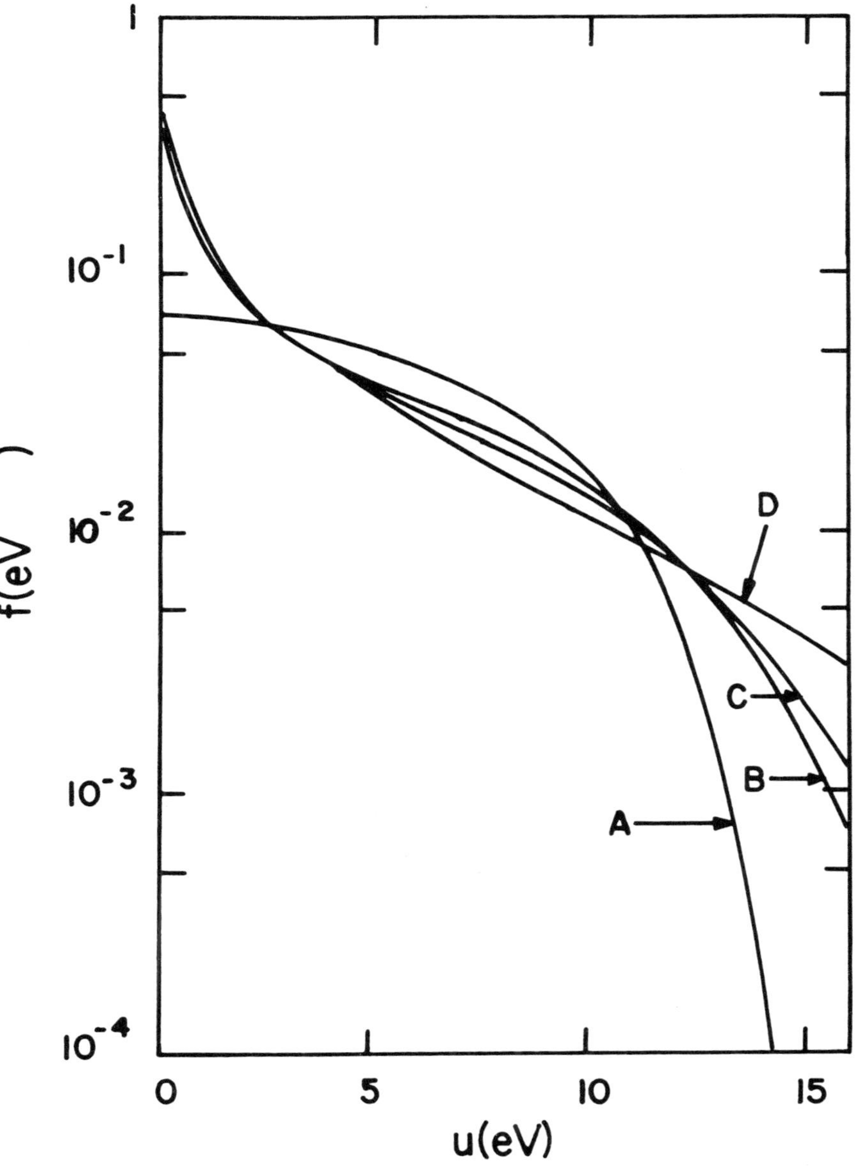

Fig. 10

NONLINEAR INTERACTION OF MAGNETOSTATIC WAVES WITH ULTRASONIC AND ELECTROMAGNETIC WAVES

by

Yu. V. GULYAEV; S.A. NIKITOV
Institute of Radiotechnics and Electronics,
Academy of Sciences, Moscow, USSR

The paper is devoted to a theoretical investigation of propagation of magnetostatic waves (MSW) with ultrasonic (UW) and electromagnetic (EMW) waves. The velocities of the waves depend from the amplitudes of the waves. This effect leads to the effect of nonlinear interaction of the waves. This nonlinear interaction is defined from the second members of distribution of Hamiltonian of magnetoelastic or magnetooptical interactions. The system of parabolic equations is removed in the work. The analytical decisions of this system received for the spherical and cylindrical waves and for the plan waves. The effect of interfocusing is promised in the first case and the effect of stratification is promised in the second case. It was present estimations of the power of UW and MSW, MSW and EMW, which is necessary for the effects of interfocusing and stratification.

It was also present estimations of the power of MSW for these effects, when the effect of automodulation is neglected.

PROPAGATION OF PULSES IN SOLIDS AND PLASMAS

P. Halevi

Departamento de Física
Instituto de Ciencias de la Universidad Autónoma de Puebla
Apdo. Postal J-48, Puebla, Pue. 72570, México

and

J. A. Gaspar-Armenta

Centro de Investigaciones en Física
de la Universidad de Sonora
Apdo. Postal A 088, Hermosillo, Son. 83190, México

I. Introduction

We present a review of limited scope that focuses on certain aspects of transient wave-phenomena studied recently. The discussion is of interest to gaseous plasmas and solids, however the emphasis is on the latter. The typical problem is the response of a material medium to a pulse that is Gaussian in time. We consider both unbounded and bounded media.

In the past much effort was devoted to arguments about the respective merits of various definitions of the velocity of light; in 1977 Bloch[1] counted no less than eight of them: phase velocity, group velocity, energy velocity, signal velocity, relativistic velocity constant, ratio - of - units velocity, centrovelocity, and correlation velocity. In solids experimental work was hampered by the lack of techniques for measurement of extremely short propagation times. It should be realized that it takes $\sim$ 10 psec for light to traverse 1 mm. This situation has been dramatically changed with the advent of picosecond spectroscopy.[2] In a typical experiment a laser beam is split into a "probe beam" that passes through the sample and a "reference beam" that propagates in air. The time t_o that it takes for the pulse to traverse the sample is measured by means of a cross-correlation technique that utilizes second-harmonic generation.[3] If D is the distance traversed then a correlation velocity, $v_{cor} = D/t_o$, may be defined.

The correlation velocity is a valid concept even in the case of a complicated transient response, for example when a single incident pulse splits into several peaks. If the emerging light consists of a single pulse then the correlation velocity may be interpreted as the "pulse velocity". In simple cases it is not expected to differ substantially from the peak velocity v_{peak}. Thus one may look at the time that it takes for the peak of the pulse to traverse a given distance. This "temporal definition" of pulse velocity seems to be gaining acceptance.[4,5]

In general it has been found that the pulse velocity is the same as the group velocity $v_g = d\omega/dk_r$, where ω is the circular frequency and k_r is the real part of the wavevector. This is true for amplifying, as well as absorptive media. For instance, Casperson and Yariv[6] found that ultrashort pulses in a high-gain infrared xenon laser propagate with the velocity v_g, which may be a factor 2.5 smaller than c. Similar results have been obtained by Ulbrich and Fehrenbach[7] who propagated pulses of 12 psec of duration through a 3.7μm thick GaAs crystal. The experiment was performed in the near infrared, an electron being excited from the valence band to the conduction band of the semiconductor. Thus an "exciton-polariton" is formed; the corresponding group velocities ranged from c/3.6 to c/2,000. The experiment was performed by utilizing a cross-correlation technique. In another semiconducting crystal, CuCl, Segawa et al.[8] observed an "anomalously slow" group velocity, however this was qualitatively explained by taking into account the dispersive effect of a neighboring exciton resonance.

In a particularly interesting experiment Chu and Wong[9] have measured the transit time of pulses of light through GaP films of thicknesses of 76μm and 9.5μm. The samples were strongly doped and the measurements were taken in the immediate vicinity of an absorption line. This is the region of anomalous dispersion: the group velocity may become greater than c, equal to $\pm\infty$, and negative. Quite often one encounters the comment that, in such cases, the group velocity is not a meaningful concept.[10] Nevertheless Chu and Wong[9] found that the correlation velocity, i.e. the pulse velocity is given by v_g even in these "anomalous" cases. Such a behavior, for Gaussian pulses in an unbounded medium, has been actually predicted by Garrett and McCumber.[11] This is subject to two conditions, namely the distance traversed by the pulse must not be too large, and the frequency spread of the pulse $\Delta\omega(\approx 1/\tau$ where τ is the temporal width) must be much smaller than the damping frequency. Other authors[4,5,12,13] have also reached similar conclusions.

Recently Segard and Macke[14] made an experimental study of pulse propagation in the region $v_g < 0$ in a gaseous plasma. They used a circular waveguide of internal diameter 60mm and 24m long. The carrier frequency was about 109.5 GHz, corresponding to a vacuum wavelength of 2.74mm, and the pulse width was about 4μs. The absorption in the gas was caused by rotational transitions of molecules corresponding, principally, to the $J = 8 \rightarrow J = 9$ OCS line. The pressure in the waveguide was sufficiently low, so that the collisional time was about 0.35μs. The pulse distortion was "quite moderate". The authors observed pulse advances of $\sim$ 1μs compared with the propagation times in vacuum. In fact, the measured pulse advances did not depart significantly from the expression based on the theoretical group velocity for a gas absorber. Thus pulse propagation with velocities $v_{peak} = v_g < 0$ has been verified experimentally in both solid state[9] and in gaseous[14] plasmas.

In section 2 we will give a simplified proof of the relation $v_{peak} = v_g$ for an unbounded medium, and we will discuss and interprete this result. We will also review higher-order effects such as "chirping", and changes in the width and the shape of the pulse. It should be stressed that, in the presence of strong distortion of the pulse, "peak-velocity" is not a meaningful concept. Also, in any experiment, the propagation is limited by the finite extension of the apparatus (film thickness, discharge-tube length, etc.). The effect of boundaries (in the presence of strong absorption)[15] is studied in section 3.

Section 4 deals with the other extreme of a transparent (i.e. absorptionless) medium of finite width D. On the basis of analytic[16] and numerical[17] calculations we will show that the pulse velocity exhibits a rather unexpected behavior if $D/v_g \ll \tau$.

In direct-gap semiconductors the maximum of the valence band and the minimum of the conduction band are both located at the origin of the Brillouin zone. The momentum of the incident photon is transmitted to the exciton and an exciton-polariton with kinetic energy is formed. This, in turn, gives rise to a wavevector - dependent dielectric function, that is, to spatial dispersion (nonlocal effects). For a given ω two values of k are now possible which means that the $\omega(k)$ diagram has two branches. Pulse propagation corresponding to both branches has been observed in GaAs[7] and, with particular clarity, in CuCl by Masumoto et al.[18] (however no effects of spatial dispersion for CuCl have been reported in reference 8). The group velocities calculated from the lower and upper polariton branches account quite well for the measured cross-correlation velocities.[18] Also, for the lower polariton branch the relation $v_g \approx \hbar k/m$ permitted to determine the exciton mass m.

The lower polariton branch exhibits an inflection point, corresponding to a minimum of v_g for a certain frequency

$\omega = \omega^*$. In the vicinity of ω^* the usual derivation breaks down, as pointed out by Linton Johnson.[19] Then the pulse that emerges from the plate will oscillate according to the properties of the Airy function. This effect has not been observed, however Itoh et al.[20] have noticed that the pulses change their shape near to the exciton resonance frequency (which is rather near to ω^*). The conclusions of reference 19 are based on analytical approximations; in section 5 we present numerical results[21] on pulse propagation in nonlocal semiconductors at $\omega \sim \omega^*$.

This review deals with smooth pulses that have no beginning and no end. It has been known for a long time[22] that, for pulses with an abrupt beginning, "precursors" precede the main body of the pulse. Precursors have been observed in the microwave region by Pleshko and Palócz.[23] In addition to the "Somerfeld precursor" and the "Brillouin precursor"[22] a spatially dispersive medium can also support a "nonlocal precursor", as shown by Frankel and Birman.[24] The corresponding transient optical reflectivity has been studied by Agrawal et al.[25] The reader is referred to the original articles and a review article by Birman[26] for additional information on precursors.

2. Propagation in an absorptive unbounded medium

In this section we are concerned with unlimited propagation in one direction, z. At an arbitrary point z and a time t the electric field of the pulse is

$$E(z,t) = \int_{-\infty}^{\infty} d\omega S_0(\omega) e^{i[k(\omega)z-\omega t]} \tag{1}$$

The spectral function $S_o(\omega)$ is centered at the carrier frequency (or laser frequency) $\bar{\omega}$ and has a width $\sim 1/\tau$. The response of the medium is given by the functional dependence $k(\omega)$. The wavevector may be expanded arround $\bar{\omega}$ as follows:

$$k(\omega) = k(\bar{\omega}) + k'(\bar{\omega})(\omega-\bar{\omega}) + \frac{1}{2} k''(\bar{\omega})(\omega-\bar{\omega})^2 + \ldots \qquad (2)$$

where the primes denote differentiation. The medium is dissipative, and therefore $k(\bar{\omega})$, $k'(\bar{\omega}),\ldots$ are in general all complex quantities.

From the mathematical point of view it is convenient to limit the discussion to Gaussian pulses. These are characterised by the spectral function

$$S_o(\omega) = \exp[-(\omega-\bar{\omega})^2\tau^2/2] \qquad (3)$$

If the pulse is sharp in frequency, that is $\tau^{-1} << \nu$, where ν is the phenomenological damping frequency, then for sufficiently small propagation distances, $z << \tau^2/|k''(\bar{\omega})|$, the third and higher-order terms in eq. (2) may be neglected. Then eq. (1) becomes

$$E(z,t) \cong \int_{-\infty}^{\infty} d\omega e^{-(\omega-\bar{\omega})^2\tau^2/2} e^{i[k(\bar{\omega})z+k'(\bar{\omega})(\omega-\bar{\omega})z-\omega t]} \qquad (4)$$

This is easily integrated to give

$$E(z,t) = \frac{\sqrt{2\pi}}{\tau} e^{-[t-k'(\bar{\omega})z]^2/2\tau^2} e^{i[k(\bar{\omega})z-\bar{\omega}t]} \qquad (5)$$

If we write $k(\bar{\omega}) = k_r+ik_i$ and $k'(\omega) = k_r'+ik_i'$ this expression may be expressed in the form

$$E(z,t) = \frac{\sqrt{2\pi}}{\tau} e^{i(k_r - k_r' k_i' \tau^{-2} z)z} e^{-i(\bar{\omega} - k_i' \tau^{-2} z)t}$$

$$e^{-[\underline{k}_i - \frac{1}{2}(k_i')^2 \tau^{-2} z]z} e^{-(t-k_r' z)^2/2\tau^2} \tag{6}$$

There are four exponential factors. The parenthesis in the exponents of the first three factors define, respectively, the effective wavector, the effective carrier frequency, and one-half of the effective absorption coefficient. These quantities all depend on the pulse-width τ and on the position z. The terms proportional to $\tau^{-2}z$ are corrections to k_r, $\bar{\omega}$, and k_i that may be neglected because our assumption of small propagation distances z.

The interpretation of z as the propagation distance makes sense if we look at the form of the pulse at the origin. Substituting z = 0 in eq. (6) we get

$$E(0,t) = \frac{\sqrt{2\pi}}{\tau} e^{-i\bar{\omega}t} e^{-t^2/2\tau^2} \tag{7}$$

We see that the maximum of the pulse appears at z = 0 at the instant t = 0. Therefore the origin z = 0 corresponds to the "launching point" of the pulse. Similarly, the last exponential factor of eq.(6) gives the instant at which the maximum arrived at the point z. This time is $t = k_r' z$ and so the velocity of the peak of the pulse is

$$v_{peak} = \frac{x}{t} = \frac{1}{k_r'}$$

$$= (\frac{dk_r}{d\bar{\omega}})^{-1} = v_g(\bar{\omega}) \tag{8}$$

The last expression is the group velocity (defined for a

dissipative medium) evaluated at the carrier frequency.

Note that nowhere did we assume that $0 < v_g < c$. Thus we see that the peak of the pulse propagates at the group velocity even when v_g has "anomalous" values.[11] In order to reconcile this strange result with casuality one should realize that a Gaussian pulse (in time) has no beginning and no end. In the presence of strong absorption the medium causes a redistribution of the energy in such a way that the peak of the pulse at some pint z has been actually formed from the early part ($t < 0$) of the pulse that was launched at $z = 0$. Therefore causality is not violated. As pointed out by Crisp,[27] "more energy is absorbed from the trailing half of the pulse than from the leading half and this results in a motion of the pulse maximum at a velocity greater than the phase velocity. Under certain conditions this asymmetric absorption of energy occurs in such a way that the transmitted pulse has the same shape as the incident pulse". This behavior may be attributed to the finite response time $1/\nu$ of the absorbing particles in the material medium.

It is instructive to study the behavior of a modified Gaussian pulse, given by a Gaussian function that has been curtailed for negative times.[28] Then eq. (7) is replaced by

$$E(0,t) = \begin{cases} \dfrac{\sqrt{2\pi}}{\tau}\, e^{-i\omega t}\, e^{-(t-T)^2/2\tau^2}, & t \geq 0 \\ 0, & t \leq 0 \end{cases} \qquad (9)$$

Here T is a suitably chosen time that gives the maximum of the pulse. This pulse <u>does</u> have a beginning (at $t = 0$), however it has still no end. The numerical results[28] for

$0 < v_g < \infty$ show that, for finite z, a precursor is formed. The precursor has the form of a sharp peak that propagates at the vacuum-speed of light c. This part of the pulse stems from the high-frequency discontinuity at t = 0. As for the main body of the pulse, its peak propagates at the group velocity v_g. It is plausible that similar results will obtain for $v_g < 0$. In this case we expect that the precursor will move in the positive t-direction (as before), while the peak will more in the negative t-direction. There can be no response before the arrival of the precursor because this would be a violation of causality. Such a conclusion was, in fact, reached for pulses that have the form of the positive half-period of a cosine function.[5,13]

We wish to stress that the proof of eq.(6) does not depend on the specific form of the dispersion relation $k = k(\omega)$. For this reason our conlusions are valid for bulk waves, surface waves, waves in complex geometries, magnetoplasma waves, etc.

It is interesting to inquire into the shape of the pulse, as a function of z, at a given instant of time t. Rearranging eq.(6) the electric field is found to be proportional to the expression

$$E(z,t) \propto e^{-(z-v_g t + k_i v_g^2 \tau^2)^2/2v_g^2 \tau^2} \tag{10}$$

Thus the pulse is Gaussian in space, as well as in time. The spatial width is $|v_g|\tau$, and is typically much greater than an absorption length $1/2\,k_i$. The position of the peak is

$$z_{peak} = v_g t - k_i v_g^2 \tau^2 \tag{11}$$

Therefore, at the time $t = 0$ the peak of the Gausssian is <u>not</u> located at $z = 0$, but rather at the point $-k_i v_g^2 \tau^2$! The proportionality to k_i indicates that this is an effect of absorption. We also see that the peak is located at the origin at the time $k_i v_g \tau^2$ which is negative for $v_g < 0$. This shows that, for anomalous dispersion, the main body of the pulse entered the region $z > 0$ before the instant $t = 0$. This is another way to understand the negative transit times obtained for $v_g < 0$.

Eq.(6) is based on the assumption that $k''(\bar{\omega})$ and higher derivatives of k are all negligible. While this is a good approximation for $|k''(\bar{\omega})| z << \tau^2$ and $\nu\tau >> 1$ it is interesting to consider the effects of the term $1/2\, k''(\bar{\omega})(\omega-\bar{\omega})^2$ in eq.(2). For a Gaussian function $S_0(\omega)$ the integration in eq.(1) may be readily performed. The resulting pulse $E(z,t)$ is still Gaussian as a function of t, and the peak still propagates at the group velocity $v_g(\bar{\omega})$. However, the width of the Gaussian is $(\tau^2 + k_i'' z)^{1/2}$, which means that the pulse expands or is compressed as it propagates, depending if k_i'' is positive or negative. In fact, Katz and Alfano[29] called attention to a compression of the pulses observed by Chu and Wong.[9] In their response Chu and Wong[30] show that k_i'' is negative (and inversely proportional to ν^3) corresponding, indeed, to a compression.

Another interesting effect of keeping k'' finite is the appearance of an additional factor in eq.(6), namely

$$\exp\left[- i k_r'' z(t - k_r' z)^2/2\tau^4\right] \tag{12}$$

This is an oscillatory effect that gives rise to a time-dependent shift $\delta\omega$ in the carrier frequency $\overline{\omega}$:

$$\delta\omega = k_r'' z(t - k_r' z)/2\tau^4 \tag{13}$$

If we assume that $k_r'' > 0$ and observe the pulse at a fixed z then for $t < z/v_g$ (that is before the peak arrives) the effective carrier frequency is lower than $\overline{\omega}$. When the peak arrives at $t = z/v_g$ then $\delta(\omega) = 0$ and the carrier frequency is $\overline{\omega}$. For $t > z/v_g$ (that is after the peak has passed the point z) the frequency shift is positive and the effective carrier frequency is greater than $\overline{\omega}$. On an audio transducer one would hear a gradually rising pitch as the intensity of the pulse increases to a maximum and then decreases. The effect is a chirping sound and one refers to a time-dependent frequency shift, e.g. eq.(13), as to "chirping". If $k_r'' < 0$ then the pitch will fall as the pulse passes by. The chirping effect increases with z, however we must keep z rather small so as not to violate our assumption that $|k''(\overline{\omega})|z << \tau^2$.

Crisp[27] comments that even if the pulse is distorted its maximum may still propagate with the group velocity. For a smooth pulse such as a Gaussian, distortion may be expected for $z \gtrsim \tau^2/|k''(\overline{\omega})|$, and the distortion will increase with z.[11,13] In a numerical simulation corresponding to the experiment of Chu and Wong,[9] Segard and Macke[14] find that, for $2k_i(\overline{\omega})z = 5.7$, an incident Gaussian pulse becomes strongly distorted: in addition to the principal peak two smaller peaks develop at later times. In such a situation the concept of "peak velocity" obviously loses its meaning. Moreover, a second-order autocorrelation scan[9] gives no information as to pulse shape.[29,14]

Even for small propagation distances, if the pulse is not smooth distortions will rapidly develop. The requirement of distortionless propagation would impose severe conditions on the continuity of the pulse envelope and its successive derivatives.[31] The slightest anomaly in the original pulse corresponds to a high-frequency transient whose front propagates with the velocity c.[14] As a result "wiggles" appear, and their size increases with z. For a large propagation distance the pulse becomes completely distorted and bears no semblence to the original pulse. Clearly, in such cases it is best to look at the complete shape E(z,t).

3. Transmission through an absorptive slab

We are concerned with a dielectric slab of thickness D bounded by vacuum. The dielectric medium is characterised by a dielectric constant $\varepsilon(\omega)$ and a magnetic permeability $\mu(\omega)$ that are, in general, complex quantities so as to allow for absorption. The wave-vector in the vacuum is $k_o = \omega/c$ and the wave-vector in the medium is $k = k_o[\varepsilon(\omega)\mu(\omega)]^{1/2}$. We will assume normal incidence, so the electric and magnetic fields are both parallel to the slab. For an incident monochromatic wave of unit amplitude the electric fields in the three media are

$$E(z,t) = \begin{cases} (e^{ik_o z} + re^{-ik_o z})\, e^{-i\omega t}, & z \le 0 \\ (a_+ e^{ikz} + a_- e^{-ikz})\, e^{-i\omega t}, & 0 \le z \le D \\ te^{ik_o z}\, e^{-i\omega t}, & z \ge D \end{cases} \tag{14}$$

Here r and t are the amplitudes of the reflected and the transmitted waves and a_+ and a_- are the amplitudes of the two waves in the slab. Using Faraday's law we may also write down the magnetic field:

$$H(z,t) = \begin{cases} (e^{ik_o z} - re^{-ik_o z})\, e^{-i\omega t}, & z \leq 0 \\ (k/k_o\mu)(a_+e^{ikz} - a_-e^{-ikz})\, e^{-i\omega t}, & 0 \leq z \leq D \\ te^{ik_o z}\, e^{-i\omega t}, & z \geq D \end{cases} \tag{15}$$

Utilizing the continuity of E(z,t) and H(z,t) at the interfaces z = 0 and z = d the following solutions are found for r and for t:

$$r = \frac{1-Q^2}{1+Q^2+2iQ \cot kD} \tag{16}$$

$$t = \frac{2iQ \exp(-ik_oD)}{(1+Q^2)\sin kD+2iQ \cos kD} \tag{17}$$

We have abbreviated, $Q = k/k_o\mu$. Being interested in the propagation of pulses we only pursue the behavior of the transmitted fields. From eqs.(14), (15), and (17), for $z \geq D$ these fields are

$$E(z,t) = H(z,t) = \frac{2iQ}{(1+Q^2)\sin kD+2iQ \cos kD}\, e^{ik_o(z-D)-i\omega t} \tag{18}$$

This may be expressed in the form

$$E(z,t) = \frac{4\mu n}{(n+\mu)^2}\, \frac{\exp(ikD)}{1-[(n-\mu)/(n+\mu)]^2 \exp(2ikD)}\, e^{ik_o(z-D/-i\omega t}$$

$$= \frac{4\mu n}{(n+\mu)^2} \sum_{m=0}^{\infty} \left(\frac{n-\mu}{n+\mu}\right)^{2m} e^{i(1+2m)kD} e^{ik_o(z-D)-i\omega t} \tag{19}$$

where

$$n(\omega) = k/k_o = \left[\varepsilon(\omega)\mu(\omega)\right]^{1/2} \tag{20}$$

is the index of refraction. Eq.(19) has a simple interpretation in terms of the transmission and reflection coefficients for a single interface. These coefficients, for transmission from vacuum into the medium, and from the medium into vacuum, are respectively

$$t_1(\omega) = \frac{2\mu(\omega)}{n(\omega)+\mu(\omega)} \tag{21}$$

$$t_1'(\omega) = \frac{2n(\omega)}{n(\omega)+\mu(\omega)} \tag{22}$$

The reflection coefficient for a medium-vacuum interface is

$$r_1(\omega) = \frac{\mu(\omega)-n(\omega)}{\mu(\omega)+n(\omega)} \tag{23}$$

Thus, from eq.(19)

$$E(z,t) = \sum_{m=0}^{\infty} t_1(\omega) r_1^{2m}(\omega) t_1'(\omega) e^{i(1+2m)kD} e^{ik_o(z-D)-i\omega t} \tag{24}$$

The electric field has been expressed in terms of an infinite number of partial waves m = 0, 1, 2, ... For a given value of m the wave undergoes a transmission at the vacuum-medium interface, then 2m internal reflections (m "zig-zags"), and finally another transmission at the medium-vacuum interface. These three processes are described by the algebraic factor $t_1 r_1^{2m} t_1'$

in eq.(24). The total path traversed by the m'th wave inside the plate is (1+2m)D; the first exponential factor gives the corresponding phase change. The wave propagates an additional distance z-D in vacuum, as expressed by the second exponential factor in eq.(24). These considerations establish the transmitted field as a superposition of waves resulting from m = 0, 1, 2, ... double internal reflections in the plate. Eq.(24) is suitable for the discussion of interference phenomena.[10]

Consider now a Gaussian pulse, whose spectral function is given by eq.(3). Moreover, we are interested in the temporal behavior of the outcoming pulse at z = D, so eq.(24) gives rise to the following wave packet:

$$E(D^+,t) = \sum_{m=0}^{\infty} \int_{-\infty}^{\infty} d\omega\, e^{-(\omega-\bar{\omega})^2\tau^2/2}\, t_1(\omega)\, r_1^{2m}(\omega)\, t_1'(\omega)\, e^{i(1+2m)k(\omega)D-i\omega t} \qquad (25)$$

We assume that the pulse is very sharp in frequency, that is the pulse width τ must be much greater than any characteristic time such as the collisional time $(1/\nu)$ and the period of the light $(2\pi/\bar{\omega})$. Then we may expand $k(\omega)$ as in eq.(2). Keeping only the first two terms in the expansion we get

$$E(D^+,t) \cong \sum_{m=0}^{\infty} t_1(\bar{\omega})\, r_1^{2m}(\bar{\omega})\, t_1'(\bar{\omega})\, e^{i(1+2m)k(\bar{\omega})D-i\bar{\omega}t}$$

$$\int_{-\infty}^{\infty} d\omega\, e^{-(\omega-\bar{\omega})^2\tau^2/2+i[(1+2m)k'(\bar{\omega})D-t](\omega-\bar{\omega})}$$

The integration gives

$$E(D^+,t) \cong \frac{\sqrt{2\pi}}{\tau} \sum_{m=0}^{\infty} t_1(\bar{\omega}) r_1^{2m}(\bar{\omega}) t_1'(\bar{\omega}) e^{-[t-(1+2m)k'(\bar{\omega})D]^2/2\tau^2}$$

$$e^{i[(1+2m)k(\bar{\omega})D-\bar{\omega}t]} \tag{26}$$

In the presence of damping the wave-vector is a complex quantity, $k = k_r + ik_i$. Then the ratio of succesive terms in the last equation is of the order of

$$|r_1(\bar{\omega})|^2 e^{-2k_i(\bar{\omega})D} << 1$$

where the inequality holds if

$$\alpha D = 2k_i(\bar{\omega})D \gtrsim 2 \tag{27}$$

If this is the case then, in lowest order, we may keep only the term $m = 0$ in eq.(26), giving

$$E(D^+,t) \cong \frac{\sqrt{2\pi}}{\tau} t_1(\bar{\omega}) t_1'(\bar{\omega}) e^{-[t-k'(\bar{\omega})D]^2/2\tau^2} e^{i[k(\bar{\omega})D-\bar{\omega}t]} \tag{28}$$

This is the same as eq.(5), obtained for propagation in an unbounded medium, except for the factor $t_1(\bar{\omega})t_1'(\bar{\omega})$ in eq.(28). This factor describes the effect of transmission through the plate without any internal reflections. Thus we are back at the result that the peak of the pulse propagates at the group velocity. It may be shown[15] that this conclusion is independent of the particular form of the spectral function $S_0(\omega)$, provided that this is symmetric with respect to the laser frequency $\bar{\omega}$. Then, neglecting $k_i'(\bar{\omega})$,

$$|E(D^+,t)| \cong t_1(\bar{\omega})t_1'(\bar{\omega})e^{-1/2\,\alpha D}|E_o(0^-,t-\frac{D}{v_g})| \tag{29}$$

$E_o(z,t)$ is the incident electric field and α and v_g are the absorption coefficient and the group velocity evaluated at $\omega = \bar{\omega}$.

We conclude that the boundaries play no important role provided that the condition (27) holds. That is, the damping must be so strong that the amplitude of the wave that is reflected at the second interface is negligibly small. Then there are no Fabry-Perot oscillations and one has essentially propagation in the positive z direction.

Garrett and McCumber[11] are concerned with an infinite medium; however they speculate that, for a finite medium, "the slab must be thinner than something like $\nu^2\tau^2$ exponential-decay (or growth) lengths". Because $\nu\tau >> 1$ we see that this is an understatement: the slab must be thinner than <u>just two</u> exponential-decay (or growth) lengths $1/\alpha$.

In the experimental configuration of Chu and Wong[9] the waves traversed a thick, undoped substrate before leaving the plate. Therefore, they suffered little reflection at the interface between the epilayer and the substrate. For the same reason the waves reflected back at the substrate-air interface were transmitted back into the epilayer with a small change of amplitude. Our derivation based on the vacuum-medium-vacuum configuration is still applicable, approximately, if we allow for a phase change corresponding to an increased (epilayer + substrate) thickness D. So let us look at Fig. 1, taken from Chu and Wong.[9] The dashed curve is the

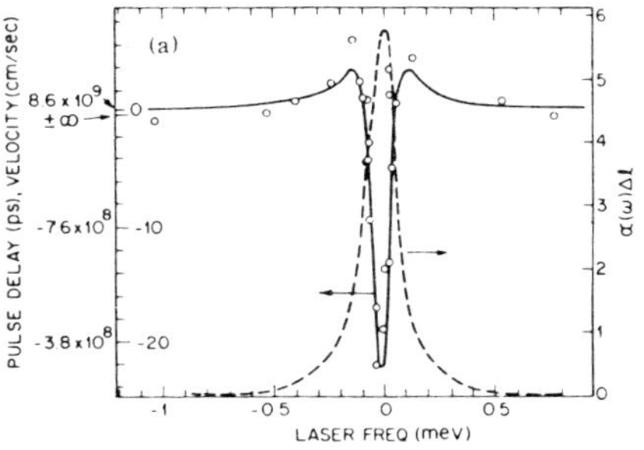

Fig. 1

The pulse delay t_0 (in ps) as a function of frequency (in meV, measured relative to the center of the absorption line) for transmission through a 76μm thick GaP film doped with nitrogen impurities (concentration $1.5 \times 10^{17}\ cm^{-3}$). The corresponding velocity D/t_0 (in cm/s) is also given. The solid line is the theoretical expression $D/v_g(\omega)$. The dashed curve is the measured absorption curve $\alpha(\omega)D$. Note that the agreement is very good near the center of the absorption line, however is quite poor in the wings of this line. From Chu and Wong.[9]

---

quantity $\alpha(\omega)D$, and we observe that, for $\alpha(\omega)D \gtrsim 2$ there is an excellent agreement between the experimentally measured transit times (circles) and the theoretical curve $D/v_g(\omega)$ (solid line). By eq.(23) this is as it should be.

We also notice from Fig. 1 that the agreement between the experimental times and the curve $D/v_g(\omega)$ is poor for $\alpha(\omega)D < 2$, that is in the wings of the absorption line. In this region of relatively small absorption we believe that the wave reflected at the second interface of the film may not be neglected. In other words, higher-order terms ($m \geq 1$) should be included in the sum of eq.(26). For $\alpha D \sim 1$ it is reasonable to calculate the correction to eq.(28) due to the term $m = 1$. By eq.(26) this term gives a contribution

$$\Delta E(D^+,t) \cong \frac{\sqrt{2\pi}}{\tau} t_1(\bar{\omega}) r_1^2(\bar{\omega}) t_1'(\bar{\omega}) e^{-[t-3k'(\bar{\omega})D]^2/2\tau^2}$$

$$\exp\{i[3k(\bar{\omega})D-\bar{\omega}t]\} \qquad (30)$$

Obviously this corresponds to the pulse that suffers one double reflection (one zig-zag) in the film. The peak of this pulse would emerge from the film at a time $t = 3k_r'(\bar{\omega})D = 3D/v_g$ if it did not overlap with the main pulse, $m = 0$. In fact, because our assumption of a very wide pulse, the overlap is so strong that a single pulse will emerge. However, the correction eq.(30) causes a small shift in the transit time of the peak. It may be shown[15] that this shift is

$$t_o - D/v_g \cong 2|r(\bar{\omega})|^2 \exp(-\alpha D)\cos(2k_r D)D/v_g \qquad (31)$$

The sign of this expression depends on the phase $2k_r(\bar{\omega})D$ and may cause an interesting variation of behavior from one sample to another. As for the order of magnitude, $|t_o - D/v_g| \sim 1$ psec for $\alpha D \sim 1$, in agreement with the deviations, from the theoretical curve, of the two uppermost circle data in Fig. 1. The "counterintuitive singularities"[9] in the region $\alpha(\omega)D << 1$ (not covered by our approximation) may have a similar origin.

4. Transmission through a transparent slab

This is the opposite extreme of the case studied in section 3. Of course, eq.(27) is not valid, in fact we assume that $\nu = 0$, and, as a result, $k_i(\omega) = 0$ and $\alpha(\omega) = 0$. The multiple interference effects are extremely important and the sum over m in eq.(26) may not be curtailed. Substituting eqs.(21)-(23) and $k'(\bar{\omega}) = 1/v_g$ in eq.(26) we obtain that

$$E(D^+,t) = \frac{\sqrt{2\pi}}{\tau} \frac{4\mu n}{(n+\mu)^2} e^{-(t-D/v_g)^2/2\tau^2} e^{i[k(\bar{\omega})D-\bar{\omega}t]}$$

$$\sum_{m=0}^{\infty} (\frac{n-\mu}{n+\mu})^{2m} e^{-2(mD/v_g\tau)^2+i2mk(\bar{\omega})D} e^{2m(D/v_g)(t-D/v_g)/\tau^2} \qquad (32)$$

The pulse is appreciable only for times such that $|t-D/v_g| \lesssim \tau$. The corresponding variation of the time-dependent exponent in the sum is $\lesssim 2mD/v_g\tau$. If this quantity is small then we may expand the last exponential function in eq.(32) in a power series. (We need not worry that the approximation is violated for high values of m because the factor $[(n-\mu)/(n+\mu)]^{2m}$ takes care of quite rapid convergence). Let us assume then that

$$D/v_g << \tau \qquad (33)$$

This being satisfied we may also write $\exp[-2(mD/v_g\tau]^2 \cong 1$, so that the sum in eq.(32) may be approximated by the expression

$$\sum_{m=0}^{\infty} \cong \sum_{m=0}^{\infty} \left\{ (\frac{n-\mu}{n+\mu} e^{ik(\bar{\omega})D})^{2m} \left[1 + \frac{2mD}{v_g\tau^2}(t-\frac{D}{v_g})\right]\right\} \qquad (34)$$

If we abbreviate

$$\frac{n-\mu}{n+\mu} e^{ik(\bar{\omega})D} = \rho \qquad (35)$$

then we see that the first term of eq.(34) is a geometric series in ρ and that the second term is proportional to the derivative (with respect to ρ) of the same series. Eqs.(32), (34), and (35) then lead to

$$E(D^+,t) \cong \frac{\sqrt{2\pi}}{\tau} \frac{4\mu n}{(n+\mu)^2} e^{i[k(\bar\omega)D-\bar\omega t]} \frac{1}{1-\rho^2} e^{-(t-D/v_g)^2/2\tau^2}$$

$$\left[1 + \frac{2D}{v_g\tau^2} \frac{\rho^2}{1-\rho^2}(t-\frac{D}{v_g})\right] \quad (36)$$

Let us write the last factor in the form

$$\left[1+ \frac{2D}{v_g\tau^2} \frac{\rho^2}{1-\rho^2}(t-\frac{D}{v_g})\right] \equiv \left[1 + \alpha(t-\frac{D}{v_g})+i\beta(t-\frac{D}{v_g})\right] \quad (37)$$

Here α and β are real quantities that may be calculated by taking the real and imaginary parts of $\rho^2/(1-\rho^2)$. We find that

$$\alpha = \frac{2D}{v_g\tau^2} \frac{N^2\cos[2k(\bar\omega)D]-1}{N^4-2N^2\cos[2k(\bar\omega)D]+1} \quad (38)$$

$$\beta = \frac{2D}{v_g\tau^2} \frac{N^2\sin[2k(\bar\omega)D]}{N^4-2N^2\cos[2k(\bar\omega)D]+1} \quad (39)$$

$$N = (n+\mu)/(n-\mu) \quad (40)$$

Because of the assumption eq.(33) both terms on the RHS of eq.(37) must be small; therefore this expression may be replaced by $\exp[(\alpha+i\beta)(t-D/v_g)]$. After some algebra eq.(36) may be recast in the form

$$E(D^+,t) = \frac{\sqrt{2\pi}}{\tau} \frac{2i\mu n}{(n^2+\mu^2)\sin[k(\bar\omega)D]+2i\mu n\cos[k(\bar\omega)D]}$$

$$e^{-(t-D/v_g)^2/2\tau^2+\alpha(t-D/v_g)} e^{-i(\bar\omega-\beta)t} \quad (41)$$

In eq.(41) the quantity β, given by eq.(39) describes a frequency shift $\Delta\omega = -\beta$. This shift may be positive or negative, it depends on D in a complicated way, however it is very small:

$$\frac{\Delta\omega}{\bar{\omega}} \sim \frac{D}{v_g \tau} \frac{1}{\bar{\omega}\tau} << 1 \tag{42}$$

The quantity α in eq.(41) determines the position of the maximum of the pulse. Clearly, our pulse is still approximately Gaussian, the peak being located at

$$t_o = D/v_g + \alpha\tau^2 \tag{43}$$

Substituting eqs.(38) and (39), after some algebra we get

$$\begin{aligned} v_{peak} &= D/t_o \\ &= v_g(\bar{\omega}) \left[\frac{2\mu n}{n^2+\mu^2} \cos^2 k(\bar{\omega}) D + \frac{n^2+\mu^2}{2\mu n} \sin^2 k(\bar{\omega}) D\right] \end{aligned} \tag{44}$$

Note that the pulse parameters have dropped out of this result; the same formula for v_{peak} is, in fact, obtained if $S_o(\omega)$ is an arbitrary function (however symmetric about $\bar{\omega}$).[16] The important conclusion is that the pulse velocity v_{peak} is a periodic function of the plate thickness, as shown in Fig.2. The period is $1/2\bar{\lambda}$, where $\bar{\lambda} = 2\pi c/(\bar{\omega} n)$ is the wavelength in the medium. The minimum and maximum velocities are given by the coefficients of the $\cos^2$ and $\sin^2$ terms, respectively:

$$v_m = v_g \frac{2\mu n}{n^2+\mu^2}, \quad V_M = v_g \frac{n^2+\mu^2}{2\mu n} \tag{45}$$

It is seen that always $v_m < v_g$ and $v_M > v_g$; interestingly it follows from eq.(45) that v_g is the geometric average of v_m

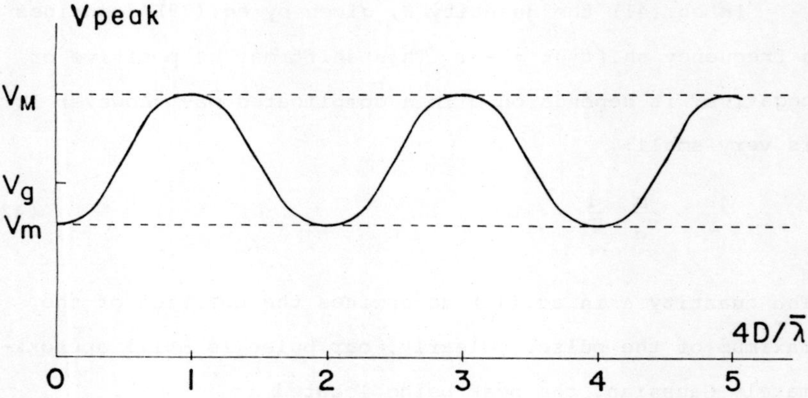

Fig. 2

The peak-velocity v_{peak} of a pulse as a function of $4D/\bar{\lambda}$ where D is the thickness of the plate and $\bar{\lambda}$ is the wavelength in the plate. The period of the oscillation is $1/2\bar{\lambda}$. The minimum and maximum velocities are defined in the text.

---

and v_M:

$$\sqrt{v_m v_M} = v_g \tag{46}$$

The only assumptions involved in the derivation are $1/\bar{\omega} \ll D/v_g \ll \tau$. An exact numerial calculation[17] confirms the periodic behavior of D. In Fig. 3 we plot the transmitted energy, normalized to the incident energy at z = 0 and t = 0, as a function of time. The parameters are $\mu = 1$, $n = 2$, $\lambda_o = 0.4\mu m$, and $\tau = 100$ ps. The thickness D is roughly 1mm, however the plate corresponding to Fig. 3(b) has a thickness $0.05\mu m$ greater than the plate corresponding to Fig. 3(a). This difference is just 1/4 of a wavelength in the plate.

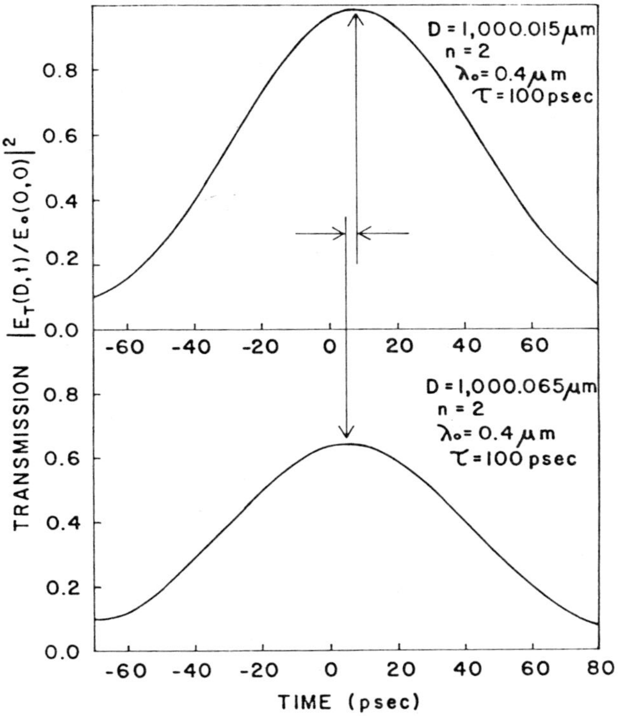

Fig. 3

The energy transmitted through a transparent plate at $z = D^+$, normalized to the incident energy at $z = 0^-$ and $t = 0$, as a function of time t. Note that the transit time of the maximum is smaller for the thicker plate (b) than it is for the thinner plate (a). The height of the maximum is also smaller for the thinner plate.

In both figures the transmitted pulses are approximately Gaussian. In (b) the peak emerges about 3 ps <u>before</u> the emergence of the peak in (a) - in spite of the fact that (b) corresponds to the thicker plate! This fact alone makes it perfectly clear that, in a transparent material, the transit time of a pulse, t_o, is not proportional to the thickness D. If we plot D/t_o, that is v_{peak}, as a function of D then Fig. 4 is obtained.

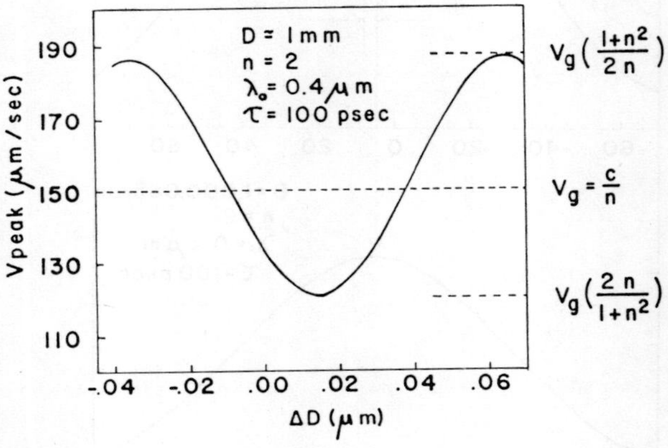

Fig. 4

The velocity of the maximum of a pulse transmitted through a transparent plate, as a function of the thickness of the plate (relative to a thickness of 1mm). Two points in this graph have been determined from Figs. 3(a) and (b), and other points have been found in a similar way. Note that this calculation, based on the exact formula for the transmitted fields, reproduces well the result of an approximate calculation given in Fig. 2.

So we see that the periodic effect of Fig. 2 is well reproduced by the numerical calculation.[17]

For a nonmagnetic glass (μ = 1, n = 1.5) v_{peak} oscillates between $v_m \simeq 0.6\ c$ and $v_M \simeq 0.7\ c$. It should not be difficult to observe this effect with pulses in the visible and a width $\tau \sim 50$ ps. In order to satisfy the condition $D \ll v_g\tau$ the plate should then be chosen to have a thickness $\sim$ 1mm, and a matching liquid could be used in order to increase D gradually. Note that $v_g\tau$ is just the coherence length of the light in the material medium; by eq.(33) this is much greater than D, so no problems of coherence are expected. On the other hand, for incoherent light v should be averaged over D. It is seen that this gives $\bar{v} = 1/2(v_m + v_M)$, which is greater than v_g, independently of D.

Going back to Fig. 3 we observe that the transmitted energy in (b) is smaller than in (a). This behavior stems from the algebraic prefactor in eq.(41). It has the same period $(1/2\bar{\lambda})$ as the peak velocity, Fig. 2, and gives rise to an oscillatory dependence of the transmittivity as a function of D. Such a behavior of the transmittivity (or reflectivity) has been known[32] for a long time. It is not difficult to show that the maxima of the transmittivity coincide with the minima of v_{peak}, and vice versa. So, depending on the value of D, either much light comes through the plate - but it is slow in coming - or else little light goes through, rapidly... Both oscillatory effects are consequences of the interference of light in the plate. The assumption $D/v_g \ll \tau$ is crucial; in the opposite case $(D/v_g \gg \tau)$ one would of course observe a series of pulses emerging from the plate at times D/v_g, $3D/v_g$, $5D/v_g, \ldots$

5. Transmission through spatially dispersive semiconductors

In this section we turn our attention to a special class of semiconductors whose dielectric response is characterised by spatial dispersion. These are semiconductors of the group II - VI with a direct gap, such as CdS, ZnO, etc. We are concerned with the optical excitation of an electron from the valence band into the conduction band and the formation of an electron-hole pair. Such a bound state is called an exciton. The energy levels are similar to those of the Bohr model, except that the background dielectric constant ε_o modifies the interaction between the electron and the hole, and that the mass of the electron must be replaced by the effective mass $m_e m_h/(m_e + m_h)$ of the electron and the hole. Considering a transition to a particular quantum level n of the "exciton-atom" a photon of certain energy $\hbar\omega_T$ (at least) is required. (The subscript "T" refers to transverse-mode vibrations, as will be revealed presently). Now, at the microscopic level, spatial dispersion is manifested by the fact that the exciton possesses a kinetic energy which is <u>not</u> negligible in comparison to the potential energy $\hbar\omega_T$. If the wave-vector of the light is $\vec{k}$ then the conservation of momentum requires that this be also the momentum of the exciton. Thus the total energy of the exciton is

$$\hbar\omega_T + \frac{(\hbar k)^2}{2(m_e+m_h)} \simeq \hbar(\omega_T^2 + \Delta k^2)^{1/2} \tag{47}$$

where $\Delta = \hbar\omega_T/(m_e+m_h)$, and the approximation is valid for sufficiently small values of k. Eq.(47) suggests that, in an oscillator model of the susceptibility function, the resonant frequency ω_T must be replaced by $(\omega_T^2 + \Delta k^2)^{1/2}$. This gives rise to the Hopfield - Thomas model[33]

$$\chi(\omega,k) = \frac{\varepsilon_o - 1}{4\pi} + \frac{\omega_P^2/4\pi}{\omega_T^2 + \Delta k^2 - \omega^2 - i\nu\omega} \tag{48}$$

Here ν is a phenomenological damping frequency and ω_P^2 is proportional to the oscillator-force of the excitonic transition. Formally ω_P plays a role that is similar to the plasma frequency, however it is <u>not</u> related to the presence of charge carriers; in fact we are assuming that the semiconductor is "intrinsic", that is undoped, so it really behaves like an insulator, rather than a conductor. Moreover, ω_P is extremely small, $\omega_P << \omega_T$, where ω_T is typically in the visible range of the spectrum. So in eq.(48) the second term gives rise to a polarisation of the medium exclusively due to the excitonic transition "n" in question, while the "tails" of all the other transitions are lumped together into a background susceptibility $(\varepsilon_o - 1)/4\pi$ (assumed to be independent of ω).

This is not yet a complete description of the dielectric response, the reason being that, in Maxwell's equations, one needs a model for the polarisation vector $P(\vec{r},t)$ <u>in real space</u>. Confining our attention to an isotropic linear response we may write, quite generally,

$$P(\vec{r},t) = \int \chi(\vec{r},\vec{r}')E(\vec{r},t)d^3r' \tag{49}$$

The question is: what is the relation between the real-space susceptibility function $\chi(\vec{r},\vec{r}')$ and the phase-space susceptibility function $\chi(\omega,k)$, eq.(48)? Clearly, there is no obvious prescription. The presence of the surface of the semiconductor turns out to be extremely important, so we cannot consider a bulk response. Taking the z coordinate to be perpendicular to the surface and eliminating in eq.(49) the dependence on t, x, and y we assume - as a part of our model - that the following relation is valid:[34]

$$\chi(z,z') = \chi(z-z') + U\chi(z+z') \tag{50}$$

Here $\chi(z \pm z')$ is the Fourier transform of eq.(48) and U is a phenomenological parameter. The first term in eq.(50) describes the bulk response because it is independent of the position of the surface. The second term gives the surface response, that is it describes in a simplified way the interaction between the exciton and the surface. The choice of a numerical value for U is directly related to the choice of a, so-called "Additional Boundary Condition" (ABC).[34] Such a condition is necessary for the following reason. The dispersion relation for transverse modes is given by

$$\varepsilon(\omega,k) = 1 + 4\pi\chi(\omega,k) = c^2k^2/\omega^2 \tag{51}$$

where $\chi(\omega,k)$ is given by eq.(48). For a given value of ω this equation has two solutions, say $q_1(\omega)$ and $q_2(\omega)$. The continuity of the electromagnetic fields E(z) and H(z) at the surface is not sufficient in order to determine the amplitudes of both transverse modes. That is why an ABC is needed.

There has been much discussion in the literature[26] just what is the "correct" ABC. A comparison between reflectivity experiments on CdS and corresponding calculations for a series of values of U reveals that the value U = -1 gives a very good description of the experimental observations.[35] It is not difficult to prove that, for U = -1, the excitonic part of the polarisation vector vanishes at the surface, that is[34]

$$\left| P_{ex}(\vec{r},t) \right|_{z=0^+} = 0 \qquad (52)$$

This is the so-called "Pekar ABC".[36]

In what follows we assume that the Pekar ABC is valid, thus U = -1 in eq.(50). A further improvement of the model is obtained by taking eq.(52) to hold at a depth $z = \ell^+$, rather than at the surface $z = 0^+$. This corresponds to the introduction of an exciton-free ("dead") surface layer. Such a layer, typically of thickness $\ell \sim 100$ Å, has been observed in many experiments.

Now we are ready to calculate the transmittivity of a thin semiconducting film of thickness D. There is a dead layer at both surfaces of the film and the dielectric response of the spatially dispersive "bulk" is given by eqs.(48) and (50) with U = -1. This problem was recently solved by Gaspar-Armenta and Halevi,[37] who utilized a surface-impedance approach. The reflectivity and the transmittivity were calculated by expressing the surface impedance Z(z) at a given interface in terms of the surface impedance at the preceding interface. Thus $Z(0^+)$, $Z(\ell^+)$, $Z(D-\ell^+)$, and $Z(D^+)$ were involved in the calculation. Reference 37 is limited to monochromatic waves and focuses on spectra $R(\omega)$ and $T(\omega)$. These spectra are dominated by standing-wave resonances due to the interference of four plane waves with wavevectors $\pm k_1(\omega)$ and $\pm k_2(\omega)$ in the thin film.

Now let us turn our attention to the subject of this review, that is propagation of pulses. So we allow for a spectral function $S_0(\omega)$ which we choose to be Gaussian, eq.(3). We wish to calculate the temporal shape of the pulse, as it emerges from the film at the point $z = D^+$, that is the quantity

$$E_T(D^+, t) = \int_{-\infty}^{\infty} d\omega e^{-(\omega-\overline{\omega})^2\tau^2/2} E_T(\omega) e^{i(k_0 D-\omega t)} \quad (53)$$

where $E_T(\omega)$ is the spectral amplitude of the transmitted electric field. This quantity is given, of course, by an extremely complicated set of formulas derived in reference 37, as explained above. Corresponding computations of the integral, eq.(53), have been also performed and will be the subject of a future publication.[21] We focus on a particularly interesting spectral region, namely we choose the carrier frequency $\overline{\omega}$ of the pulse to be very near to the frequency ω^* at which the lower polariton branch, say $k_1(\omega)$, exhibits a minimum of the group velocity $v_{g1}(\omega) = [d(\mathrm{Re}k_1)/d\omega]^{-1}$. At $\omega = \omega^*$ the function $\mathrm{Re}k_1(\omega)$ has an inflection point; $d(\mathrm{Re}k_1)/d\omega$ has a maximum; and $d^2(\mathrm{Re}k_1)/d\omega^2$ vanishes. Then if $|\overline{\omega} - \omega^*| \lesssim 1/\tau$ and if damping effects are small it may be important to allow for cubic terms $(\omega-\omega^*)^3$ in the expansion of $q(\omega)$ in a Taylor series, as pointed out by Johnson.[19] On the basis of several approximations this author[19] shows that the intensity of the emergent light oscillates before decaying, and this behavior is described by the Airy function. In what follows we present exact numerical results for the A(n = 1) exciton of CdS.[21] The Poynting vector of the transmitted light is proportional to the absolute value squared of the right-hand side of eq.(53). This is normalized by the maximum value of the Poynting vector of the incident light. So we are plotting the "transmission" $|E_T(D^+,t)/E_o(0^-,0|^2$ as a function of time.

In Fig. 5 we show the result of the computation for a film of thickness D = 1μm and a pulse of width τ = 3.5 ps centered at $\bar{\omega} = \omega^*$. The oscillations seen for t $\gtrsim$ 35 ps correspond, approximately, to the absolute value squared of the Airy function, as predicted by Johnson.[19] The earlier times are governed by Fourier components of the pulse that have group velocities $v_{g1}(\omega)$ greater than the minimum velocity $v_{g1}(\omega^*)$. Now an inspection of the spectral transmittivity reveals that $E_T(\omega)$ has a minimum at $\omega \simeq \omega^*$ and that it increases more rapidly for $\omega < \omega^*$ than for $\omega > \omega^*$. Thus the interference process becomes less effective as we move away from ω^*. This is our interpretation of the reason that no oscillations are seen for t < 35 ps. The position t $\simeq$ 46 ps of the dominant peak is roughly given by $D/v_g(\omega^*)$. In our case $v_g(\omega^*) = 2.06 \times 10^4$m/s, four orders of magnitude less than the speed of light in vacuum! It has been shown by Puri and Birman[38] that such a slowing down also occurs for the energy-transport velocity in GaAs. The signal that precedes the main peak describes a transient response (rather than "precursor") which is a direct consequence of spatial dispersion

This effect has not yet been observed, possibly because of the small values of the transmission, $\sim 10^{-5}$.

We have repeated the calculation leading to Fig. 5

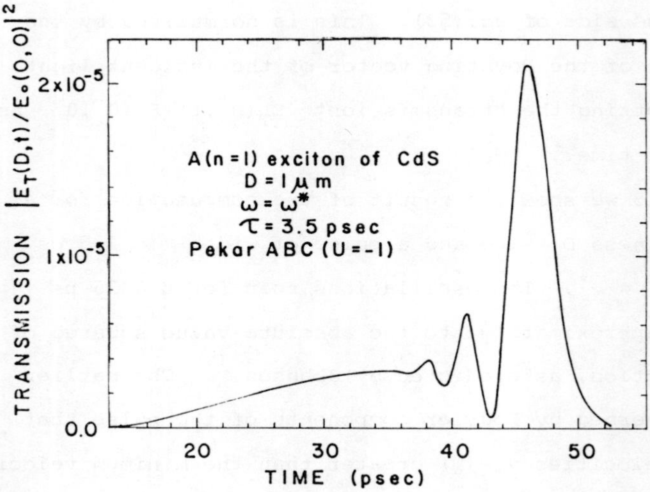

Fig. 5

Transmission of a Gaussian pulse through a thin film of CdS as a function of time. The carrier frequency is centered at ω^*, at which frequency the group velocity for the lower exciton branch has a minimum. The exciton parameters are quoted in reference 35. From reference 21.

---

for a smaller thickness, $D = 0.5\mu m$. The transmission increased by two orders of magnitude, to $\sim 10^{-3}$. Now, however, the oscillations were reduced to only two peaks. We speculate that, for $D = 0.5\mu m$, the wave $k_2(\omega)$ is not yet damped out and makes an appreciable (destructive) contribution for earlier times of arrival.

If the carrier frequency is sufficiently detuned from ω^* we may expect that the oscillatory effect will disappear. Indeed, for $\bar{\omega} - \omega^* = \pm 0.5$meV and $\tau \geq 3.5$ ps only a single peak is transmitted. For $\tau = 15$ ps and $\tau = 30$ ps a single peak emerges even for $\bar{\omega} = \omega^*$. These pulses are approximately Gaussian and v_{peak} is roughly equal to $v_g(\bar{\omega})$. However, the peak velocity somewhat depends on τ.

The oscillations reappear in the case of a very brief pulse, with $\tau = 1.8$ ps, as seen in Fig. 6. Now the spectral

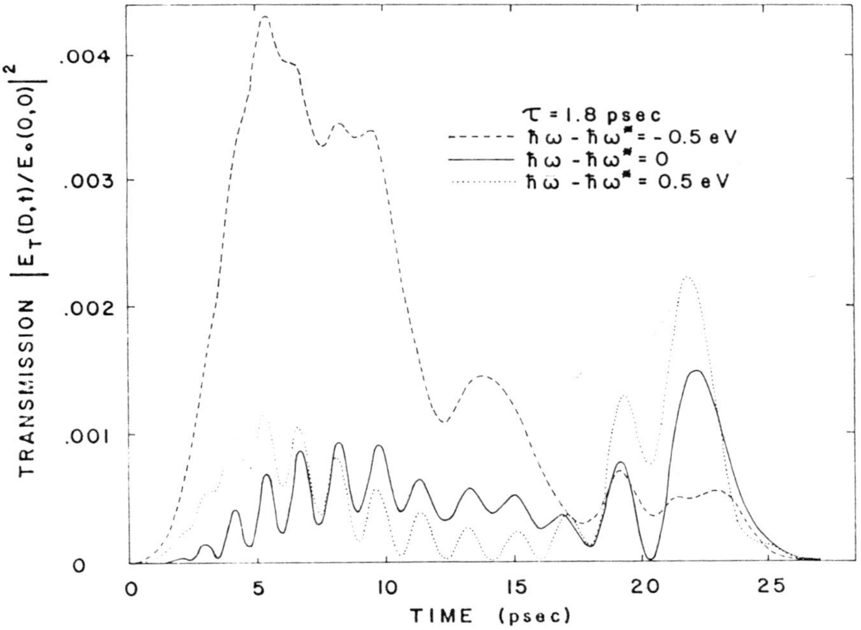

Fig. 6. As in Fig. 5, however $D = 0.5\mu$m and $\tau = 1.8$ ps. The transmission is plotted for three values of the carrier frequency. From reference 21.

width of the pulse is great enough so that oscillations are obtained even for $\bar{\omega} - \omega^* = \pm 0.5$meV. The asymmetry stems from the fact that the spectral transmission increases more rapidly for $\omega < \omega^*$ than for $\omega > \omega^*$, as mentioned before. The oscillatory behavior described by Johnson[19] may be observed for $t \gtrsim 17$ ps in the cases $\bar{\omega} = \omega^*$ and $\bar{\omega} = \omega^* + 0.5$meV. For $\bar{\omega} = \omega^* - 0.5\text{meV} = \omega_T$ the oscillations are irregular. A new type of oscillations, with a period ~ 1.4 ps, appears for $t < 17$ ps. We expect that this behavior should be readily observable with a resolution $\simeq 1$ ps.

Acknowledgements

The research leading to this paper was partially supported by a cooperative program between the authors' universities. The first author also wishes to acknowledge support by CONACyT.

REFERENCES

1. S.C. Bloch, Amer. J. Phys., 45 538 (1977)

2. See, for example, S.L. Shapiro (editor), Ultrashort Light Pulses, Springer, Berlin 1977

3. E.P. Ippen and C.V. Shank, in ref. 2, p. 83

4. D. Anderson, J. Askne, and M. Lisak, Phys. Rev., A 12 1546 (1975)

5. L.A. Vainshtein, Usp. Fiz. Nauk, 118 339 (1976); English translation: Sov. Phys. Usp., 19 189 (1976)

6. L. Casperson and A. Yariv, Phys. Rev. Lett., 26 293 (1971)

7. R.G. Ulbrich and G.W. Fehrenbach, Phys. Rev. Lett., 43 963 (1979)

8. Y. Segawa, Y. Aoyagi, and S. Namba, Solid State Commun., 32 229 (1979)

9. S. Chu and S. Wong, Phys. Rev. Lett., 48 738 (1982)

10. M. Born and E. Wolf, Principles of Optics, 5th ed., Pergamon 1975 (see comment at the end of sec. 1.3.4)

11. C.G.B. Garrett and D.E. McCumber, Phys. Rev. A, 1 305 (1970)

12. D.G. Anderson and J.I.H. Askne, Proc. IEEE, 62 1518 (1974)

13. E.S. Birger and L.A. Vainshtein, Zh. Tekh. Fiz., 43 2217 (1973); English translation: Sov. Phys. Tech. Phys., 18 1405 (1974)

14. B. Segard and B. Macke, Phys. Lett., 109A 213 (1985)

15. P. Halevi and R. Fuchs, Phys. Rev. Lett., 55 338 (1985)

16. P. Halevi, to be published

17. J.A. Gaspar-Armenta and P. Halevi, to be published

18. Y. Masumoto, Y. Unuma, Y. Tanaka, and S. Shionoya, J. Phys. Soc. Japan, 47 1844 (1979)

19. D.L. Johnson, Phys. Rev. Lett., 41 417 (1978)

20. T. Itoh, P. Lavallard, J. Reydellet, and C. Benoit à la Guillaume, Solid State Commun., 37 925 (1981)

21. J.A. Gaspar-Armenta and P. Halevi, to be published

22. L. Brillouin, Wave Propagation and Group Velocity, Academic, New York 1962

23. P. Pleshko and I. Palócz, Phys. Rev. Lett., 22 1201 (1969)

24. M. Frankel and J.L. Birman, Opt. Commun., 13 303 (1975) and Phys. Rev. A, 15 2000 (1977)

25. G. P. Agrawal, J.L. Birman, D.N. Pattanayak, and A. Puri, Phys. Rev. B, 25, 2715 (1982)

26. J.L. Birman, in Excitons, ed. E.I. Rashba and M.D. Sturge, North-Holland 1982, p. 27

27. M.D. Crisp, Phys. Rev. A, 4 2104 (1971). See also, ibid 1 1604 (1970)

28. R. Fuchs and P. Halevi, unpublished

29. A. Katz and R.R. Alfano, Phys. Rev. Lett., 49 1292 (1982)

30. S. Chu and S. Wong, Phys. Rev. Lett., 49 1293 (1982)

31. B. Macke, Optics Commun., 49 307 (1984)

32. R. Messner, Zeiss Nachr., 4 (H9) 253 (1943)

33. J.J. Hopfield and D.G. Thomas, Phys. Rev., 132 563 (1963)

34. P. Halevi and R. Fuchs, J. Phys. C: Solid State Phys., 17 3869 (1984)

35. P. Halevi, Phys. Rev. Lett., 48 1500 (1982)

36. S.I. Pekar, Sov. Phys. JETP, 6 785 (1958); 7 813 (1958); 9 314 (1959); J. Phys. Chem. Solids 5 11 (1958)

37. P. Halevi, G. Hernández-Cocoletzi, and J.A. Gaspar-Armenta, Thin Solid Films, 89 271 (1982) and J. A. Gaspar Armenta and P. Halevi, to be published

38. A. Puri and J.L. Birman, Phys. Rev. Lett. 47, 173 (1981)

Magnetohydrodynamic Surface Waves

Akira Hasegawa
AT&T Bell Laboratories
Murray Hill, New Jersey 07974

I. General Feature of Magnetohydrodynamic (MHD) Surface Waves

In a homogeneous plasma with magnetic field, three types of magnetohydrodynamic waves are known to exist. One, called the Alfvén wave,[1] is a shear wave where energy propagates only in the direction of the magnetic field and the others, called the fast and the slow magnetosonic waves, are compressional waves.

When the plasma is bounded for example by a magnetic pressure, a surface wave can propagate along the boundary. For a wave to be localized near the boundary surface, it is required that the wave energy does not propagate away from the surface. This requirement can be fulfilled either by a choice of wave frequency below the cut off frequency of the body wave or by exciting a perturbation which does not propagate perpendicular to the boundary. Hence the Alfvén wave is an obvious candidate to form a MHD surface wave. Assuming incompressibility, the surface wave dispersion relation which propagates along the boundary between two plasmas with the mass density ρ_I and ρ_{II}, and magnetic flux density B_{0I} and B_{0II} are given by[1]

$$\epsilon_I + \epsilon_{II} = 0 \qquad (1)$$

where

$$\epsilon_j = \omega^2 \rho_j - \frac{(\mathbf{k} \cdot \mathbf{B}_{0j},)^2}{\mu_0}$$

$$\equiv \rho_j \left[\omega^2 - (\mathbf{k} \cdot \mathbf{v}_{Aj})^2 \right] , \tag{2}$$

and $\mathbf{v}_{Aj} \left[= \mathbf{B}_{0j}/(\mu_0 \rho_j)^{1/2} \right]$ is the Alfvén speed. In particular when ϵ_{II} represents vacuum, $\rho_{II} = 0$ and the dispersion relation becomes

$$\omega = \sqrt{2}\, k_{\|} v_A \tag{3}$$

where $k_{\|}$ is the wave number in the direction of the magnetic field. We note that the surface Alfvén wave has a frequency larger than the body Alfvén frequency ($= k_{\|} v_A$) but smaller than the vacuum Alfvén frequency ($= \infty$).

Since the Alfvén wave can propagate only in the direction of the magnetic field, the disturbance is localized at the surface and forms a surface wave.

II. Instabilities

II.1 Kelvin-Helmholtz Instability

MHD surface waves become unstable in the manner similar to hydrodynamic surface waves.

When the medium I has a flow parallel to the boundary with a velocity given by $\mathbf{v}_0$, ϵ_I in Eq. (12) is modified by the Doppler effect to

$$\epsilon_I = \rho_{0I}(\omega - \mathbf{k} \cdot \mathbf{v}_0)^2 - \frac{(\mathbf{k} \cdot \mathbf{B}_0)_I^2}{\mu_0}$$

$$= \rho_{0I}\left[(\omega - \mathbf{k} \cdot \mathbf{v}_0)^2 - (\mathbf{k} \cdot \mathbf{v}_A)_I^2\right] . \tag{4}$$

The modified ϵ_I produces a complex ω when substituted in Eq. (1). The instability is the Kelvin-Helmholtz instability of the incompressable MHD wave. In particular when $\rho_I = \rho_{II}$, the instability condition is given by

$$\mathbf{k} \cdot \mathbf{v}_0 > 2\, \mathbf{k} \cdot \mathbf{v}_A . \tag{5}$$

We note that the threshold flow speed goes to zero if $\mathbf{k} \cdot \mathbf{v}_A = 0$, namely for a wave with the wave vector perpendicular to the magnetic field.

II.2 Rayleigh-Tayler Instability.

Rayleigh-Tayler instability is an instability of a gravity wave caused by an upside-down situation, namely when the pressure gradient is anti-parallel to the gravity force. For a plasma, the natural gravity is generally small compared with other forces and the instability is excited by an effective gravity g_{eff} which is produced by the centrifugal force on plasma particles moving along curved magnetic field lines, $g_{eff} \simeq m\, v_T^2/R$, where m is the effective mass of plasma particles, v_T is the thermal speed and R is the radius of curvature of the magnetic line of force. The gravity wave dispersion relation is modified also by the magnetic tension and is given by

$$\omega^2 = |k_y|\, g_{eff} + (\mathbf{k} \cdot \mathbf{v}_A)^2 \tag{6}$$

where k_y is the component of the wave vector in the direction tangential to the boundary, which is normal to the plane of $\mathbf{R}$ and $\mathbf{B}_0$.

When two plasmas with density ρ_I and ρ_{II}, pressure p_I and p_{II} are bounded by a curved field line with the radius of curvature R, pointing to the medium I (the centrifugal force pointing to medium II) g_{eff} can be written as[3]

$$g_{eff} = \frac{2}{R} \frac{p_{II}-p_I}{\rho_{II}+\rho_I} \begin{cases} > \\ < \end{cases} 0 \tag{7}$$

The instability occurs when

$$-|k_y| g_{eff} > (\mathbf{k} \cdot \mathbf{v}_A)^2 \tag{8}$$

where $(\mathbf{k} \cdot \mathbf{v}_A)$ is the value at the boundary. Unlike the hydrodynamic Rayleigh-Tayler instability where the instability occurs when $\nabla\rho \cdot \underset{\sim}{g} < 0$, in MHD the magnetic field tension helps to stabilize the instability. Inequality (8) shows that the instability occurs when the gravity force overcomes the magnetic field tension.

III. Effect of Compressibility

When compressibility is introduced the surface wave becomes dispersive. Using adiabatic equation of state in the linearized ideal MHD equations, the wave equation for the plasma displacement ξ_x in the direction of inhomogenity can be shown to satisfy the following wave equation[4]

$$\frac{d}{dx}\left(\frac{\epsilon\,\alpha B_0^2\,/\,\mu_0}{\alpha k_\perp^2 B_0^2/\mu_0-\epsilon}\right)\frac{d\xi_x}{dx}-\epsilon\,\xi_x=0\ , \tag{9}$$

where

$$\alpha(x)=1+\frac{\omega^2 v_s^2}{v_A^2[(\omega^2-(\mathbf{k}\cdot\mathbf{v}_s)^2]}\ , \tag{10}$$

$$\epsilon(x)=\rho(x)\omega^2-[\mathbf{k}\cdot\mathbf{B}_0(x)]^2/\mu_0\ , \tag{11}$$

and v_s is the sound speed. In the low β limit $\alpha\simeq 1$, the compressible dispersion relations for the surface wave with a sharp boundary can be expressed

$$\frac{\epsilon_I}{k^2-\omega^2/v_{AI}^2}+\frac{\epsilon_{II}}{k^2-\omega^2/v_{AII}^2}=0\ . \tag{1'}$$

Comparing (1') with (1), we see now that the surface wave frequency depends on the perpendicular wave number $k_\perp\left(=\sqrt{k^2-k_\parallel^2}\right)$. In particular when $\rho_{II}=0$, the dispersion relation becomes

$$\omega\simeq\sqrt{2}\,\mathbf{k}\cdot\mathbf{v}_A\Big(1-\frac{k_\parallel}{4\sqrt{2}k_\perp}\Big)\ . \tag{12}$$

IV. Effect of A Smooth Boundary

When the boundary is smooth such that the background plasma parameter varies gradually at the boundary, the surface wave dissipates due to the resonant absorption. For example, if we take an incompressibility limit in Eq. (9) the wave equation becomes

$$\frac{1}{\epsilon}\frac{d}{dx}\left[\epsilon\frac{d\xi_x}{dx}\right]-k_\perp^2\xi_x=0\,. \tag{13}$$

This equation is singular at the point $x = x_0$ where $\epsilon(x_0)$ vanishes for $\omega = \omega_s(= \sqrt{2}\mathbf{k}\cdot\mathbf{v}_A)$ and the solution near the singular point can be expressed

$$\begin{aligned}\xi_x &= \ln(x-x_0)\,, &\quad x > x_0\\ &= \ln|x-x_0| + i\pi\,, &\quad x < x_0\end{aligned} \tag{14}$$

When the Poynting flux is evaluated across the singular point, the $i\pi$ term in Eq. (15) produces discontinuity in the flux indicating the local absorption of the energy.[4,5] At $x = x_0$ the surface eigen value $\omega = \omega_s$ resonates with the local Alfvén frequency $\mathbf{k}\cdot\mathbf{v}_A(x_0)$ and the energy is dissipated by this local resonance.

V. Kinetic Effects

The singularity which appears in the wave equation Eq. (9) is a consequence of the ideal MHD assumption in which the microscopic scale such as the Larmor radius or the Debye lengths are assumed to be zero. Hence a correction to the "fluid" approximations is expected to remove the singularity. To take into account of these kinetic effects, the Vlasov equation is used to derive the modified dispersion relation for the Alfvén wave. The dispersion relation then reads[6,7]

$$\omega_r^2 = (\mathbf{k}\cdot\mathbf{v}_A)^2\,(1+k_\perp^2\bar{\rho}^2) \tag{15}$$

$$\text{where}\quad \bar{\rho}^2 = \rho_i^2\left(\frac{3}{4}+\frac{T_e}{T_{||}}\right)$$

$$\omega_i = -\sqrt{\frac{\pi}{2}}\,\frac{\omega_r^2}{k_{||}v_e}\,k_\perp^2\bar{\rho}^2 \quad \text{for} \quad v_{Te} > v_A$$

where ω_r and ω_i are the real and the imaginary part of the frequency, ρ_i is the ion Larmor radius, T_e and T_i are the electron and ion temperatures v_{Te} is the electron thermal speed and $k_{||}$ is the wave vector in the direction of the magnetic field. The imaginary part of omega is a consequence of the electron Landou damping which originates from the parallel wave electric field (to the ambient magnetic field) which is produced due to the finite Larmor radius correction. The wave given by Eq. (15) is called the kinetic Alfvén wave. The dispersion relation is modified if $v_{Te} < v_A$ and is given by

$$\omega^2 = \frac{(\mathbf{k}\cdot\mathbf{v}_A)^2\,(1+k_\perp^2\rho_i^2)}{1+k_\perp^2 c^2/\omega_{\rho e}^2}\,. \tag{16}$$

VI. Resonant Mode Conversion

The singularity associated with the local resonance of the Alfvén wave can be eliminated if the finite Larmor radius effect is included in the wave equation. The resultant wave equation expressed in terms of ξ_x becomes[7]

$$\bar{\rho}^2\,\frac{d^4\xi_x}{dx^4} + \frac{d}{dx}\left(\epsilon\,\frac{d\xi_x}{dx}\right) - k_y^2\epsilon\xi_x = 0\,. \tag{17}$$

We see, by comparing this expression with Eq. (13), the leading term which is proportional to the square of the ion Larmor radius is added. If this equation is solved with an appropriate boundary condition, the solution expresses the mode coupling between the MHD surface wave and the kinetic Alfvén wave derived in the previous chapter. This means that the surface wave energy is converted to the kinetic Alfvén wave energy at the resonant surface and carried away to the higher (lower) density side if $v_{Te} > v_A$ ($v_{Te} < v_A$). Namely the resonant absorption is a manifestation of the energy absorption due to the mode coupling to the kinetic Alfvén wave.

VII. Effect of Field Aligned Current and Magnetic Field Curvature.

In the presence of a stationary current in the direction of the ambient magnetic field in an inhomogeneous plasma with curved magnetic field, the surface eigen frequency is modified. Take for example a cylindrical plasma with magnetic field parallel to the axis. The surface Alfvén frequency is still given by $\sqrt{2}\, k_{\|} v_A$ if there exists no current in the axial (z) direction. However if there exists a current, the dispersion relation is modified.

For simplicity, let us take the current to flow only at the surface of the cylindrical plasma with radius a. The dispersion relation for such a geometry has been derived by Kruskal[8] and Shafranov[9] and is given by

$$\omega^2 = k_z^2 v_A^2 \left[1 + (1 + \frac{mB_{0\theta}}{ak_zB_{0z}})^2 - \frac{mB_{0\theta}^2}{a^2k_z^2B_{0z}^2} \right]$$

where m(= 0, ±1, ±2...) is the azimuthal mode number and $B_{0\theta}$ is the azimuthal magnetic field which is produced by the surface current. We see that for m=1 and for $B_{0\theta}/(ak_zB_{0z}) > 1$, $\omega^2 < 0$ and the instability appears (Kruskal-Shafranov instability). However, if the parallel current is smaller than this threshold, ω^2 becomes positive, yet can take a value smaller than the minimum value of k_zv_A in the plasma. This means that the surface Alfvén frequency can become smaller than the minimum local Alfvén frequency hence the resonant absorption does not occur for such a case.[10]

VIII. SUMMARY

Various properties of surface waves in MHD frequency range are presented. Since the (bulk) Alfvén wave has an anisotropic velocity of propagation which does not allow its energy to propagate perpendicular to the ambient magnetic field, it provides an ideal medium of forming a surface wave.

Because of its richness, the MHD surface waves present several unique physics such as the resonant mode conversion to the kinetic (Alfvén) waves which can propagate either to a higher or to a lower density side, or the production of a surface eigen mode free from a resonant absorption.

The MHD surface wave has wide areas of applications which include heating and instability of fusion plasmas, wave excitation and heating of magnetospheric and solar plasmas.

References

1. H. Alfvén, Nature *150*, 406 (1942).

2. A. Hasegawa and C. Uberoi, *The Alfvén Wave*, Technical Information Center, US Department of Energy, DE802001702 (DOE/TIC-11197), National Technical Information Service US Dep. of Commerce, Springfield, VA 22161.

3. Yu. P. Maltsev and R. Lyatsky, Planet. Space Sci. *32*, 1547 (1984).

4. A. Hasegawa and L. Chen, Phys. Fluids *17*, 1399 (1974).

5. D. J. Southwood, Planet, Space Sci. *22*, 483 (1974).

6. K. N. Stepanov, Sov. Phys.-JETP (English Transl) *34*, 892 (1958).

7. A. Hasegawa and L. Chen, Phys. Fluids 19, 1924 (1976).

8. M. Kruskal and M. Schwarzschild, Proc. Roy. Soc. Landou, Ser. A *223*, 348 (1954).

9. V. D. Shafranov, Sov. J. At. Energy (English Transl) *1*, 709 (1976).

10. K. Appert, R. Gruber, F. Trayon and J. Vaclavik, Plasma Phys. *24*, 1147 (1982).

SURFACES IN THE INTERACTION OF INTENSE LONG WAVELENGTH LASER LIGHT WITH PLASMAS

ROGER D. JONES

Los Alamos National Laboratory
Los Alamos, New Mexico 87545 USA

ABSTRACT

The role of surfaces in the interaction of intense CO_2 laser light with plasmas is reviewed. The collisionless absorption of long wavelength light is discussed. Specific comments on the role of ponderomotive forces and profile steepening on resonant absorption are made. It is shown that at intensities above $10^{15} W/cm^2$ the absorption is determined by ion acoustic-like surface modes. It is demonstrated experimentally that harmonics up to the forty-sixth can be generated in steep density profiles. Computer simulations and theoretical mechanisms for this phenomena are presented. The self generation of magnetic fields on surfaces is discussed. The role these fields play in the lateral transport of energy, the insulation of the target from hot electrons, and the acceleration of fast ions is discussed.

I. INTRODUCTION

The industrialized countries of this planet are being forced to become less reliant on petroleum. Table I is taken from a recent issue of "Science News".[1] It can be seen that at present rates of comsumption the United States will have used its oil reserves in about nine years. The Soviet Union will have used its domestic supply in fourteen years. The world's oil supplies will be consumed in thirty-four years. As supplies dwindle, the industrialized nations will increasingly become affected by volatile political conditions in the Middle East. A cheap and abundant supply of energy could diffuse this situation. There is an additional concern over atmospheric changes that may be induced by the burning of fossil fuels. A clean energy supply may alleviate this situation.

Table I

	Annual Production	Reserves	Reserves/Production
	(billion barrels)		(ratio)
Kuwait	0.36	90.0	250
Iraq	0.43	44.5	104
Saudi Arabia	1.71	169.0	99
United Arab Emirates	0.44	31.9	73
Iran	0.80	48.5	61
Libya	0.40	21.1	52
Mexico	1.10	48.6	45
Venezuela	0.68	25.8	38
Nigeria	0.51	16.7	33
United Kingdom	0.94	13.6	14
Soviet Union	4.53	63.0	14
United States	3.79	34.5	9
World Total	21.10	707.2	34

There are a number of alternate energy sources which have been proposed as candidates to replace petroleum. Perhaps our most abundant energy source and the one requiring the least investment in new technologies is coal. The burning of coal, however, causes unacceptable environmental modifications. Fission plants suffer from small uranium reserves. Additionally, they generate dangerous and long lived wastes. Breeder reactors can overcome the uranium shortage. They unfortunately generate large amounts of high grade plutonium, thus raising concerns over nuclear proliferation. Solar and geothermal energy cannot economically meet existing demands.

These concerns permit fusion to be a viable candidate. There is an essentially infinite supply of deuterium in the oceans, no fissionable byproducts are generated, and chemical production may be limited to helium. The two fusion approaches are magnetic confinement and inertial confinement. Magnetic confinement schemes suffer from the proximity of delicate and complicated coils to the burning plasma. This is irrespective of the technological problems involved in achieving the burning plasma. In inertial fusion it is possible to stand off the delicate parts of the apparatus from the plasma. However, the drivers in inertial fusion are much less efficient than in magnetic fusion. Projected efficiencies for KrF excimer lasers are in the range of 10%, for example. Projected efficiencies for heavy ion drivers are in the 25% range. Engineering contraints[2] require the product of driver efficiency and target gain to be greater than 10. Thus inertial fusion targets must have yields of 40 to 100 times breakeven to be commercially acceptable. This is to be compared with 10 to 20 times breakeven for magnetic fusion.

In this paper we will be concerned with the problem of achieving high gains with a particular inertial fusion driver system, the CO_2 laser. For this driver system, surfaces in the target plasma play an important role in the laser-target coupling and consequently in the target gain. For reasons that will be discussed in this paper the CO_2 laser is no longer a leading candidate for a reactor driver. Surfaces have played an important role in this decision.

At the start of the inertial fusion program in the United States, there were two laser systems that were thought to be potentially powerful enough to drive fusion capsules, the Nd-glass laser and the CO_2 laser. The CO_2 laser had the advantages of being capable of the high repetition rates required in inertial fusion and of having a relatively high efficiency (about 10%). Because of the time required to relax the solid state crystals, Nd-glass lasers are incapable of high repetition rates. Additionally, they have yet to be able to demonstrate efficiencies as high as 1%. The key disadvantage of CO_2 lasers is their wavelength. The wavelength of CO_2 lasers is 10 microns and of of Nd-glass is 1 micron. An important quantity is the ratio of the oscillatory energy of an electron in the laser field to the thermal energy of the electron, v_0^2/v_e^2. Here, v_0 is the oscillatory speed of an electron and v_e is the thermal speed. This ratio determines the role of collisionless, nonlinear effects. A high ratio is an indication of strong effects. But v_0 is proportional to the laser wavelength. Therefore we would expect nonlinear effects to be more important in the CO_2 laser-plasma interaction than in the Nd-glass. An unfortunate feature of collisionless nonlinear plasma physics is that energy absorbed from the laser is deposited in a hot tail of the electron distribution function. These hot electrons can preheat the cold fuel core of the target making it difficult to achieve high compressions necessary for ignition. Additionally, energy deposited in hot electrons couples poorly to ablation, once again making high compressions difficult. Therefore the long wavelength of the CO_2 laser is a serious disadvantage. The present thrust of the inertial fusion program is toward even shorter wavelengths than the 1 micron of the Nd-glass laser. This is being done in two ways, by frequency multiplying Nd-glass lasers and by developing the KrF laser which has a wavelength of 0.25 microns. The KrF laser has the advantages of the CO_2 laser but with a shorter wavelength.

The organization of the paper is as follows. In Section II, the role of surfaces in the absorption of CO_2 laser light will be discussed. In Section III, we discuss the generation of high laser harmonics at the critical surface. Self generated magnetic fields are discussed in Section IV. These fields play an important role in the insulation of hot electrons from the cold fuel and in the lateral transport of energy. At very high intensities the fields mediate the transfer of the laser energy into a very fast ion blowoff. This is discussed in Section V. Many of the topics in this paper have been reviewed elsewhere.[3,4]

II. ABSORPTION OF CO_2 LASER LIGHT

To achieve high gain the laser light must be absorbed efficiently. Figure 1 is a schematic of the density profile. The laser light is absorbed at or before the critical surface (the density at which the plasma frequency equals the laser frequency). A thermal front is generated which eats into and ablates the cold plasma. The reaction to the ablation causes the inner core to compress. At high intensities and long wavelengths the absorption is collisionless. The principal absorption mechanism is resonant absorption of the light into plasma waves at the critical surface (see Fig. 2). These waves Landau damp and deposit their energy into superthermal electrons.

It may seem surprising that resonant absorption is the dominant mechanism. There are a number of other mechanisms that might seem to be more likely candidates. In the underdense plasma there is stimulated Raman and Brilluoin absorption as well as two plasmon decay. There is also the oscillating two stream instability and the parametric decay instability. These mechanisms are less important than resonant absorption because the ponderomotive force of the laser modified the density profile[5] (see Fig. 2). A sharp surface is formed with the lower density shelf at about 0.1 times the critical density. This density is too low for subcritical absorption processes to occur. Pressure balance at the surface is determined by bulk electrons balancing the ponderomotive force. Energy balance is maintained by the superthermal electrons.[3] The superthermal electron energy depends on the details of the surface profile (see Fig. 3). Self consistent computer simulations[3] indicate a scaling for the superthermal electron temperature as

$$T_h = 7(T_c I \lambda^2/10^{15})^{1/3}$$

where T_h is the superthermal electron temperature[6] in KeV, T_c is the bulk electron temperature, I is the intensity in W/cm^2, and λ is the laser wavelength in microns. Typically the absorption is 30-40%. This is valid for intensities up to about $10^{15} W/cm^2$. For higher intensities other mechanisms take over, increasing the absorption (see Fig. 4). Absorption as high as 70% has been measured. The phenomena, not observed at low intensity for p-polarization in a sharp density profile, is surface rippling. In these high intensity simulations, strong plasma rippling is observed. Simulations at intensities of 2 x $10^{16} W/cm^2$ show 25% absorption at early times. This absorption increases to more than 60% after 5 psec, apparently due to increased surface roughness resulting from the radiative decay instability. At the same time, T_h, increases from about 200 KeV to 900 KeV. A typical snapshot from the simulation showing surface roughness is shown in Fig. 5. Figure 5 is a density contour plot of the critical surface. The incident wave vector is indicated. The critical density is displayed as a dashed line. This simulation is for p-polarized light incident on a sharp density profile at fifteen times critical density. Other parameters are $v_0 = 1.5c$ and $m_i = 1836 m_e$, and $T_e/T_i = 10$.

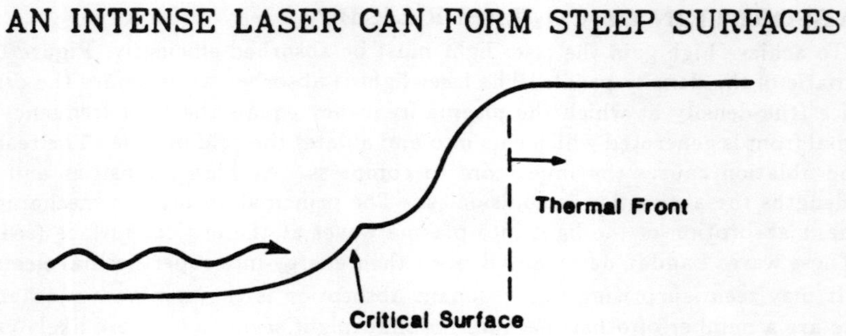

Figure 1.

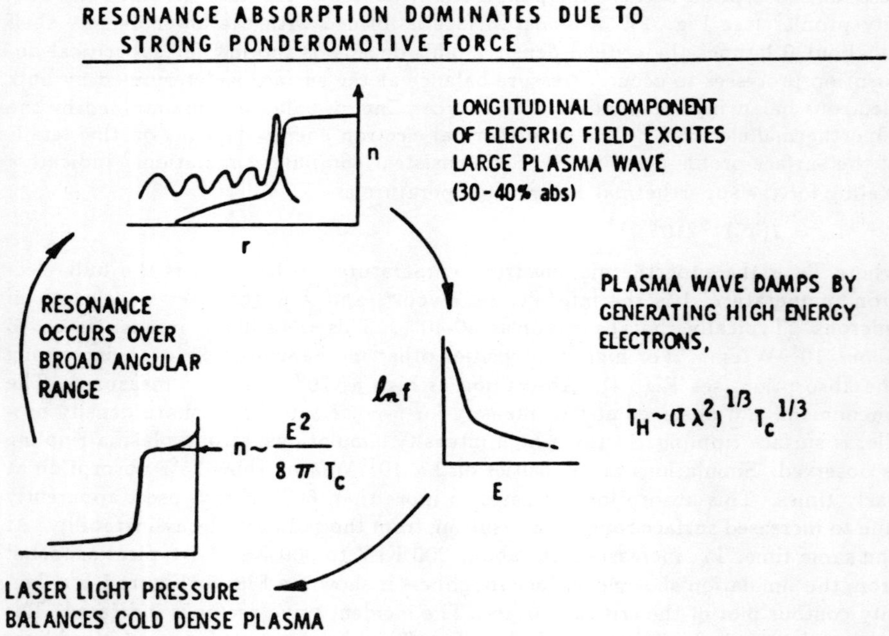

Figure 2.

HIGH ENERGY ELECTRON COMPONENT CONTROLLED BY SURFACE STRUCTURE

ENERGY OF ELECTRONS DEPENDS ON AMPLITUDE AND WIDTH OF LARGE AMPLITUDE PLASMA WAVE

Energy = eE δ

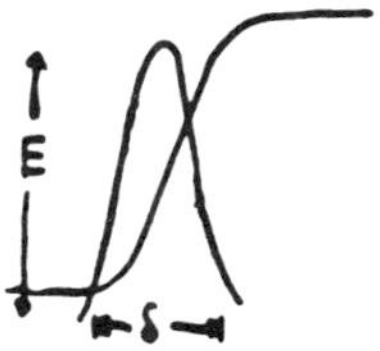

WIDTH OF WAVE δ CONTROLLED BY SHARP DENSITY GRADIENT LENGTH

Figure 3.

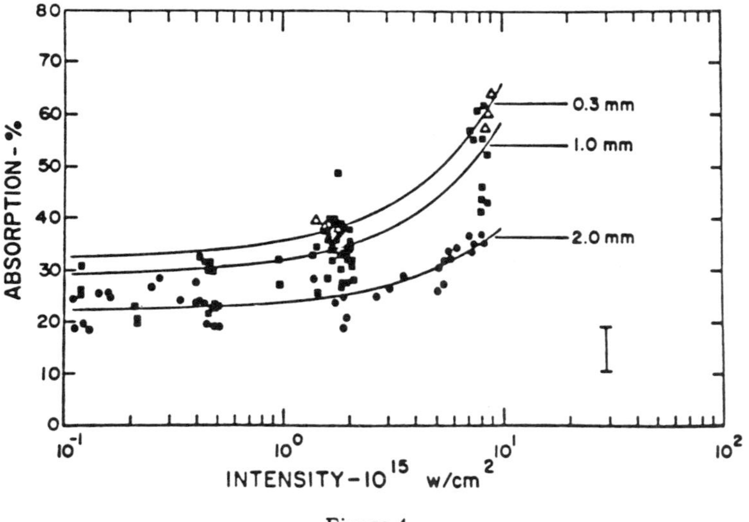

Figure 4.

III. HIGH HARMONIC GENERATION

An interesting feature of irradiation at high intensities is the generation of tens of laser harmonics.[6] Figure 6 is a display of the observed harmonic spectrum in the visible. Up to 46 harmonics have been observed. The spectrum is flat over most of its range. Conversion efficiencies of 10^{-4} - 10^{-5} per order have been seen. Duration of the harmonics is about 200 psec. The polarization is the same as the incident light. Also displayed in Fig. 6 are harmonic spectra taken from computer simulations. Simulations at three different intensities are shown. The plasma frequency corresponding to the upper density shelf are indicated.

The mechanism for generating high harmonics[7] is shown in Fig. 7. Imagine an electron fluid element close to the critical surface. If an oscillating laser field is applied, the fluid element will tend to oscillate in the electric field. If the fluid element is displaced to the right a distance δ, it will feel a restoring force of ${\omega_{pu}}^2\delta$, where ω_{pu} is the plasma frequency corresponding to the upper density shelf. If the fluid element is displaced to the left, a restoring force of ${\omega_{pl}}^2\delta$ is felt, where ω_{pl} is the plasma frequency corresponding to the lower density shelf. A trajectory of the fluid element is displayed in Fig. 8. It can be seen that whenever the fluid element collides with the surface, it experiences a sharp acceleration. In this respect, the fluid element resembles a basketball bouncing on the floor. If the basketball is charged, it will radiate every time it strikes the floor. The radiated frequency will be in harmonics of the driver frequency and the harmonic spectrum will be determined by the duration of the collision. The collision time for the fluid element is π/ω_{pu}. Therefore, from the uncertainty principle, the harmonic spectrum will be flat to ω_{pu} as seen in the experiment. In general, adjacent fluid elements will radiate coherently, but we can obtain a lower limit on the power emission by assuming incoherent emission,

$$P = \frac{2}{3}\frac{e^2}{c^3}\ddot{\delta}^2$$

This gives conversion efficiencies of

$$\frac{I_h}{I_c} \approx 10^{-5} - 10^{-6}$$

about an order of magnitude lower than that seen in experiment. A more detailed analysis of the fluid trajectories has been performed by Grebogi et al.,[8] but a proper treatment of coherence has yet to be made.

IV. SELF GENERATED SURFACE MAGNETIC FIELDS

Historically there have been a number of transport mysteries. The heat flux into targets has been much less than the classical value. Lateral energy transport has been much larger than that predicted by fluid codes. The partitioning of energy into fast ions has been inexplicably large. And in large laser spots, filaments in the blowoff plasma have been seen. Many of these issues were resolved when self generated magnetic fields were seen in particle code simulations.[9] Figure 9 is a plot of magnetic field contours on a sharp surface. The width of the field is of the order

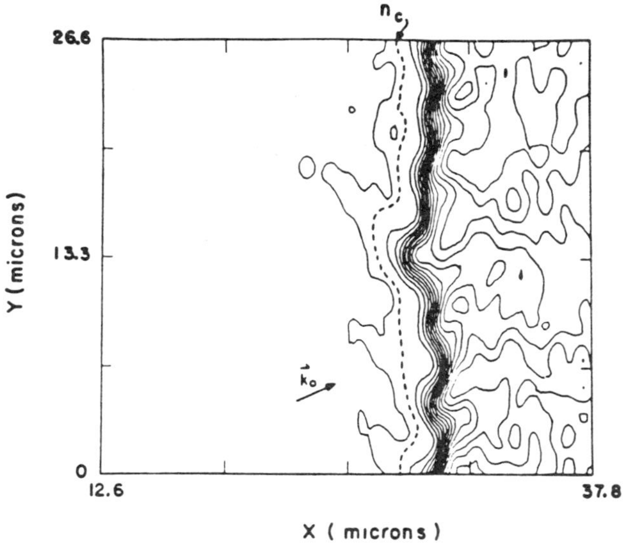

Figure 5.

The spectrum is flat up to ω_{pu}.

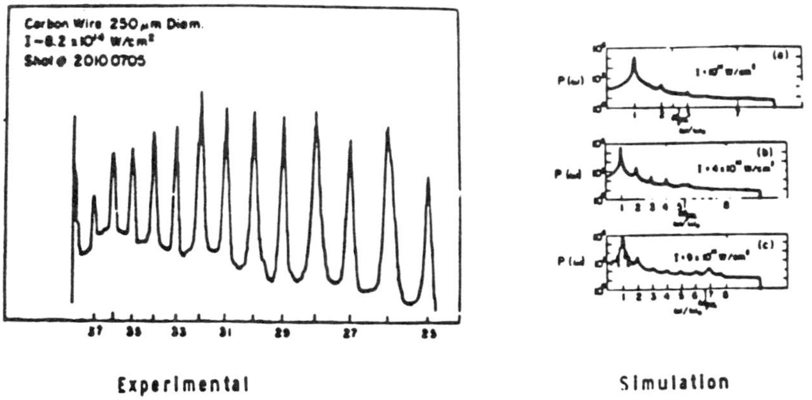

Figure 6.

THE SIMPLE FLUID MODEL

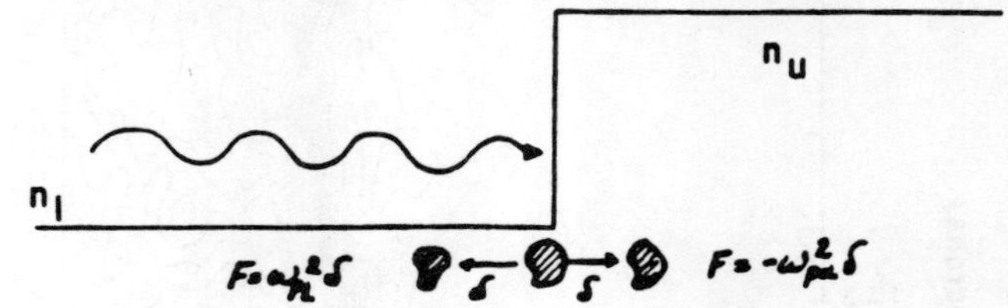

- Cold, fixed ions in an infinitely sharp density gradient

  $n_u >> n_c > n_l$

- Cold electrons

Figure 7.

ACCELERATION OF FLUID ELEMENT IS IMPULSE-LIKE AND PERIODIC

- MOTION IS APPROXIMATELY THAT OF AN OSCILLATOR WITH PERIOD $\frac{2\pi}{\omega_{pu}}$ IN OVERDENSE REGION WHICH LEADS TO STRONG ACCELERATION IN THIS SHORT TIME

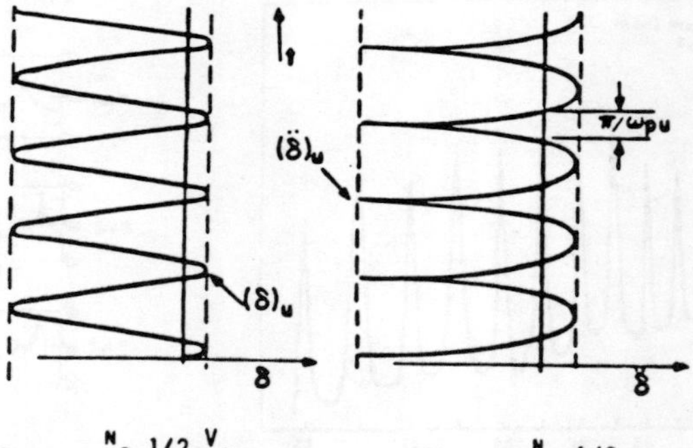

$(\delta)_u = (\frac{N_c}{N_u})^{1/2} \frac{V_o}{W}$ $\qquad$ $(\ddot{\delta})_u = (\frac{N_u}{N_c})^{1/2} \omega V_o$

Figure 8.

of a collisionless skin depth. The maximum field amplitude is about a megagauss. The laser spot is located at y = 1.8. It can be seen that the fields coat the surface of the target. Hot electrons are localized to the region of the field. The fields thus mediate the lateral transport of energy. Transport of energy into the target is retarded. The mechanism for this transport may be thermal-magnetic surface waves.[10]

Evidence of lateral energy transport has been seen in x-ray images of multiple laser spots.[11] Adjacent laser spots generate a region of null field between the spots. Hot electrons are no longer successfully insulated from the target in this region thus permitting enhanced x-ray emission (Fig. 10).

For larger laser spots qualitatively new physics emerges. Figure 11 pictures three dimensional magnetic field contours for the simulation of Fig. 9. Also displayed is the same run but with the larger spot indicated. Superimposed on the large scale fields are small scale filamentary structures. These structures seem to be related to the Weibel instability in that they depend on how isotropically superthermal electrons are emitted from the laser spot.

V. FAST ION PRODUCTION

The magnetic insulation of hot electrons at the critical surface generates a large space charge. Large electrostatic fields are produced quite far from the laser spot. These large fields accelerate the cold plasma ions to kinetic energies greater than the hot electron temperature.[12] This can be seen in the charge cup data displayed in Fig. 12.

At high laser intensities all of the absorbed energy can be partitioned into fast ions. This is seen in the calorimeter data in Fig. 13. The total absorption is indicated by dashed curves and the absorption into fast ions is indicated by solid curves. T_{Apache} is the hot electron temperature as measured by the x-ray bremsstrahlung spectrum. Intensities of 10^{14} and 10^{16} W/cm^2 are indicated.

An interesting feature of the blowoff is that it is well collimated. Figure 14 is a plot of the angular ion distribution. The squares are experimental points and the solid curve is computer simulation. Most of the energy in fast ions is emitted within a 10 degree cone. Jet like features from the neighborhood of the laser spot have also been seen in simulations.

It has been suggested[14] that the fast ion blowoff can be used to drive inertial fusion targets as well as to fuel magnetic fusion devices with high energy fuel.[15]

VI. CONCLUDING REMARKS

In conclusion, we have reviewed the critical role surfaces play in the coupling of intense long wavelength lasers to plasmas. The absorption at intensities below 10^{15}W/cm^2 is due to resonant absorption in steep density profiles. The profiles are generated by the ponderomotive force of the laser. Above 10^{15}W/cm^2 the absorption is mediated by rippling of the critical surface. At early times many harmonics of the laser frequency are generated. It was shown that this is due to the

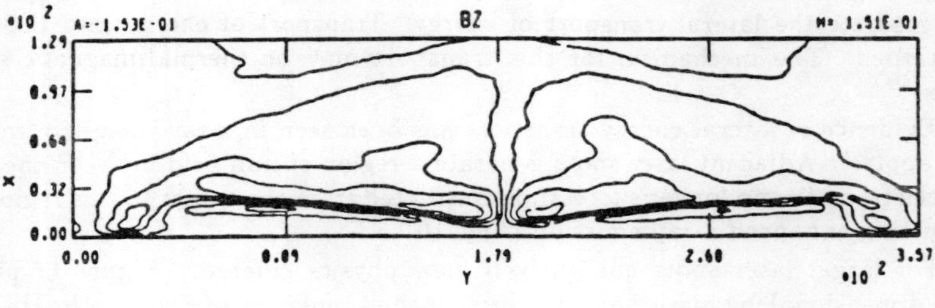

Figure 9.

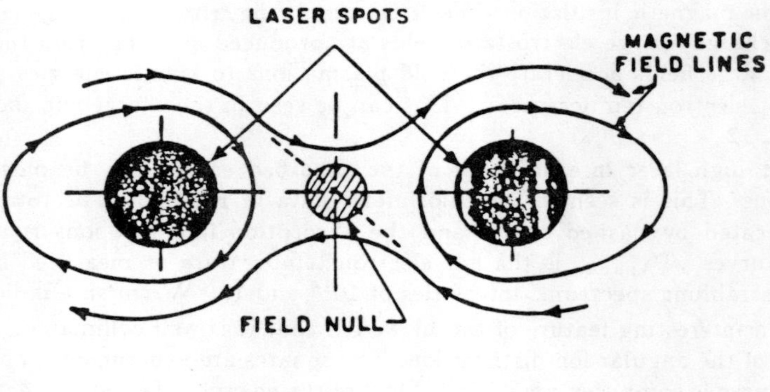

Figure 10.

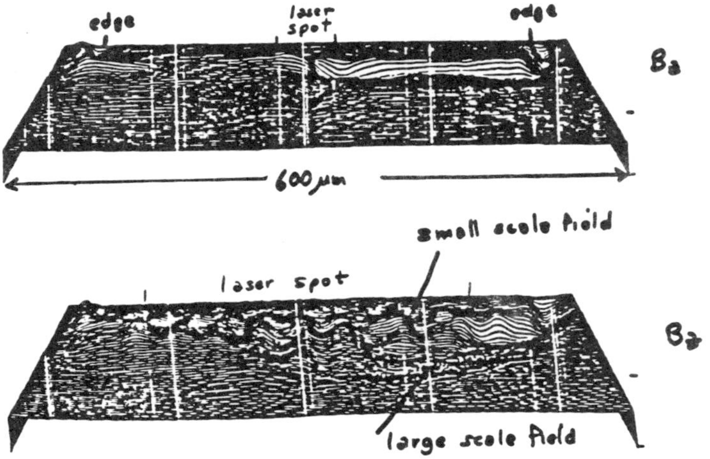

Figure 11.

ION ENERGY IS A STRONG FUNCTION OF T_{HOT}

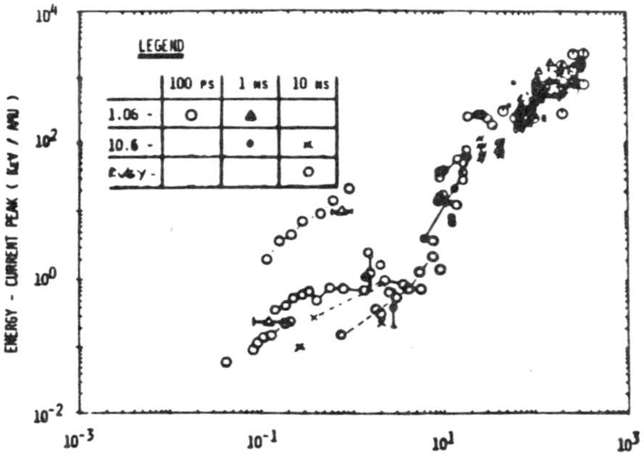

Figure 12.

FAST ION PRODUCTION EFFICIENCY INCREASES WITH T_{HOT}

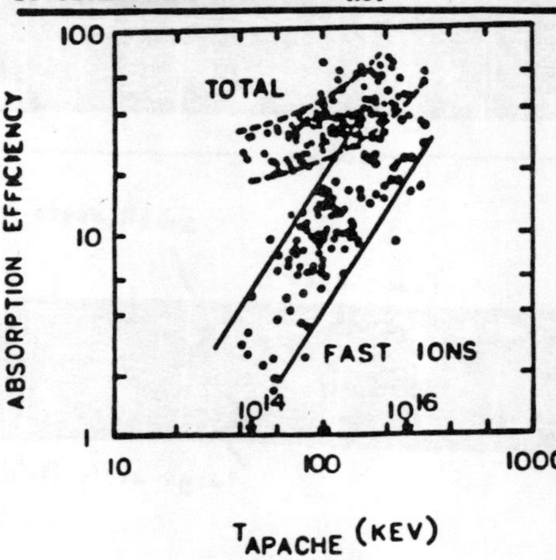

Figure 13.

THEORETICAL MODELING GIVES ANGULAR DISTRIBUTION OF FAST IONS CONSISTENT WITH EXPERIMENT

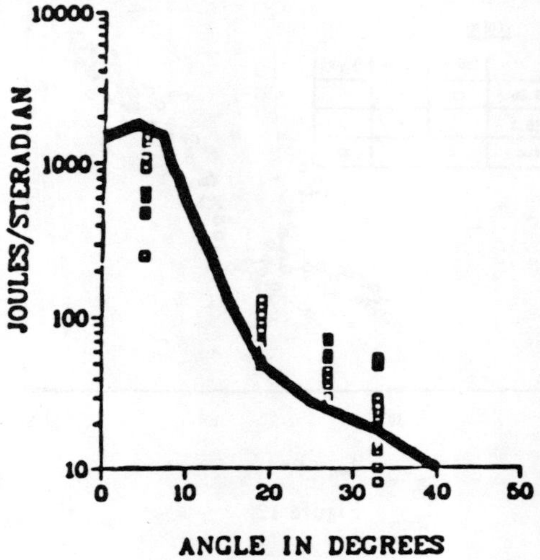

Figure 14.

sharp boundary at the critical surface. A surface layer of magnetic fields is generated at the critical surface. These fields induce lateral energy transport, insulation of hot electrons from the cold target, and the generation of a very fast ion blowoff. At intensities greater than $10^{15} W/cm^2$ most of the absorbed energy ends up as ion kinetic energy. The physics has turned out to be quite interesting and exotic. However, the fact that the energy is deposited in a small number of high energy electrons and fast ions, and that the energy deposition is along the target surface rather than into the interior is fatal for CO_2 lasers as fusion drivers.

This work was supported by the United States Department of Energy.

REFERENCES

1. Science News, 128 68 (1985).
2. J. J. Duderstadt and G. A. Moses, Inertial Confinement Fusion (Wiley Interscience, New York, 1982).
3. D. W. Forslund and P. D. Goldstone, Los Alamos Science 12, 2 (1985).
4. R. D. Jones and J. M. Kindel, Radiation in Plasmas, Ed. by B. McNamara (World Scientific, Singapore, 1984) p. 423.
5. D. W. Forslund, J. M. Kindel, and K. Lee, Phys. Rev. Lett. 39 284 (1977).
6. B. Bezzerides, S. J. Gitomer, and D. W. Forslund, Phys. Rev. Lett. 44 651 (1980); B. Bezzerides and S. J. Gitomer, Phys. Rev. Lett. 46 593 (1981).
7. R. L. Carman, D. W. Forslund, and J. M. Kindel, Phys. Rev. Lett. 46 29 (1981).
8. B. Bezzerides, R. D. Jones, and D. W. Forslund, Phys. Rev. Lett. 49 202 (1982).
9. C. Grebogi, V. K. Tripathi, and H-H. Chen, Phys. Fluids 26 1904 (1983).
10. D. W. Forslund and J. U. Brackbill, Phys. Rev. Lett. 48 1614 (1982).
11. R. D. Jones, Phys. Rev. Lett. 51 1269 (1983).
12. M. A. Yates, D. B. vanHulsteyn, H. Rutkowski, G. A. Kyrala, and J. U. Brackbill, Phys. Rev. Lett. 49 1702 (1982).
13. S. J. Gitomer, R. D. Jones, F. Begay, A. W. Ehler, J. F. Kephart, and R. Kristal, "Fast Ions and Hot Electrons in the Laser-Plasma Interaction," in preparation; S. J. Gitomer, R. D. Jones, J. Kephart, F. Begay, and A. W. Ehler, IEEE 1983 International Conference on Plasma Science, San Diego, Calif. and Proceedings of the 13th Annual Anomalous Absorption Conference, 1983, Banff, Alberta.
14. C. W. Barnes, D. Bach, H. Rutkowski, and G. Kyrala, Bull. Am. Phys. Soc. 27 1042 (1982)
15. F. Mayer, 1985 Sherwood Theory Meeting, Madison, Wisc.

NONRECIPROCAL EFFECTS IN SURFACE WAVE PROPAGATION

B. Lüthi, L. Remer

Physikalisches Institut, Universität Frankfurt,
P.O. Box 11 19 32, Frankfurt, FRG

Introduction We consider surface wave propagation in the continuum limit, i.e. the wavelength of the excitation $\lambda >> a$ with a equal the lattice constant. In this limit surface wave excitations have been calculated for surface acoustic waves[1] (SAW), ferromagnetic surface spin waves[2], antiferromagnetic spin waves[3], surface plasmons[4] and surface polaritons[5]. In the presence of a magnetic field nonreciprocal propagation effects for these surface excitations can occur. We will review these effects emphasising the physics which leads to these properties and giving simple coupling constant arguments. In addition we shall discuss nonreciprocal reflectivity experiments which result from these nonreciprocal propagation effects. We will end with a generalization of Kirchhoff's law of radiation.

In the next section we introduce the formal symmetry argument leading to nonreciprocal behaviour. Next we discuss several examples of nonreciprocal propagation experiments. In the final section we discuss the reflectivity experiments and its generalizations.

Symmetry argument We consider a solid bounded by an infinite half plane (Fig. 1). The reflection symmetries on the planes xy, S_{xy}, on the plane xz, S_{xz} and on the plane yz, S_{yz} shall be allowed symmetry operations in the solid. Other symmetry elements are unimportant for the argument. For the polar wave vector k_x and the axial magnetic field vector B_y we have $S_{xy}(k_x,B_y) = (k_x,-B_y)$, $S_{xz}(k_x,B_y) = (k_x,B_y)$ and $S_{yz}(k_x,B_y) = (-k_x,-B_y)$. For a bulk excitation frequency $\omega(k_x,B_y)$ it follows that $\omega(k_x,B_y) = \omega(k_x,-B_y) = \omega(-k_x,B_y)$, a well known fact that volume excitation frequencies are even functions of k_x and B_y. For the surface excitation the reflection operation S_{xy} is no longer a symmetry operation. We therefore can only conclude for the excitation frequency $\omega_s(k_x,B_y) = \omega_s(-k_x,-B_y)$ but in general $\omega_s(k_x,B_y) \neq \omega_s(-k_x,B_y)$ and $\omega_s(k_x,B_y) \neq \omega_s(k_x,-B_y)$. This symmetry property was noted before[2] and formulated as a symmetry principle for the case of magnetoelastic surface waves[6]. Analogously one can show that corresponding nonreciprocal behaviour does not exist for other field directions (B_x,B_z). One can visualize this nonreciprocity for surface wave propagation rather easily by noting that surface waves in the continuum limit have particle- or field- or spin- polarizations which are not linear but elliptical (figure 1). Therefore the rotation sense is different for $+k_x$ and $-k_x$ propagation. We shall discuss this fact in a number of examples below for propagation along symmetry directions. In the notation of figure 1 a general surface wave propagation in x-direction can be written as $u_n = U_n e^{\alpha z} e^{i(kx-\omega t)}$ where for $z<o$ $\alpha>o$.

<u>Examples of nonreciprocal surface wave propagation</u> Here we discuss briefly some nonreciprocal effects which have been realized also experimentally.

a) <u>Nonreciprocal dipolar surface spin waves</u> The first nonreciprocal surface wave effect has been discussed for ferromagnetic surface spin waves[2]. It has been observed experimentally by means of Brillouin scattering in a number of cases: EuO[7], Ni, Fe[8]. Theory and experiment for this case have

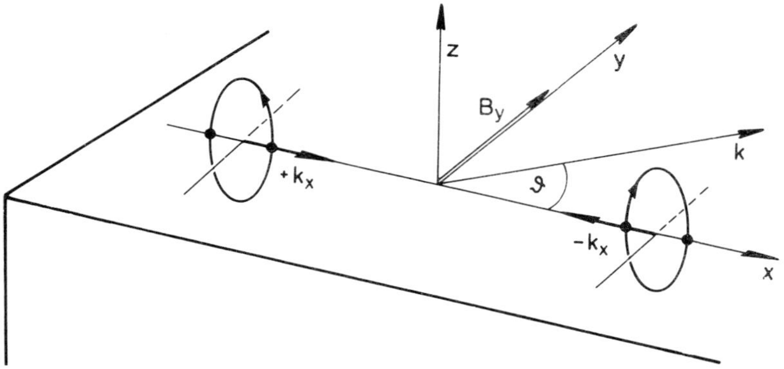

Fig.1 Geometrical arrangement for surface wave propagation

been reviewed recently[9]. Therefore we make just a few brief remarks. In the dipolar limit where the exchange contributions can be neglected in the spinwave dispersion, the surface spin wave frequency reads (with angle Θ in the xy plane defining the propagation, fig. 1)

$$\omega_{s}/\gamma = -\frac{1}{2}[H/\cos\Theta + (H+4\pi M)\cos\Theta] \qquad (1)$$

with stable solutions occuring only for $\alpha > 0$, $\cos\Theta < \cos\Theta_c = \sqrt{H/(H+4\pi M)}$. In figure 2a the bulk and surface spinwaves are given as a function of Θ for M = 480 G (Ni). It is seen that the stable surface spin waves are confined to a region $0 < \Theta < 75^{\circ}$ for H = 0.4 kG. For $\Theta = 75^{\circ}$ the SSW touches the bulk continuum and in the damping term α becomes negative which means that the SSW are no longer stable and decay into volume modes. In the stable region ω_s is larger than ω_{bulk} because the positive magnetostatic energy contribution is larger for SSW due to the dipolar fields extending into the region z>0. The nonreciprocal feature is easily understood because for $\gamma < 0$ only $k_x > 0$ SSW have the correct rotation sense for the magnetization components (M_x, M_z) in the presence of a magnetic field.

Experimentally this nonreciprocal effect can be clearly seen in light scattering (fig. 2b), where in the geometry of fig. 1 the Stokes component is absent and only the anti-Stokes component due to SSW shows up. Further details can be found in ref. 9.

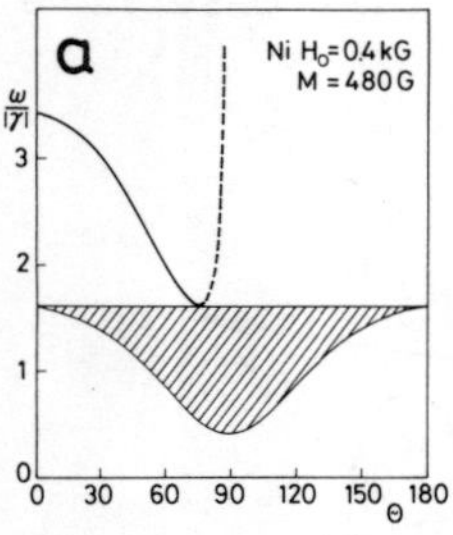

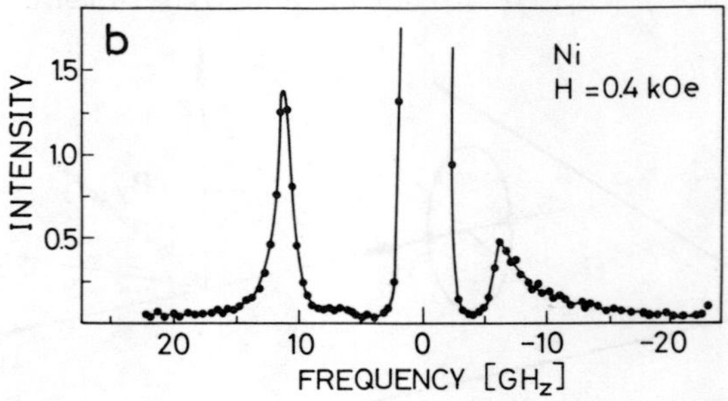

Fig. 2 a) calculated ω_s and ω_{bulk} for Ni using eq. (1)
b) Brillouin scattering spectra for Ni, adapted from ref. 8.

Besides dipolar ferromagnetic spinwaves there are also dipolar antiferromagnetic spinwaves and magnetostatic modes[10]. Dipolar antiferromagnetic SSW have been calculated[3], but not observed yet. Their frequency region of stability is rather limited. An interesting way of observing them has been recently shown[11] and will be discussed below.

b) Nonreciprocal surface plasma polaritons The coupled modes of plasmons and electromagnetic waves, the socalled plasma polaritons were discussed in detail for various magnetic field geometries[12]. For nonreciprocal effects the geometry of figure 1 is again appropriate. One can understand the nonreciprocal feature of the plasma polaritons rather easily by noting that the surface mode described by P_x+iP_z has a different dispersion law than the P_x-iP_z mode[5,12]. The dispersion equation for the polariton reads

$$\sqrt{k_x^2-\omega^2\varepsilon_v/c^2}+\varepsilon_v\sqrt{k_x^2-\omega^2/c^2}+k_x\varepsilon_2/\varepsilon_1=0 \qquad (2)$$

where $\varepsilon_v=\varepsilon_1-\varepsilon_2^2/\varepsilon$, is the Voigt dielectric constant and ε_1, $\pm i\varepsilon_2$ denote the diagonal and off-diagonal tensor components of the (Drude)-dielectric tensor. Since ε_2 is linear in B the nonreciprocity arises from the last term in (2) which is linear in k and B. Calculated dispersion curves are shown in figure 5 below.

Experimentally nonreciprocal features in propagation of surface polaritons have been observed[13,14] using the technique of attenuated total reflection[15]. Figure 3 shows an example of nonreciprocal surface plasma polaritons measured for InSb. Again the criterion $\alpha>0$ for $z<0$ gives the region of stability, which is limited as in the case of spinwaves by the bulk polariton spectra.

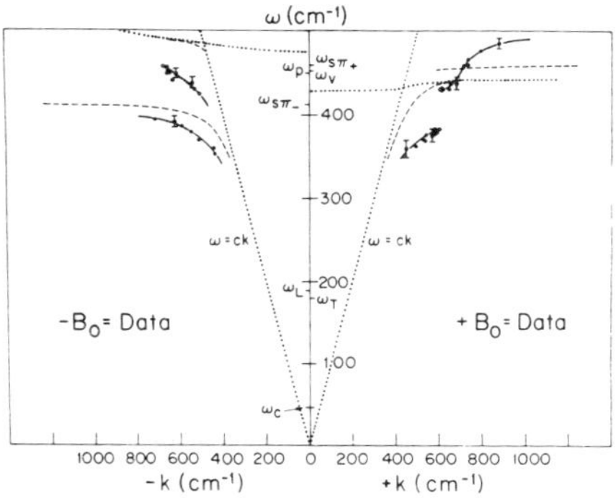

Fig. 3 Measured and calculated nonreciprocal surface polariton for InSb, adapted from ref. 13

In the case of optical phonon polaritons[15] no such nonreciprocal effects exist because the optical phonons do not couple to the magnetic field.

c) Surface acoustic waves in metals An intriguing nonreciprocal effect exists in acoustic surface wave propagation in pure metals[16]. Since SAW have a particle motion on elliptical paths[1,5] they couple differently to electrons on cyclotron orbits for propagation direction in $+k_x$ or $-k_x$ directions respectively. The velocity effects are of the order of .1%. The effect is largest for $\omega_c\tau \sim 1$ and it disappears for $\omega_c\tau \ll 1$ and $\omega_c\tau \gg 1$ (ω_c cyclotron frequency, τ electron relaxation time). The boundary condition $\sigma_{zi} = 0$ for a free surface is fulfilled by noting that the stress tensor is composed of an elastic one and a kinetic stress tensor for the electrons. This latter one shows the nonreciprocal behaviour $t_{ik}(B) \neq t_{ik}(-B)$. For $\omega_c\tau \sim 1$ this kinetic stress tensor is nonzero, whereas for $\omega_c\tau \gg 1$ it becomes small because the momentum transfer through a given surface approaches zero. It is important to note that in this coupled system acoustic surface wave-conduction electrons the nonreciprocal nature occurs only in the presence of damping. This is somewhat similar to the nonreciprocal reflection of electromagnetic waves (see below).

An interesting application is the acoustic analogue of r-f size effect. (Ghantmahker effect) where excitation and/or detection could be done with SAW. Such an experiment has not been performed yet.

d) Nonreciprocal magnetoelastic waves Nonreciprocity for magnetoelastic waves have been calculated for ferromagnets[6] and paramagnets[17]. In paramagnetic rare earth compounds with crystal field split rare earth ions the nonreciprocal effect on surface acoustic wave propagation due to magnetoelastic coupling is very small unless the wave frequency is equal to the crystal field splitting[18]. In $CeAl_2$ for acoustic frequencies of the order of 100 MHz the effect is unobservable[19]. On the other hand a SAW attenuation experiment in Yttrium Gallium Iron Garnet showed nonreciprocal behaviour[20]. In this case one can have a resonant coupling between SAW and the spinwaves. However disturbing domain effects cannot be excluded.

Nonreciprocal Reflectivity As a consequence of nonreciprocal surface wave propagation the reflectivity of e.g. an electromagnetic wave can also be nonreciprocal. The first investigation on nonreciprocal reflection was a theoretical study of acoustic wave reflection on a ferromagnetic surface[21]. Here we briefly discuss nonreciprocal reflectivity for a magnetoplasma[22] and an antiferromagnet[11] with electromagnetic waves.

a) Nonreciprocity in the optical reflection of magnetoplasmas For the configuration given in Fig. 4 one can expect a nonreciprocal effect in reflection. This feature is directly related to the nonreciprocal behaviour of surface polaritons shown in Fig. 5. In this figure we have plotted as a function of ω/ω_p the dispersion relation for surface polaritons and the calculated reflectivity in the presence of an applied magnetic field. ω_p denotes the plasma frequency. It is seen that nonreciprocal reflectivity occurs in the frequency region where nonreciprocal features of the polaritons occur. The amount of nonreciprocity depends on the damping factor $\omega\tau$ with τ electron relaxation time. The nonreciprocal effect disappears without damping. This is easily seen by looking at Fig. 5 where vacuum light cannot couple to surface polaritons. This is the reason for introducing attenuated total reflection techniques[15].

For the reflection coefficient one obtains[22]:

$$1-r/1+r = -\varepsilon_1\varepsilon_v\cos\alpha/(i\varepsilon_2\sin\alpha+\varepsilon_1\kappa) \quad (3)$$

with $\kappa = -\sqrt{\varepsilon_v - \sin^2\alpha}$ Again the nonreciprocity arises from the term $\varepsilon_2\sin\alpha$. Experimental results[22] for n-type InSb are shown in Fig. 6. Two frequencies are shown, one near the

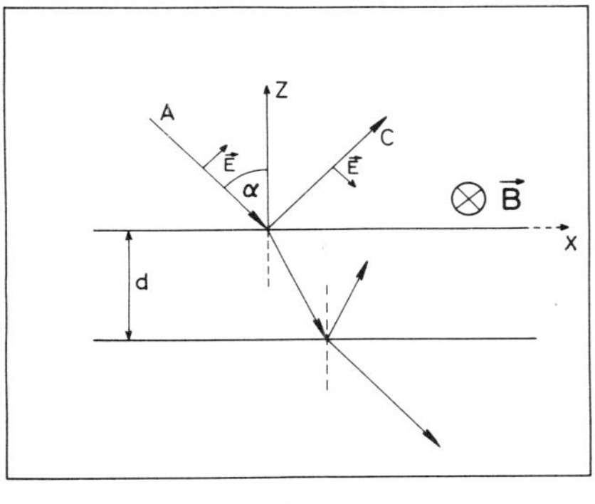

Fig. 4 Schematic plot of the reflection geometry

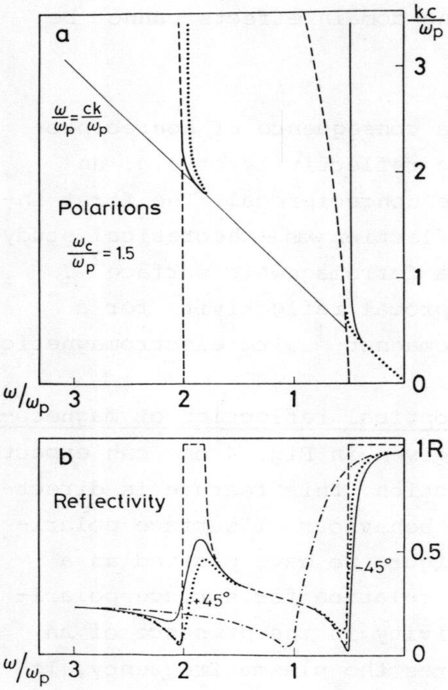

Fig. 5 Calculated dispersion relation (a) and reflectivity (b) for plasmon polaritons, adapted from ref. 22.

plasma edge and the other near the collective cyclotron resonance. The agreement between the calculated and experimentally determined reflectivity is very good. In addition to this nonreciprocal effects one observes also thickness interferences for a plane parallel plate. Again this effect can be interpreted satisfactorily.

b) Nonreciprocity in the optical reflection of antiferromagnets Analogous to the case of magnetoplasmas one can observe nonreciprocal effects in the optical reflection of antiferromagnets. If one considers an uniaxial antiferromagnet

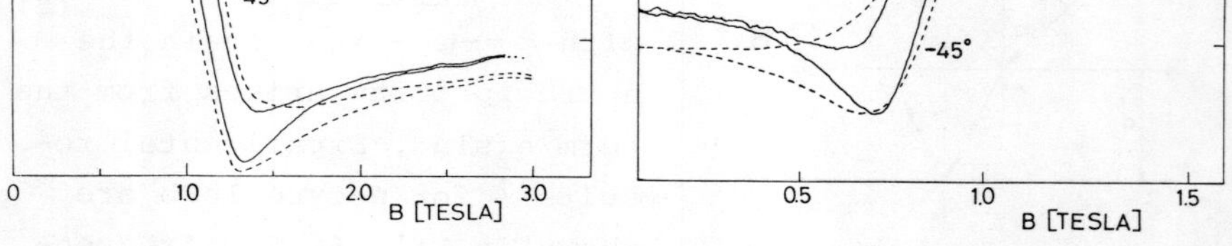

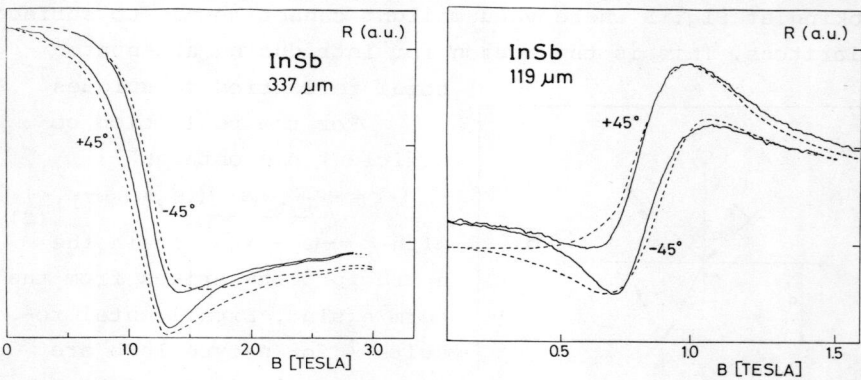

Fig. 6 Measured and calculated reflectivity for two laser frequencies at T=8oK, adapated from ref. 22

with a spinwave energy gap close to the laser frequency one can treat the problem in an analogous way to the case of magnetoplasmas. The coupling of spinwaves to electromagnetic waves gives rise to magnetic polaritons. The magnetic surface polaritons again exhibit nonreciprocal behaviour[11] in close analogy to the nonreciprocal behaviour of antiferromagnetic spinwaves mentioned in section a or to the plasma polaritons treated in section b.

A reflectivity experiment using far infrared laser frequencies has been performed recently[11] in CoF_2. Since one couples in this case to the dipolar magnetic polaritons the electric field vector has to be perpendicular to the plane of incidence (Fig. 4). The agreement between theory and experiment is in this case not perfect, but the salient features are clearly seen.

c) Nonreciprocity in the optical reflection of ferromagnets In analogy to the dipolar surface spin waves treated above one also has ferromagnetic polaritons and one can expect nonreciprocal reflectivity in this case. This experiment has not been done yet.

However a related experiment, the socalled equatorial Kerr effect, has been studied a long time ago[23,24]. The difference to our nonreciprocal effects lies in the different geometry of E and B vectors. For coupling to dipolar spinwaves or polaritons the B vector has to lie in the plane of incidence, whereas in the equatorial Kerr effect the E vector is in this plane. For optical transitions in the visible one has not magnetic dipole transitions but electric ones (in Fe, Ni from 3d shells). In other words instead of probing the dipolar type magnetostatic modes, one deals with stronger electric dipole quasi-atomic transitions.

Nonreciprocity in Emission and Absorbtion: Generalization of Kirchhoffs law of radiation Let us turn back to the case of nonreciprocal reflection in a magnetoplasma (section above). We introduce the concepts of emittance E, absorbtance A and reflectance R in the usual way[22]. Numerical calculations of reflection and transmission of an electromagnetic wave through a plate (Fig. 4) show that for $B \neq 0$ and $(\omega_c \tau)^{-1} \neq 0$ one has $R(\alpha) \neq R(-\alpha)$, nonreciprocal and $T(\alpha) = T(-\alpha)$ reciprocal. From $R(\alpha) + T(\alpha) + A(\alpha) = 1$ (energy conservation) follows that $A(\alpha) \neq A(-\alpha)$; and $R(\alpha) = R(-\alpha)$ for $A = 0$. With $R(-\alpha) + T(\alpha) + E(\alpha) = 1$ (again energy conservation) follows the nonreciprocity of emission $E(\alpha) \neq E(-\alpha)$ and $E(\alpha) = A(-\alpha)$, the generalization of Kirchhoffs law of radiation[25] to the case of magnetic fields.

These findings can be formulated in a more general way using Onsagers reciprocal relations[26].

<u>References</u>

1. J. W. Rayleigh, The Theory of Sound, New York, Dover Publications

2. R.W. Damon, J.R. Eshbach, J. Phys. Chem. Solids, <u>19</u> 3o8 (198o)

3. B. Lüthi, D.L. Mills, R.E. Camley, Phys. Rev. B<u>28</u> 1475 (1983)

4. R.H. Ritchie, Surface Science <u>34</u> 1 (1973)

5. A.A. Maradudin, in Festkörperprobleme Vol. XXI, ed. J. Treusch, Vieweg Braunschweig, p. 25 (1981)

6. R.Q. Scott and D.L. Mills, Phys. Rev. B<u>15</u> 3545 (1977)

7. P. Grünberg and F. Metawe, Phys. Rev. Lett. <u>39</u> 1561 (1977)

8. J.R. Sandercock and W. Wettling, J. Appl. Phys. <u>5o</u> 7784 (1979)

9. D.L. Mills, in Surface Excitations, ed. V.M. Agranovich and R. London, North-Holland Amsterdam chapter 3 (1983)

1o. J.P. Kotthaus, AIP Conference Proceedings, <u>1o</u> Part 1 57 (1973)

11. L. Remer, B. Lüthi, R.E. Camley, to be published

12. R.F. Wallis, J.J. Brion, E. Burstein, A. Hartstein, Phys. Rev. B<u>12</u> 3186 (1975)

13. A. Hartstein et al., Solid State Comm. <u>14</u> 1223; Phys. Rev. B<u>12</u> 3186 (1975)

14. V.S. Ambrazeviciene and R.S. Brazis, Solid State Comm. 18 415 (1976)

15. A. Otto in Festkörperprobleme Vol. XIV ed. by H.Queisser, Stuttgart

16. J. Heil, I. Kouroudis, B. Lüthi, P. Thalmeier J. Phys. C 17 2433 (1984)

17. R.E. Camley, P. Fulde Phys. Rev. B 23 2614 (1981)

18. R.E. Camley, P. Fulde Phys. Rev. B 30 4137 (1984)

19. C. Lingner, B. Lüthi Phys. Rev. 23 256 (1980)

20. M.F. Lewis and E. Patterson Appl. Phys. Lett. 20, 276 (1972)

21. B. Laks and D.L. Mills Phys. Rev. B, 22 4445 (1980)

22. L. Remer, E. Mohler, W. Grill and B. Lüthi Phys. Rev. B 30 3277 (1984)

23. L.R. Ingersoll, Phys. Rev. 35 312 (1912)

24. D. H. Martin, K.F. Neal, T.J. Dean, Proc. Phys. Soc. 86 605 (1965)

25. G. Kirchhoff, reprinted in ABHANDLUNGEN ÜBER EMISSION UND ABSORPTION, edited by M. Planck (Akademische, Verlagsgesellschaft, Leipzig, 1921)

26. L. Remer, doctoral thesis, Universität Frankfurt 1985

Surface Waves on Rough Surfaces

Alexei A. Maradudin
Department of Physics
University of California*
Irvine, CA 92717, U.S.A.

and

Max Planck Institut für Festkörperforschung
Heisenbergstrasse 1
7000 Stuttgart 80
Federal Republic of Germany

Abstract

A survey will be given of recent work on the propagation of acoustic, electromagnetic, and magnetostatic surface waves across a rough surface. The roughness to be considered is of three kinds: (a) an isolated protruberance or indentation on an otherwise planar surface; (b) a periodically corrugated surface; and (c) a randomly rough surface. The first kind of roughness gives rise to surface shape resonances, i.e. to modes that are spatially localized in the vicinity of the surface perturbation and that are characterized by discrete frequencies. Waves localized to the surface can propagate across surfaces of the second and third kinds. The effects of the roughness on their dispersion curves will be described. In addition, the localization of surface electromagnetic waves by random surface roughness, and its experimental consequences, will be discussed.

*Permanent address.

1. Introduction

Why study surface waves on rough surfaces? There are several answers that can be given to this question. The first is that all real surfaces, even the most carefully prepared ones, possess some degree of roughness, and it is of interest to know how their properties are modified by that roughness. For example, a surface wave propagating across a randomly rough surface can be scattered by the hills and valleys on the surface into other surface waves and into bulk waves. The surface wave is attenuated thereby, even if it is propagating on the surface of a lossless medium. It can be important for applications of surface waves to know what degree of such surface roughness is tolerable before the attenuation to which it gives rise reaches an unacceptable level. In the same vein, the propagation of surface waves across periodically corrugated surfaces gives rise to the phenomenon of wave slowing, to the appearance of pass bands and stop bands in their dispersion curves that can be useful in the design of filters for surface waves, and to the possibility of the controlled conversion (transduction) of surface waves into bulk waves. A second reason for studying surface waves on rough surfaces is that surface roughness gives rise to new types of surface waves that cannot exist in the absence of roughness. This enriches the taxonomy of this field of science, and adds to the armamentarium of those interested in applications of surface waves.

When one speaks of rough surfaces one instinctively thinks of randomly rough surfaces. However that is only one type of rough surface. In this talk I will consider three types of rough surfaces each of which would be planar in the absence of the roughness. They are all defined by an equation for the position of the surface of the form $x_3 = \zeta(\vec{x}_\parallel)$, where $\vec{x}_\parallel = \hat{x}_1 x_1 + \hat{x}_2 x_2$ and $\hat{x}_1$ and $\hat{x}_2$ are unit vectors in the x_1- and x_2-directions, respectively. They differ in the nature of the surface profile function $\zeta(\vec{x}_\parallel)$. For the first type of rough surface $\zeta(\vec{x}_\parallel)$ is deterministic and is nonzero only for a finite range of values of x_1 and x_2. It describes an isolated protuberance or indentation on an otherwise planar surface. For the second type of rough surface $\zeta(\vec{x}_\parallel)$ is again deterministic, but is a periodic function of x_1 and/or x_2. It describes a classical grating or a bigrating on a material surface. For the third type of rough surface $\zeta(\vec{x}_\parallel)$ is a stochastic function of $\vec{x}_\parallel$, and describes a randomly rough surface. The restriction of the rough surfaces I will consider here to those that would be planar in the absence of the roughness is both a reflection of the relative paucity of work on surface waves on rough, nonplanar surfaces and a consequence of restrictions imposed by both time and space.

To present the discussion from a unified point of view I will base it on the Rayleigh hypothesis,[1] i.e. on the assumption that the solutions of the equations of motion outside the selvedge region (the region of the surface between the minimum and maximum excursions of the surface from flatness) that satisfy the boundary conditions at infinity can be continued in to the surface and used to satisfy the boundary conditions there. Despite the well-known limits of validity of this hypothesis[2-5], we will find that it yields convergent results even for strongly corrugated surfaces, which convergence may have an asymptotic nature.[4,6]

2. Surface Waves on Elastic Media

Historically, the first kind of surface wave to be studied was the surface acoustic wave on the planar, stress-free surface of a semi-infinite, isotropic elastic medium, now called the Rayleigh wave after its discoverer.[7] The study of the propagation of Rayleigh waves and other types of surface acoustic waves across rough surfaces, however, is of comparatively recent origin. In this section we survey briefly what has been learned from such studies.

A. Scattering of a Rayleigh Wave by a Surface Perturbation

We begin by considering the behavior of surface acoustic waves on the planar, stress-free surface of a semi-infinite, isotropic elastic medium that is perturbed by an isolated protuberance or indentation of finite extent. We assume that the medium occupies the region $x_3 > \zeta(\vec{x}_\parallel)$. Then, since the surface perturbation is static, we can work with a single frequency Fourier component of the elastic displacement field in the medium. If we write the latter in the form $u_\alpha(\vec{x},t) = u_\alpha(\vec{x}|\omega)\exp(-i\omega t)$, an application of Green's theorem yields the following integral equation for the amplitude $u_\alpha(\vec{x}|\omega)$ [8]:

$$u_\alpha(\vec{x}|\omega) = u_\alpha^{(0)}(\vec{x}|\omega) + \frac{1}{\rho}\sum_\gamma \int d^2y_\parallel\, D_{\alpha\gamma}(\vec{x};\vec{y}|\omega)\Big|_{y_3=0}$$

$$\times \sum_{\mu\nu}\Big[\sum_{\delta=1}^{2} C_{\gamma\delta\mu\nu}\frac{\partial\zeta(\vec{y}_\parallel)}{\partial y_\delta}\frac{\partial u_\mu(\vec{y}|\omega)}{\partial y_\nu} -$$

$$- C_{\gamma 3\mu\nu}\,\zeta(\vec{y}_\parallel)\frac{\partial^2 u_\mu(\vec{y}|\omega)}{\partial y_3\partial y_\nu}\Big]_{y_3} = 0, \tag{2.1}$$

to first order in the surface profile function $\zeta(\vec{x}_\parallel)$. In this equation ρ is the mass density of the medium, and $C_{\alpha\beta\mu\nu}$ is an element of the elastic modulus tensor, which for an isotropic medium has the form

$$C_{\alpha\beta\mu\nu} = \rho\,(c_\ell^2 - 2c_t^2)\delta_{\alpha\beta}\delta_{\mu\nu} + \rho c_t^2(\delta_{\alpha\mu}\delta_{\beta\nu} + \delta_{\alpha\nu}\delta_{\beta\mu}), \qquad (2.2)$$

where c_ℓ and c_t are the speeds of longitudinal and transverse sound waves in the medium, respectively. $D_{\alpha\beta}(\vec{x};\vec{y}|\omega)$ is the dynamical Green's tensor for a semi-infinite, isotropic elastic medium occupying the half-space $x_3 > 0$ and bounded by a planar, stress-free surface at $x_3 = 0$.[9] $u_\alpha^{(o)}(\vec{x}|\omega)$ is the amplitude of the incident Rayleigh wave (on a planar surface). The wave scattered by the protuberance or indentation is therefore given by the remaining terms on the right hand side of Eq. (2.1).

We can assume with no loss of generality that the incident Rayleigh wave propagates in the $+x_1$-direction. The displacement field $u_\alpha^{(o)}(\vec{x}|\omega)$ thus has the nonzero components

$$u_1^{(o)}(\vec{x}|\omega) = A\Big[e^{-\beta_\ell x_3} - \Big(1 - \frac{c_R^2}{2c_t^2}\Big)e^{-\beta_t x_3}\Big]e^{ik_o x_1} \qquad (2.3a)$$

$$u_3^{(o)}(\vec{x}|\omega) = i\Big(1 - \frac{c_R^2}{c_\ell^2}\Big)^{1/2} A\Big[e^{-\beta_\ell x_3} - \frac{e^{-\beta_t x_3}}{1 - c_R^2/2c_t^2}\Big]e^{ik_o x_1}, \qquad (2.3b)$$

where C_R is the speed of Rayleigh waves, and

$$\beta_\ell = k_o\Big(1 - \frac{c_R^2}{c_\ell^2}\Big)^{1/2}, \quad \beta_t = k_o\Big(1 - \frac{c_R^2}{c_t^2}\Big)^{1/2}. \qquad (2.4)$$

When Eq. (2.1) is solved in the first Born approximation with the use of Eqs. (2.3)-(2.4) and the results of Ref. 9, it is found that the incident Rayleigh wave is scattered into (a) bulk longitudinal waves; (b) bulk transverse waves of p-polarization; (c) bulk transverse waves of s-polarization; and (d) other Rayleigh waves. The displacement field associated with the scattering process (d) is particularly interesting. It is given in the far field region by

$$u_\alpha^{(s)}(\vec{x},t) \sim i\,\frac{Ae^{-i\frac{\pi}{4}}}{(2\pi)^{1/2}}\,\frac{c_R^2}{2Rc_t^2}\left(\frac{\omega}{c_R}\right)^{5/2}\hat{\zeta}(\vec{k}_R - \vec{k}_\parallel^{(o)}) \times$$

$$\times\,\frac{(\cos\phi_s - 1)(\cos\phi_s - \tfrac{1}{2}(\xi^2-1))}{(1-\tfrac{1}{2}\xi^2)(1-\xi^2\lambda^2)^{\frac{1}{2}}}\,\frac{e^{i\frac{\omega}{c_R}x_\parallel - i\omega t}}{x_\parallel^{1/2}} \times$$

$$\times\left\{(\hat{x}_1\cos\phi_s + \hat{x}_2\sin\phi_s)\left[e^{-\beta_\ell x_3} - (1-\tfrac{1}{2}\xi^2)e^{-\beta_t x_3}\right] + \right.$$

$$\left. + i\hat{x}_3(1-\xi^2\lambda^2)^{\frac{1}{2}}\left[e^{-\beta_\ell x_3} - \frac{e^{-\beta_t x_3}}{1-\tfrac{1}{2}\xi^2}\right]\right\}. \qquad (2.5)$$

In writing Eq. (2.5) we have introduced the notation

$\xi = c_R/c_t$, $\lambda = c_t/c_\ell$, $\vec{k}_R = k_o(\cos\phi_s, \sin\phi_s, 0)$,
$\vec{k}_\parallel^{(o)} = k_o(1, 0, 0)$, $k_o = \omega/c_R$, and

$$R = \frac{1}{(1-\tfrac{1}{2}\xi^2)^5}\left[\xi^4(1+2\lambda^2) - \xi^2(9-5\lambda^2) + 8(1-\lambda^2)\right] \qquad (2.6a)$$

$$\hat{\zeta}(\vec{k}_\parallel) = \int d^2x_\parallel\,\zeta(\vec{x}_\parallel)\,e^{-i\vec{k}_\parallel\cdot\vec{x}_\parallel}. \qquad (2.6b)$$

We note from Eq. (2.5) the interesting result that for values of the azimuthal scattering angle ϕ_s given by

$$\cos\phi_s = 1, \quad \cos\phi_s = \tfrac{1}{2}(\xi^2-1) \qquad (2.7)$$

the scattered Rayleigh surface wave vanishes identically, irrespective of the shape of the protuberance on the surface. Of course, this result had been obtained in the first Born approximation for the scattered wave, and is unlikely to persist in higher approximations. It has been obtained earlier by Polevoi[10].

B. Gratings

The propagation of surface acoustic waves across surfaces that are periodically corrugated in one direction (classical gratings) has been studied by several authors for reasons that will become apparent from the results to be described below.

If we assume an isotropic elastic medium occupies the region $x_3 > \zeta(x_1)$, where $\zeta(x_1)$ is a periodic function of x_1 with period a, and assume an elastic displacement field polarized in the x_1x_3-plane (sagittal polarization),

$$\vec{u}(\vec{x},t) = (u_1(x_1,x_3|\omega),\ 0,\ u_3(x_1x_3|\omega))e^{-i\omega t}, \qquad (2.8)$$

the time-independent equations of motion of the medium take the form

$$\left(\omega^2 + c_\ell^2 \frac{\partial^2}{\partial x_1^2} + c_t^2 \frac{\partial^2}{\partial x_3^2}\right) u_1 + (c_\ell^2 - c_t^2) \frac{\partial^2}{\partial x_1 \partial x_3} u_3 = 0 \qquad (2.9a)$$

$$(c_\ell^2 - c_t^2) \frac{\partial^2}{\partial x_1 \partial x_3} u_1 + \left(\omega^2 + c_\ell^2 \frac{\partial^2}{\partial x_3^2} + c_t^2 \frac{\partial^2}{\partial x_1^2}\right) u_3 = 0. \qquad (2.9b)$$

The stress-free boundary conditions on the surface $x_3 = \zeta(x_1)$ become

$$\left\{-\zeta'(x_1)\left[c_\ell^2 \frac{\partial u_1}{\partial x_1} + (c_\ell^2 - 2c_t^2) \frac{\partial u_3}{\partial x_3}\right] + c_t^2\left[\frac{\partial u_1}{\partial x_3} + \frac{\partial u_3}{\partial x_1}\right]\right\}_{x_3 = \zeta(x_1)} = 0$$

(2.10a)

$$\left\{-\zeta'(x_1) c_t^2\left[\frac{\partial u_1}{\partial x_3} + \frac{\partial u_3}{\partial x_1}\right] + \left[c_\ell^2 \frac{\partial u_3}{\partial x_3} + (c_\ell^2 - 2c_t^2) \frac{\partial u_1}{\partial x_1}\right]\right\}_{x_3 = \zeta(x_1)} = 0.$$

(2.10b)

If on the other hand, we assume the elastic displacement field is polarized perpendicular to the x_1x_3-plane (shear horizontal polarization),

$$\vec{u}(\vec{x},t) = (0,\ u_2(x_1x_3|\omega),\ 0)e^{-i\omega t}, \qquad (2.11)$$

the time independent equation of motion of the medium is

$$\left(\omega^2 + c_t^2 \frac{\partial^2}{\partial x_1^2} + c_t^2 \frac{\partial^2}{\partial x_3^2}\right) u_2 = 0. \qquad (2.12)$$

The stress-free boundary condition in this case is

$$\left[-\zeta'(x_1)\frac{\partial u_2}{\partial x_1}+\frac{\partial u_2}{\partial x_3}\right]_{x_3=\zeta(x_1)} = 0. \tag{2.13}$$

We consider each of these cases in turn.

i. <u>Sagittal</u> <u>Polarization</u>(11)

In the case of sagittal polarization we write $u_{1,3}(x_1x_3|\omega)$ in the region $x_3 > \zeta(x_1)_{max}$ in the forms

$$u_1(x_1x_3|\omega) = \sum_{m=-\infty}^{\infty} e^{ik_mx_1}\{A_m^{(\ell)}e^{-\alpha_\ell(k_m\omega)x_3} +$$

$$+ A_m^{(t)}e^{-\alpha_t(k_m\omega)x_3}\} \tag{2.14a}$$

$$u_3(x_1x_3|\omega) = \sum_{m=-\infty}^{\infty} e^{ik_mx_1}i\{\frac{\alpha_\ell(k_m\omega)}{k_m}A_m^{(\ell)}e^{-\alpha_\ell(k_m\omega)x_3} +$$

$$+ \frac{k_m}{\alpha_t(k_m\omega)}A_m^{(t)}e^{-\alpha_t(k_m\omega)x_3}\} , \tag{2.14b}$$

where $k_m = k+(2\pi m/a)$, with k the wave vector of the surface acoustic wave, and

$$\alpha_{\ell,t}(k_m\omega) = \left(k_m^2 - \frac{\omega^2}{c_{\ell,t}^2}\right)^{1/2} \quad k_m^2 > \frac{\omega^2}{c_{\ell,t}^2} \tag{2.15a}$$

$$= -i\left(\frac{\omega^2}{c_{\ell,t}^2} - k_m^2\right)^{1/2} \quad k_m^2 < \frac{\omega^2}{c_{\ell,t}^2}. \tag{2.15b}$$

These expressions satisfy the equations of motion, the boundary conditions at infinity, and the Bloch theorem. When Eqs. (2.14) are substituted into the boundary conditions (2.10), the following pair of coupled equations for $A_m^{(\ell,t)}$ is obtained:

$$\sum_{m=-\infty}^{\infty} \left\{ I_{p-m}(\alpha_{\ell m}) \frac{\omega^2 (k_p - k_m) + 2c_t^2 \alpha_{\ell m}^2 k_p}{c_t^2 k_m \alpha_{\ell m}} A_m^{(\ell)} + \right.$$

$$\left. + I_{p-m}(\alpha_{tm}) \frac{2c_t^2 k_p k_m - \omega^2}{c_t^2 \alpha_{tm}} A_m^{(t)} \right\} = 0 \qquad p = 0, \pm 1, \pm 2, \ldots$$

(2.16a)

$$\sum_{m=-\infty}^{\infty} \left\{ I_{p-m}(\alpha_{\ell m}) \frac{2c_t^2 k_p k_m - \omega^2}{c_t^2 k_m} A_m^{(\ell)} + \right.$$

$$\left. + I_{p-m}(\alpha_{tm}) \frac{c_t^2 (k_m^2 + \alpha_{tm}^2) k_p - k_m \omega^2}{c_t^2 \alpha_{tm}^2} A_m^{(t)} \right\} = 0 \qquad p = 0, \pm 1, \pm 2, \ldots$$

(2.16b)

where

$$I_n(\alpha) = \frac{1}{\alpha} \int_{-\frac{a}{2}}^{\frac{a}{2}} dx_1 e^{-i \frac{2\pi n}{a} x_1} e^{-\alpha \zeta(x_1)} . \qquad (2.17)$$

The dispersion curve for Rayleigh waves on a grating is obtained by equating to zero the determinant of the matrix of coefficients in Eq. (2.16). In actual calculations the matrix of coefficients is made to have a finite size by restricting the sums on m in Eqs. (2.14) to run from -N to +N, and increasing N until convergent results for the frequencies are obtained.

The results of the numerical calculations of the dispersion curves can be analyzed in terms of the flat-surface dispersion curves in both the extended- and reduced-zone schemes. In the extended-zone scheme of Fig. 1(a) the straight line $\omega = c_R k$ (shown with a solid and then a dotted segment) is the dispersion curve for a Rayleigh wave on a flat surface (c_R is the speed of such a wave), and the dashed lines are the lines $\omega = ct|k+(2\pi/a)m|$, where either m = 0 (the transverse sound line) or m = ±1, ±2,... . Since any point (ω,k) in the region above the dashed lines, $\omega > c_t|k+(2\pi/a)m|$, will correspond to an imaginary $\alpha_t(k_m\omega)$, the displacement fields at such points will have radiative components. Thus, a Rayleigh wave initially on a flat surface with its (ω,k) on the dotted part of the dispersion curve in Fig. 1(a) will begin to radiate into the bulk

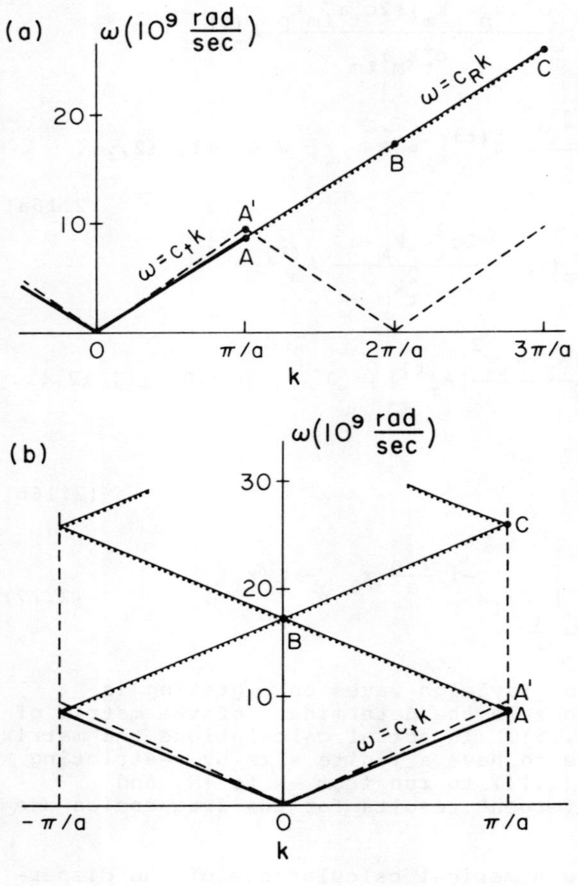

Fig. 1. The flat surface Rayleigh wave dispersion curve $\omega = c_R k$ in the nonradiative region (——) and in the radiative region (·····) defined by the boundary lines (- - -) $\omega = c_t|k+(2\pi/a)m|$, where $m = 0, \pm1, \pm2, \ldots$, where the grating period is $a = 1 \times 10^6$m, and where $c_\ell = 5 \times 10^3$m/sec and $c_t = 3 \times 10^3$m/sec. (a) The extended-zone and (b) the reduced-zone scheme.

at the moment the grating, however weak, is turned on. If a Rayleigh wave with (ω,k) on this radiative part of the dispersion curve is launched across the grating surface, it will decay with some characteristic lifetime or, equivalently, will acquire a finite mean free path. We will return to this point below in discussing the attenuation of Rayleigh waves on a grating. Gratings thus can act as transducers converting Rayleigh waves into bulk waves in a controlled manner. The Bloch-type surface waves that are true eigenmodes for the surface with a grating (i.e. possess an infinite lifetime) will thus be found only with (ω,k) beneath the dashed lines in

Fig. 1(a). In other words, they originate out of the solid line part of the flat surface Rayleigh wave dispersion curve as the grating is turned on.

The flat surface dispersion curve with its solid and dotted line segments is folded back into the domain $-\pi/a \leq k \leq \pi/a$ (the first Brillouin zone for the grating) in the reduced zone scheme of Fig. 1(b). The nonradiative region is now beneath the single dashed line $\omega = c_t|k|$. At points like A, B, C,... on the flat surface dispersion curve for Fig. 1(a) we have two degenerate modes separated in wave vector by a multiple of $2\pi/a$, i.e. by a reciprocal lattice vector of the grating. If the Fourier coefficients $\zeta(n)$ of the grating profile function that correspond to these reciprocal lattice vectors are nonzero, then the degeneracy of the modes at the points A, B, C,... will be lifted in the first order of perturbation theory in the surface profile function, and gaps will occur in the dispersion curves at these points in both the extended and reduced zone schemes.

From Eqs. (2.16) it is easy to show that this dispersion curve is an even function of k, and is periodic in k with period $2\pi/a$. Thus we confine our attention to the region $0 \leq k \leq \pi/a$, i.e. to half the one-dimensional first Brillouin zone defined by the periodicity of the grating.

The preceding qualitative results are what is observed in a typical dispersion curve, such as the one plotted in Fig. 2 for a sinusoidal surface profile function $\zeta(x_1) = \zeta_0 \cos 2\pi x_1/a$ with $\zeta_0/a = 0.3$. In the long wavelength limit the dispersion curve osculates the dispersion curve for Rayleigh waves on a flat surface. With increasing k the curve bends away from the latter curve, into the nonradiative region. Both its group and phase velocity decrease below c_R. This is the phenomenon of wave slowing that occurs for Rayleigh waves on a grating. Wave slowing occurs because in contrast with the dispersion curve for Rayleigh waves on a flat surface, which has a constant slope for all k, the dispersion curve for a Rayleigh wave on a grating is required to have zero slope at the Brillouin zone boundary $k = \pi/a$ due to its periodicity and continuity. This forces it to bend over into the nonradiative region of the (ω,k)-plane reducing its group velocity thereby. At the zone boundary a gap opens up in the dispersion curve, which defines a stop band for the propagation of Rayleigh waves on a grating. Knowledge of where this stop band occurs can be useful for the design of filters for Rayleigh waves, finally, there is a high frequency branch to the dispersion curve above the gap, that terminates when it intersects the transverse sound line.

We will see some of these features of the dispersion curve for Rayleigh waves on a grating reflected in the dispersion curve for Rayleigh waves on a randomly rough surface.

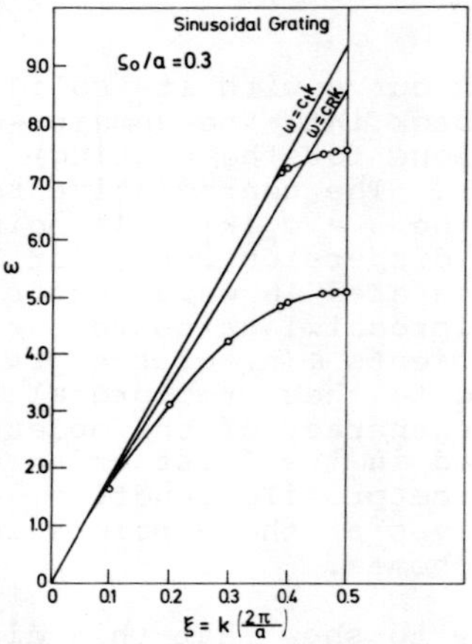

Fig. 2. Dispersion curve for Rayleigh waves propagating across a sinusoidal grating with corrugation strength $\zeta_o/a = 0.3$ on a medium characterized by $c_\ell = 5 \times 10^3$ m/sec and $c_t = 3 \times 10^3$ m/sec.

The attenuation of Rayleigh waves propagating across a grating can also be studied exactly by the same approach.[12] It is only necessary to make the wave vector k in Eqs. (2.14)-(2.15) complex, with ω real, $k = k_R + ik_I$, and to solve Eqs. (2.16) for k_R and ω in the radiative region of the (ω, k)-plane. Both k_R and k_I must be positive in order that Eqs. (2.14)-(2.15) describe a wave propagating in the x_1-direction and being attenuated in the direction of propagation. The attenuation length $\ell(\omega)$ for the Rayleigh wave, viz. the distance over which the energy in the wave decreases to 1/e of its initial value, is then given by $\ell^{-1}(\omega) = 2k_I(\omega)$.

When complex values of k are used in Eqs. (2.15) one must be careful in defining how the square roots are to be defined. It turns out that in order to describe waves that radiate energy into the interior of the solid ($\mathrm{Im}\alpha_{\ell,t}(k\omega) < 0$), propagate in the x_1-direction, and are attenuated in the direction of propagation, one should choose the branch cut in evaluating the square roots to lie along the positive imaginary axis, and work on the upper sheet of the corresponding Riemann surface[12]. However, with this choice of branch cut $\mathrm{Re}\alpha_{\ell,t}(k\omega) < 0$, i.e. the amplitude of the Rayleigh wave <u>grows</u> exponentially with increasing distance into the medium from the surface. This behavior of the displacement field is characteristic of leaky surface acoustic waves[13], and is intimately connected with the exponential growth of the amplitude of such an attenuating wave on the surface of the grating as one proceeds to $-\infty$ along the negative x_1-axis.

Results of such calculations are depicted in Figs. 3(a)

and 3(b), where we have plotted the attenuation coefficients

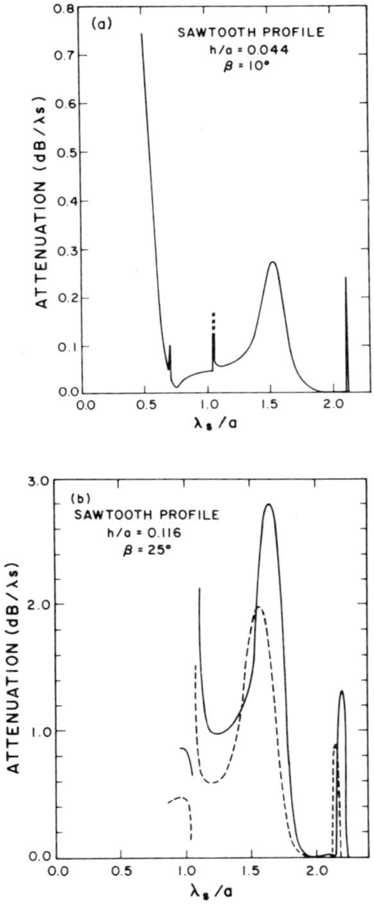

Fig. 3. Attenuation as a function of shear wavelength λ_s for Rayleigh waves on sawtooth gratings of corrugation strengths (a) h/a = 0.044 (c_t/c_ℓ = 0.3) and (b) h/a = 0.1166 [c_t/c_ℓ = 0.3 (---) and c_t/c_ℓ = 0.5 (——)].

for Rayleigh waves propagating across a symmetric sawtooth grating, defined by the surface profile function

$$\zeta(x_1) = \begin{cases} h + \frac{4h}{a} x_1 & -\frac{a}{2} \leq x_1 \leq 0 \\ h - \frac{4h}{a} x_1 & 0 \leq x_1 \leq \frac{a}{2} \end{cases} \tag{2.18}$$

ruled on an isotropic elastic medium, chosen to represent

aluminum. Two sets of theoretical curves are presented for the second grating, corresponding to the values $c_t/c_\ell = 0.3$ and 0.5, which span the range of values of this ratio displayed by aluminum, an anisotropic solid. These results can be compared with the experimental data of Rischbieter[14] for such gratings presented in Fig. 4, together with results obtained on the basis of first order perturbation theory in the surface profile function.[15,16] Good qualitative and quantitative agreement between the exact results and experiment is observed.

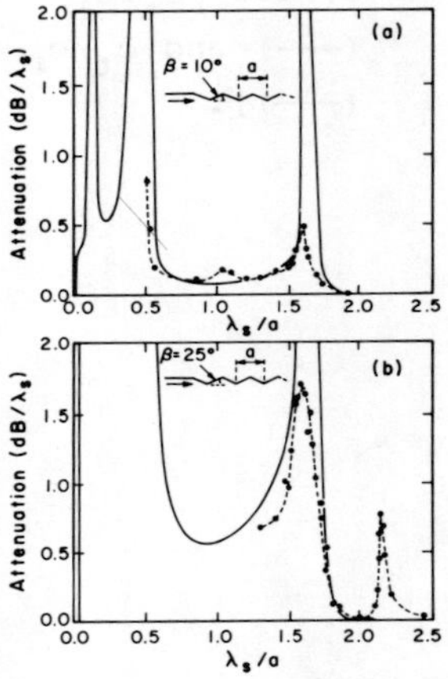

Fig. 4. Experimental (crosses) and theoretical (full line) results for the attenuation of Rayleigh waves propagating across a symmetric sawtooth profile ruled on an aluminum surface. λ and λ_s are the period of the grating and the wavelength of shear waves, respectively. (a) $h/a = 0.044$; (b) $h/a = 0.116$.

ii. <u>Shear Horizontal Polarization</u>[17]

In the case of shear horizontal polarization we write the solution of Eq. (2.12) that satisfies the boundary condition at infinity and the Bloch theorem in the form

$$u_2(x_1x_3|\omega) = \sum_{m=-\infty}^{\infty} e^{ik_mx_1} A_m e^{-\alpha_t(k_m\omega)x_3} \qquad x_3 > \zeta(x_1)_{max} \ . \tag{2.19}$$

Substitution of this expansion into the boundary condition (2.13) yields the following set of homogeneous linear equations for the coefficients $\{A_m\}$:

$$\sum_{m=-\infty}^{\infty} \frac{k_pk_m-(\omega^2/c_t^2)}{\alpha_t(k_m\omega)} I_{p-m}(\alpha_{tm})A_m = 0 \qquad p = 0, \pm 1, \pm 2, \ldots \tag{2.20}$$

The dispersion relation for shear horizontal surface acoustic waves on a grating is obtained by equating to zero the determinant of the coefficients in Eq. (2.20).

Dispersion curves for such waves are plotted in Fig. 5 for a sinusoidal grating defined by $\zeta(x_1) = \zeta_o \cos(2\pi x_1/a)$. We see that in the long wavelength limit these curves are tangent to the dispersion curve for bulk transverse waves for the same wave vector, $\omega = c_t k$ (the "sound line") (unlike the situation for sagittal polarization there are no surface wave solutions of Eqs. (2.12)-(2.13) for a planar surface, i.e. when $\zeta(x_1) \equiv 0$). As the grating strength ζ_o/a increases, the dispersion curves bend away from the sound line at progressively lower frequencies, or smaller wave vectors, and display the phenomenon of wave slowing. For the values of ζ_o/a indicated in Fig. 5 there is only a single branch to the dispersion curve. Thus the entire range of frequencies above the zone boundary frequency is a stop band for these waves.

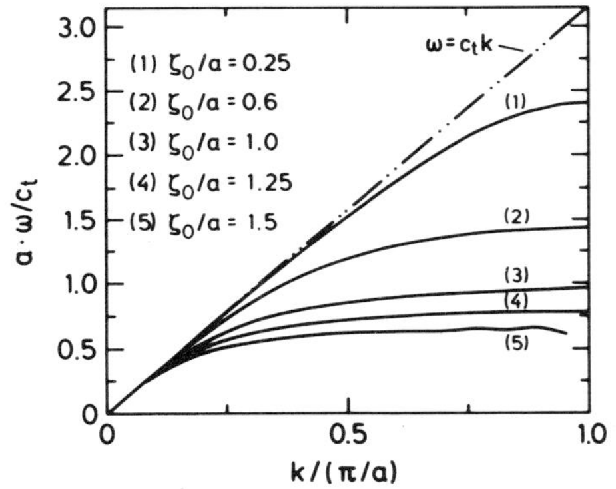

Fig. 5. The dispersion curves for shear horizontal surface acoustic waves propagating across a sinusoidal grating with a period of 500Å and $c_t = 3 \times 10^3$m/sec, for several values of ζ_o/a.

We have here an example of surface roughness giving rise to surface acoustic waves that cannot exist in the absence of surface roughness.

The determination of the attenuation of shear horizontal surface acoustic waves on a grating on the basis of Eq. (2.20) has not been carried out yet.

C. Randomly Rough Surfaces

In studying surface acoustic waves on a randomly rough surface the physical situation we consider is that of an isotropic elastic medium occupying the region $x_3 > \zeta(\vec{x}_\parallel)$ in contact with vacuum in the region $x_3 < \zeta(\vec{x}_\parallel)$. The surface x_3

$= \zeta(\vec{x}_\parallel)$ is stress-free. The surface profile function $\zeta(\vec{x}_\parallel)$ is assumed to be a stationary stochastic process. The probability distribution function of $\zeta(\vec{x}_\parallel)$ is not known in general, but for our purposes its first two moments suffice. They are given by

$$\langle\zeta(\vec{x}_\parallel)\rangle = 0 \tag{2.21a}$$

$$\langle\zeta(\vec{x}_\parallel)\zeta(\vec{x}_\parallel')\rangle = \delta^2 W(|\vec{x}_\parallel - \vec{x}_\parallel'|). \tag{2.21b}$$

The angular brackets in Eqs. (2.21) denote an average over the ensemble of realizations of the surface profile function. The first of these just tells us that the mean surface is the plane $x_3 = 0$. In the second $\delta^2 = \langle\zeta^2(\vec{x}_\parallel)\rangle$ is the mean square departure of the surface from flatness. The correlation function $W(|\vec{x}_\parallel|)$ will be assumed to have a Gaussian form in what follows, viz. $W(|\vec{x}_\parallel|) = \exp(-x_\parallel^2/a^2)$. The characteristic length a over which $W(|\vec{x}_\parallel|)$ decays to 1/e of its initial value of unity is called the <u>transverse correlation length</u> of the surface roughness. It is a measure of the distance between consecutive peaks and valleys on the surface.

Since the surface profile function is time-independent, we can work with a single frequency Fourier component of the displacement field in the medium, and we write the latter in the form $\vec{u}(\vec{x};t) = \vec{u}(\vec{x}|\omega)\exp(-i\omega t)$. The equations of motion satisfied by $\vec{u}(\vec{x}|\omega)$ are

$$\sum_\mu \left(\omega^2\delta_{\alpha\mu} + \frac{1}{\rho}\sum_{\beta\nu} C_{\alpha\beta\mu\nu}\frac{\partial^2}{\partial x_\beta \partial x_\nu}\right) u_\mu(\vec{x}|\omega) = 0,$$

$$\alpha = 1, 2, 3;\; x_3 > \zeta(\vec{x}_\parallel), \tag{2.22}$$

where the elastic modulus tensor has the form given by Eq. (2.2). The stress-free boundary conditions at the surface $x_3 = \zeta(\vec{x}_\parallel)$ are

$$\sum_\beta T_{\alpha\beta}(\vec{x}|\omega)\hat{n}_\beta\Big|_{x_3=\zeta(\vec{x}_\parallel)} = 0 \quad \alpha = 1, 2, 3, \tag{2.23}$$

where $\hat{n}$ is the unit vector normal to the surface at each point,

$$\hat{n} = [1 + (\nabla\zeta(\vec{x}_\parallel))^2]^{-1/2}\left(-\frac{\partial\zeta(\vec{x}_\parallel)}{\partial x_1}, -\frac{\partial\zeta(\vec{x}_\parallel)}{\partial x_2}, 1\right). \tag{2.24}$$

The stress tensor $T_{\alpha\beta}(\vec{x}|\omega)$ for an isotropic elastic medium is given by

$$T_{\alpha\beta}(\vec{x}|\omega) = \rho(c_\ell^2 - 2c_t^2)\delta_{\alpha\beta}\nabla\cdot\vec{u}(\vec{x}|\omega) + + \rho c_t^2\left(\frac{\partial u_\alpha(\vec{x}|\omega)}{\partial x_\beta} + \frac{\partial u_\beta(\vec{x}|\omega)}{\partial x_\alpha}\right). \tag{2.25}$$

The solution of the equations of motion (2.22) in the region $x_3 > \zeta(\vec{x}_\parallel)_{max}$ that vanish as $x_3 \to \infty$ or satisfy a radiation condition in that limit is

$$\vec{u}(\vec{x}|\omega) = \int\frac{d^2q_\parallel}{(2\pi)^2} e^{i\vec{q}_\parallel\cdot\vec{x}_\parallel}\Big\{(\hat{q}_\parallel + i\hat{x}_3\frac{\alpha_\ell(q_\parallel\omega)}{q_\parallel}A_1(\vec{q}_\parallel\omega) \times$$

$$\times e^{-\alpha_\ell(q_\parallel\omega)x_3} + (\hat{q}_\parallel \times \hat{x}_\parallel)_3 A_2(\vec{q}_\parallel\omega)e^{-\alpha_t(q_\parallel\omega)x_3} +$$

$$+ \left(\hat{q}_\parallel + i\hat{x}_3\frac{q_\parallel}{\alpha_t(q_\parallel\omega)}\right)A_3(\vec{q}_\parallel\omega)e^{-\alpha_t(q_\parallel\omega)x_3}\Big\}, \tag{2.26}$$

where

$$\alpha_{\ell,t}(q_\parallel\omega) = \left(q_\parallel^2 - \frac{\omega^2}{c_{\ell,t}^2}\right)^{1/2} \qquad q_\parallel^2 > \frac{\omega^2}{c_{\ell,t}^2} \tag{2.27a}$$

$$= -i\left(\frac{\omega^2}{c_{\ell,t}^2} - q_\parallel^2\right)^{1/2} \qquad q_\parallel^2 < \frac{\omega^2}{c_{\ell,t}^2}. \tag{2.27b}$$

When Eq. (2.26) is substituted into the boundary conditions (2.23)-(2.25), the equations for the amplitudes $\{A_\alpha(\vec{q}_\parallel\omega)\}$ can be put in the form[8,18]

$$\sum_\beta M_{\alpha\beta}^{(o)}(q_\parallel|\omega)A_\beta(\vec{q}_\parallel\omega)$$

$$= \sum_\beta\int\frac{d^2k_\parallel}{(2\pi)^2}\hat{\zeta}(\vec{q}_\parallel - \vec{k}_\parallel)M_{\alpha\beta}^{(1)}(\vec{q}_\parallel;\vec{k}_\parallel|\omega)A_\beta(\vec{k}_\parallel\omega), \tag{2.28}$$

in the small roughness limit, where

$$\hat{\zeta}(\vec{Q}_{\parallel}) = \int d^2x_{\parallel}\, e^{-i\vec{Q}_{\parallel}\cdot\vec{x}_{\parallel}}\, \zeta(\vec{x}_{\parallel}). \tag{2.29}$$

The matrix $M^{(1)}_{\alpha\beta}(\vec{q}_{\parallel};\vec{k}_{\parallel}|\omega)$ depends on $\vec{q}_{\parallel}$ and $\vec{k}_{\parallel}$ only through their magnitudes and the cosine and sine of the angle between them.

Equation (2.28) is a stochastic, matrix integral equation because of the presence of $\hat{\zeta}(\vec{q}_{\parallel}-\vec{k}_{\parallel})$ in its kernel. The amplitude $A_{\alpha}(\vec{q}_{\parallel}\omega)$ is therefore a stochastic quantity also. The problem of obtaining the complete probability distribution function of $A_{\alpha}(\vec{q}_{\parallel}\omega)$ is a formidable one. Fortunately, it is not necessary for us to know it. The first moment of this distribution function, $\langle A_{\alpha}(\vec{q}_{\parallel}\omega)\rangle$, suffices for our purposes since it describes the mean wave propagating along the surface.

To obtain from Eq. (2.28) the equations satisfied by $\langle A_{\alpha}(\vec{q}_{\parallel}\omega)\rangle$ we proceed as follows. To the equation written in operator form,

$$(H_o - V)A = 0, \tag{2.30}$$

where H_o is a nonstochastic operator while V is stochastic, let us apply the operators P and Q in turn, where P is the smoothing operator that projects out the average of anything it operates on, $Pf = \langle f\rangle$, while $Q = 1-P$. The results are

$$H_oPA - PV(PA+QA) = 0 \tag{2.31a}$$

$$H_oQA - QV(PA+QA) = 0. \tag{2.31b}$$

When QA is eliminated from this pair of equations the resulting equation for $\langle A\rangle$ takes the form

$$(H_o - \langle M\rangle)\langle A\rangle = 0, \tag{2.32}$$

where

$$\langle M\rangle = \langle (I-VH_o^{-1}Q)^{-1}V\rangle. \tag{2.33}$$

If, as in the present case, $\langle V\rangle = 0$ (see Eq. (2.21a)), then to lowest nonvanishing order in V, $\langle M\rangle = \langle VH_o^{-1}V\rangle$.

Applied to Eq. (2.28) the preceding results have the

consequence that the equation for $\langle A_\alpha(\vec{q}_\parallel \omega)\rangle$ in the small roughness limit has the form

$$\sum_\beta \{M^{(o)}_{\alpha\beta}(q_\parallel \omega) - \Delta M_{\alpha\beta}(q_\parallel|\omega)\}\langle A_\beta(\vec{q}_\parallel|\omega)\rangle = 0, \tag{2.34}$$

where

$$\Delta M_{\alpha\beta}(q_\parallel|\omega) = \delta^2 \int \frac{d^2k_\parallel}{(2\pi)^2} g(|\vec{q}_\parallel - \vec{k}_\parallel|) \times$$

$$\times \sum_{\mu\nu} M^{(1)}_{\alpha\mu}(\vec{q}_\parallel;\vec{k}_\parallel|\omega) M^{(o)-1}_{\mu\nu}(k_\parallel|\omega) M^{(1)}_{\nu\beta}(\vec{k}_\parallel;\vec{q}_\parallel|\omega), \tag{2.35a}$$

with

$$g(Q_\parallel) = \int d^2x_\parallel e^{-i\vec{Q}_\parallel \cdot \vec{x}_\parallel} W(|\vec{x}_\parallel|). \tag{2.35b}$$

The fact that Eq. (2.34) is a matrix equation rather than a matrix integral equation, as was the starting equation, Eq. (2.28), and the fact that $\Delta M_{\alpha\beta}(q_\parallel|\omega)$ depends on $\vec{q}_\parallel$ only through its magnitude, are consequences of the restoration of inifinitesimal translational invariance and of isotropy in the plane $x_3 = 0$ to our system as a consequence of the averaging process. They follow from the form of Eq. (2.21b). In particular, the restoration of isotropy in the plane $x_3 = 0$ has the consequance that Eqs. (2.34) break up into two sets of equations of the form

$$\begin{pmatrix} m_{11}(q_\parallel|\omega) & m_{13}(q_\parallel|\omega) \\ m_{31}(q_\parallel|\omega) & m_{33}(q_\parallel|\omega) \end{pmatrix} \begin{pmatrix} \langle A_1(\vec{q}_\parallel|\omega)\rangle \\ \langle A_3(\vec{q}_\parallel|\omega)\rangle \end{pmatrix} = 0 \tag{2.36a}$$

and

$$m_{22}(q_\parallel|\omega) \ \langle A_2(q_\parallel|\omega)\rangle = 0. \tag{2.36b}$$

The first of these is associated with the propagation of a surface acoustic wave polarized in the sagittal plane; the second with a surface acoustic wave polarized perpendicular to the sagittal plane. The dispersion relation for each of these two types of surface acoustic waves are obtained by equating to zero the determinant of the matrix of coefficients.

i. Sagittal Polarization

The frequency of a Rayleigh wave propagating across a randomly rough surface that is obtained from Eq. (2.36a) can be written in the form[8]

$$\omega(q_\parallel) = c_R q_\parallel\left[1 + \frac{\delta^2}{a^2}\omega_1(aq_\parallel) - i\frac{\delta^2}{a^2}\omega_2(aq_\parallel)\right]. \quad (2.37)$$

The functions $\omega_1(aq_\parallel)$ and $\omega_2(aq_\parallel)$ are plotted in Figs. 6 and 7, respectively.[8] $\omega(q_\parallel)$ is complex because as the Rayleigh

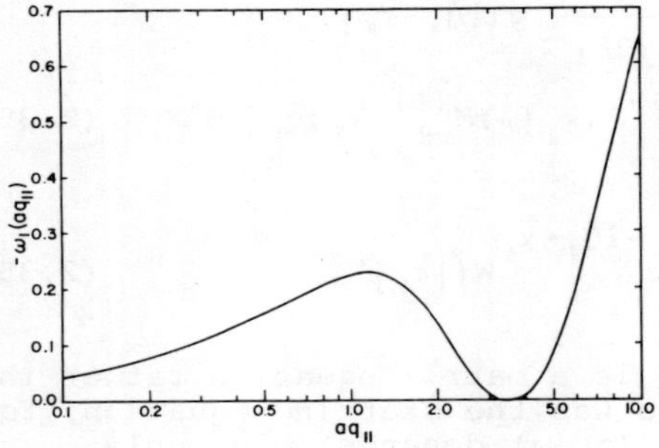

Fig. 6. The shift in the frequency of a Rayleigh wave due to surface roughness as a function of $aq_\parallel$. The actual shift is given by $c_R q_\parallel(\delta^2/a^2)\omega_1(aq_\parallel)$.

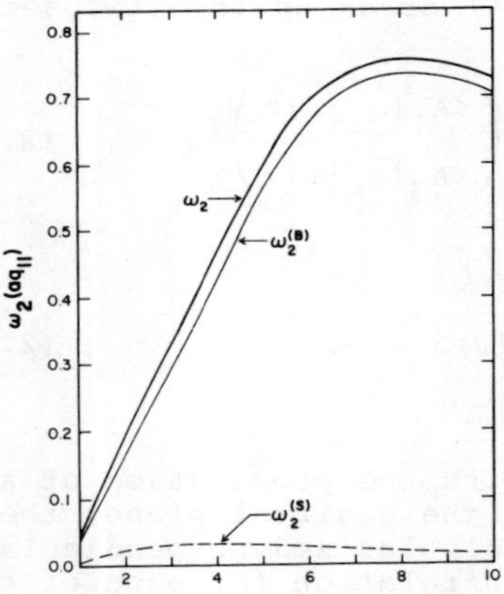

Fig. 7. The dimensionless imaginary part $\omega_2(aq_\parallel)$ of the frequency of a Rayleigh wave on a randomly rough surface as a function of $aq_\parallel$. The figure shows that the contribution to $\omega_2(aq_\parallel)$ from decay of the Rayleigh wave into bulk modes ($\omega_2^{(B)}$) dominates that due to decay into other Rayleigh waves ($\omega_2^{(s)}$).

wave propagates along the rough surface it is scattered elastically by the hills and valleys on the surface into other Rayleigh waves and into bulk waves. Energy is removed from the incident beam thereby, and the wave is attenuated. It acquires a finite lifetime that is given by $\tau(q_\parallel) = (2\mathrm{Im}\,\omega(q_\parallel))^{-1}$. It is found that the dominant attenuation mechanism is the scattering into bulk waves. This is shown explicitly in Fig. 7. The situation here is similar to the attenuation of a Rayleigh wave on a grating by the grating induced conversion of the surface wave into bulk waves. The chief difference between the two cases is that unlike the situation for a grating, there is no threshold in frequency or wave vector for attenuation to begin in the case of a randomly rough surface. This is because the latter can be viewed as a superposition of gratings with different amplitudes, periods, and orientations. In particular there are contributions to the profile from gratings with such large periods that their Brillouin zones shrink to zero in width, and the attenuation commences already at $q_\parallel = 0$. The roughness-induced shift in the frequency of the Rayleigh wave is negative. This is analogous to the wave slowing that occurs on a grating. It is seen from Fig. 6 that when $aq_\parallel \approx 1$, i.e. when the wavelength of the Rayleigh wave becomes comparable to the transverse correlation length of the roughness, there is a kind of geometric resonance in the interaction of the wave with the roughness that leads to a maximum in $\omega_1(aq_\parallel)$.

If we choose the values $\delta/a = 0.3$, $\omega = 10^8\ \text{sec}^{-1}$, and $c_R = 3 \times 10^5$cm/sec, the relative downward shift in the frequency of the Rayleigh wave $(\delta^2/a^2)\omega_1(aq_\parallel)$ is 0.4%, 2%, and 5.8%, for $aq_\parallel = 0.1$, 1, and 10, respectively. The corresponding mean free paths of the Rayleigh wave are 2.6×10^3cm, 0.42cm, and 0.02cm, respectively.

ii. Shear Horizontal Polarization

The frequency of a surface acoustic wave on a randomly rough surface that is polarized perpendicular to the sagittal plane is obtained from Eq. (2.36b) in the form[19]

$$\omega(q_\parallel) = c_t q_\parallel - \frac{\delta^4}{a^4}\left(\omega_1(aq_\parallel) + i\omega_2(aq_\parallel)\right). \tag{2.38}$$

The real part of this frequency is plotted in Fig. 8. There is no surface wave of this polarization in the absence of roughness, so the real part of the frequency approaches $c_t q_\parallel$ as $q_\parallel = 0$, in which limit it becomes a surface skimming bulk transverse wave. For larger values of $q_\parallel$ the dispersion curve bends over into the nonradiative region $\omega < c_t|q_\parallel|$ and exhibits wave slowing. The imaginary part of the frequency is due to the scattering of the wave into bulk waves. Because $\omega(q_\parallel)$ departs from $c_t q_\parallel$ only by terms of $O(\delta^4/a^4)$ in this case, it is a very weakly bound surface wave, i.e. it penetrates very deeply into the medium.

The wave discussed here is another example of a wave

bound to a surface by roughness that has no counterpart on a flat surface.

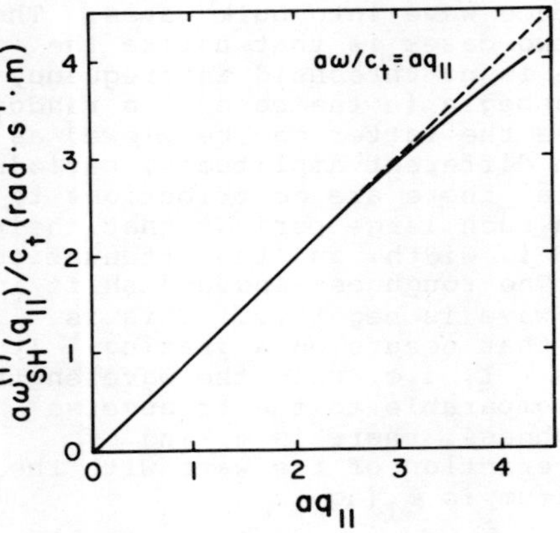

Fig. 8. The real part of the frequency of a shear horizontal surface acoustic wave on a randomly rough surface as a function of $aq_{\|}$.

3. <u>Surface Excitations on Dielectric Media</u>

A wide variety of electrostatic and electromagnetic excitations localized at dielectric surfaces has been studied theoretically and experimentally by now. In this section we survey some of the results obtained.

The physical system considered in each case is a dielectric medium characterized by an isotropic, frequency-dependent, dielectric constant $\varepsilon(\omega)$ in the region $x_3 < \zeta(\vec{x}_{\|})$, and vacuum in the region $x_3 > \zeta(\vec{x}_{\|})$. In order for surface excitations to exist in this system, the dielectric constant must be negative. In all cases we will assume it to have the simple free electron metal form $\varepsilon(\omega) = 1-(\omega_p^2/\omega^2)$, where ω_p is the plasma frequency of the electrons in the bulk of the metal. However, the formal results obtained do not depend on the specific form of $\varepsilon(\omega)$, although the numerical results do.

If one works in the unretarded limit, and writes the electrostatic scalar potential in the vacuum region in the form

$$\phi^{>}(\vec{x}|\omega) = \int \frac{d^2k_{\|}}{(2\pi)^2} A(\vec{k}_{\|}\omega)\, e^{i\vec{k}_{\|}\cdot\vec{x}_{\|} - k_{\|}x_3} \qquad x_3 > \zeta(\vec{x}_{\|})_{max}\ , \tag{3.1}$$

where a time dependence form $\exp(-i\omega t)$ has been assumed, then the use of the Rayleigh hypothesis, Green's theorem, and the extinction theorem results in the following homogeneous integral equation for $A(\vec{k}_{\parallel}\omega)$ [20]

$$\frac{\varepsilon(\omega)+1}{\varepsilon(\omega)-1} A(\vec{q}_{\parallel}\omega) = \int \frac{d^2p_{\parallel}}{(2\pi)^2} J(q_{\parallel}-p_{\parallel}|\vec{q}_{\parallel}-\vec{p}_{\parallel})(1 - \hat{q}_{\parallel}\cdot\hat{p}_{\parallel})p_{\parallel}A(\vec{p}_{\parallel}\omega), \tag{3.2}$$

where

$$J(\alpha|\vec{Q}_{\parallel}) = \int d^2x_{\parallel}\, e^{-i\vec{Q}_{\parallel}\cdot\vec{x}_{\parallel}} \frac{e^{\alpha\zeta(\vec{x}_{\parallel})}-1}{\alpha}. \tag{3.3}$$

The solvability condition for this equation yields the frequencies of electrostatic modes localized in the vicinity of the surface $x_3 = \zeta(\vec{x}_{\parallel})$.

In the case of a one-dimensional surface profile $\zeta(x_1)$, the corresponding equations are [21]

$$\frac{\varepsilon(\omega)+1}{\varepsilon(\omega)-1} A(q\omega) = \int \frac{dp}{2\pi} J(|q|-|p||q-p) \times$$

$$\times\ (1- \mathrm{sgn}q\,\mathrm{sgn}p)|p|A(p\omega), \tag{3.4}$$

with

$$J(\alpha|Q) = \int dx_1 e^{-iQx_1} \frac{e^{\alpha\zeta(x_1)}-1}{\alpha}. \tag{3.5}$$

In electromagnetic theory, we can write the solution of Maxwell's equations for the electric field in the vacuum region that satisfies the boundary conditions at infinity in the form

$$\vec{E}(\vec{x}|\omega) = \int \frac{d^2q_{\parallel}}{(2\pi)^2} \{\frac{c}{\omega}[i\hat{q}_{\parallel}\beta_o(q_{\parallel}\omega) - \hat{x}_3q_{\parallel}]A_{\parallel}(\vec{q}_{\parallel}\omega) +$$

$$+ (\hat{x}_3\times\hat{q}_{\parallel})_3 A_{\perp}(\vec{q}_{\parallel}\omega)\} e^{i\vec{q}_{\parallel}\cdot\vec{x}_{\parallel}-\beta_o(q_{\parallel}\omega)x_3} \qquad x_3 > \zeta(\vec{x}_{\parallel}), \tag{3.6}$$

where

$$\beta_o(q_{\parallel}\omega) = (q_{\parallel}^2 - \frac{\omega^2}{c^2})^{1/2} \qquad q_{\parallel}^2 > \frac{\omega^2}{c^2} \tag{3.7a}$$

$$= -\, i(\frac{\omega^2}{c^2} - q_{\parallel}^2)^{1/2} \qquad q_{\parallel}^2 < \frac{\omega^2}{c^2}. \tag{3.7b}$$

$A_{\parallel,\perp}(\vec{q}_\parallel\omega)$ are the amplitudes of the p- and s-polarized components of the field, respectively. The use of the Rayleigh hypothesis, the vectorial equivalent of the Kirchhoff integral, and the extinction theorem yields the following pair of coupled equations for these amplitudes[21]

$$\begin{pmatrix} \frac{\varepsilon(\omega)\beta_o(q_\parallel\omega)+\beta(q_\parallel\omega)}{\varepsilon(\omega)-1} & 0 \\ 0 & \frac{\beta_o(q_\parallel\omega)+\beta(q_\parallel\omega)}{\varepsilon(\omega)-1} \end{pmatrix} \begin{pmatrix} A_\parallel(\vec{q}_\parallel\omega) \\ A_\perp(\vec{q}_\parallel\omega) \end{pmatrix}$$

$$= \int \frac{d^2p_\parallel}{(2\pi)^2} J(\beta(q_\parallel\omega) - \beta_o(p_\parallel\omega)|\vec{q}_\parallel - \vec{p}_\parallel) \times$$

$$\times \begin{pmatrix} q_\parallel p_\parallel - \beta(q_\parallel\omega)(\hat{q}_\parallel\cdot\hat{p}_\parallel)\beta_o(p_\parallel\omega) & -i\frac{\omega}{c}\beta(q_\parallel\omega)(\hat{q}_\parallel\times\hat{p}_\parallel)_3 \\ i\frac{\omega}{c}(\hat{q}_\parallel\times\hat{p}_\parallel)_3\beta_o(p_\parallel\omega) & \frac{\omega^2}{c^2}(\hat{q}_\parallel\cdot\hat{p}_\parallel) \end{pmatrix} \begin{pmatrix} A_\parallel(\vec{p}_\parallel\omega) \\ A_\perp(\vec{p}_\parallel\omega) \end{pmatrix} \tag{3.8}$$

where

$$\beta(q_\parallel\omega) = \left(q_\parallel^2 - \varepsilon(\omega)\frac{\omega^2}{c^2}\right)^{1/2} \quad \mathrm{Re}\beta(q_\parallel\omega) > 0 \quad \mathrm{Im}\beta(q_\parallel\omega) < 0. \tag{3.9}$$

The solvability condition for this pair of equations yields the frequencies of electromagnetic excitations localized to the surface $x_3 = \zeta(\vec{x}_\parallel)$.

In the case of a one-dimensional surface profile $\zeta(x_1)$ it is more convenient to work with the single nonzero component of the p-polarized magnetic field in the vacuum. Its Fourier coefficient satisfies the homogeneous integral equation[21].

$$\frac{\varepsilon(\omega)\beta_o(q\omega)+\beta(q\omega)}{\varepsilon(\omega)-1} A(q\omega) = \int \frac{dp}{2\pi} J(\beta(q\omega)-\beta_o(p\omega)|q-p) \times$$

$$\times [qp - \beta(q\omega)\beta_o(p\omega)] A(p\omega). \tag{3.10}$$

We now proceed to apply these equations to the cases of surface shape resonances, surface waves on gratings, and surface waves on randomly rough surfaces.

3.A. Surface Shape Resonances

Electrostatic and electromagnetic surface shape resonances are solutions of Laplace's and Maxwell's equations, respectively, that are spatially localized in the vicinity of a protuberance or indentation on the otherwise planar surface of a dielectric medium in contact with vacuum. The frequencies of these modes, and the field distributions associated with them, depend on the shape of the protuberance or indentation.(22-29)

Such surface shape resonances are of interest because their excitation by an electromagnetic field incident on the surface of the dielectric from the vacuum side leads to a significant enhancement of the total electromagnetic field in the vicinity of the surface perturbation that gives rise to them. This in turn gives rise to a significant enhancement of a variety of physical processes associated with solid surfaces, e.g. surface enhanced Raman scattering from molecules adsorbed on solid surfaces(30), and the enhancement of second harmonic generation in the reflection of light from a metal surface(31). At the same time, certain kinds of rough surfaces can be viewed as consisting of a random array of indentical protuberances or indentations on a planar surface. An understanding of the resonances associated with an isolated protuberance should aid in understanding the surface excitations of such rough surfaces.

Equations (3.2) and (3.8) have been solved in the case that the surface profile function $\zeta(\vec{x}_\parallel)$ has the Gaussian form $\zeta(\vec{x}_\parallel) = A\exp(-x_\parallel^2/R^2)$.(29) When $A > 0$ this profile function describes a protuberance on the surface $x_3 = 0$; when $A < 0$ it describes an indentation. In both the electrostatic and electromagnetic calculations only the frequencies of the totally cylindrically symmetric modes were studied. In Fig. 9 we have plotted the frequencies of the three lowest frequency (for $A > 0$) electrostatic surface shape resonances as functions of A/R. There is in fact an infinity of such modes with frequencies that approach the frequency $\omega_p/\sqrt{2}$ of surface plasmons on a flat surface with increasing mode index. From this figure we see that surface shape resonances associated with a protuberance on the surface have frequencies that are lower than the frequency of surface plasmons on a flat surface. The opposite is the case for surface shape resonances associated with an indentation.

When the effects of retardation are included in the theory of surface shape resonances, a feature is introduced into the results that is absent in the electrostatic approximation, viz. the damping of these localized modes by radiation damping. As a result, the frequencies of surface shape resonances become complex, $\omega = \omega_R - i\omega_I$ $(\omega_R, \omega_I > 0)$, with the

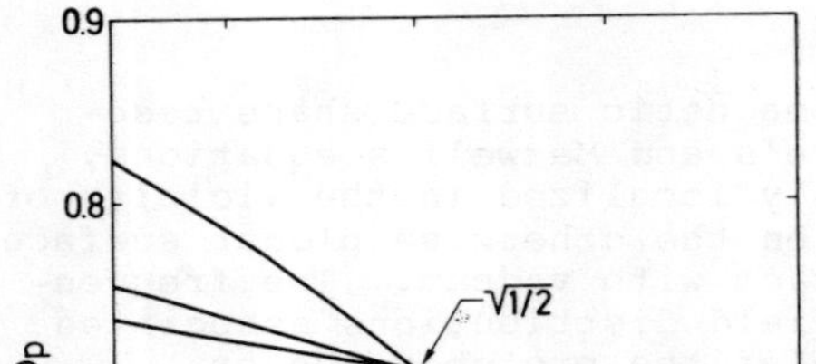

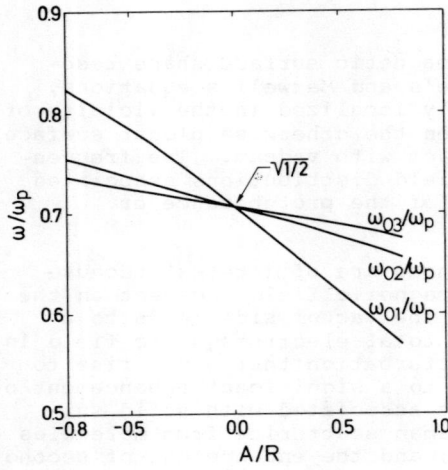

Fig. 9. The frequencies of the first three electrostatic surface shape resonances as functions of A/R, for the Gaussian profile $\zeta(\vec{x}_\parallel) = A\exp(-x_\parallel^2/R^2)$.

imaginary part of the frequency a measure of the finite lifetime of the mode caused by the radiation of energy into the vacuum above the surface perturbation. In Fig. 10 are plotted the real parts of the frequencies of the p-polarized electromagnetic surface shape resonances that correspond to the two

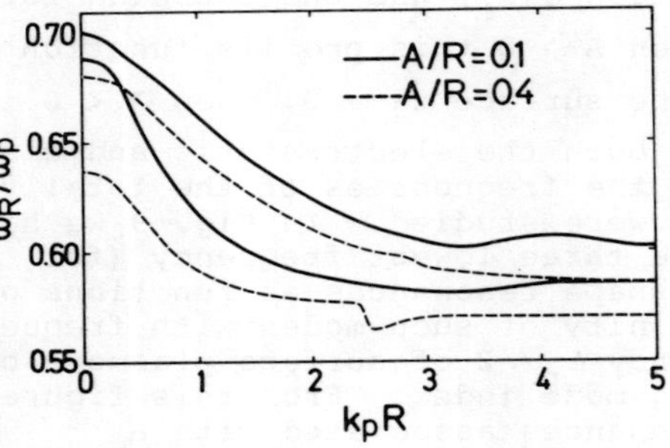

Fig. 10. The real parts of the frequencies of the first two electromagnetic surface shape resonances, as functions of k_pR, where $k_p = \omega_p/c$, for the Gaussian profile $\zeta(\vec{x}_\parallel) = A\exp(-x_\parallel^2/R^2)$. The frequencies have been calculated for A/R = 0.1 and 0.4.

lowest frequency modes in the electrostatic case depicted in Fig. 9, for two protuberances, with A/R = 0.1 and 0.4, as functions of k_pR, where $k_p = \omega_p/c$. For silver, $h\omega_p$ = 3.78eV, and k_pR = 2 if R = 104nm. The frequencies are substantially lowered as the scale of the protuberance increases relative to the wavelength of the radiation.

The imaginary parts of these frequencies are small. For example, for the mode with $\omega_R/\omega_p = 0.58$ at $k_pR = 2(A/R = 0.4)$, $\omega_I/\omega_p = 4.52 \times 10^{-5}$. The corresponding mode for the protuberance with $A/R = 0.1$ at $k_pR = 2$ with $\omega_R/\omega_p = 0.59$ has $\omega_I/\omega_p = 1.36 \times 10^{-5}$. Thus ω_I increases with increasing A/R as would be expected on physical grounds.

3.B. Gratings

Grating surfaces, whether periodic in one direction (classical gratings) or in two noncollinear directions (bigratings), constitute an important category of rough surfaces. They are of interest for several reasons. They are widely used in optical systems. They provide an experimental method for coupling volume electromagnetic waves into surface polaritons that can then be used in devices. This same coupling makes it possible to measure the dispersion curves of surface polaritons from the dips in the reflectivity from a grating that occur when the conditions for exciting surface polaritons through the grating are satisfied. It is then of interest to know how the dispersion curves of a surface polariton is affected by the very grating that makes it observable. Finally, gratings are rough surfaces whose profiles are known. They therefore provide an excellent testing ground for the predictions of theories of rough surface phenomena.

The propagation of electromagnetic surface waves across grating surfaces has been studied with the neglect of retardation[34,35] and with its inclusion.[36-38] For the sake of brevity we will consider here only results based on electromagnetic theory with the inclusion of the effects of retardation. This further restricts us to the case of classical gratings, because the propagation of surface polaritons across bigratings has not been studied theoretically or experimentally up to the present time.

The starting point for obtaining the dispersion relation for surface polaritons propagating in the x_1-direction across a grating surface defined by the equation $x_3 = \zeta(x_1)$ is Eq. (3.10). It is now assumed that $\zeta(x_1)$ is a periodic function of x_1 with a period a. We make the following Ansatz for the amplitude $A(q\omega)$:

$$A(q\omega) = \sum_{m=-\infty}^{\infty} A_m(k\omega)\, 2\pi\delta(q-k_m) \tag{3.11}$$

where

$$k_m = k + \frac{2\pi m}{a}, \tag{3.12}$$

and k is the wave vector of the surface polariton. The form (3.11) is required in order that the corresponding electromagnetic field satisfy the Bloch condition that is a consequence of the grating structure. When Eq. (3.11) is substituted into Eq. (3.10) the following equation for the expansion coefficients $\{A_m(k\omega)\}$ is obtained:

$$\sum_{n=-\infty}^{\infty} \frac{k_m k_n - \beta(k_m\omega)\beta_o(k_n\omega)}{\beta(k_m\omega) - \beta_o(k_n\omega)} \frac{1}{a}\int_0^a dx_1\, e^{-i\frac{2\pi}{a}(m-n)x_1} \times$$

$$\times\, e^{(\beta(k_m\omega)-\beta_o(k_n\omega))\zeta(x_1)} A_n(k\omega) = 0 \qquad m = 0, \pm1, \pm2, \ldots \tag{3.13}$$

The dispersion relation sought is obtained by equating to zero the determinant of the matrix of coefficients in this equation. In practice the infinite set of equations (3.13) is made finite by restricting m and n in it to run from -N to +N, and increasing N until the frequencies obtained converge.

One can show on the basis of Eq. (3.13) that the frequencies $\omega(k)$ that are the solutions of this equation are even functions of k and are also periodic functions of k with a period given by $2\pi/a$. We can therefore confine k to the interval $(0, \pi/a)$ to obtain the distinct solutions.

What should we expect these solutions to look like? To answer this question in the upper half of Fig. 11 we have plotted the dispersion curve of a surface polariton on the planar surface of a metal described by a free electron dielectric constant. Also indicated on this figure are the boundaries of the first, second, third,... Brillouin zones of the grating structure. In addition, we have drawn the light lines $\omega = \pm ck$, and the light lines displaced by integer multiples of $(2\pi/a)$. The significance of these light lines is that in the region of the (ω,k)-plane above them, $\beta_o(k_m\omega)$ is pure imaginary for some m, and the solutions of Eq. (3.13) correspond to radiative modes rather than to the nonradiative surface polaritons. Two portions of the dispersion curve that fall into the radiative region are indicated by being hatched. It is only the portions of the dispersion curve that fall below the light lines and the displaced light lines that can belong to the dispersion curve of a surface polariton on a grating.

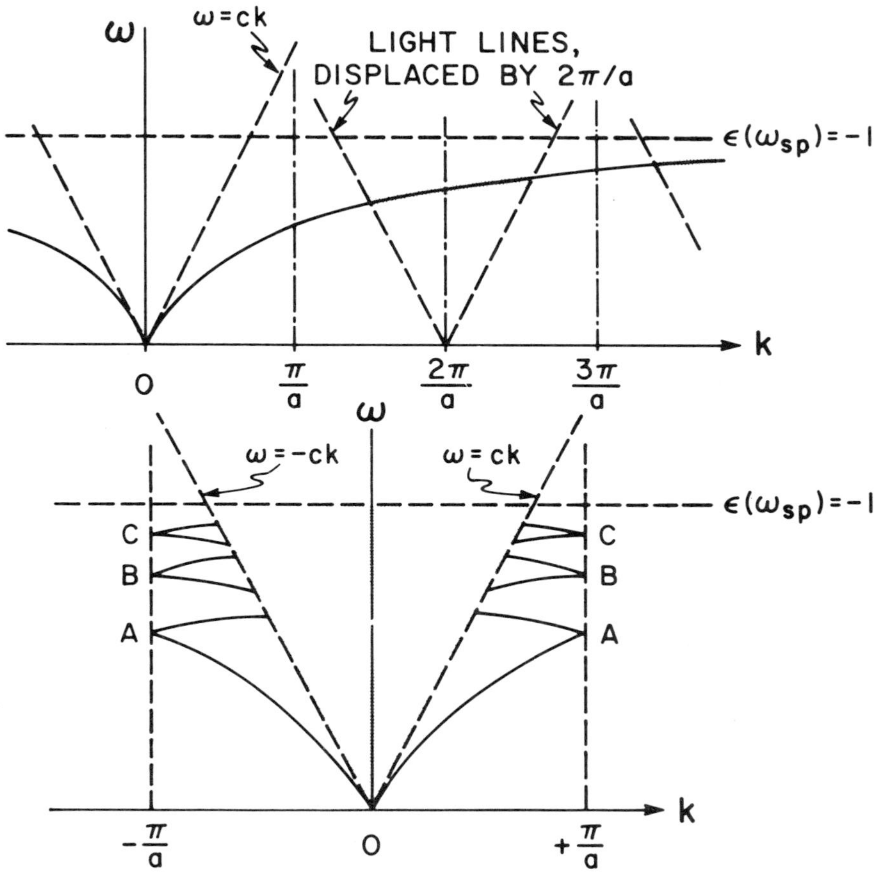

Fig. 11. (a) The dispersion curve for surface polaritons on a flat metal surface in the extended-zone scheme. The hatched portions of the curve become unstable with respect to radiation into the vacuum above the metal when the periodic grating is introduced. (b) The nonradiative portions of the flat surface dispersion curve have been translated back into the first Brillouin zone, to form the reduced-zone scheme appropriate to the present problem.

In the lower part of Fig. 11 we have shown the result of bringing the portions of the dispersion curve for surface polaritons on a planar surface that lie outside the first Brillouin zone into the first Brillouin zone by displacing them to the right or left by suitable integer multiples of $(2\pi/a)$. The radiative region is shown hatched here. The resulting dispersion curve now consists of several branches that extend from the light lines $\omega = \pm ck$ to the boundaries of the first Brillouin zone, $k = \pm\pi/a$, where the upper and lower signs go together. The points marked A, B, C... on this figure are of particular interest. At these points two points on the flat surface dispersion curve, separated by a grating reciprocal lattice vector, are degenerate. When the grating profile is turned on, if the surface profile function $\zeta(x_1)$ has a nonzero Fourier coefficient corresponding to this grating reciprocal lattice vector, we know from degenerate perturbation theory that the degeneracies at the points A, B, C,..., will be lifted and gaps will open up in the dispersion curve for surface polaritons at these points. Thus stop bands for surface polariton propagation are created.

This is indeed what is observed. In Fig. 12 we have plotted the dispersion curves for surface polaritons on a grating, defined by the sinusoidal profile function $\zeta(x_1) = \zeta_o\cos(2\pi x_1/a)$, ruled on the surface of a free electron metal.

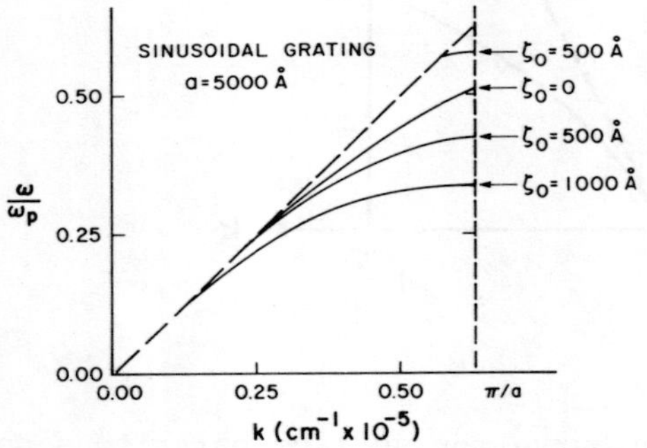

Fig. 12. The dispersion curves for surface polaritons on a grating defined by the profile function $\zeta(x_1) = \zeta_o\cos(2\pi x_1/a)$ for $a = 5000Å$ and three values of ζ_o.

The value of $\hbar\omega_p$ was chosen to be 2eV. The period a was assumed to be 5000Å, and dispersion curves for ζ_o = 0, 500Å, and 1000Å are presented. It is seen that for ζ_o = 500Å the dispersion curve consists of two branches with a gap at $k = \pi/a$. Any other branches in this case have higher frequencies and lie in the radiative region. When ζ_o = 1000Å the gap in the dispersion curve at the zone boundary has become so large that the upper branch is in the radiative region. The phenomenon of wave slowing is well displayed by the dispersion curves depicted in Fig. 12. Both the group velocity and the phase velocity of surface polaritons on the lowest branch are smaller in the presence of the grating than on the planar surface, for the same value of k. Related to this is the result that the frequency of the surface polariton on the lowest branch of the dispersion curve is lower in the presence of the grating than on the planar surface, for the same value of k.

The attenuation of surface polaritons on a grating due to their conversion into volume electromagnetic waves in the vacuum can also be studied on the basis of Eq. (3.13). It is only necessary to choose k complex and ω real, or _vice versa_, and to work in the radiative region of the (ω,k)-plane. The same choice for the branch cut in the definitions of $\beta(k_m\omega)$ and $\beta_o(k_m\omega)$ has to be made as in the corresponding calculation for surface acoustic waves described in Section 2. The resulting surface electromagnetic waves are again leaky waves. The attenuation of surface polaritons on a symmetric sawtooth grating has been calculated by Glass _et al._[38] by a method that goes beyond the Rayleigh hypothesis. Their results are plotted in Fig. 13.

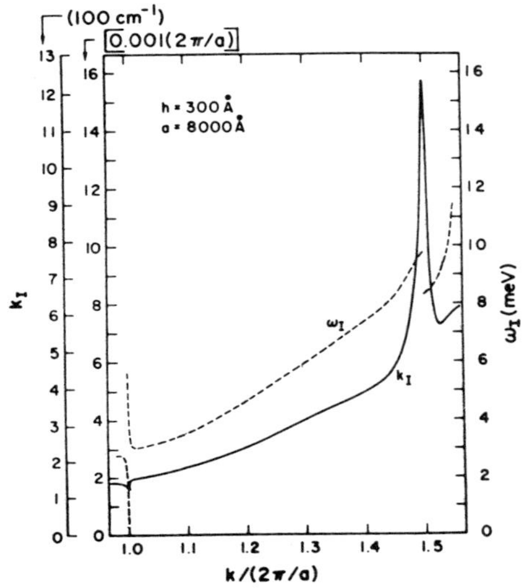

Fig. 13. Imaginary part of the complex wave vector solution, k_I, and the imaginary part of the complex frequency solution, ω_I, both as functions of real (or real part of) k, for surface polaritons on a symmetric sawtooth grating on Ag. The solid and dashed lines show the exact solutions for k_I and ω_I, respectively.

3.C. Randomly Rough Surfaces

In studying the propagation of surface electromagnetic waves across a randomly rough surface our starting equations are Eqs. (3.2) and (3.8) when the effects of retardation are neglected and taken into account, respectively. However, the surface profile function $\zeta(\vec{x}_\parallel)$ appearing in these equations must now be regarded as a stationary stochastic process with the properties expressed by Eqs. (2.21). Consequently, the solution $A(\vec{q}_\parallel\omega)$ of Eq. (3.2), and the solutions $A_{\parallel,\perp}(\vec{q}_\parallel\omega)$ of Eq. (3.8), are random functions. Just as in Section 2.C we will seek the first moments $\langle A(\vec{q}_\parallel\omega)\rangle$ and $\langle A_{\parallel,\perp}(\vec{q}_\parallel\omega)\rangle$ of the probability distribution function of these solutions, which describe the propagation of the corresponding mean wave across a randomly rough surface.

As in Section 2.C we confine ourselves to the small roughness limit that is defined by the following approximation to $J(\alpha|\vec{Q}_\parallel)$

$$J(\alpha|\vec{Q}_\parallel) \cong \hat{\zeta}(\vec{Q}_\parallel), \tag{3.14}$$

and the approximation $\langle VH_o^{-1}V\rangle$ to $\langle M\rangle$ in Eq. (2.32).

i) Surface Plasmons[20]

When the effects of retardation can be neglected the electromagnetic surface waves become electro-static surface waves, that we call surface plasmons. In the small roughness limit the dispersion relation for these surface excitations is the solvability condition for the integral equation

$$\frac{\varepsilon(\omega)+1}{\varepsilon(\omega)-1}A(\vec{q}_\parallel\omega) = \int\frac{d^2p_\parallel}{(2\pi)^2}\hat{\zeta}(\vec{q}_\parallel-\vec{p}_\parallel)(1-\hat{q}_\parallel\cdot\hat{p}_\parallel)\,p_\parallel A(\vec{p}_\parallel\omega). \tag{3.15}$$

When we apply the smoothing method to this equation we find that the equation satisfied by $\langle A(\vec{q}_\parallel\omega)\rangle$ is the algebraic equation

$$\left[\frac{\varepsilon(\omega)+1}{\varepsilon(\omega)-1} - M(q_\parallel\omega)\right]\langle A(\vec{q}_\parallel\omega)\rangle = 0, \tag{3.16}$$

where

$$M(q_\parallel\omega) = \frac{\varepsilon(\omega)-1}{\varepsilon(\omega)+1}\int\frac{d^2p_\parallel}{(2\pi)^2}\delta^2 g(|\vec{q}_\parallel-\vec{p}_\parallel|)\,q_\parallel(1-\hat{q}_\parallel\cdot\hat{p}_\parallel)^2\,p_\parallel. \tag{3.17}$$

If we introduce the representations

$$\vec{q}_\parallel = \vec{q}_\parallel(\cos\theta, \sin\theta), \quad \vec{p}_\parallel = p_\parallel(\cos\theta', \sin\theta'), \tag{3.18}$$

we find that

$$g(|\vec{q}_\parallel - \vec{p}_\parallel|) = \sum_{\ell=-\infty}^{\infty} g_\ell(q_\parallel|p_\parallel) e^{i\ell(\theta-\theta')}, \tag{3.19}$$

where

$$g_\ell(q_\parallel|p_\parallel) = 2\pi\int_0^\infty dx_\parallel x_\parallel W(|\vec{x}_\parallel|) J_\ell(q_\parallel x_\parallel) J_\ell(p_\parallel x_\parallel), \tag{3.20}$$

with $J_\ell(x)$ a Bessel function. The angular integration can now be carried out in Eq. (3.17) with the result that

$$M(q_\parallel \omega) = \delta^2 \frac{\varepsilon(\omega)-1}{\varepsilon(\omega)+1} q_\parallel \int_0^\infty \frac{dp_\parallel}{2\pi} G(q_\parallel|p_\parallel) p_\parallel^2 \tag{3.21}$$

where

$$G(q_\parallel|p_\parallel) = \frac{3}{2} g_0(q_\parallel|p_\parallel) - 2g_1(q_\parallel|p_\parallel) + \frac{1}{2} g_2(q_\parallel|p_\parallel). \tag{3.22}$$

The dispersion relation for surface plasmons on a randomly rough surface in the small roughness limit therefore becomes

$$\left(\frac{\varepsilon(\omega)+1}{\varepsilon(\omega)-1}\right)^2 = \delta^2 q_\parallel \int_0^\infty \frac{dp_\parallel}{2\pi} G(q_\parallel|p_\parallel) p_\parallel^2.$$

or

$$\frac{\varepsilon(\omega)+1}{\varepsilon(\omega)-1} = \pm\delta\left[q_\parallel \int_0^\infty \frac{dp_\parallel}{2\pi} G(q_\parallel|p_\parallel) p_\parallel^2\right]^{1/2}. \tag{3.23}$$

Thus, we see that to every solution of the equation $\varepsilon(\omega) + 1 = 0$, which is the dispersion relation for surface plasmons on a planar surface, there correspond two solutions of the dispersion relation for surface plasmons on a randomly rough surface.

This roughness-induced splitting of the surface plasmon dispersion curve has been studied theoretically by several authors[20,39,40] and has been observed experimentally by several groups[41-44]. The results of a recent experimental study of this effect are shown in Fig. 14, in which differential reflectivity spectra at normal incidence, $\Delta R/R = 2(R'-R)/(R'+R)$, are plotted for quenched silver deposits of four different thicknesses on silver surfaces. Here R' and R are the reflectivities with and without the quenched films.

In Fig. 14 the minima assigned to surface plasmons are pointed out by arrows. Their separation in energy gives the magnitude of the surface plasmon splitting, which is seen to increase with increasing film thickness, perhaps because the roughness increases with increasing film thickness.

The physical reason for the splitting of the surface plasmon dispersion curve is the following.[39] A rough surface can be regarded as a superposition of diffraction gratings, each with its own spacing, amplitude, and orientation in the x_1x_2-plane, which vary continuously from one grating to the next. Each grating can split the surface

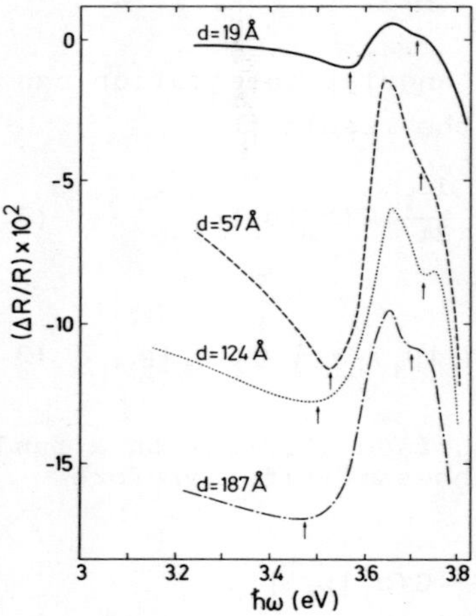

Fig. 14. Differential reflectivity spectra at normal incidence for Ag deposits of different thicknesses on an Ag substrate at 140K.

plasmon dispersion relation. This splitting occurs at a given frequency on the dispersion curve if two degenerate surface plasmons with different wave vectors can couple through the wave vector of the grating. Since the dispersion curve for a surface plasmon obtained from the equation $\varepsilon(\omega)+1 = 0$ is flat, i.e. it depends neither on the magnitude nor on the direction of the wave vector $\vec{k}_\parallel$, all wave vectors entering the Fourier decomposition of $\zeta(\vec{x}_\parallel)$ couple two degenerate surface plasmons with different wave vectors, and split the dispersion curve thereby. The fact that we obtain only two branches to the dispersion curve is due to our retaining only the term linear in $\zeta(\vec{x}_\parallel)$ in the integral for $J(\alpha|\vec{Q}_\parallel)$ in going from Eq. (3.4) to Eq. (3.15). If higher order terms were kept, more branches would be obtained.

Recently it has been shown that an infinite subset of terms in the expansion of $\langle M \rangle$, Eq. (2.33), in powers of V has the same order of magnitude for frequencies in the vicinity of the solution of $\varepsilon(\omega)+1 = 0$ as the leading term $\langle VH_o^{-1}V\rangle$.[20]

An approximate summation of these terms yields some quantitative changes in the splitting of the surface plasmon dispersion curve, but essentially no qualitative changes.

ii) Surface Polaritons

In obtaining the equations satisfied by $\langle A_{\parallel,\perp}(\vec{q}_{\parallel}\omega)\rangle$ from Eqs. (3.8) we will also make the small roughness approximation. In this approximation Eq. (3.8) becomes

$$\begin{pmatrix} \frac{\varepsilon(\omega)\beta_o(q_{\parallel}\omega)+\beta(q_{\parallel}\omega)}{\varepsilon(\omega)-1} & 0 \\ 0 & \frac{\beta_o(q_{\parallel}\omega)+\beta(q_{\parallel}\omega)}{\varepsilon(\omega)-1} \end{pmatrix} \begin{pmatrix} A_{\parallel}(\vec{q}_{\parallel}\omega) \\ A_{\perp}(\vec{q}_{\parallel}\omega) \end{pmatrix}$$

$$= \int \frac{d^2p_{\parallel}}{(2\pi)^2}\, \hat{\zeta}(\vec{q}_{\parallel}-\vec{p}_{\parallel}) \times$$

$$\times \begin{pmatrix} q_{\parallel}p_{\parallel}-\beta(q_{\parallel}\omega)(\hat{q}_{\parallel}\cdot\hat{p}_{\parallel})\beta_o(p_{\parallel}\omega) & -i\frac{\omega}{c}\beta(q_{\parallel}\omega)(\hat{q}_{\parallel}\times\hat{p}_{\parallel})_3 \\ i\frac{\omega}{c}(\hat{q}_{\parallel}\times\hat{p}_{\parallel})_3\beta_o(p_{\parallel}\omega) & \frac{\omega^2}{c^2}(\hat{q}_{\parallel}\cdot\hat{p}_{\parallel}) \end{pmatrix} \begin{pmatrix} A_{\parallel}(\vec{p}_{\parallel}\omega) \\ A_{\perp}(\vec{p}_{\parallel}\omega) \end{pmatrix}. \tag{3.24}$$

When the smoothing method is applied to this pair of equations the equations for $\langle A_{\parallel}(\vec{q}_{\parallel}\omega)\rangle$ and $\langle A_{\perp}(\vec{q}_{\parallel}\omega)\rangle$ decouple due to the restoration of isotropy in the plane $x_3 = 0$ by the averaging process, and become algebraic rather than integral equations due to the restoration of infinitesimal translational invariance by the average process. The dispersion relation for p-polrized surface polaritons then takes the form

$$F(q_{\parallel}\omega) = \delta^2(1-\varepsilon(\omega))^2 G(q_{\parallel}\omega), \tag{3.25}$$

where

$$F(q_{\parallel}\omega) = \varepsilon(\omega)\beta_o(q_{\parallel}\omega) + \beta(q_{\parallel}\omega) \tag{3.26a}$$

$$G(q_\parallel\omega) = \int\frac{d^2p_\parallel}{(2\pi)^2}\,\frac{g(|\vec{q}_\parallel-\vec{p}_\parallel|)}{\varepsilon(\omega)\beta_o(p_\parallel\omega)+\beta(p_\parallel\omega)}\times$$

$$\times\ [q_\parallel p_\parallel\cos(\theta-\theta')-\beta_o(q_\parallel\omega)\beta(p_\parallel\omega)][p_\parallel q_\parallel\cos(\theta-\theta')-\beta_o(p_\parallel\omega)\beta(q_\parallel\omega)],$$

$$= G^{(1)}(q_\parallel\omega)-iG^{(2)}(q_\parallel\omega), \qquad (3.26b)$$

and the angles θ and θ' have been defined in Eq. (3.18). If we denote the solution of the equation $F(q_\parallel\omega) = 0$ by $\omega_o(q_\parallel)$, the frequency of a surface polariton on a flat surface, the solution of Eq. (3.25) for the frequency of a surface polariton on a randomly rough surface can be written in the form

$$\omega(q_\parallel) = \omega_o(q_\parallel)+\Delta(q_\parallel)-i\Gamma(q_\parallel), \qquad (3.27)$$

where

$$\Delta(q_\parallel) = \delta^2(1-\varepsilon(\omega_o(q_\parallel)))^2\,\frac{G^{(1)}(q_\parallel,\omega_o(q_\parallel))}{[dF(q_\parallel\omega)/d\omega]_{\omega=\omega_o(q_\parallel)}} \qquad (3.28a)$$

$$\Gamma(q_\parallel) = \delta^2(1-\varepsilon(\omega_o(q_\parallel)))^2\,\frac{G^{(2)}(q_\parallel,\omega_o(q_\parallel))}{[dF(q_\parallel\omega)/d\omega]_{\omega=\omega_o(q_\parallel)}}\,. \qquad (3.28b)$$

The imaginary part of $\omega(p_\parallel)$ arises from the fact that in the presence of surface roughness a surface polariton is attenuated even if the dielectric constant $\varepsilon(\omega)$ is real. The physical reason for this is the roughness-induced scattering of the surface polaritons into radiative modes in the vacuum, and into other surface polariton modes. Both scattering processes remove energy from the incident beam, and the surface polariton is attenuated thereby. This is the situation we would like to focus our attention on. Therefore, to separate the attenuation of surface polaritons that has its origin in surface roughness from that which is due to the dissipative processes present in the bulk of the material, we assume $\varepsilon(\omega)$ to be real everywhere in Eq. (3.25), except in the denominator $\varepsilon(\omega)\beta_o(p_\parallel\omega)+\beta(p_\parallel\omega)$ of the integrand on the right hand side of Eq. (3.26b). The retention of the small, positive imaginary part of the dielectric constant here serves only to define the manner in which the pole in the integrand that occurs when $\varepsilon(\omega)\beta_o(p_\parallel\omega)+\beta(p_\parallel\omega) = 0$ is to be treated in the evaluation of the integral over $p_\parallel$.

The attenuation length of the surface polariton $\ell(q_\parallel)$, is the distance over which the energy of the polariton decays to 1/e of its initial value. It is given by

$$\ell(q_\parallel) = \frac{V_E(q_\parallel)}{2\Gamma(q_\parallel)} . \qquad (3.29)$$

Here $V_E(q_\parallel)$ is the energy transport velocity of the surface polariton and, in the absence of damping, is equal to the group velocity of the surface polariton,

$$V_E(q_\parallel) = c^2 \frac{q_\parallel}{\omega_o(q_\parallel)} \left. \frac{[\varepsilon(\omega)+1]^2}{\varepsilon(\omega)[\varepsilon(\omega)+1] + 1/2\, \omega\varepsilon'(\omega)} \right|_{\omega=\omega_o(q_\parallel)} . \qquad (3.30)$$

Numerical results for $\Delta(q_\parallel)$, $\Gamma(q_\parallel)$, and $\ell^{-1}(q_\parallel)$ for aluminum are shown in Fig. 15. The dielectric constant was chosen to have the free electron form with $\hbar\omega_p = 1.2018 \times 10^5 cm^{-1}$. The values of δ and a used were δ = a = 2500Å, so that the surface is a very rough one. From the figure we see that for a surface polariton wavelength of 2.1μm ($q_\parallel = 3\times10^{-4}$ Å^{-1}), $\ell^{-1}(q_\parallel) \cong 4 \times 10^2 cm^{-1}$, so that $\ell(q_\parallel) \cong 2.5 \times 10^{-3}$cm.

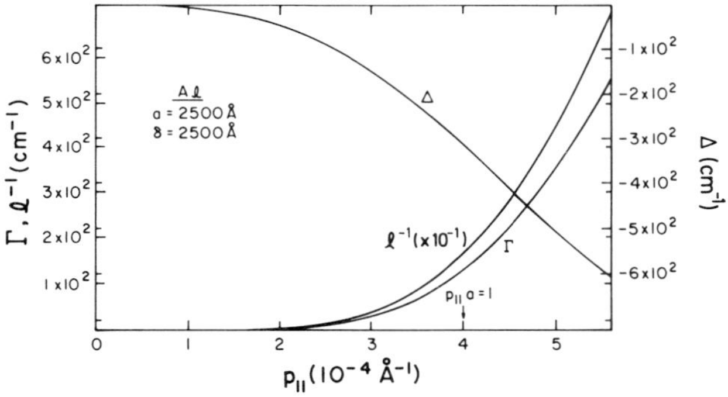

Fig. 15. The functions $\Delta(p_\parallel)$, $\Gamma(p_\parallel)$ and $\ell^{-1}(p_\parallel)$ as functions of $p_\parallel$ for surface polaritons on a randomly rough Aℓ surface. Note that the graphical value of $\ell^{-1}(p_\parallel)$ is to be multiplied by 10.

In all cases studied $\Delta(q_\parallel)$ was found to be negative, i.e. surface roughness depresses the frequency of a surface polariton below its value on a planar surface. This is reminiscent of the depression of the frequency of (the lowest branch of) a surface polariton propagating across a grating by the periodic corrugations of the latter surface.

The numerical calculations also show that the dominant contribution to $G^{(1)}(q_\parallel\omega)$ and $G^{(2)}(q_\parallel\omega)$ comes from the part of the integral in Eq. (3.26b) that comes from the region $q_\parallel > (\omega/c)$, and corresponds to the roughness-induced scattering into other surface polariton states. In a separate calculation Mills[46] has shown that a surface polariton is damped predominantly by scattering into other surface polariton states when $cq_\parallel >> \omega$ so that $|\varepsilon(\omega)| \approx 1$, and $q_\parallel a << 1$, which conditions are reasonably well satisfied for the situation depicted in Fig. 15. When $\omega << \omega_p$, so that $\omega \approx cq_\parallel$ the most effective mechanism for attenuating the surface polariton is the roughness induced radiation into the vacuum above the material.

3.D. Localization of Surface Polaritons by Surface Roughness

It is now well known that when a wave propagates in a random medium the interference between the scattered waves can lead to localization phenomena. For example, in the case of an electron moving in a random potential this kind of interference leads to Anderson localization, viz. the renormalized diffusion coefficient of the electron vanishes[47-50]. The localization of elastic waves in disordered media has been discussed recently[51-53], and it has been suggested that a mobility edge exists for an electromagnetic wave propagating in a disordered dielectric[54,55].

It has recently been shown[56-58] that a surface polariton can be localized parallel to the surface by the presence of random roughness on that surface. The theory of this effect is outlined in this section. We follow Refs. 57 and 58, since an operational method for observing effects of this localization is presented there.

We consider the scattering of p-polarized light, incident from the vacuum side on a one-dimensional grating whose profile is given by the equation $x_3 = \zeta(x_1)$. The plane of incidence is the x_1x_3-plane. The region $x_3 > \zeta(x_1)$ is vacuum, while the region $x_3 < \zeta(x_1)$ is filled by an isotropic dielectric medium characterized by the complex dielectric constant $\varepsilon(\omega) = \varepsilon_1(\omega)+i\varepsilon_2(\omega)$, with $\varepsilon_1(\omega) < -1$ and $\varepsilon_2(\omega) << |\varepsilon_1(\omega)|$ at the frequency of the incident light.

The surface profile function $\zeta(x_1)$ is assumed to be a stationary stochastic process with the following properties:

$$\langle\zeta(x_1)\rangle = 0 \tag{3.31a}$$

$$\langle\zeta(x_1)\zeta(x_1')\rangle = \delta^2 W(|x_1-x_1'|). \tag{3.31b}$$

The angular brackets again denote an average over the ensemble of realizations of $\zeta(x_1)$, and $\delta^2 = \langle\zeta^2(x_1)\rangle$ is the mean square departure of the surface from flatness. In evaluating higher order moments of the probability distribution of $\zeta(x_1)$ we assume that $\zeta(x_1)$ is a Gaussianly distributed random variable. In numerical work we assume a Gaussian form for $W(|x_1-x_1'|)$,

$$W(|x_1-x_1'|) = \exp(-|x_1-x_1'|^2/a^2). \tag{3.32}$$

The magnetic field vector in this problem has the form $\vec{H}(\vec{x},t) = (0, H_2(x_1x_3|\omega), 0)\exp(-i\omega t)$, and the expression for $H_2(x_1x_3|\omega)$ in the region $x_3 > \zeta(x_1)_{max}$ that satisfies the boundary conditions at $x_3 = \infty$ is

$$H_2(x_1x_3|\omega) = e^{ikx_1-i\alpha_o(k\omega)x_3} + \int\frac{dq}{2\pi}R(q|k)e^{iqx_1+i\alpha_o(q\omega)x_3}, \tag{3.33}$$

where

$$\alpha_o(q\omega) = \left(\frac{\omega^2}{c^2} - q^2\right)^{1/2} \qquad q^2 < \frac{\omega^2}{c^2} \tag{3.34a}$$

$$= i\left(q^2 - \frac{\omega^2}{c^2}\right)^{1/2} \qquad q^2 > \frac{\omega^2}{c^2}, \tag{3.34b}$$

while $\alpha_o(k\omega)$ is real. The use of the Rayleigh hypothesis, Green's theorem, and the extinction theorem shows that the amplitude of the reflected field is the solution of

$$\int\frac{dq}{2\pi}\frac{I(\alpha(p\omega)-\alpha_o(q\omega)|p-q)}{\alpha(p\omega)-\alpha_o(q\omega)}\left[\alpha(p\omega)\alpha_o(q\omega)+pq\right]R(q|k)$$

$$= \frac{I(\alpha(p\omega)+\alpha_o(k\omega)|p-k)}{\alpha(p\omega)+\alpha_o(k\omega)}\left[\alpha(p\omega)\alpha_o(k\omega)-pk\right], \tag{3.35}$$

where $\alpha(p\omega) = \left[\varepsilon(\omega)(\omega^2/c^2)-p^2\right]^{1/2}$, with $\mathrm{Re}\alpha(p\omega) > 0$, $\mathrm{Im}\alpha(p\omega) > 0$, while

$$I((\alpha|Q) = \int dx_1 e^{-iQx_1-i\alpha\zeta(x_1)}. \tag{3.36}$$

We seek a solution of Eq. (3.35) in the form

$$R(p|k) = 2\pi\delta(p-k)R_o(k) - 2iG_o(p)T(p|k)G_o(k)\alpha_o(k\omega) \tag{3.37}$$

where

$$R_o(p) = \frac{\varepsilon(\omega)\alpha_o(p\omega)-\alpha(p\omega)}{\varepsilon(\omega)\alpha_o(p\omega)+\alpha(p\omega)} \tag{3.38}$$

is the Fresnel coefficient for the reflection of p-polarized light from a flat dielectric surface, and

$$G_o(p) = \frac{i\varepsilon(\omega)}{\varepsilon(\omega)\alpha_o(p\omega)+\alpha(p\omega)} \tag{3.39}$$

is the Green's function for a surface polariton on a flat surface. The scattering matrix $T(p|k)$ is postulated to satisfy the equation

$$T(p|k) = V(p|k) + \int\frac{dq}{2\pi} V(p|q)G_o(q)T(q|k). \tag{3.40}$$

Equations (3.35), (3.37), and (3.40) in fact define the scattering potential $V(p|k)$. To first order in $\zeta(x_1)$, which is all that we consider here, it is given by

$$V(p|k) = \frac{\varepsilon(\omega)-1}{\varepsilon^2(\omega)}\hat{\zeta}(p-k)\left[\varepsilon(\omega)pk-\alpha(p\omega)\alpha_o(k\omega)\right]. \tag{3.41}$$

The diffuse scattering efficiency (the diffuse part of the total scattered flux in the x_3-direction normalized by the total incident flux in the $-x_3$-direction), averaged over the ensemble of realizations of $\zeta(x_1)$ is[57]

$$I(p|k)_{diff} = \frac{4\alpha_o(p\omega)\alpha_o(k\omega)}{L}\langle|G(p|k)|^2\rangle_{diff} \tag{3.42}$$

where L is the length of the surface $x_3 = 0$ in the x_1-direction, and $G(p|k)$ is the Green's function for a surface polariton on the randomly rough surface, and is the solution of

$$G(p|k) = 2\pi\delta(p-k)G_o(k) + G_o(p)\int\frac{dq}{2\pi} V(p|q)G(q|k) \tag{3.43a}$$

$$= 2\pi\delta(p-k)G_o(k) + G_o(p)T(p|k)G_o(k). \tag{3.43b}$$

In Eq. (3.42) $\langle|G(p|k)|^2\rangle_{diff}$ is that contribution to

$\langle|G(p|k)|^2\rangle$ that contains no factor of $2\pi\delta(p-k)$.

To obtain the scattered intensity (3.42) we study the two polariton Green's function

$$L\phi_{pp'}(q\omega_o) = -\frac{1}{2\pi i}\langle G(p_+p'_+|\omega+\omega_o)G^*(p_-p'_-|\omega)\rangle, \qquad (3.44)$$

where we have indicated the frequency dependence of $G(p|k)$ explicitly, and have defined $p_+ = p+(q/2)$ and $p_- = p-(q/2)$. In the limit that $q \to 0$, $\omega_o \to 0$, $\phi_{pp'}(q\omega_o)$ is proportional to $\langle|G(p|p')|^2\rangle$.

$\phi_{pp'}(q\omega_o)$ satisfies the Bethe-Salpeter equation

$$\phi_{pp'}(q\omega_o) = G(p_+,\omega+\omega_o)G^*(p_-,\omega)[\,i\delta(p-p')+$$

$$+\int\frac{dp''}{2\pi}U_{pp''}(q\omega_o)\phi_{p''p'}(q\omega_o)], \qquad (3.45)$$

where $U_{pp'}(q\omega_o)$ is the irreducible vertex function. The function $G(k\omega)$ appearing in Eq. (3.45) is defined in terms of the surface polariton Green's function $G(p|k)$ by $\langle G(p|k)\rangle = 2\pi\delta(p-k)G(k\omega)$. We approximate $G(k\omega)$ by

$$G(k\omega) \cong \frac{C_1}{k-k_{sp}-i\Delta_{tot}} - \frac{C_1}{k+k_{sp}+i\Delta_{tot}}, \qquad (3.46)$$

where

$$k_{sp}(\omega) = \frac{\omega}{c}\left[\frac{\varepsilon_1(\omega)}{\varepsilon_1(\omega)+1}\right]^{1/2} \qquad (3.47)$$

$$C_1(\omega) = \frac{\varepsilon_1(\omega)\sqrt{-\varepsilon_1(\omega)}}{1-\varepsilon_1^2(\omega)}, \qquad (3.48)$$

and $\Delta_{tot} = \Delta_\varepsilon + \Delta_{sp}$, with

$$\Delta_\varepsilon(\omega) = \frac{\varepsilon_2(\omega)k_{sp}(\omega)}{2\varepsilon_1(\omega)[\varepsilon_1(\omega)+1]} \qquad (3.49)$$

$$\Delta_{sp}(\omega) = 2\pi^{1/2}a\delta^2C_1^2\left[\frac{\varepsilon_1(\omega)-1}{\varepsilon_1(\omega)}\right]^2 k_{sp}^4\exp(-a^2k_{sp}^2). \qquad (3.50)$$

$\Delta_\varepsilon(\omega)$ describes the damping of the surface polariton on a flat surface due to the imaginary part of $\varepsilon(\omega)$; $\Delta_{sp}(\omega)$ describes the damping of the surface polariton by its roughness-induced conversion into radiative modes in the vacuum. The form (3.46) follows from an approximate self-consistent solution of Eq. (3.43a).

In what follows we will be interested in the case in which $\Delta_\varepsilon << \Delta_{sp}$, and consequently where $\Delta_\varepsilon << \Delta_{tot}$. The somewhat simpler case in which $\Delta_\varepsilon >> \Delta_{sp}$ is discussed in Ref. 57.

With the use of Eq. (3.46) it is found that the solution of Eq. (3.45) can be written in the form

$$\phi_{pp'} = iG(p_+,\omega+\omega_o)G^*(p_-,\omega)\{\delta(p-p') + \frac{1}{2\pi} U_{pp'}G(p'_+,\omega+\omega_o) \times$$

$$\times\, G^*(p'_-,\omega) - \frac{i}{2}[(U_{p,k_{sp}} + U_{p,-k_{sp}})\hat{\phi}_{p'} + (U_{p,k_{sp}} - U_{p,k_{sp}})\hat{\phi}_{jp'}] \times$$

$$\times\, G(p'_+,\omega+\omega_o)G^*(p'_-,\omega)\}, \tag{3.51}$$

where

$$\hat{\phi}_{p'} = \int\frac{dp}{2\pi}\,\phi_{pp'} \cong \frac{1}{2\gamma_\varepsilon+M}\,\frac{(\omega_o+iM)N_{p'}+c((1+\varepsilon_1)/\varepsilon_1)^{1/2}qN_{jp'}}{-i(\omega_o+i\frac{2\gamma_\varepsilon M}{2\gamma_\varepsilon+M}) + Dq^2} \tag{3.52a}$$

$$\hat{\phi}_{jp'} = \int\frac{dp}{2\pi}\,\frac{p}{|p|}\,\phi_{pp'} \cong \frac{1}{2\gamma_\varepsilon+M}\,\frac{c((1+\varepsilon_1)/\varepsilon_1)^{1/2}qN_{p'}+(\omega_o+2i\gamma_\varepsilon)N_{jp'}}{-i(\omega_o + i\frac{2\gamma_\varepsilon M}{2\gamma_\varepsilon M}) + Dq^2}\,. \tag{3.52b}$$

In these expressions

$$\gamma_\varepsilon = c((1+\varepsilon_1)/\varepsilon_1)^{1/2}\Delta_\varepsilon \tag{3.53}$$

$$M = 2c((1+\varepsilon_1)/\varepsilon_1)^{1/2}\Delta_{tot} + M_o \tag{3.54}$$

$$M_o = \frac{1}{2} c\left((1+\varepsilon_1)/\varepsilon_1\right)^{1/2} \int\frac{dp}{2\pi}\int\frac{dp'}{2\pi} \Delta G_p \frac{p}{|p|} U_{pp'} \frac{p'}{|p'|} \Delta G_{p'} \quad (3.55)$$

$$\Delta G_p = G^*(p_-,\omega) - G(p_+,\omega+\omega_o) \quad (3.56)$$

$$N_{p'} = \frac{cC_1^2}{2\pi}\left((1+\varepsilon_1)/\varepsilon_1\right)^{1/2}\left[U_{k_{sp},p'} + U_{-k_{sp},p'}\right] \quad (3.57)$$

$$N_{jp'} = \frac{cC_1^2}{2\pi}\left((1+\varepsilon_1)/\varepsilon_1\right)^{1/2}\left[U_{k_{sp},p'} - U_{-k_{sp},p'}\right] \quad (3.58)$$

$$D = \frac{c^2}{2\gamma_\varepsilon + M}\,\frac{1+\varepsilon_1}{\varepsilon_1} \, . \quad (3.59)$$

The quantity D is the diffusion constant for surface polariton motion in the x_1-direction along the interface.

In the limit $q \to 0$, $\omega_o \to 0$ we have the approximations

$$\Delta G_p \cong -2iC_1\left[\frac{\Delta_{tot}}{(p-k_{sp})^2 + \Delta_{tot}^2} + \frac{\Delta_{tot}}{(p+k_{sp})^2 + \Delta_{tot}^2}\right] \quad (3.60a)$$

$$\cong -2\pi i C_1\left[\delta(p-k_{sp}) + \delta(p+k_{sp})\right], \quad (3.60b)$$

that can be used in appropriate circumstances.

To obtain the lowest order approximation in $\zeta(x_1)$ to M, and hence to D, we use the lowest order approximation to $U_{pp'}$ in $\zeta(x_1)$, viz

$$U_{pp'}^{(o)} = \langle V(p|p')V^*(p|p')\rangle_o \, , \quad (3.61)$$

where we have used the definition

$$\langle V(p|q)V^*(r|s)\rangle = 2\pi\delta(p-q-r+s)\langle V(p|q)V^*(r|s)\rangle_o \, . \quad (3.62)$$

$U^{(o)}_{pp'}$ is slowly varying in the vicinity of $p \cong \pm k_{sp}$ and $p' \cong \pm k_{sp}$, so we can use Eqs. (3.60) and (3.47) in Eq. (3.55) to obtain

$$M_o^{(o)} = cC_1^2((1+\varepsilon_1)/\varepsilon_1)^{1/2} U^{(o)}_{-k_{sp},k_{sp}} \tag{3.63}$$

and

$$U^{(o)}_{-k_{sp},k_{sp}} \cong \frac{2}{C_1^2} \Delta_{sp}. \tag{3.64}$$

The corresponding approximations to M and D from Eqs. (3.54) and (3.59) are

$$M^{(o)} = 2c((1+\varepsilon_1)/\varepsilon_1)^{1/2}[\Delta_{tot}+\Delta_{sp}] \tag{3.65}$$

$$D_o = \frac{c((1+\varepsilon_1)/\varepsilon_1)^{1/2}}{4\Delta_{tot}} . \tag{3.66}$$

To incorporate the effects of localization into D we must include the contribution to $U_{pp'}$ from the maximally crossed diagrams:

$$U_{pp'} = U^{(o)}_{pp'} + \Lambda_{pp'} . \tag{3.67}$$

$\Lambda_{pp'}(q\omega_o)$ is obtained from the relation[49]

$$\Lambda_{pp'}(q\omega_o) = \tilde{\Gamma}_{\frac{1}{2}(p-p'+q),\, \frac{1}{2}(p'-p+q)}(p+p',\omega_o), \tag{3.68}$$

where $\tilde{\Gamma}_{pp'}(q\omega)$ is defined by the pair of equations

$$\Gamma_{pp'}(q\omega_o) = \langle V(p_+|p'_+)V^*(p_-|p'_-)\rangle_o +$$

$$+ \int\frac{dk}{2\pi} \langle V(p_+|k_+)V^*(p_-|k_-)\rangle_o\, G(k_+,\omega+\omega_o)G^*(k_-,\omega)\Gamma_{kp'}(q\omega_o) \tag{3.69a}$$

$$\equiv \langle V(p_+|p'_+)V^*(p_-|p'_-)\rangle_o + \tilde{\Gamma}_{pp'}(q\omega_o) . \tag{3.69b}$$

$\Gamma_{pp'}(q\omega_o)$ represents the sum of the so-called ladder diagrams.

Just as in the two-dimensional problem of electrons in a random potential[49-50], $\Lambda_{pp'}$ is largest for $p \cong -p'$. Therefore, from Eqs.(3.55) and (3.60b) we see that the dominant contribution to M_o from $\Lambda_{pp'}$ comes from $p \cong \pm k_{sp}$ and $p' \cong \mp k_{sp}$. To obtain $\Lambda_{pp'}$ for $p = -k_{sp}$ and $p' \cong k_{sp}$ we write the product of Green's functions in Eq. (3.69a) in the form

$$G(k_+,\omega+\omega_o)G^*(k_-,\omega) \cong \frac{\pi C_1^2}{\Delta_{tot}}\left[A_+(q\omega_o\omega)\delta(k+k_{sp}) + \right.$$

$$\left. + A_-(q\omega_o\omega)\delta(k-k_{sp})\right] \tag{3.70}$$

where

$$A_\pm(q\omega_o\omega) = \left(1 \pm \frac{iq}{2\Delta_{tot}} - \frac{q^2}{4\Delta_{tot}^2}\right) + i\frac{\omega_o}{c}\frac{(\varepsilon_1/(\varepsilon_1+1))^{1/2}}{2\Delta_{tot}} . \tag{3.71}$$

Equation (3.69a) reduces to a set of algebraic equations for $\tilde{\Gamma}_{\pm k_{sp},\pm k_{sp}}(q\omega_o)$ which upon solution yield

$$\tilde{\Gamma}_{-k_{sp},k_{sp}}(q\omega_o) \cong \frac{2\Delta_{tot}^2}{C_1^2}\frac{c((1+\varepsilon_1)/\varepsilon_1)^{1/2}}{-i(\omega_o+2i\gamma_\varepsilon)+D_oq^2} \tag{3.71}$$

and hence

$$\Lambda_{pp'}(q\omega_o)\Big|_{\substack{p\,=\,-k_{sp}\\ p\,\cong\,k_{sp}}} \cong \frac{2\Delta_{tot}^2}{C_1^2}\frac{c((1+\varepsilon_1)/\varepsilon_1)^{1/2}}{-i(\omega_o+2i\gamma_\varepsilon)+D_o(p+p')^2} . \tag{3.72}$$

In the limit as $\omega_o \to 0$ this result becomes

$$\Lambda_{pp'}\Big|_{\substack{p\,\cong\,-k_{sp}\\ p'\,\cong\,k_{sp}}} = \frac{8\Delta_{tot}^2}{C_1^2}\frac{\Delta_{tot}}{8(\Delta_\varepsilon\Delta_{tot}) + (p+p')^2} \tag{3.73}$$

which is more sharply peaked in the vicinity of $p = -p$ than ΔG_p and $\Delta G_{p'}$ are in the vicinity of $p = \pm k_{sp}$ and $p' = \pm k_{sp}$, respectively, when $\Delta_\varepsilon << \Delta_{tot}$. We therefore use Eqs. (3.60a) and (3.67) in Eq. (3.55), and the equation for M becomes

$$M = 2c((1+\epsilon_1)/\epsilon_1)^{1/2}[\Delta_{tot} + \Delta_{sp} + c((1+\epsilon_1)/\epsilon_1)^{1/2}\Delta_{tot} \times$$

$$\times \int \frac{dQ}{2\pi} \frac{1}{-i(\omega_o + 2i\gamma_\epsilon) + D_o Q^2}] \tag{3.74}$$

$$\underset{\omega_o \to 0}{=} 2c((1+\epsilon_1)/\epsilon_1)^{1/2}[\Delta_{tot} + \Delta_{sp} + \Delta_{tot}(\frac{\Delta_{tot}}{2\Delta_\epsilon})^{1/2}]. \tag{3.75}$$

When $\Delta_\epsilon << \Delta_{tot}$ this result differs significantly from its bare value $M^{(o)}$, Eq. (3.65), and so, therefore, does D. We must therefore follow Vollhardt and Wolfle[49,50] in replacing D_o in Eq. (3.74) by the fully renormalized D. We also replace $2\gamma_\epsilon$ by its renormalized value $2\gamma_\epsilon M/(2\gamma_\epsilon + M)$, and obtain the following self-consistent equation for M:

$$M = 2c((1+\epsilon_1)/\epsilon_1)^{1/2}[\Delta_{tot} + \Delta_{sp} +$$

$$+ c((1+\epsilon_1)/\epsilon_1)^{1/2}\Delta_{tot} \int \frac{dQ}{2\pi} \frac{1}{-i(\omega_o + i\frac{2\gamma_\epsilon M}{2\gamma_\epsilon + M}) + DQ^2}]. \tag{3.76}$$

The solution of this equation in the limit that $|M| >> 2\gamma_\epsilon$, an assumption that is confirmed by the result, is

$$M = ic^2\Delta_{tot}^2 \frac{1+\epsilon_1}{\epsilon_1} \frac{1}{\omega_o + 2i\gamma_\epsilon}. \tag{3.77}$$

The diffusion constant of the surface polariton, from Eq. (3.59), is therefore

$$D = \frac{-i(\omega_o + 2i\gamma_\epsilon)}{\Delta_{tot}^2 + \frac{2\gamma_\epsilon(\omega_o + 2i\gamma_\epsilon)}{c^2(1+\epsilon_1)/\epsilon_1}}, \tag{3.78}$$

and goes to zero as $\omega_o + 2i\gamma_\epsilon$ goes to zero. This indicates that the surface polaritons are localized by the surface roughness. The localization length ℓ is then given by[49,50]

$$\frac{1}{\ell^2} = \lim_{\omega_o + 2i\gamma_\epsilon \to 0} [\frac{-i(\omega_o + 2i\gamma_\epsilon)}{D}] = \Delta_{tot}^2, \tag{3.79}$$

so that

$$\ell = \frac{1}{\Delta_{tot}} \cong \frac{1}{\Delta_{sp}} . \tag{3.80}$$

Thus the localization length is equal to the mean free path for surface polariton propagation along the interface. A similar relation is found between the localization length and the mean free path of electrons in one-dimensional systems.[58] If we use the value of the dielectic constant of silver at the frequency corresponding to a vacuum wavelength of light of $\lambda = 4579Å$, viz. $\varepsilon = -7.5+i0.24$[60], then for a random grating for which $\delta = 50Å$ and $a = 1000Å$ we find that $\ell = 1.2 \times 10^6 Å$.

The implications of the preceding results for the diffraction of light from a random grating follow from Eqs. (3.42) and (3.44), which tell us that

$$I(p|k)_{diff} = 4\alpha_o(p\omega)\alpha_o(k\omega) \lim_{\substack{q\to 0\\ \omega_o\to 0}} [-2\pi i\phi_{pk}(q\omega_o)]_{diff}, \tag{3.81}$$

and from Eq. (3.51), which tells us that

$$\lim_{\substack{q\to 0\\ \omega_o\to 0}} [-2\pi i\phi_{pk}(q\omega_o)]_{diff} = |G(p\omega)|^2|G(k\omega)|^2 \times$$

$$\times \{U_{pk} - \pi i[U_{p,k_{sp}} + U_{p,-k_{sp}}]\hat{\phi}_k -$$

$$- \pi i[U_{p,k_{sp}} - U_{p,-k_{sp}}]\hat{\phi}_{jk}\}_{\substack{q=0\\ \omega_o=0}} . \tag{3.82}$$

If we use the preceding results and recall that p and k are now in the radiative region ($|k|,|p| < \omega/c$), rather than in the nonradiative region, we are led to the result that when $|M| >> 2\gamma_\varepsilon$ as before

$$I(p|k)_{diff} = 4\alpha_o(p\omega)\alpha_o(k\omega)|G(p\omega)|^2|G(k\omega)|^2 \times$$

$$\times \{K(p,k) + \frac{C_1^2}{4\Delta_\varepsilon}[K(p,k_{sp})+K(p,-k_{sp})][K(k,k_{sp})+K(k,-k_{sp})]$$

$$+ \frac{C_1^2}{4\Delta_\varepsilon}\frac{\Delta_{tot}^2}{\Delta_{tot}^2+(p+k)^2}[K(p,k_{sp})+K(p,-k_{sp})]^2\}, \tag{3.83}$$

where

$$K(p,k) = \pi^{1/2} \alpha\delta^2 \left[\frac{\varepsilon_1(\omega)-1}{\varepsilon_1(\omega)}\right]^2 |\varepsilon_1(\omega) pk - \alpha(p\omega)\alpha(k\omega)|^2 \times$$

$$\times \exp(-a^2(p-k)^2/4). \tag{3.84}$$

The presence of the denominator $\Delta^2_{tot} + (p+k)^2$ in Eq. (3.83) gives a maximum in the antispecular direction in the angular distribution of the scattered intensity.

The differential reflection coefficient $\partial R/\partial\theta$ is obtained from $I(p|k)_{diff}$ according to

$$\frac{\partial R}{\partial \theta} = \frac{\omega}{2\pi c} \cos\theta \; I(p|k)_{diff} , \tag{3.85}$$

where θ is the angle between the scattered wave vector and the outward normal to the mean surface. We have plotted in Fig. 16 the differential reflection coefficient for light of wavelength λ = 4579Å incident at two angles onto a random silver grating whose dielectric constant at this wavelength is ε = -7.5+i0.24. The parameters characterizing the roughness of the surface are δ = 50Å and a = 1000Å. The peak in the antispecular direction is clearly seen. Its angular width is estimated to be < 0.05° in the plot given by Fig. 16. The observation of this peak would provide experimenal evidence for the one-dimensional localization of surface polaritons on a randomly rough grating by the random roughness of the grating.

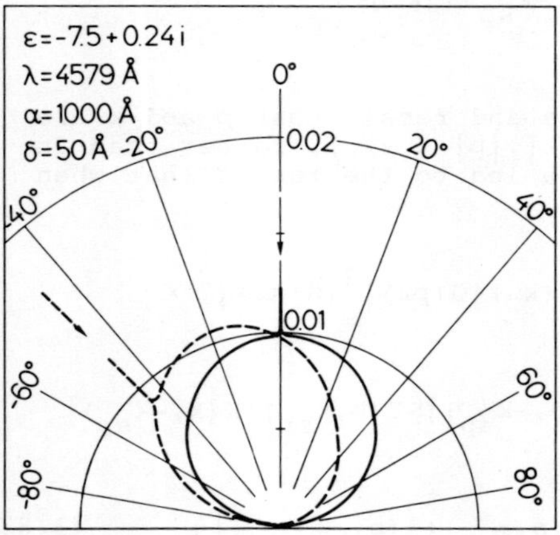

Fig. 16. A plot of the differential reflection coefficient $\partial R/\partial\theta$ in (radians)$^{-1}$ as a function of θ for light of wavelength λ = 4579Å incident on a random silver grating whose dielectric constant is $\varepsilon(\omega)$ = -7.5 +i0.24, and whose roughness is characterized by the parameters δ = 50Å and a = 1000Å. Curves for normal incidence and for an angle of incidence of 45° are shown.

4. Surface Waves on Magnetic Media

In contrast with the great deal of work that has been done on surface waves on rough elastic and dielectric surfaces, comparatively little is known about such waves on magnetic surfaces. The only results of this kind of which I am aware in which approximations are not made are for a magnetostatic surface wave (MSSW) propagating across a periodically corrugated surface of a semi-infinite magnetic insulator(61). In other, approximate, work the propagation of MSSW on an epitaxial yttrium-iron-garnet (YIG) film grooved periodically on one side and flat on the other has been studied theoretically and experimentally(62). The propagation of MSSW through a YIG slab, both of whose surfaces are periodically corrugated in such a way that the midplane of the slab is a plane of reflection symmetry for the structure has been studied theoretically(63), as have the filter characteristics of MSSW propagating in a layered YIG film periodically corrugated on one surface but not on the other(64). The interest in such systems is due to the possibility of using MSSW in devices useful in microwave signal processing applications. In what follows, a brief discussion of the propagation of MSSW across a periodically corrugated, semi-infinite, ferromagnetic insulator, rather than a film of finite thickness, will be discussed(61). This discussion is prefaced by an outline of the theory of magnetostatic surface shape resonances, and is followed by a discussion of MSSW on a random grating.

The physical system assumed here consists of a ferromagnetic medium occupying the region $x_2 > \zeta(x_1)$ and vacuum occupying the region $x_2 < \zeta(x_1)$. An external, static magnetic field H_o is directed along the x_3-axis, as is the saturation magnetization M_s. We seek straightcrested MSSW propagating in the x_1-direction.

We introduce a magnetic scalar potential $\phi(\vec{x},t) = \phi(x_1x_2|\omega)\exp(-i\omega t)$, in terms of which the Fourier coefficient of the demagnetizing field in the system is given by $\vec{h}(\vec{x}_1\vec{x}_2|\omega) = -\nabla\phi(x_1x_2|\omega)$. The dynamic magnetic induction in the medium is related to the magnetic field by $\vec{b}(x_1x_2|\omega) = \overleftrightarrow{\mu}(\omega)\vec{h}(x_1x_2|\omega)$, where the magnetic permeability tensor $\overleftrightarrow{\mu}(\omega)$ has the form

$$\overleftrightarrow{\mu}(\omega) = \begin{pmatrix} \mu_1(\omega) & i\mu_2(\omega) & 0 \\ -i\mu_2(\omega) & \mu_1(\omega) & 0 \\ 0 & 0 & 1 \end{pmatrix} . \tag{4.1}$$

In this equation $\mu_1(\omega) = 1+4\pi\omega_H\omega_M/(\omega_H^2-\omega^2)$, $\mu_2(\omega) = 4\pi\omega_M\omega/(\omega_H^2-\omega^2)$, where $\omega_H = \omega H_o$, $\omega_M = \gamma M_s$, and γ (<0) is the gyromagnetic ratio.

The magnetic scalar potential satisfies Laplace's equation in each medium, and outside the selvedge region can be written in the forms

$$\phi^>(x_1x_2|\omega) = \int\frac{dq}{2\pi}\,A(q\omega)e^{iqx_1-|q|x_2} \qquad x_2 > \zeta_{max} \tag{4.2a}$$

$$\phi^<(x_1x_2|\omega) = \int\frac{dq}{2\pi}\,B(q\omega)e^{iqx_1+|q|x_2} \qquad x_2 < \zeta_{min}. \tag{4.2b}$$

The boundary conditions in the problems are

$$\phi^>(x_1,\zeta(x_1)|\omega) = \phi^<(x_1,\zeta(x_1)|\omega) \tag{4.3a}$$

$$\hat{n}\cdot\vec{b}^>(x_1,\zeta(x_1)|\omega) = \hat{n}\cdot\vec{b}^<(x_1,\zeta(x_1)|\omega), \tag{4.3b}$$

where $\hat{n} = [1+(\zeta'(x_1))^2]^{-1/2}(-\zeta'(x_1),1,0)$ is the unit vector normal to the surface at each point. We invoke the Rayleigh hypothesis and use Eqs. (4.2) in the boundary conditions (4.3). When this is done we obtain a pair of coupled integral equations for $A(q\omega)$ and $B(q\omega)$:

$$\int\frac{dq}{2\pi}\,e^{iqx_1}[e^{-|q|\zeta(x_1)}A(q\omega)-e^{|q|\zeta(x_1)}B(q\omega)] = 0 \tag{4.4a}$$

$$\int\frac{dq}{2\pi}\,e^{iqx_1}(\{\zeta'(x_1)[iq\mu_1(\omega)-i|q|\mu_2(\omega)] + [-q\mu_2(\omega) + |q|\mu_1(\omega)]\}e^{-|q|\zeta(x_1)}A(q\omega) - [\zeta'(x_1)iq-|q|]e^{|q|\zeta(x_1)}B(q\omega)) = 0. \tag{4.4b}$$

At this point we simplify the problem by eliminating $B(q\omega)$ from the problem, to obtain only a single equation, for $A(q\omega)$. We do this by multiplying Eq. (4.4a) by

$[ip\zeta'(x_1)+|p|]\exp[-ipx_1+|p|\zeta(x_1)]$, Eq. (4.4b) by $\exp[-ipx_1+|p|\zeta(x_1)]$, integrating the resulting equations over all x_1, and adding the results. The equation for $A(q\omega)$ obtained this way can be written

$$D(p\omega)A(p\omega) = \int\frac{dq}{2\pi}\, J(|p|-|q||p-q)(1-\text{sgn}p\,\text{sgn}q)|q|A(q\omega), \tag{4.5}$$

where

$$D(p\omega) = \frac{1+\mu_1(\omega)-\text{sgn}p\,\mu_2(\omega)}{1-\mu_1(\omega)-\text{sgn}p\,\mu_2(\omega)}$$

$$= \frac{1+\mu_1(\omega)-\mu_2(\omega)}{1-\mu_1(\omega)-\mu_2(\omega)} \equiv D_+(\omega) \quad p > 0 \tag{4.6a}$$

$$= \frac{1+\mu_1(\omega)+\mu_2(\omega)}{1-\mu_1(\omega)+\mu_2(\omega)} \equiv D_-(\omega) \quad p < 0. \tag{4.6b}$$

Equations (4.5)-(4.6) provide the starting point for the discussion in the remainder of this section. I will use them to discuss magnetostatic surface shape resonances and the propagation of MSSW across periodically corrugated and randomly rough surfaces.

4.A. Magnetostatic Surface Shape Resonances

Equation (4.5) is not in the form of a standard eigenvalue problem because of the p-dependence of the factor $D(p\omega)$ appearing on its left hand side. However, it can be transformed into an eigenvalue problem. Because of the factor $(1-\text{sgn}p\,\text{sgn}q)$ in the integrand on the right hand side of Eq. (4.5), we see that p and q must have opposite signs in order that the integral not vanish. We can use this fact and Eqs. (4.6) to obtain the equation satisfied by $A(p\omega)$ for $p > 0$:

$$D_+(\omega)D_-(\omega)A(p\omega) = \int_0^\infty dq\; K(p|q)qA(q\omega), \tag{4.7}$$

where

$$K(p|q) = \frac{1}{\pi^2}\int_0^\infty dk J(p-k|p+k)kJ(k-q|-k-q) \quad p > 0,\ q > 0. \tag{4.8}$$

In a similar fashion we can obtain the equation satisfied by $A(p\omega)$ when $p < 0$:

$$D_+(\omega)D_-(\omega)A(p\omega) = \int_{-\infty}^{0} dq\ \overline{K}(p|q)|q|A(q\omega), \tag{4.9}$$

where

$$\overline{K}(p|q) = \frac{1}{\pi^2}\int_0^\infty dk\ J(|p|-k|p-k)kJ(k-|q||k-q) \qquad p < 0,\ q < 0. \tag{4.10}$$

Thus, from Eq. (4.7) we see that if λ_n is the nth eigenvalue of the kernel $K(p|q)q$, the frequency of the nth magnetostatic surface shape resonance is given by

$$D_+(\omega_n)D_-(\omega_n) = \frac{1}{(2\pi\omega_m)^2}\left[(\omega_H+2\pi\omega_M)^2 - \omega_n^2\right] = \lambda_n \tag{4.11}$$

which yields the simple relation

$$\omega_n = \left[\omega_{DE}^2 - (2\pi\omega_M)^2\lambda_n\right]^{1/2}, \tag{4.12}$$

where $\omega_{DE} = -\omega_H - 2\pi\omega_M$ is the frequency of the Damon-Eshbach magnetostatic surface wave on the planar surface of a semi-infinite magnetic insulator(65) (recall that γ, and hence ω_H and ω_M, are negative).

In the absence of a numerical determination of the $\{\lambda_n\}$ we can carry out an analytic determination of these eigenvalues in the small amplitude limit, in the case that the surface profile is given by

$$\zeta(x_1) = \frac{AR^2}{x_1^2+R^2}. \tag{4.13}$$

In this case

$$K(p|q) \cong \frac{1}{\pi^2}\int_0^\infty dk\ \hat{\zeta}(p+k)k\hat{\zeta}(-k-q)$$

$$= \frac{A^2}{4}\, e^{-(p+q)R}. \tag{4.14}$$

The eigenvalue equation becomes

$$\lambda_n a_n(p) = \frac{A^2}{4} e^{-pR} \int_0^\infty dq\, e^{-qR}\, q\, a_n(q). \tag{4.15}$$

Since the kernel of this integral equation is separable, we find readily that there is only one eigenvalue, and it has the value

$$\lambda_1 = \frac{A^2}{16R^2}. \tag{4.16}$$

The corresponding surface shape resonance frequency is

$$\omega_1 = [\omega_{DE}^2 - \omega_M^2 (\frac{\pi A}{2R})^2]^{1/2}. \tag{4.17}$$

It is lower than the frequency of the Damon-Eshbach mode, and in this approximation is the same for a ridge ($A < 0$) as for an indentation ($A > 0$).

The result given by Eq. (4.17) is also obtained if we start from Eqs. (4.9)-(4.10), as must be the case.

More modes, closer in frequency to ω_{DE}, are obtained if higher order terms in the expansion of $J(\alpha|Q)$ in powers of $\zeta(x_1)$ are kept in obtaining $K(p|q)$.

4.B. A Grating Surface

To study the propagation of a MSSW across a grating surface, we return to Eqs. (4.5)-(4.6) and set

$$A(p\omega) = 2\pi \sum_{n=-\infty}^{\infty} A_n(k\omega)\delta(p-k_n), \tag{4.18}$$

where $k_n = k+(2\pi n/a)$, a is the period of the grating, and k is the wave vector of the MSSW. The form (4.18) is required in order that $\phi^>(x_1x_2|\omega)$ given by Eq. (4.2a) have the Bloch form as required by the periodicity of the grating. The equation for $A_n(k\omega)$ obtained in this way is

$$D(k_m\omega)A_m(k\omega) = \sum_{n=-\infty}^{\infty} M_{mn}(k)A_n(k\omega) \qquad m = 0, \pm 1, \pm 2, \ldots \tag{4.19}$$

where

$$M_{mn}(k) = \frac{|k_m||k_n|-k_m k_n}{|k_m|(|k_m|-|k_n|)} \frac{1}{a}\int_0^a dx_1\, e^{-i\frac{2\pi}{a}(m-n)x_1} \times$$

$$\times\, e^{(|k_m|-|k_n|)\zeta(x_1)} \qquad m \neq n \tag{4.20a}$$

$$= 0 \qquad m = n. \tag{4.20b}$$

Equation (4.17) is invariant if k is replaced by $k+(2\pi/a)$. Thus, the frequencies $\omega(k)$ that are the solutions of Eq. (4.19) are periodic functions of k with period $2\pi/a$, and we can restrict k to the interval $-\pi/a < k \leqslant \pi/a$.

Equation (4.19) can be transformed into a standard eigenvalue problem in exactly the same way this was done in the preceding subsection. The result, which is the analogue of Eqs. (4.9)-(4.10), is for $k > 0$

$$D_+(\omega)D_-(\omega)A_m(k\omega) = \sum_{n=-\infty}^{-1} N_{mn}(k)A_n(k\omega) \qquad m < 0, \tag{4.21}$$

where

$$N_{mn}(k) = \sum_{p=0}^{\infty} M_{mp}(k)M_{pn}(k) \qquad m < 0,\ n < 0\ . \tag{4.22}$$

Exactly the same equation is obtained if it is assumed that $k < 0$. Thus the frequencies $\omega(k)$ that are the solutions of Eq. (4.21) are also even functions of k, and we can further restrict k to the interval $0 \leqslant k \leqslant \pi/a$. If we denote the nth eigenvalue of $\overleftrightarrow{N}(k)$ by $\lambda_n(k)$, the dispersion relation for the corresponding MSSW is given as before by

$$\omega_n(k) = [\omega_{DE}^2 - (2\pi\omega_M)^2\lambda_n(k)]^{1/2}. \tag{4.23}$$

The fact that the dispersion curves for MSSW on a grating are found to be reciprocal, i.e. $\omega_n(-k) = \omega_n(k)$, is in contrast to the result for a planar surface, on which in the present geometry no magnetostatic surface wave can propagate for negative k.

In Fig. 17 are plotted the frequencies of the first four branches of the dispersion curve for a MSSW on a sinusoidal grating defined by $\zeta(x_1) = \zeta_0\cos(2\pi x_1/a)$, with $\zeta_0/a = 0.5$.[61] The frequencies are measured from the flat surface Damon-Eshbach frequency ω_{DE}, and are normalized by $|\omega_H|$ in the case $M_s/H_o = 7/20$ that is appropriate to YIG. The higher frequency modes are too close to ω_{DE} to be resolved in this figure.

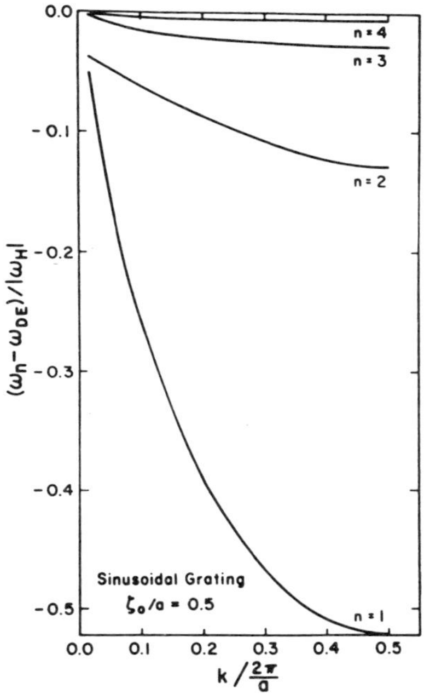

Fig. 17. The dispersion curves, for branches n = 1→4, for a magnetostatic surface wave on a sinusoidal grating with a corrugation strength $\zeta_o/a = 0.5$. The frequencies are plotted as measured from the flat surface Damon-Eshbach frequency ω_{DE} and normalized by $|\omega_H| = |\gamma|H_o$, in the case that $M_s/H_o = 7/20$, corresponding to YIG.

The multiplicity of branches in this dispersion curve can be understood in the following way. We have seen that an isolated ridge on an otherwise planar ferromagnetic surface can support an infinite number of surface shape resonances whose frequencies lie below the frequency of the Damon-Eshbach mode. When an infinite number of such ridges are brought together to produce a grating, the mutual interaction among the surface shape resonances spread these discrete frequencies into bands.

4.C. A Random Grating

The propagation of a MSSW across a random grating can be studied conveniently by starting from Eqs. (4.7)-(4.8) in the small roughness limit. If we apply the smoothing operator P and the complementary operator Q to Eq. (4.7) in turn we find that

$$D_+(\omega)D_-(\omega)PA(p\omega) = \int_0^\infty dq\ PK(p|q)q(PA(q\omega)+QA(q\omega)) \qquad (4.24a)$$

$$D_+(\omega)D_-(\omega)QA(p\omega) = \int_0^\infty dq\ QK(p|q)q(PA(q\omega)+QA(q\omega)), \quad (4.24b)$$

where now

$$K(p|q) \cong \frac{1}{\pi^2}\int_0^\infty dk\ \hat{\zeta}(p+k)k\ \hat{\zeta}(-k-q) \quad p > 0,\ q > 0. \quad (4.25)$$

Since we see from Eqs. (4.24b) and (4.25) that $QA(p\omega)$ is of $O(\zeta^2)$, we can neglect it on the right hand side of Eq. (4.24a), which then becomes

$$D_+(\omega)D_-(\omega)\langle A(p\omega)\rangle = \int_0^\infty dq\langle K(p|q)\rangle q\langle A(q\omega)\rangle. \quad (4.26)$$

It follows from Eqs. (4.25), (2.35), and (3.31b) that

$$\langle K(p|q)\rangle = \frac{1}{\pi^2}\int dk\ k\langle\hat{\zeta}(p+k)\hat{\zeta}(-k-q)\rangle$$

$$= \frac{2\delta^2}{\pi}\delta(p-q)f(p) \quad (4.27)$$

where

$$f(p) = \int_0^\infty dk\ kg(|k+p|). \quad (4.28)$$

The use of Eq. (4.27) in Eq. (4.26) immediately yields the dispersion relation

$$D_+(\omega)D_-(\omega) = \frac{\omega_{DE}^2-\omega^2}{(2\pi\omega_M)^2} = \frac{2\delta^2}{\pi}pf(p), \quad (4.29)$$

whence it follows that

$$\omega(p) = [\omega_{DE}^2 - \omega_M^2 8\pi\delta^2 pf(p)]^{1/2} \quad p > 0. \quad (4.30)$$

The same result is found when $p < 0$. Thus the dispersion curve for MSSW on a random grating is reciprocal, $\omega(-p) = \omega(p)$, just as it is in the case of a periodic grating.

In the case that the surface structure factor $g(|k|)$ has the Gaussian form,

$$g(|k|) = \pi^{1/2} a e^{-1/4 k^2 a^2}, \quad (4.31)$$

the function $f(p)$ is given by

$$f(p) = \frac{\sqrt{\pi}}{a} e^{-\frac{1}{4}\xi^2} F(\xi) \qquad (4.32)$$

where $\xi = ap$, and

$$F(\xi) = \int_0^\infty du\, u e^{-\frac{1}{2}\xi u - \frac{1}{4}u^2}$$

$$= 2\left[1 - \tfrac{1}{2}\sqrt{\pi}\,\xi + \tfrac{1}{2}\xi^2 - \tfrac{1}{8}\sqrt{\pi}\,\xi^3 + \ldots\right] \qquad \xi \ll 1 \qquad (4.33a)$$

$$= \frac{4}{\xi^2}\left[1 - \frac{6}{\xi^2} + \frac{60}{\xi^4} - \ldots\right] \qquad \xi \gg 1 . \qquad (4.33b)$$

The frequency $\omega(p)$, Eq. (4.30), then takes the form

$$\omega(p) = \left[\omega_{DE}^2 - \omega_M^2\, 8\pi^{3/2} \frac{\delta^2}{a^2} e^{-\frac{1}{4}\xi^2} \xi F(\xi)\right]^{\frac{1}{2}} \quad \xi > 0. \quad (4.34)$$

To obtain more branches to the dispersion curve we need to keep higher order terms in $\zeta(x_1)$ in the expression for $K(p|q)$, Eq. (4.8).

The effects of roughness disappear both when $\xi \to 0$ and when $\xi \to \infty$. In the former case this is because the wavelength of MSSW becomes much longer than the transverse correlation length of the surface roughness and it sees a flat mean surface. In the latter case the wavelength is much shorter than the transverse correlation length and it sees a locally flat surface. For intermediate values of ξ, when the wavelength of the MSSW is comparable with the transverse correlation length, the effect of roughness on the frequency of the MSSW is largest.

5. Some Directions for Future Work

In concluding this paper I should like to point to several directions in which it seems to me studies of surface waves on rough surfaces could go in the next few years.

The study of acoustic surface shape resonances would seem to be worthwhile. These are vibrational modes that are spatially localized in the vicinity of a protuberance or indentation on an otherwise planar surface of an elastic medium. Like their electromagnetic and magnetostatic counterparts they are characterized by discrete frequencies that are complex. The imaginary part of these frequencies arises because they fall in the range of frequencies allowed the vibrational modes of the unperturbed system, viz. a semi-infinite elastic medium bounded by a planar, stress-free surface. The surface shape resonances can therefore decay into the latter modes and acquire a finite lifetime as a result. The excitation of surface shape resonances, either by a volume elastic wave or by a surface acoustic wave impinging on the surface perturbation responsible for them, could give rise to an enhancement of the dynamic stress field in the vicinity of the surface perturbation that could be important for chemical processes occurring at surfaces.

There has been very little work done to date on the propagation of surface waves of any nature across doubly periodic gratings (bigratings). Such gratings are better deterministic models of randomly rough surfaces than are classical gratings, and at the same time the dispersion curves of surface waves on such surfaces should display a richer structure than is found in the case of classical gratings.

A study of the localization of Rayleigh waves propagating across a randomly rough surface, that parallels the recent studies of the localization of surface polaritons by surface roughness, would also be of interest. The localization of bulk elastic waves in disordered media has been discussed recently[51-53], and the analogue of this effect for surface acoustic waves should certainly exist as well.

Much of the work on surface waves to date has been concerned with surface waves on linear media. Interest is now growing in surface waves and surface shape resonances in systems that include nonlinear media. The propagation of surface acoustic, electromagnetic, and magnetostatic waves across gratings on nonlinear media, or on linear media in contact with nonlinear media, would be of interest, because the positions of the pass bands and stop bands for these waves would be modulated by changing the intensity of these waves. Similarly, the frequencies of electrostatic surface shape resonances associated with a protuberance or indentation on a metal film deposited on a nonlinear dielectric substrate would depend on the intensity of the probe used to excite them, e.g. a beam of electrons moving along a trajectory in the vicinity of the surface perturbation supporting these resonances. The structure proposed seems to be necessary for the realization of nonlinear electrostatic surface shape resonances because metals, which have the negative dielectric constants needed for the existence of such localized excitations, are notoriously linear, while nonlinear media required for the existence of nonlinear effects usually do not have negative dielectric constants.

Magnetostatic surface waves on randomly rough surfaces have been studied very little up to the present time. The brief discussion of MSSW on a random grating presented in the preceding section is the only such work known to me. The extension of that work to randomly rough surfaces characterized by a two-dimensional surface profile function should lead to some interesting results, particularly in connection with the effects of roughness on the nonreciprocity of these waves.

Finally, the effects of surface roughness on excitations localized to surfaces that are not planar in the absence of the roughness, but instead are cylindrical or spherical, for example, would be of interest in technological contexts such as the propagation of optical pulses through optical fibers, and in physical contexts such as in the optical properties of small particles of one dielectric medium embedded in a matrix of a second. Very little work on such systems has been carried out up to now.[66,67]

Acknowledgements

The work described in this paper was carried out during the past few years in collaboration with R. E. Camley, V. Celli, L. Dobrzynski, A. G. Eguiluz, N. E. Glass, O. Hardouin Duparc, B. Laks, R. Loudon, A. M. Marvin, A. R. McGurn, D. L. Mills, T. S. Rahman, T. P. Shen, W. M. Visscher, and W. Zierau. I am grateful to all of them. I am also grateful to Professors H. Bilz and W. Hanke for the hospitality of the Max Planck Institut für Festkörperforschung, Stuttgart, where this paper was written. This work was supported in part by ARO(D) Grant DAAG 29-85K-0025.

References

1. Lord Rayleigh, Phil. Mag. 14, 70 (1907); Theory of Sound, 2nd ed. (Dover, New York, 1945), vol. II, p. 89.
2. R. Petit and M. Cadilhac, C. R. Acad. Sci. B262, 468 (1966).
3. R. F. Millar, Proc. Camb. Phil. Soc. 69, 175, 217 (1971); Radio Sci. 8, 785 (1973).
4. N. R. Hill and V. Celli, Phys. Rev. B17, 2478 (1978).
5. P. M. van den Berg and J. T. Fokkema, J. Opt. Soc. Am. 69, 27 (1979); Radio Sci. 15, 723 (1980).
6. F. O. Goodman, J. Chem. Phys. 66, 976 (1977).
7. Lord Rayleigh, Proc. Lond. Math. Soc. 17, 4 (1887).
8. A. G. Eguiluz and A. A. Maradudin, Phys. Rev. B28, 728 (1983).
9. A. A. Maradudin and D. L. Mills, Ann. Phys. (N.Y.) 100, 262 (1976).
10. V. G. Polevoi, Akust. Zhur. 29, 91 (1983) [Soviet Physics-Acoustics 29, 52 (1983)].
11. N. E. Glass, R. Loudon, and A. A. Maradudin, Phys. Rev. B24, 6843 (1981).
12. N. E. Glass and A. A. Maradudin, J. Appl. Phys. 54, 796 (1983).
13. K. A. Ingebrigtsen and A. Tonning, Phys. Rev. 184, 942 (1969).
14. F. Rischbieter, Acustica 16, 75 (1965).

15. L. M. Brekhovskikh, Akust. Zhur. 5, 288 (1959) [Soviet Physics-Acoustics 5, 288 (1960)].
16. P.V.H. Sabine, Electronics Lett. 6, 149 (1970).
17. N. E. Glass and A. A. Maradudin, Electronics Lett. 17, 773 (1981).
18. A. A. Maradudin, in Nonequilibrium Phonon Dynamics, ed. W. E. Bron (Plenum, New York, 1985), p. 395.
19. O. Hardouin Duparc and A. A. Maradudin, J. Electron. Spectr. and Rel. Phenom. 30, 145 (1983).
20. G. A. Farias and A. A. Maradudin, Phys. Rev. B28, 5675 (1983).
21. A. A. Maradudin (unpublished work).
22. R. W. Rendell, D. J. Scalapino, and B. Mühlschlegel, Phys. Rev. Lett. 41, 1746 (1978).
23. R. W. Rendell and D. J. Scalapino, Phys. Rev. B24, 3276 (1981).
24. D. W. Berreman, Phys. Rev. 163, 855 (1967).
25. R. Ruppin, Solid State Commun. 39, 908 (1981).
26. J. I. Gersten and A. Nitzan, J. Chem. Phys. 75, 1139 (1981).
27. P. C. Das and J. I. Gersten, Phys. Rev. B25, 6281 (1982).
28. A. G. Mal'shukov and Sh. A. Shekhmamet'ev, Fiz. Tver. Tela 25, 2623 (1983) [Soviet Physics-Solid State 25, 1509 (1983)].
29. A. A. Maradudin and W. M. Visscher, Z. Physik B (to appear).
30. D. L. Jeanmaire and R. P. van Duyne, J. Electroanal. Chem. 84, 1 (1977); see, also, E. Burstein, S. L. Lundqvist, and D. L. Mills, in Surface Enhanced Raman Scattering, R. K. Chang and T. E. Furtak, eds. (Plenum, New York, 1982), p. 67.
31. C. K. Chen, A.R.B. de Castro, and Y. R. Shen, Phys. Rev. Lett. 46, 145 (1981); R. Reinisch and M. Neviere, Phys. Rev. B28, 1870 (1983); G. A. Farias and A. A. Maradudin, Phys. Rev. B30, 3002 (1984).
32. I. Pockrand, Opt. Commun. 13, 311 (1975).
33. I. Pockrand and H. Raether, Opt. Commun. 17, 353 (1976); 18, 395 (1976).
34. N. E. Glass and A. A. Maradudin, Phys. Rev. B24, 595 (1981).
35. N. E. Glass, A. A. Maradudin, and V. Celli, Phys. Rev. B26, 5357 (1982).
36. P. M. van den Berg and J.C.M. Borburgh, Appl. Phys. 3, 55 (1974).
37. B. Laks, D. L. Mills, and A. A. Maradudin, Phys. Rev. B23, 4965 (1981).
38. N. E. Glass, M. Weber, and D. L. Mills, Phys. Rev. B29, 6548 (1984).
39. E. Kretschmann, T. L. Ferrel, and J. C. Ashley, Phys. Rev. Lett. 42, 1312 (1979).
40. T. S. Rahman and A. A. Maradudin, Phys. Rev. B21, 2137 (1980).
41. R. E. Palmer and S. E. Schnatterly, Phys. Rev. B4, 2329 (1971).

42. R. Kötz, M. J. Lewerenz, and E. Kretschmann, Phys. Lett. 70A, 452 (1979).
43. M. W. Williams, J. C. Ashley, E. Kretschmann, T. A. Calcott, M. S. Chung, and E. T. Arakawa, Phys. Lett. 73A, 231 (1979).
44. T. Lopez-Rios and Y. Borensztein, Phys. Rev. B31, 5507 (1985).
45. A. A. Maradudin and W. Zierau, Phys. Rev. B14, 484 (1976).
46. D. L. Mills, Phys. Rev. B12, 4036 (1975).
47. E. Abrahams, P. W. Anderson, D. C. Licciardello, and T. V. Ramakrishnan, Phys. Rev. Lett. 42, 673 (1979).
48. P. W. Anderson, E. Abrahams, and T. V. Ramakrishnan, Phys. Rev. Lett. 43, 718 (1979).
49. D. Vollhardt and P. Wölfle, Phys. Rev. B22, 4666 (1980).
50. D. Vollhardt and P. Wölfle, in Anderson Localization, eds. Y. Nagaoka and H. Fukuyama (Springer-Verlag, New York, 1982), p. 26.
51. H. Levine and J. F. Willemsen, J. Acoust. Soc. Am. 73, 32 (1983).
52. S. John, H. Sompolinsky, and M. J. Stephen, Phys. Rev. B27, 5592 (1983). S. John and M. J. Stephen, Phys. Rev. B28, 6358 (1983).
53. C. A. Condat and T. R. Kirkpatrick, Phys. Rev. B32, 495 (1985).
54. S. John, Phys. Rev. Lett. 53, 2169 (1984).
55. S. John, Phys. Rev. B31, 304 (1985).
56. K. Arya, Z. B. Su, and J. L. Birman, Phys. Rev. Lett. 54, 1559 (1985).
57. A. R. McGurn, A. A. Maradudin, and V. Celli, Phys. Rev. B31, 4866 (1985).
58. V. Celli, A. A. Maradudin, A. M. Marvin, and A. R. McGurn, J. Opt. Soc. Am. (to appear).
59. N. F. Mott and E. A. Davis, Electronic Processes in Non-Crystalline Materials, 2nd ed. (Clarendon Press, Oxford, 1979), pp. 62-64.
60. P. B. Johnson and R. W. Christy, Phys. Rev. B6, 4370 (1972).
61. R. E. Camley, N. E. Glass, and A. A. Maradudin, J. Appl. Phys. 53, 3170 (1982).
62. C. G. Slykes, J. D. Adam, and J. H. Collins, Appl. Phys. Lett. 29, 388 (1976).
63. M. Tsutsumi, Y. Sakaguchi, and N. Kumagai, Appl. Phys. Lett. 31, 779 (1977).
64. N. S. Chang and Y. Matsuo, Appl. Phys. Lett. 35, 352 (1979).
65. R. W. Damon and J. R. Eshbach, J. Phys. Chem. Solids 19, 308 (1961).
66. P. Mazur and D. L. Mills, J. Appl. Phys. 54, 3735 (1983).
67. R. G. Lee, Jr., G. A. Farias, A. A. Maradudin, and B. Mühlschlegel (unpublished work).

SURFACE MAGNETOSTATIC WAVES IN PERIODIC STRUCTURES

by

S.A. NIKITOV; Yu. V. GULYAEV
IRE Academy of Science, Moscow, USSR

The paper is devoted to a theoretic investigation of propagation of surface magnetostatic waves (MSSW) in a plates (films) of ferromagnets, one of the surface of which has a periodic roughness or a system of thin metallic conductor or semiconductor. The main method, which was used in the work, is the method of two coupled waves. According to this method the wave function in the region of roughness has a form of Bloh function. Fourier-series of a periodic member of the Bloh function contains only two of the first members. These waves have a fase velocity directed along and opposite axe of propagation. The use of this method gives a possibility to obtain the analytical decisions for the coefficient of attenuation of waves near a periodic part of surface of plate, coefficient of reflection, and other coefficient.

The different physics of Bragg reflection MSSW from the periodic roughness part of surface and from the periodic system of thin metallic conductor (semiconductor) leads to the qualitative and quantitative results. The fase velocity of the wave is modulated in the case of periodic roughness. The coefficient of absorption of the wave is modulated in the case of periodic system of conductor (semiconductor).

The theory of resonator Fabri-Perot is developed on the MSSW in the work.

EXCITATION OF SURFACE ELECTROMAGNETIC WAVES BY REB IN COAXIAL PLASMA RESONATOR

Rukhadze A.A.
General Physics Institute USSR
Academy of Sciences, Moscow

1. In the spatially bounded plasma there excist not only volume but also surface electromagnetic waves. The field of surface waves decreases with distances from the plasma boundary and the decrease rate dependences on the plasma density. Therefore it is possible to excitate the plasma surface waves by a thin annular electron beam passing near the plasma boundary. By varieing the distance between the beam and plasma boundary one can easily change the plasma-beam coupling and investigate both regimes of the surface waves excitation the one-particle (Compton) when this distance is small as well as the collective (Raman) when this distance is sufficiently large.

In the long wavelength limit the phase velocity of the surface waves is close to the velocity of light. Therefore such waves are highly nonpotential and easily radiated out of the plasma. Moreover for sufficiently dense plasma the dispersion of the surface waves is linear which opens the possibility for the effective energy transformation from REB into the electromagnetic radiation in the wide range of wave frequency.

The above advantages stimulated the investigations [1,2] in the General Physics Institute Academy of Sciences of USSR where was realized the wide frequency range surface wave generator exciteded by REB. This is the first successful experimental realization of powerful UHF generator using the mechanism of stimulated Cherenkov radiation in a plasma, predicted by soviet and americal physicists A.I.Akhiezer, Ya.B.Fainberg [3] and D.Bohm, E.Gross [4] in 1949.

The previous experimental attempts on electromagnetic wave excitation by nonrelativistic electron beams up to 70-th were unsuccessful. That may be due to the small phase velocity and strong potentiality of the field of electromagnetic waves excitated by the nonrelativistic beams. Such waves were really excitated very effectively in a plasma. But the problem of taking this radiation out of the plasma remined unsolved.

The summary of the experimental results on nonrelativistic plasma electronics was made in ref. [5,6,7] which sounded as a minore finale in this field of UHF electronics. The success of the experiments [1,2] was provided by using REB and exciting the nonpotential surface plasma waves. The relativistic phase velocity of such waves solved the problem of radiation energy output from the plasma.

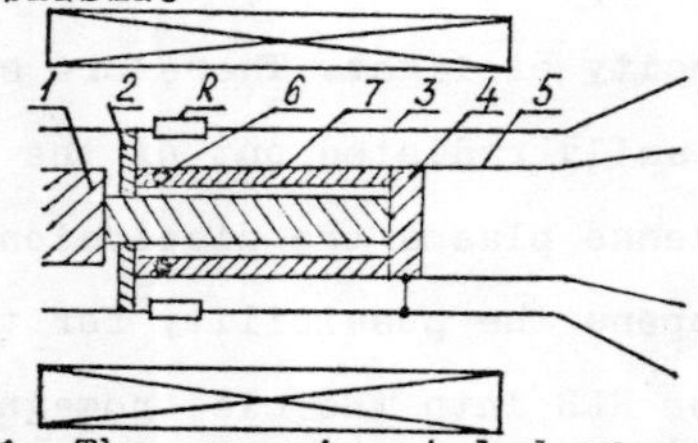

Fig.1. The experimental device for excitation of a plasma surface wave: 1 - kathod, 2 - anod, 3 - cylindrical waveguide, 4 - collector, 5 - central waveguide, 6 - plasma source, 7 - annular plasma

2. Fig. 1 shows a principle scheme of experimental device. REB formed in the "Terek-II" accelerator with an energy of electrons $\mathcal{E}$ = 0,3-0,5 MeV, current $\mathcal{J}_b$ = 1-10 kA, pulse duratinn τ = 40 nsec is injected into the cylindrical resonator

with the radius R_0 = 1.5 cm and L = 10-30 cm in length which was preliminary filled with a annular plasma with the radius r_p = 0.7 cm and thickness $\Delta_p \approx 1$ mm produced by a low energy beam ($\mathcal{E} \approx$ 1.5 keV, $J_b \approx 10$ A). REB is injected either inside the plasma cylinder or out of it blowing over the plasma and exciting the main mode of a plasma surface wave (TEM mode of a cable wave). Besides REB configuration and current a change was mode in angular spread of beam electrons φ = 0 - 20^0, system length L = 10-30 cm, plasma density $n_p \simeq 10^{12}$-10^{14} cm^{-3} and induction of an external magnetic field B_0 = 15-30 kGauss. During the experiments the wavelength (frequency), power and structure of the field excited by REB emerging from the plasma of electromagnetic radiation were measured depending on the above parameters.

Fig.2 shows field structure of the mine mode of a plasma surface wave. Field components E_z , E_r and B_φ differs from zero [8]

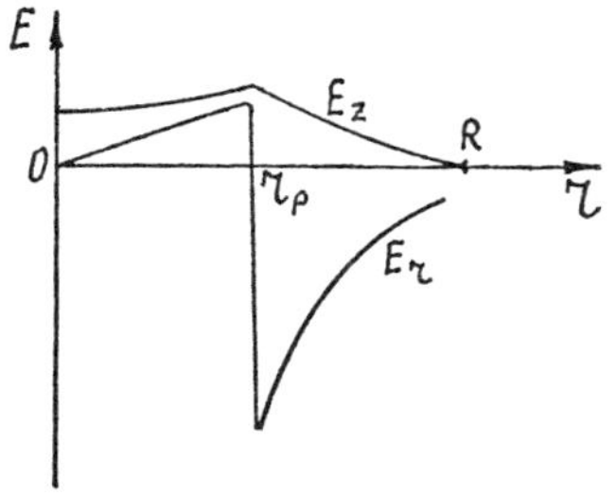

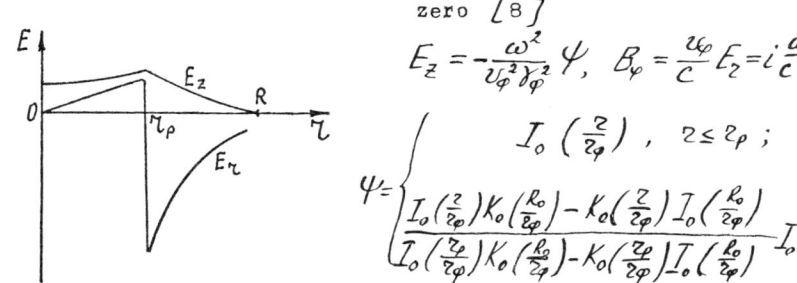

$$E_z = -\frac{\omega^2}{v_\varphi^2 \gamma_\varphi^2}\Psi, \quad B_\varphi = \frac{v_\varphi}{c}E_r = i\frac{\omega}{c}\frac{\partial \Psi}{\partial r}, \tag{1}$$

$$\Psi = \begin{cases} I_0\left(\frac{r}{r_\varphi}\right), & r \le r_p ; \\ \dfrac{I_0\left(\frac{r}{r_\varphi}\right)K_0\left(\frac{R_0}{r_\varphi}\right) - K_0\left(\frac{r}{r_\varphi}\right)I_0\left(\frac{R_0}{r_\varphi}\right)}{I_0\left(\frac{r_p}{r_\varphi}\right)K_0\left(\frac{R_0}{r_\varphi}\right) - K_0\left(\frac{r_p}{r_\varphi}\right)I_0\left(\frac{R_0}{r_\varphi}\right)} I_0\left(\frac{r_p}{r_\varphi}\right), & r_p < r. \end{cases}$$

Fig.2. The structure of the cable mode of a plasma surface wave.

Here $r_\varphi = \frac{v_\varphi \gamma_\varphi}{\omega}$, $v_\varphi = \frac{\omega}{K_z}$, ω - is frequency and K_z - is longitudinal wave number, $\gamma_\varphi = \left(1 - v_\varphi^2/c^2\right)^{-1/2}$. Just this mode of a plasma surface wave is excited by REB in the experiments [1,2] due the Cherenkov resonance when $v_\varphi = u$ and $\gamma_\varphi = \gamma$. One mode excitation in the considered system by REB is possible only under the condition

$$\frac{2u^2\gamma^2}{r_p \Delta_p}\left(1-\frac{r_p^2}{R_0^2}\right) > \omega_p^2 > \frac{u^2\gamma^2}{r_p \Delta_p\left(1+\ln\frac{R_0}{r_p}\right)}, \qquad (2)$$

where $\omega_p = \sqrt{4\pi e^2 n_p / m}$ — is Langmuir frequency of plasma electrons.

The main mode of a plasma surface wave has two very importent advantages. The first is that in the case of Cherenkov resonance the beam electrons interact with the E_z -component of wave field. It can be seen in Fig.2 that this field component is large and practically constant in the inner part of the plasma cylinder. As a result if REB is injected into this inner part then its coupling with the wave field is strong and the Compton regime of wave excitation takes place. But if annular REB is injected out of the plasma cylinder were E_z -component of wave field is small then the Ruman regime of wave excitation will be realized. This it is possible to investigate experimentally the both regimes of wave excitation and to compare their efficiency.

The second advantage is that the transverse field components E_z and B_φ are large as it can be seen from Fig.2 only in the outside part of the plasma cylinder. Therefore almost the full flux of electromagnetic energy is localized in this part of the resonator and it is no problem of effective transformation of the excited mode of surface wave into the output electromagnetic radiation by a coaxial metalic waveguide (horn).

It must be noticed that this two advantages can be useful

if the phase velocity of excited waves is close to the velocity of light. The reflection coefficient of such waves from the horn is of the order of

$$\varkappa = \frac{1 - v_\varphi/c}{1 + v_\varphi/c} \approx \frac{1}{4\gamma^2}, \qquad (3)$$

when $\gamma^2 \gg 1$ then $\varkappa \ll 1$ and the efficiency of radiation output from the plasma is very high.

The relativistic phase velocity of surface waves provide the wide frequency range in which the effective amplification by REB is possible. It is seen from Fig.3 that the dispersion curve $\omega = \omega(k_z)$ of the main mode of the plasma surface wave is linear in the frequency range $\omega < \omega_p$ (in the Fig.3 are shown the lines $\omega = k_z c$ and $\omega = k_z u$) and the resonance amplification takes place for the surface waves with

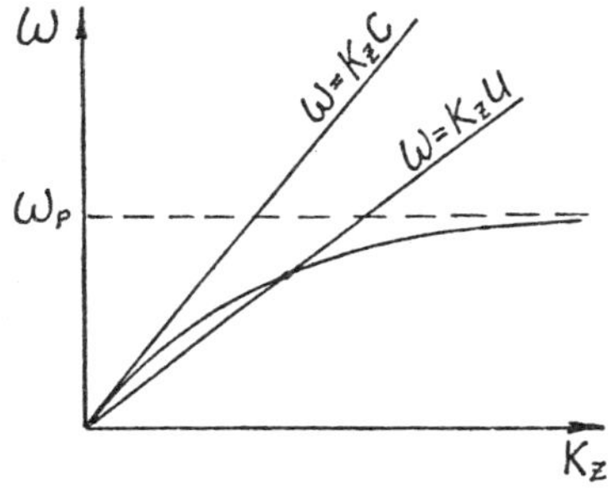

Fig.3. The dispersion curve for a plasma surface wave.

$$\omega = \sqrt{\omega_p^2 - \frac{u^2\gamma^2}{r_p \Delta_p (1 + \ln R_0/r_p)}} \,. \qquad (4)$$

This formula shows that it is possible of increasing the radiation frequency by increasing the plasma density.

Finally we'l write the condition when the relations given above are valid. This condition shows that the external longitudinal magnetic field must be sufficiently strong such as

$$\Omega^2 \gg (\gamma^2 - 1)^2 \omega_p^2, \qquad (5)$$

where $\Omega = eB_0/mc$ - is the electron Larmoure frequency. For $\gamma = 3$ and $\omega = 10^{11}$ sec^{-1} (i.e. $n_p \approx 3.10^{12}$cm^{-3}) it follows from (5) that $B_0 >$ 50 kGauss. In the experimental conditions [1,2] $\gamma \leq 2$ and therefore $B_0 \gtrsim$ 15 kGauss is sufficient.

3. Let us now begin to discuss the experimental investigations [1,2] . Without considering the experimental measurements in detail we'l give here only the results and their explanation.

Ref. [1] describes experiments on injection of a solid REB with a radius r_b = 0.5 cm inside a plasma cylinder in the case of $\mathcal{E}$ = 480 keV and B_0 = 22.5 kBauss. Fig.4 shows the power of total radiation versus plasma density at I_b = 0.9 kA and L = 24 cm. The radiation is seen to occure when $n_p >$ 5.10^{12}cm^{-3}, then sharply increase and begin to fall slowly when $n_p > 2\cdot 10^{13}$cm^{-3} which agree with theory discussed above.

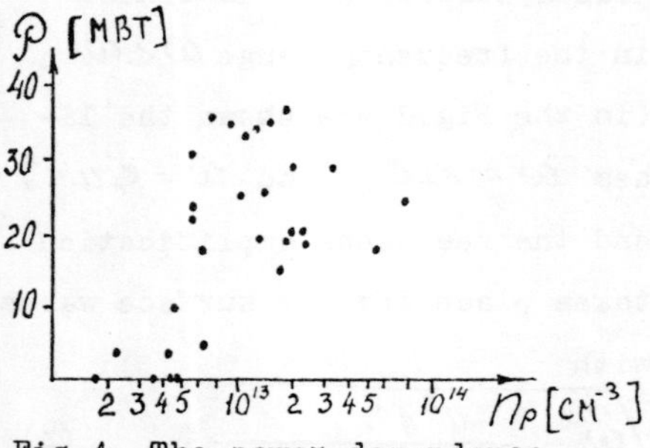

Fig.4. The power dependence on the plasma density.

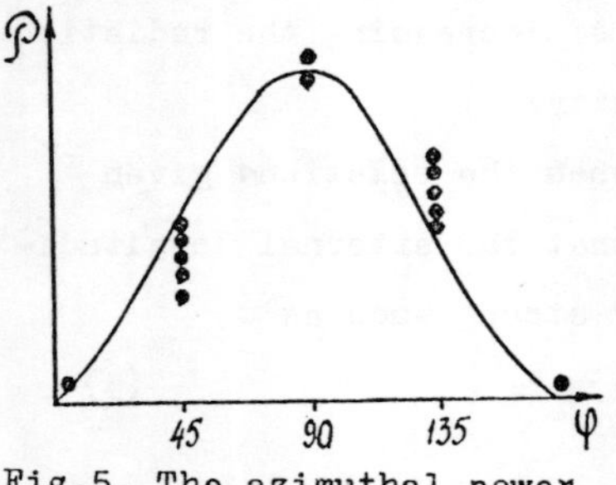

Fig.5. The azimuthal power dependence.

The stable distribution of emerging radiation at the horn aperture is observed when a beam collector is electrically connected to the waveguide walls by means of two radial plates

(Fig.1). The azimutal power distribution for this case is shown in Fig.5. This distribution together with the radial dependence $\mathcal{P} \sim 1/r^2$ shows that the radiation corresponds to the main mode of coaxial waveguide with parpartiations and it is close to the mode H_{10} of a rectangular waveguide with the dimentions $[\pi R, R_2 - R_1]$, where $R = \frac{1}{2}(R_1 + R_2)$, R_1 and R_2 are internal and external radiuses of the coaxial waveguide. These results confirmed that in our system it really excitates the main mode of the annular plasma surface waves.

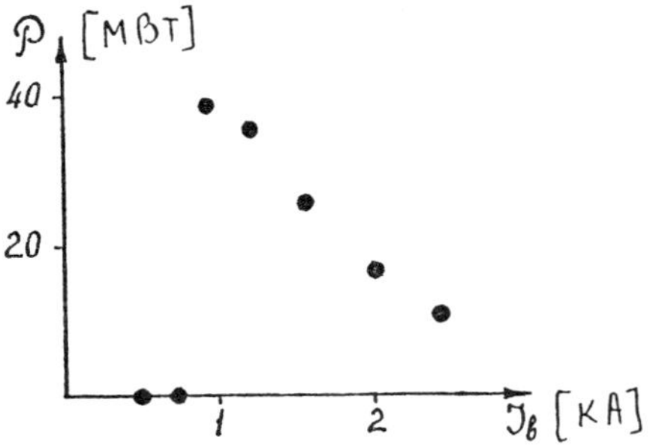

Fig.6. The power dependence on the beam current.

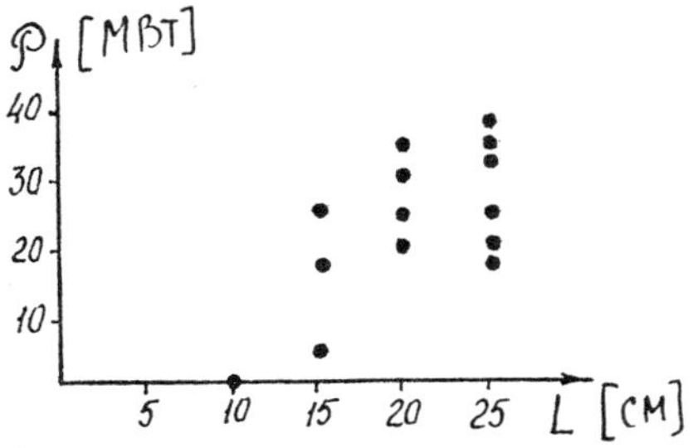

Fig.7. The power dependence on the resonator length.

The power dependences on the beam current or on the resonator length provide also evidence for the nature of observed radiation. These dependences are given in Fig.6 (in the case of L = 24 cm and n_p = 5.10^{12}-3.10^{13}cm^{-3}) and in Fig.7 (in the case of J_b = 0.9 kA and $n_p \approx$ 5.10^{12} - 3.10^{13}cm^{-3}). It can be seen that the radiation power increases sharply at $J_b \approx 0.9$ kA and then slowly falls. Similar power dependence is observed when the resonator length increases - radiation ocurres if $L > 15$ cm and it increases up to $L \approx 25$ cm. Maximal radiation in the wavelength range $\lambda \simeq 1.8 - 3.2$ cm is $\mathcal{P} \simeq$ 30-35 MW what corresponds to the maximal efficiency of wave generation $\eta \simeq 8\%$. The relative width of radiation lines does not exceed 20%.

We also note that the radiation wavelength is retuned by changing plasma density in the range of $5.10^{12}\,cm^{-3} < n_p < 3.10^{13}\,cm^{-3}$ and it decreases when the plasma density increases. This conclusion is in a good agreement with theory. It follows from the condition (2) that the radiation must occure in the considered system at $n_p \simeq 6.10^{12}\,cm^{-3}$ and if $n_p > 2.10^{13}\,cm^{-3}$ its intensity must fall as a result of excitation of two or more modes of the plasma oscillations. The sufficiently large E_φ -component of exciting field is observed experimentally when $n_p > 2.10^{13}$ cm^{-3} which corresponds to the H-tipe plasma oscillations in the output radiation.

Under the conditions when the solid REB is injected inside a plasma cylinder the generation regime is always oneparticle (Compton). If the REB current is lower then the vacuum limiting current J_0 (for considered system $J_0 \simeq 3$ kA) the theoretical value of REB threshold current and the radiation effeciency in this case are equal [1,8]

$$J_{th} \simeq 1{,}75 \cdot 10^{-2} S_p \frac{\omega_p^2}{c^2} \gamma(\gamma^2-1)^2 \left(\frac{\lambda}{L} \ln \frac{3}{\varkappa}\right)^3 \text{ kA},$$
$$\eta \simeq 0{,}7 \left(4\gamma \frac{S_b}{S_p} \frac{n_b}{n_p}\right)^{1/3} \qquad (6)$$

Here S_b and S_p are the cross-section squires of beam and plasma correspondingly, $\lambda = \frac{2\pi c}{\omega}$ - is the radiation wavelength and $\varkappa$ is given in exp. (3). The generation occures if $J > J_{th} =$

$= J_{th}$ = 0.5 kA and its efficiency increases with REB current reaching $\eta_{max} \simeq$ 11% at $J_b \simeq$ 1.5 kA and then it falls as J_b^{-1} (see curve 1 in Fig.8). The threshold current and the maximal efficiency are in a good agreement with the results of the described experiments. But the radiation power which must be constant when J_b > 1.5 kA really falls slowly. May be it is the result of influence beam space charge effects because the beam current is close to the vacuum limiting current J_0 = 3 kA.

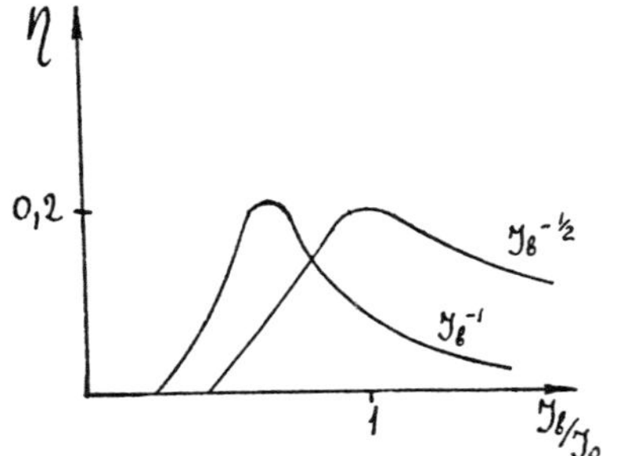

Fig.8. The efficiency dependence on the beam current:
1 - Compton regime, 2 - Ruman regime.

In a quantitative agreement with theory are also the experimental results for radiation power dependence on the resonator length. As Fig.7 shows the radiation occures when L > 15 cm and then its power increases up to L = 25 cm. The theoretical value of the threshold length is just equal L_{th} = 15 cm when $J_b \simeq$ 1 kA.

4. An attempt to increase radiation power by increasing beam current in the considered plasma configuration is described in ref. [2]. Experiments with a annular REB blowing up the plasma cylinder were carryed out in order to achieve high beam currents and to increase radiation power. As has already b een mentioned in this case both Compton and Ruman operation regimes of a plasma generator are possible. If

$$G = \frac{\ln R_0/r_p}{\ln R_0/r_b} > 64\gamma \frac{S_b n_b}{S_p n_p} \tag{7}$$

then the Compton regime takes place and in the contrary limit of an intense beam the Ruman regime may be realized. For the Ruman regime the threshold current and the generation efficiency are given by the expressions [8]

$$J_{th} = 1{,}4\cdot 10^{-2} S_p \frac{\omega_p^2}{c^2}\gamma(\gamma^2-1)^{5/2}\frac{1}{G^3}\left(\frac{\lambda}{L}\,\mathrm{arcch}\,\frac{1}{\varkappa}\right)^4 kA,$$

$$\eta = \left(G\gamma\frac{S_b n_b}{S_p n_p}\right)^{1/2} \qquad (8)$$

In the experiments [2] the plasma cylinder with radius r_p = 0.9 cm, thickness Δ_p = 0.1 cm and length $L \approx$ 30 cm was blowed up by the annular REB with radius r_b = 1.25 cm and thickness Δ_b = 0.1 cm; the beam electron energy is $\mathcal{E} \approx$ 420 KeV. The Fig.9 shows that the radiation occures when

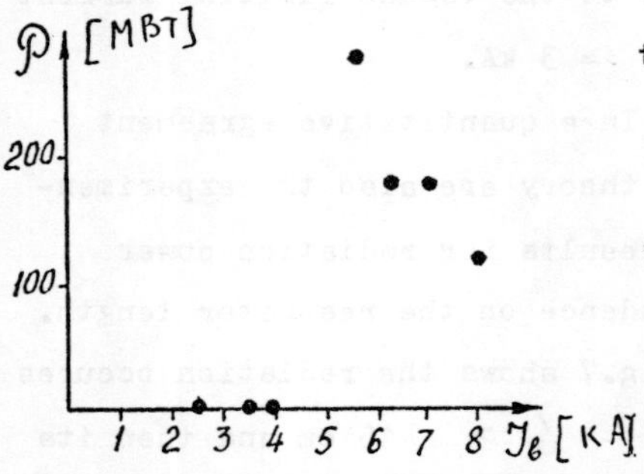

Fig.9. The power dependence on the blowing beam current.

$J_b >$ 5 kA and its power reaches $\mathcal{P}$ = 300 MW at $J_b \approx$ 6 kA (the efficiency being $\eta \approx$ 15%) and then falls up to $\mathcal{P} \approx$ 130 HW at $J_b \approx$ 8 kA. The accurate measurements of plasma density were not made, but according to the excited field structure which remined the same as in the described above experiments [1] the plasma density is apparently of the order of $n_p \approx 5.10^{12}-3.10^{13}$ cm^{-3}. In this region of plasma density the radiation wavelength varied from λ = 3.2 cm up to λ = 1.8 cm.

In the experimental conditions $\frac{n_b}{n_p} > 10^{-2}$ and therefore $G >$

> 1. This means that Ruman operation regime of plasma generator is apprently realized with the efficiency beam current dependence as is showing in Fig.8 on the curve 2. The theoretical estimations of the generation efficiency and the REB threshold current derived from exp. (8) are $\eta_{max} \approx$ 17% and $J_{th} \simeq 5$ kA which are the same as experimental ones. An appreciable decrease in radiation power which is observed in the experiments when REB current exceed 7-8 kA (see Fig.9) can be explined by the fact that is close to the vacuum limiting current $J_0 \approx 10$ kA.

From the analizes which was carried out above it follows that in the experiments [1,2] really was observed the excitation of main mode of the plasma surface waves. All experimental results demonstrate that the beam excitation of this mode is very suitable for the energy transformation from REB into the electromagnetic radiation and for the achievement of high powers of UHF radiation in the plasma surface waves generators.

REFERENCES

1. Kuzelev M.V. at all Sov. Phys. Doklady 1982 v.267 p.829; Sov. Phys. JETP 1982 v.83 p.1352.
2. Strelkov P.S., Shkvarunets A.G. The IV All-Union Seminar on relativistic UHF electronics, Moscow 1984 p.76.
3. Akhiezer A.I., Fainberg Ya.B., Sov. Phys. Doklady 1949 v.65, p.555.
4. Bohm D., Gross E., Phys. Rev. 1949, v.75, p.18151.
5. Fainberg Ya.B., Sov. Phys. Atomnaia Energia 1961 v.11 p.313; Sov. Phys. Uspekhy 1969 v.93, p.619.

6. Bernashevski G.A., Bogdanov E.V., Kislov V.Ya.,Chernov Z.S. "Plasma Amplifiers and UHF electronics", Moscow, "Radio"1965.
7. Nezlin M.V. Sov. Phys. Uspekhy 1970 v 102 p 105; "Beams dinamics in a plasma" Moscow, Atomizdat, 1979.
8. Kuzelev M.V., Mukhamedzianov F.Kh., Shkvarynets A.G. Sov. Phys. Plasma Phys. 1983 v.9 p.1137.

EXPERIMENTAL OBSERVATION OF SURFACE AND INTERFACE MODES BY LIGHT SCATTERING

J.R. Sandercock

Laboratories RCA Ltd.,
Badenerstrasse 569, CH-8048 Zurich, Switzerland

Abstract - A review is given of recent light scattering measurements at surfaces and interfaces with a view to understanding the nature of the observed phonon modes and the mechanisms by which they couple to light.

The study of surface and localised phonon modes in layered materials dates back to the investigations of Lord Rayleigh[1] into the propagation of earthquakes. More recently the development of surface acoustic wave devices lead to a resurgence of interest in these excitations. Investigations have been made using ultrasonic techniques and recently atomic and electron scattering, but the most powerful technique presently available is that of inelastic light scattering. This latter technique became possible with the development of the high contrast tandem Fabry-Perot interferometer.

This paper summarises the properties of the different surface and localised phonon states and introduces the rather complex problem of deriving the cross-section for light scattered from such states. Results of scattering measurements from a variety of different substrates and substrate/film combinations are presented. Finally, the analog problem of scattering from spin waves in magnetic substrates and films is briefly discussed.

The Light Scattering Experiment

For some years light scattering has been used as a tool to study phonons in bulk material. The interaction between the phonon and light was described by Brillouin[2] in terms of the fluctuation in the dielectric constant produced by the strain associated with the phonon. Since the development of the laser, Brillouin scattering has ripened into a useful and diversified technique.

Figure 1 illustrates the technique, where for simplicity a backscattering arrangement is shown.

A transparent sample has been assumed with the scattering volume contained within the bulk of the material. The incident light of wavevector $\underline{k}^i$ and frequency w^i is scattered by a

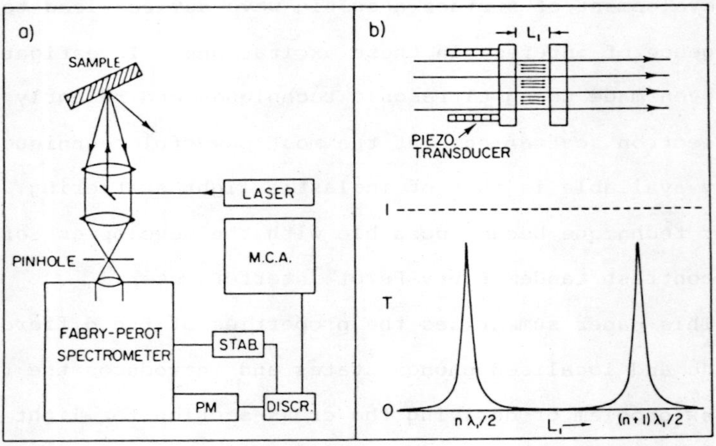

Fig. 1. a) Experimental set-up for backscattering with a Fabry-Perot Interferometer
b) Transmission of monochromatic light by scanning Fabry-Perot interferometer.

phonon $\underline{q}, \Omega$ into the state $\underline{k}^s$, w^s such that

$$\underline{k}^s - \underline{k}^i = \pm \underline{q}$$

and
$$w^s - w^i = \pm \Omega \tag{1}$$

The + sign refers to absorption of the phonon, the - sign to emission. (Respectively the anti-Stokes and Stokes processes).

For backscattering in a transparent medium of refractive index n we have $q = 2nk_o$ ($\underline{k}_o$ is the vacuum wavevector of the laser light). The scattered light contains components at the frequencies $w^i \pm \Omega$ corresponding to all excitations present having the above wavevector. A normal spectrum will therefore show peaks shifted by frequencies Ω_L, Ω_{T1} and Ω_{T2} due to scattering by the longitudinal and transverse phonon modes.

The scattered light is analysed using a Fabry-Perot interferometer as illustrated in Fig. 1b. While this instrument is adequate for the observation of scattering in transparent materials, it is quite inadequate for studying excitations near surfaces particularly in opaque materials. In this case the light scattered elastically from surface defects swamps the much weaker Brillouin components. This problem can be overcome by the use of the high contrast multipass interferometer[3] - however even then a difficulty remains, namely that neighbouring orders of interference are not widely separated from one another with the result that Brillouin spectra measured in neighbouring orders overlap. This is particularly troublesome for spectra containing many features or broad features -

typical of the spectra to be described from surface and interfaces. This problem has recently been overcome with the introduction of the tandem multipass interferometer[4, 5, 6], an instrument which combines high resolution, high contrast and a strong suppression of neighbouring interference orders. A novel scanning stage common to both interferometers automatically ensures synchronisation and forms the basis of a compact and highly stable design.

An alternative scheme relying purely on the electronic coupling of two separate interferometers[7] has also been demonstrated.

A typical spectrum taken using the tandem interferometer is shown in Figure 2. This spectrum[6] measured in backscattering on the surface of Ge, shows the depolarised light scattered from transverse phonons and demonstrates two important aspects of light scattering near surfaces.

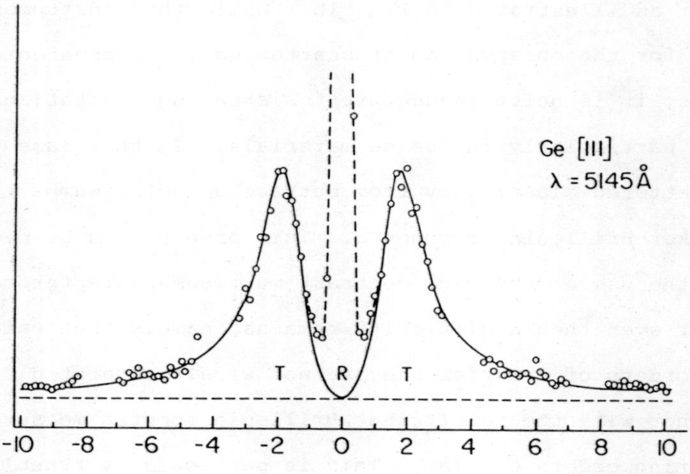

Fig. 2. Depolarised spectrum of transverse phonons in Ge.

a) The peaks are seen to be considerably broadened. This broadening is not related to the phonon lifetime, but arises due to the high optical absorption[8] which limits the penetration of the light to within a few wavelengths from the surface. The wavevector conservation condition of equation 1 only applies when all dimensions of the scattering volume are large compared to the wavelengths involved. For high optical absorption the equality is only approximate - a range of phonon wavevectors $\Delta q \sim \alpha$ (where α is the absorption coefficient) centred around $\underline{q} = \underline{k}^s - \underline{k}^i$ all contribute to the scattering and so give rise to the observed broadening effect.

b) Although the linewidth is in good agreement with simple calculations the marked asymmetry observed cannot be explained in terms of optical absorption. It was shown by Pine and Dresselhaus[9] (in an albeit incorrect calculation) that the assymetry arises from the coherent reflection of the phonons at the surface.

The above two facts, namely that the presence of the surface (and interfaces) relaxes the wavevector conservation requirement and modifies the phonon states with respect to the bulk, are of fundamental importance to an understanding of the light scattering experiments.

We consider first the modification to the phonon modes.

Phonon Modes in the Presence of Surfaces and Interfaces

The solutions of the equations of motion for an elastic continuum are the well known longitudinal and transverse acoustic excitations. In the presence of a surface, which intoduces a stress-free discontinuity, new excitations appear, namely the Rayleigh, Lamb and Love modes.

A good feeling for the nature of these modes can be obtained following the eloquent treatment of Auld[10] using the very simple transverse resonance technique applied to a slab of thickness b in the z direction. The simplest solution is a purely transverse mode polarised parallel to the surface. This mode is rather similar to a guided light wave and is known as the Love mode. More complicated solutions involve both transverse and longitudinal components. Three different frequency regimes may be described in terms of the propagation vector parallel to the surface.

a) $\omega/\beta > V_\ell$ where V_ℓ is the normal bulk longitudinal sound velocity. In this case both longitudinal and transverse wavevector components k_z and k_{zt} (parallel to surface normal) are real and the wave solutions are normal bulk modes.

b) $V_\ell > \omega/\beta > V_t$. Here, k_{zt} is real but $k_{z\ell}$ is purely imaginary. The solutions represent bulk transverse modes combined with a longitudinal component which is strongly localised at the surface. These modes are referred to as Lamb waves.

c) $\omega/\beta < V_t$. In this case, both transverse and longitudinal components are localised at the surfaces. These modes are related to the Rayleigh surface mode of a semi-inifinite solid.

In the limit $\beta b \rightarrow \infty$ the discrete Lamb and Love solutions form continua in which the Love waves are indistinguishable from normal bulk modes. The Lamb modes, on the other hand, are bulk transverse modes with an additional longitudinal component localised at the surface.

The Rayleigh mode and the surface component of the Lamb mode are localised typically within one wavelength from the surface.

The first observation[11] of these plate modes was reported for the rather trivial case of near-normal incidence scattering from thin films of collodion and from thin crystalline platelets. For this case $\beta \rightarrow 0$ with the phonons propagating backward and forward across the plate. The Lamb solution separates into a Love wave and a longitudinal wave with $k_z = m\pi/b$. The restriction of the scattering volume in the z direction leads to a wavevector uncertainty $\Delta q \sim 1/b$ with the result that typically 3 of the longitudinal modes could be observed.

A more complex example is in the spectrum of GaAs[12] shown in Figure 3. Here we see the bulk phonon peaks, absorption broadened as in the case of Ge discussed above. In addition the Rayleigh surface phonon is clearly resolved together with a broad shoulder out to about 20 GHz arising from scattering from the coninuum of Lamb waves. The shoulder extends from $\omega = q'' V_t$ to $\omega = q'' V_l$ as expected from the discussion on the nature of the Lamb wave.

A surprising feature of the spectrum however, is that the Rayleigh surface wave peak is more intense than the bulk phonon peak. Since the Rayleigh wave is localised within a wavelength from the surface, the interaction volume with the light is an order of magnitude smaller than for the bulk phonons. The origin of the strong Rayleigh wave scattering was explained by Mishra and Bray[13] who pointed out that the phonons ripple the surface and that light may be directly scattered from these

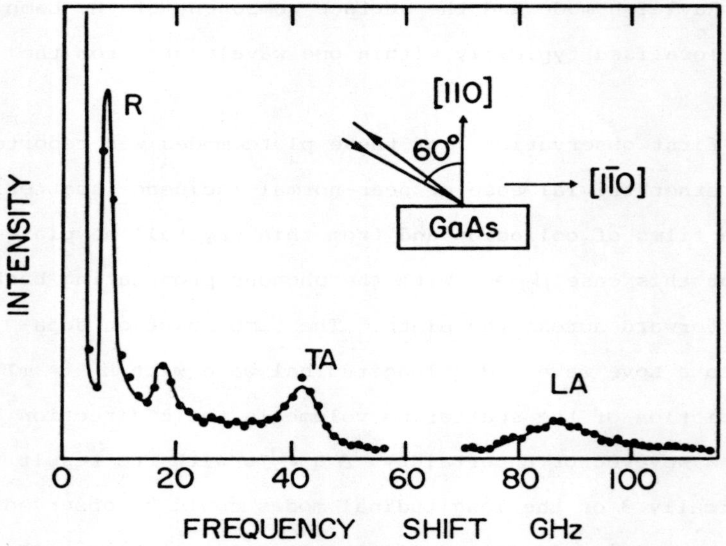

Fig. 3. Spectrum of GaAs showing scattering form both bulk (L,T) and surface modes.

ripples. The total scattered intensity must be calculated therefore in terms of both elastooptic and ripple scattering mechanisms allowing for the interference between the two. The calculations have been performed by several groups[12, 14-19] and the theory found to be in excellent agreement with the measurements. Notice that the spectrum of Figure 2 is depolarised, there being in this case no contribution from the ripple mechanism.

An important result of the calculation of the ripple scattering crossection is that in general higher reflectivity materials will scatter more strongly. In particular, phonons in metals may be easily observed via this mechanism. In Figure 4 are shown measurements on polycrystalline Al[20]. The Rayleigh wave and the Lamb continuum are clearly resolved and the inset

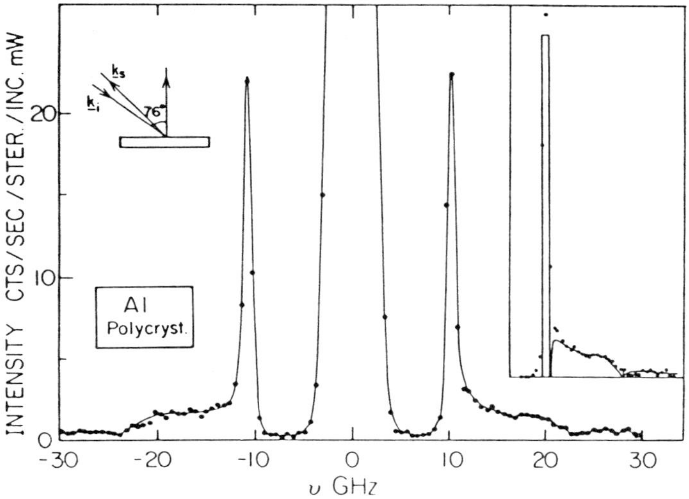

Fig. 4. Spectrum of polycrystalline Al showing Rayleigh wave and continuum of Lamb waves. Inset theoretical fit to data.

shows the excellent agreement with theory[14]. Earlier measurements by Dil and Brody[21] on liquid metals have also been satisfactorily explained in terms of the ripple mechanism[16, 22].

In addition, a variety of measurements of thin films on substrates have been performed. The substrate, of course, modifies the properties of the plate modes. Before discussing these measurements we qualitatively summarise the properties of a supported plate.

Modes of a Plate on a Half-Space

Assuming the lower surface of the plate to be the interface with the half-space, the reflection coefficient for phonons at the lower plate surface is now no longer unity. This face seriously complicates the analysis since the transmitted waves must be included. This section is restricted to a qualitative discussion of the modes with emphasis on their similarity to the modes of the unsupported plate.

The solutions of interest are those which are trapped within the plate, corresponding to total internal reflection at the interface with only an evanescent wave extending into the substrate.

Love Mode: Provided $V_t' > V_t$ where V_t' is the transverse velocity in the substrate, a purely transverse solution exists similar to the Love mode of the unsupported plate.

Generalised Lamb Mode: For $V_t' < V_t$ only one solution is possible which in the limit $\beta b \rightarrow 0$ becomes the Rayleigh wave on the substrate.

For $V_t' > V_t$ there exists a large number of solutions. As $\beta b \rightarrow 0$ only one solution is possible which becomes the Rayleigh wave of the bare substrate. As βb increases more modes become possible referred to as Sezawa modes. As $\beta b \rightarrow \infty$ the lowest Sezawa mode becomes the Rayleigh wave of the top surface of the plate.

Stoneley wave: Under certain conditions for $V_{t'} \simeq V_t$ a bound interface mode exists with velocity V_s satisfying $V_R' < V_s < V_t'$ where V_R' is the Rayleigh velocity of the bare substrate.

Following are some examples of light scattering from supported films. In general a calculation of the light scattering cross-section becomes complicated - in the most general case of a transparent film on a transparent substrate there are contributions to the scattering form both surface and interface ripples and from the elastooptic effect in both film and substrate. In addition, limitations on the scattering volume imposed by film thickness and/or optical absorption modulate the scattering cross-section as discussed above.

An example of the simplest situation in which only the surface ripple is important is shown in Figure 5 for scattering from a film of Al on Si. This spectrum should be compared with that of pure Al in Figure 4. Notice that the continuum of Lamb waves observed in Figure 4 becomes discrete modes for the thin film, The continuous line of Figure 5 is the calculated cross-section[23]. Vacher and coworkers have reported measurements of the Rayleigh wave alone on Al films on different substrates[24].

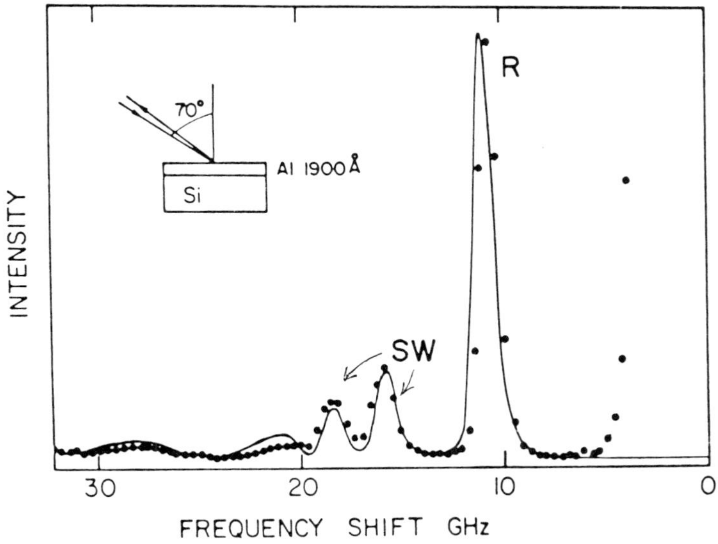

Fig. 5.. Spectrum of thin film of Al on Si showing Rayleigh and discrete Sezawa (Lamb) waves.

Measurements on SiO_2 on Si could be theoretically explained in detail[25] using the known elastooptic and elastic properties of film and substrate. On the other hand, interpretation of spectra of Au films on Si has yielded[26] the previously unknown elastooptic constants of Au.

Of interest are measurements by Dil and coworkers[7] on polycarbonate films on glass in which Love modes were successfully observed. The same authors also report a measurement of the Stoneley wave at the interface between glass and liquid Hg. The observed spectrum has been well described by Albuquerque[27] in terms of ripple scattering from the interface.

As expected for low reflectivity materials the spectra from polycarbonate/glass samples are found to be very weak. Rowell and Stegeman[28] have demonstrated that prism coupling techniques may be used to couple the light more effectively and obtained strong spectra of Rayleigh and Sezawa modes in optical waveguide structures. Interesting results for scattering from smooth Ag films on glass have recently been reported by Moretti and coworkers[29] in which coupling via the surface plasmon was used to obtain an enhancement of the intensity of the Rayleigh wave scattering by a factor of about 25. Attempts to observe enhancement from silver island films failed.

Hillebrands and coworkers[30] have investigated Au films on NaCl. Rayleigh and Sezawa waves were observed as a function of film thickness with emphasis on the thinnest films where the Au coverage becomes distontinuous. Surprisingly, in the discontinuous region, a well resolved mode was observed at an energy where trapped modes are not expected. A satisfactory explanation for these results has not yet been found.

Light Scattering from Magnetic Materials

Spinwaves in magnetic materials modulate the dielectric constant of the medium (via the spin orbit interaction) and so can scatter light. Early measurements on YIG demonstrated that the scattered intensity is comparable to that from phonons, and further that a lot of information relating to the magnetic system may be derived from the spectra. Several measurements on relatively transparent materials have been discussed in a short review by Borovik-Romanov and Kreines[31].

Two differences between phonon spectra and spinwave spectra are of particular interest:

a) Even in the region of wavevectors accessible to light scattering (less than 1% of the Brillouin zone) spinwaves show strong dispersion due to the exchange interaction. Measurements as a function of wavevector thus yield useful information - by comparison phonons show no dispersion in this range.

b) the lack of time reversal symmetry between Stokes and anti-Stokes scattering events leads to marked asymmetry in the observed spectra.

The early measurements on relatively transparent materials showed only scattering from bulk spinwaves. More recently, measurements on opaque ferromagnets have been reported in which surface features have been observed. The surface spinwave, first described by Damon und Eshbach[32] has the strange property of non-reciprocal propagation. More precisely on a surface with magnetisation in the plane of the surface a surface spinwave may only propagate from left to right across the field direction. As a result in a light scattering measurement (which involves absorption of an excitation in one direction and emission in the opposite direction) the surface spin wave peak will be observed only in the Stokes or anti-Stokes spectrum depending on the field direction. This strange asymmetry was first observed by Grunberg and Metawe[33] in EuO. Similar observations have been reported in Fe and Ni[34].

There has recently been considerable interest in scattering from thin magnetic films both theoretically[35-39] and experimentally[40-42]. The interested reader is referred to a recent review article by Grunberg[43].

Conclusions

Brillouin scattering has ripened into a powerful tool for the study of small wavevector excitations, both phonons and spinwaves, at surfaces and interfaces on materials ranging from transparent dielectrics to highly opaque metals.

References

1. Lord Rayleigh, London Math. Soc. Proc. 17, 4 (1887).
2. Brillouin, L., Ann. Phys. (Paris) 17, 88 (1929).
3. Sandercock, J.R., 2nd Int. Conf. on Light Scattering in Solids (Paris: Flammarion 1971) p. 9.
4. Sandercock, J.R., Proc. VII Int. Conf. on Raman Spectroscopy (New York: North Holland) p. 346.
5. Lindsay, S.M., Anderson, M.W., Sandercock, J.R., Rev. Sci. Instrum. 52, 1478 (1981).
6. Sandercock, J.R., Topics in Applied Physics 51, 173 (1982).
7. Dil, J.G., van Hijningen, N., van Dorst, F., Aarts, R.M., App. Opt. 20, 1374 (1981).

8. Sandercock, J.R., Phys. Rev. Lett. 28, 237 (1972).
9. Dresselhaus, G., Pine, A.S., Sol. State Comm. 16, 1001 (1975).
10. Auld, B.A., Acoustic Fields and Waves in Solids, Vol. 2 (New York: Wiley 1973).
11. Sandercock, J.R., Phys. Rev. Lett. 29, 1735 (1972).
12. Marvin, A., Bortolani, V., Nizzoli, F., Santoro, G., J. Phys. C13, 1607 (1980).
13. Mishra, S., Bray, R., Phys. Rev. Lett. 39, 222 (1977).
14. Loudon, R., Phys. Rev. Lett. 40, 581 (1978).
15. Subbaswami, K.R., Maradudin, A.A., Phys. Rev. B18, 4181 (1978).
16. Rowell, N.L., Stegeman, G.I., Phys. Rev. B18, 2598 (1978).
17. Loudon, R., Sandercock, J.R., J. Phys. C13, 2609 (1980
18. Velasco, V.R., Garcia-Moliner, F., Sol. State Comm. 33, 1 (1980).
19. Marvin, A.M., Bortolani, V., Nizzoli, F., J. Phys. C13, 299 (1980).
20. Sandercock, J.R., Sol. State Comm. 26, 547 (1978).
21. Dil, J.G., Brody, E.M., Phys. Rev. B14, 5218 (1976).
22. Dervisch, A., Loudon, R., J. Phys. C11, L291 (1978).
23. Bortolani, V., Nizzoli, F., Santoro, G., Marvin, A., Sandercock, J.R., Phys. Rev. Lett 43, 224 (1979).
24. Sussner, H., Pelous, J., Schmidt, M., Vacher, R., Sol. State Comm. 36, 123 (1980).
25. Bortolani, V., Nizzoli, F., Santoro, G., Sandercock, J.R., Phys. Rev. B25, 3442 (1982).
26. Bortolani, V., Nizzoli, F., Santoro, G., J. de Phys. C5, 45 (1984).
27. Albuquerque, E.L., Phys. Stat. Sol. (b) 118, 223 (1983).
28. Rowell, N.L., Stegeman, G.I., Phys. Rev. Lett 41, 970 (1978).
29. Moretti, A.L., Robertson, W.M., Fisher, B., Bray R., Phys. Rev. B31, 3361 (1985).
30. Hillebrands, B., Mock, R., Guntherodt, G., Bechtold, P.S., Herres, N., J. Phys. C to be published.

31. Borovik-Romanov, A.S., Kreines, N.M., J. Magn. and Magn. Mater. 15-18, 760 (1980).
32. Damon, R.W., Eshbach, J.R., J. Phys. Chem. Solids 19, 308 (1961).
33. Grunberg, P., Metawe, F., Phys. Rev. Lett. 39, 1561 (1977).
34. Sandercock, J.R., Wettling, W., J. Appl. Phys. 50, 7784 (1979).
35. Cottam, M.G., J. Phys. C11, 165 (1978).
36. Camley, R.E., Mills, D.L., Phys. Rev. B18, 4821 (1978).
37. Camley, R.E., Grunberg, P., Mayr, C.M., Phys. Rev. B26, 2609 (1982).
38. Camley, R.E., Rahman, T.S., Mills, D.L., Phys. Rev. B23, 1226 (1981).
39. Cottam, M.G., J. Phys. C16, 1573 (1983).
40. Grimsditch, M., Malozemoff, A., Brunsch, A., Phys. Rev. Lett. 43, 711 (1979),
41. Grunberg, P., Cottam, M.G., Vach, W., Mayr, C.M., Camley, R.E., J. Appl. Phys. 53, 2078 (1982).
42. Vernon, S.P., Lindsay, S.M., Stearns, M.B., Phys. Rev. B29, 4439 (1984).
43. Grunberg, P., Prog. in Surface Science 18. 1 (1985).

Numerical and Experimental Studies of Nonlinear EM Guided Waves

G.I. Stegeman and C.T. Seaton
Optical Sciences Center and Arizona Research Laboratories
Tucson, AZ 85721, USA

Abstract

The power dependence of waves guided by thin dielectric or metallic films bounded by various combinations of linear and nonlinear media are discussed and experimental evidence for their existence reviewed. Numerical examples are used extensively to illustrate the change in propagation wavevector with guided wave power for s-polarized waves guided in the presence of a single self-focussing or self-defocussing bounding medium. This includes lowest order solutions which begin to guide above a threshold power, and waves which exhibit a maximum in the power which they can transmit. The relevance of the two experiments reported to date for verifying the theoretical predictions is assessed. For two self-focussing bounding media, up to three branches with different power thresholds are obtained for each guided wave order, and it becomes possible, in principle, to switch between them. Similar behaviour is found numerically for p-polarized waves guided by thin metal films. Of particular interest is the prediction that s-polarized waves can be guided above a power threshold by a thin metal film bounded on both sides by self-focussing media.

1. Introduction

Since the advent of high power lasers in the early 1960's it has been well-known that intense optical beams can modify the refractive index of the medium through which they propagate. This effect, which is intrinsic to all materials, has been viewed as detrimental to many applications of high power lasers and has been investigated extensively for finite aperture, for example Gaussian, beams propagating in unbounded media. Materials characterized by a refractive index which increases locally with the intensity of an electromagnetic wave are called self-focussing media. Since the refractive index (and hence phase velocity) is directly (inversely) proportional to the beam intensity, a curvature is introduced into a wavefront which leads to the focussing of the optical beam into a narrow filament whose diameter is determined primarily by the balance between beam power and diffraction. This effect leads to very high power densities and hence to the onset of material damaging phenomena such as Stimulated Brillouin Scattering, or the ionization of the molecules or atoms of the medium leading to plasma creation. For materials in which the refractive index decreases with increasing intensity (self-defocussing media), the resulting wavefront curvature results in an expanding beam cross-section and hence a reduction in intensity with propagation distance. This defocussing effect is detrimental to applications of nonlinear optics because in some cases desireably large intensities cannot be maintained over usefully long distances.

Such nonlinear material properties lead to interesting phenomena when applied to waves guided by surfaces. The advantage of using a wave guided by single or multiple boundaries (versus wave propagation in unbounded media) is that the beam does not undergo diffraction in the direction normal to the interface, and hence a beam has a stable cross-section in that dimension. Therefore, the field distributions should become power dependent, but stable with respect to propagation distance. This leads to wave phenomena which have no analogs in nonlinear plane (or focussed) wave optics. Multiple guided wave branches become possible, corresponding to self-focussing in each self-focussing medium.

These effects have been investigated both theoretically and experimentally in a number of interface geometries. Light beams incident from a linear onto a nonlinear medium at an angle just greater than the critical angle for total reflection exhibit anomalous reflection and transmission properties due to the creation of "self-focussed" channels in the nonlinear transmission medium[1-9]. The finite aperture of the beams must be taken explicitly (numerically) into account when analysing this phenomenon, which makes the results vary with the details of the beam spatial profile. This phenomenon has been discussed in detail in the past and will not be pursued here. The present discussion will be restricted to guided wave geometries in which the fields are evanescent in all media at large distances from the guiding interface(s). This paper essentially continues the discussion of this phenomenon initiated in the preceding paper by Boardman and Egan[10].

The simplest case is a single interface between two semi-infinite media[11-20]: This geometry was discussed in detail by Boardman and Egan in the preceding paper[10]. For two dielectric media, at least one of the two media must exhibit a self-focussing nonlinearity for waves guided by the interface to exist above a power threshold. If one of the media is metallic, the interface can support surface plasmon polaritons for both self-focussing or self-defocussing nonlinearities[21].

The basic principles of wave guiding in the presence of one or more nonlinear media has been treated in detail in the preceding paper[10]. There the underlying assumptions were outlined, the basic formulae developed and interpreted, and some representative results discussed. In this paper we build on this theoretical exposition and examine some of the unique features predicted for a large variety of different material properties and thin film guiding geometries. In particular we examine numerical results for both self-focussing and self-defocussing media bounding thin films[22-47].

Only two experiments dealing with nonlinear guiding have been reported to date[48,49]. Both have used a single nonlinear self-focussing medium bounding a deposited dielectric film. Here we discuss the interpretation of these experiments in terms of nonlinear guided waves with power dependent field distributions.

II TE Waves Guided by Thin Films

(a) Thin Film Geometry

The geometry shown in Fig 1 consists of a thin film of refractive index n_f (relative dielectric constant $\varepsilon_f=n_f^2$) and thickness "h" ($0\geq z\geq h$), bounded by a cladding (n_c, n_{2c} and $0\geq z$) and a substrate (n_s, $\varepsilon_s=n_s^2$, n_{2s}, and $z\geq h$). The nonlinearity is of the form $n = n_\gamma + n_{2\gamma}S$ or $\varepsilon = \varepsilon_\gamma + \alpha_\gamma|E|^2$ where γ identifies the medium, S is the local intensity and $\alpha_\gamma = n_\gamma^2 c\varepsilon_0 n_{2\gamma}$.

Although the nonlinear wave equation has not yet been solved in closed form for lossy media, it is useful to estimate how the guided wave loss varies with guided wave power, based solely on power dependent changes in the fraction of power carried in the different guiding media. Assuming that the field distributions calculated on the basis of lossless media are still approximately valid in the presence of small loss, it is relatively straight-forward to calculate the power dissipated in each medium and hence an approximate attenuation coefficient. This was done for some of the curves shown in this paper in which case the effective index is written as $\beta = \beta_R + i\beta_I$.

(b) Nonlinear Cladding Medium: Dielectric Films

This case corresponds to the simplest one for nonlinear waves guided by thin films[41]. In general, for $n_{2c}>0$ ($n_{2c}<0$), the effective index increases (decreases) with increasing power. This increase (decrease) is initially proportional to the guided wave power, a result which can be simply predicted from coupled mode (perturbation) theory. When the optically induced change in the refractive index Δn in the cladding becomes comparable with the initial index difference n_f-n_c, dramatic changes in the field distributions occur and many of the guided wave parameters become strongly power dependent. It is this case which we concentrate on here.

(i) $n_{2c}>0$, $n_{2s}=0$

For self-focussing nonlinearities ($n_{2c}>0$), the dependence on power of the TE_0 (lowest order solution) is different from that of TE_m, $m\geq 1$. The field distribution is of the form

$$E_{cy}(x,z) = \frac{1}{2}\sqrt{\frac{2}{\alpha_c}}\;\frac{q}{\cosh[k_0 q(z_1-z)]}\;e^{i(\omega t-\beta k_0 x)} + \text{c.c.} \qquad n_{2c}>0 \qquad (1)$$

where $k_0=\omega/c$, $q^2 = \beta^2 - n_c^2$, and β is the effective index. The parameter z_1 depends on the guided wave power. At low powers, $z_1 \to \infty$. As power increases z_1 decreases and can become negative which corresponds to a self-focussed field in the cladding. The evolution of the TE_0 field distributions with film thickness are illustrated in Fig 2. This illustrates two of the characteristic features of nonlinear guided waves, namely power-dependent field distributions and power-dependent shifts in the position of the field maximum (beam axis). The variation in the effective index β with power (β-power plot shown in Fig 3) illustrates three additional characteristics of nonlinear guided waves, namely power-dependent effective index (wavevector), solutions which terminate at a maximum value of β in the β-power plot and lowest order solutions which degenerate into single-interface surface guided waves at high powers. For all film thicknesses, the TE_0 solution degenerates at high powers into a self-focussed surface wave guided by the nonlinear cladding-film interface alone, and hence all of the curves for different film thicknesses converge at high powers. The formation of a self-focussed peak in the cladding requires that the cladding index ($n = n_c + n_{2c}S$) at the cladding-film interface is larger than the film index n_f. For thick films, the power carried by the wave at this field value can be larger than that required for the single-interface wave, which results in a maximum in the guided wave power, as shown in Fig 3.

Self-focussing in the nonlinear cladding also occurs for higher order TE_m, $m \geq 1$, solutions. The field extremum closest to the cladding moves with increasing β into the nonlinear cladding medium, see Fig 2. Since these field solutions require at least one field extremum in the film, the field distributions must be oscillatory in the film. This implies $\beta > n_f$ and therefore these waves cannot asymptotically evolve into nonlinear single interface waves (as the TE_0 solutions do). As illustrated in Fig 4, the branches all terminate at some value of $\beta < n_f$. Therefore, for a self-focussing cladding there is an absolute maximum in the power which can be propagated in any TE_m, $m > 1$ guided wave. These waves can be used for optical limiters in a variety of applications, providing that the excitation geometry precludes the excitation of the TE_0 wave which exhibits no absolute power maximum.

All of the features discussed above are qualitatively the same for both symmetric ($n_c = n_s$) and asymmetric ($n_c \neq n_s$) waveguides. Strong power dependence of the guided wave solutions occurs when the cladding index near the boundary becomes larger than the film index. This implies that the optically induced refractive index change in the cladding must be larger than the low power index difference $n_f - n_c$.

The existence of the nonlinear cladding medium affects the cut-off conditions for a thin film waveguide. It is well-known that an asymmetric waveguide ($n_c \neq n_s$) cannot support guided waves below a critical thickness h_{co}. As the film thickness is decreased, $\beta \rightarrow$

$\max[n_C,n_S]$ and cut-off occurs when $\beta = \max[n_C, n_S]$. However, for a self-focussing cladding medium, increasing guided wave power at any given film thickness increases β and it takes the waveguide further from cut-off[40,50]. Therefore, at high powers, wave propagation can be sustained for film thicknesses $h<h_{CO}$. This is illustrated in the inset of Fig 5. The thinner the film, the higher the minimum power required for TE_0 wave propagation, as shown in Fig 5. Therefore, for asymmetric waveguides, there is a minimum power threshold for TE_0 wave propagation for film thicknesses below cut-off. This phenomenon can clearly be used as a lower power threshold device.

For each TE_m wave, $m\geq1$, wave propagation is allowed for a small range of thicknesses just below the respective cut-off thickness. The further the thickness is below the cut-off value, the smaller the range of powers over which propagation is allowed.

There are also possible applications of the TE_0 curves to optical switching and hysteresis. Whether such effects occur or not is intrinsically related to the method of excitation of the guided wave. For example, if the nonlinear waveguide is joined onto two linear waveguides as shown in Fig 6, the nonlinear waves can be excited at the transverse linear-nonlinear waveguide boundary by guided waves incident from the linear side. The waveguide geometry required is one in which the film thicknesses is (a) too small to sustain any wave other than TE_0, and (b) large enough so that there is a maximum in the β-power plot of the type shown in Fig 3. In the absence of numerical calculations to define the exact threshold conditions for the excitation of a given nonlinear wave, we also make the assumption that the nonlinear wave is excited only when the field value of the incident wave exceeds that of the nonlinear wave at every point along the transverse interface. Therefore, as the power in the linear waveguide is increased past that needed to reach the local maximum in the β-power plot, the transmitted power saturates until the incident field produces a sufficiently strong field to excite nonlinear waves on that part of the curve on which power again rises with increasing β. Assuming that further increases in incident power lead to preferential excitation of these essentially single-interface waves, the transmitted power suddenly drops, and then increases again monotonically with incident power. (There are indications that regions in which increasing β leads to decreasing power are unstable to propagation.) As the power is now reduced, the minimum in the high β part of the curve is reached and that branch can no longer be excited. The system now must switch to the low-β curve with potentially an increase in transmitted power. Determination of whether hysteresis will occur or not requires detailed numerical investigations of the field matching conditions at the transverse boundary.

(ii) $n_{2c}<0$, $n_{2s}=0$

The self-defocussing case does not lead to the rich variety of new guided wave characteristics as was the case for self-focussing. For a self-defocussing cladding, the guided wave index decreases monotonically with increasing power until waveguide cut-off is reached, asymptotically with increasing power in the case $n_s>n_c$, and at a finite power level for $n_c>n_s$. For cut-off with $\beta \to n_s$ ($n_s \geq n_c$), the guided wave field degenerates into a plane wave in the substrate and the guided wave power carried by the substrate[37,41,50]

$$P_s = \frac{\beta}{2k_0 s n_c^2 |n_{2c}| \sinh^2(k_0 q z_1)} \left[\cos(k_0 \kappa h) + \frac{q}{\kappa} \coth(k_0 q z_1) \sin(k_0 \kappa h)\right]^2 \qquad (2)$$

diverges since $s = \sqrt{\beta^2 - n_s^2} \to 0$ and $\kappa \to \sqrt{n_f^2 - n_s^2}$. For $n_c>n_s$, cut-off occurs as $\beta \to n_c$ so that P_s remains finite, and

$$P_c = \frac{\beta q}{k_0 n_c^2 |n_{2c}|} [\text{cotnh}(k_0 q z_1) - 1] \to \frac{1}{k_0^2 z_1 n_c |n_{2c}|} \qquad (3)$$

which also remains finite since $z_1>0$. This case is unique since the field in the cladding into which cut-off occurs does not degenerate into a plane wave but takes the form $E_y \propto [z_1-z]^{-1}$. The change from a plane wave (linear waveguide) at cut-off to a field with curvature (nonlinear cladding) is caused by the local decrease in the refractive index in the cladding near the cladding-film boundary. A sample calculation of the decrease in effective index with power is shown in the inset of Fig 7. The power at which cut-off occurs can be tuned by varying the film thickness, or the index difference between the film and the cladding. This phenomenon exhibits the characteristics of an optical limiter.

(c) Nonlinear Cladding and Substrate Media: Dielectric Films

The guided waves in this geometry exhibit many novel characteristics, especially for the all self-focussing case[41].

(i) $n_{2c}>0$, $n_{2s}>0$

The asymptotic ($\beta>n_f$) behaviour of the fields in this case can be understood strictly in terms of the phenomenon of self-focussing, without resort to numerical calculations of the power dependent dispersion relations. At high powers, self-focussed fields can occur in either of the two cladding media, or in both of them. Thus, three asymptotic curves are expected, which implies multiple branches, and therefore implies power thresholds for all but one branch (the one with a low power limit). As will be shown later, whether the branches are connected or not depends on the relative values of n_γ and $n_{2\gamma}$.

TE_0

First we examine the dispersion curves for a material system with complete symmetry, namely $n_c=n_s$ and $n_{2c}=n_{2s}$[41]. There are three TE_0 branches, two of which have power thresholds and are degenerate with respect to their β-power characteristics, but not their field distributions. Their evolution with film thickness is shown in Fig 8 and the variation in the field distributions with β in Fig 9. Curve A corresponds to self-focussing in both media such that the field distribution remains symmetric about the film center. The fields associated with curve B are self-focussed in either the cladding, or the substrate. Note that these waves are unique since the fields do not reflect the symmetry of the material system, that is optical intensity has been used via an intensity dependent refractive index to break material symmetry. At high powers, curve A degenerates into two single interface surface waves, one at each film-bounding medium boundary. Each of the two degenerate curves B evolves into a single interface surface wave at one film boundary. Hence in the asymptotic limit it requires twice as much power to sustain the wave on branch A as it does for a wave on either one of the B branches. Note that for thick enough films, the maxima in the β-power plot occur for the same reasons as outlined for the single nonlinear cladding case discussed in the previous section. In principle, switching, hysteresis etc. should also be possible for this case, as outlined in the preceding sections.

When either the linear or nonlinear symmetry of the sample geometry is broken ($n_c \neq n_s$, $n_{2c} \neq n_{2s}$), two disconnected branches[41] are obtained, as shown in Figs 10 and 11. We discuss first the case $n_c = n_s$ and $n_{2c} > n_{2s}$. In the limit of large β ($\beta > n_f$), the asymptotic curves in order of increasing power correspond to self-focussing in the most nonlinear medium (cladding), in the least nonlinear medium (substrate), and self-focussing in both bounding media. If in addition $n_c \neq n_s$, it becomes possible to obtain two branches to the dispersion curves which do not overlap in power. That is, there exists a range of powers over which guided waves cannot propagate. The field distributions in the large β limit are the same as those discussed for Fig 10. Note that the origin of the maxima in the branches is the same as discussed for the single nonlinear cladding in the previous section.

Therefore, <u>for two self-focussing bounding media, there are two or three TE_0 branches, one of which has a power threshold: For equal bounding media, the branches are connected and for different bounding media the branches are unconnected</u>.

There are a number of very interesting possibilities for the transmission characteristics of TE_0 nonlinear waves propagated in the linear-nonlinear-linear waveguide geometry discussed previously for Fig 6. For the dispersion curves in Fig 10, increasing guided wave power from zero to past the peak of the lower branch can only be achieved by switching to the upper branch. Conversely, decreasing the power along the upper branch below its minimum point requires switching to the lower branch. It is very likely that hysteresis will result, although the details of the response require a numerical calculation of what happens at the linear-nonlinear waveguide interfaces.

The β-power plot in Fig 11 clearly identifies that case as very promising for all-optical switching. Because of the gap in the guided wave power between the two branches, increasing the guided wave power past the maximum in the lower branch leads to no further increase in the transmitted power until the excitation field is of sufficient amplitude to excite waves on the upper branch. This results in a discontinuous jump in the guided wave power. Similarly, decreasing the guided wave power below the minimum of the upper branch should result in a discontinuous jump in the guided wave power.

$\underline{TE_1}$

The evolution of the TE_1 guided wave solutions with increasing β are shown in Figs 12 and 13. Since both field extrema can be self-focussed, one in each of the two bounding media, solutions with $\beta>n_f$ which degenerate into combinations of single-interface guided waves can and do occur. The details depend on the specific material constants.

We discuss first the case $n_c=n_s$ and $n_{2c}=n_{2s}$, illustrated in Fig 12. The corresponding field distributions are shown in Fig 14. For curve C, the field distributions retain symmetry with respect to the center of the film and the two field extrema move symmetrically into the two nonlinear bounding media where they degenerate into single-interface guided waves. D has a power threshold and corresponds to two degenerate (in the β-power plot) curves with different field distributions. These curves are similar to those labelled B for TE_0

since the field distributions are asymmetric with respect to the film center. In this case, the fields in the asymptotic limit correspond to single interface surface guided waves with peaks located at different distances from the respective film boundaries. Curve E is unique, has a power threshold, and corresponds to the condition $\kappa k_0 h = \pi$ with the field distributions shown in Fig 14. Note that it shows no dispersion in β with power.

The consequences of breaking material symmetry are shown in Fig 13. Three separate branches, two with power thresholds, are obtained, with the lowest branch always terminating at a value of $\beta < n_f$. The two upper branches have three different asymptotic field distributions, as expected.

Thus, two self-focussing bounding media exhibit three TE_1 branches which are connected only for equal cladding and substrate media: There are three asymptotic curves which correspond to different combinations of single-interface surface guided waves.

TE_m, $m \geq 2$

The cases for TE_m, $m \geq 2$ are all similar. Since at least one of the field extrema must always remain inside the film, the solutions in the film must be oscillatory and hence $n_f > \beta$. For $n_c = n_s$ and $n_{2c} = n_{2s}$, there are three branches, two of which have power thresholds. One branch is defined by the condition $\kappa k_0 h = 2\pi$. One branch has symmetric field distributions and has a low power limit. For the second there is a power threshold and the field distributions are asymmetric with respect to the center of the film.

When $n_{2c} \neq n_{2s}$ the solutions evolve into two unconnected branches, one with a power threshhold. For $n_c = n_s$, the branch which evolves from the linear case terminates with a field maximum in the more nonlinear medium of the two. The second branch starts with a field maximum in the medium with the smallest nonlinearity and terminates with field maxima in both nonlinear media. When $n_c \neq n_s$, the curves retain their basic shape but a gap in β opens up between the two curves.

(ii) $n_{2c}>0$, $n_{2s}<0$

For this case, self-focussing occurs in the cladding and self-defocussing in the substrate at high powers. The β-power plots in this case are similar to those discussed previously for just a self-focussing cladding. That is, the TE_0 solution degenerates at high powers into a single-interface nonlinear guided wave bound to the film-cladding boundary, and the TE_m branches terminate at a maximum value of β less than n_f. The only significant difference is that for self-defocussing nonlinearities larger in magnitude than the self-focussing nonlinearities, the effective index β may decrease with increasing power before the self-focussing phenomenon dominates and ultimately results in increasing β with guided wave power. (An example of this is shown in Fig 24 for guiding by a thin metal film.)

(iii) $n_{2c}<0$, $n_{2s}<0$

In this case, the effective index decreases monotonically with increasing power until cut-off is reached. Cut-off always occurs at a finite power level because the guided wave power in the two self-defocussing media takes on the form given by Eq. 3.

(d) TE Polarized Nonlinear Surface Plasmon Polaritons

It is well-known that s-polarized (TE) surface plasmon polaritons cannot be supported by the interface between a metal and a dielectric medium, or by a metal film bounded by dielectric media. Although s-polarized waves are also not supported by the interface between two linear dielectric media, such waves can be propagated provided that the guided wave power exceeds a threshold value, if one of the media has a self-focussing nonlinearity. This raises the interesting question of whether an s-polarized wave can be supported at high powers if one of the dielectric media has self-focussing properties.

The dispersion relations for waves guided by an arbitrary film bounded on one or both sides by nonlinear media have been reviewed in the preceding paper by Boardman and Egan[10]. For two self-focussing bounding media

$$\tanh(k_0\kappa h) = \frac{\kappa[q\tanh(k_0 q z_1) + s\tanh(k_0 s z_3)]}{-\kappa^2 - sq\tanh(k_0 q z_1)\tanh(k_0 s z_3)} \tag{4}$$

where $s^2=\beta^2-n_s{}^2$ and $\kappa^2=\beta^2-\varepsilon_m$ with $\varepsilon_f=\varepsilon_m$, the dielectric constant of the metal film. The parameter z_3 governs the field distribution in the substrate (same form as Eq. 1), it is related to z_1 and can be written in terms of the total guided wave power.

Numerical solutions to Eq. 4 have been reported[51] for very thin metal films bounded on both sides by nonlinear self-focussing media. The variation in guided wave power with effective index β is shown in Fig 15 for a few metal film thicknesses and for identical nonlinear media bounding the metal film (ε_m=-10.). The power threshhold at which the solutions first appear increases with film thickness. Note that for reasonable metal film thicknesses of 0.01μm or more, the threshhold value of β corresponds to a maximum index change in the bounding media in excess of 0.1 and is therefore larger than typical values of Δn_{sat} associated with real materials. The field distribution, reproduced in Fig 16a, exhibit symmetric self-focussed maxima in both bounding media for identical bounding media and these maxima move towards the metal film with increasing power. As ε_m decreases, that is $|\varepsilon_m|$ increases, both the threshhold power and minimum effective index increase rapidly. Unfortunately, for most real metals, this range of useful values for ε_m is usually characterized by large imaginary components which result in high propagation losses.

Wave solutions also exist when the bounding media have different optical properties. For example, when the value of one of the refractive indices of the bounding media is varied, the solutions persist and the field distributions become asymmetric. Two examples are shown in Figs 16b and 16c. Note that for large enough $|n_c - n_s|$, fields are obtained with maxima in only one medium. As the nonlinearity in the substrate is reduced, the threshhold guided wave power increases correspondingly. Furthermore, the field distributions become distorted with progressively more of the power carried by the medium with the smaller nonlinearity.

No s-polarized surface polaritons solutions were found numerically for any other combination of nonlinearities for the two bounding media, including the case of one self-focussing cladding medium. They also do not exist for wave propagation along the interface between nonlinear semi-infinte dielectric and semi-infinite metallic media. This can be proven directly from the dispersion relation which is especially simple for this case, namely

$$\tanh(k_0 q z_1) = -\frac{\kappa}{q} \tag{5}$$

Because $\beta^2 = q^2 + n_c{}^2 = \kappa^2 + \varepsilon_m$ for this case, $\kappa^2 > s^2$, κ/q is always larger than unity and equation (5) can never be satisfied. Therefore s-polarized waves are not possible in this case. Finally we note that the combination of the very thin metal thickness required, the losses usually associated with surface plasmons for small $|\varepsilon_m|$ and the large changes in refractive index required will probably make such waves difficult to observe experimentally.

II Experimental Status

Two experiments on nonlinear guided waves have been reported[47,48], both dealing with the case of a self-focussing cladding medium. The geometry used was essentially that shown in Fig 6. Light was coupled with a prism into a linear waveguide, the light was propagated into, through and out of the nonlinear waveguide region, and then was coupled out of the linear region with a second prism. The transmitted versus incident power was measured.

The experimental results for the liquid crystal MBBA[47] used as the nonlinear cladding are shown in Figs 17a (TE_0) and 17b (TE_1). The pertinent theoretical calculations are shown in Fig 18. For the TE_0 case, insufficient power was available to reach the maximum of the β-power curve, where-as for TE_1 the maximum was reached and surpassed.

The power-dependent field distribution in the nonlinear waveguide affects the guided wave transmission in essentially two ways. As indicated in Fig 6, the field for a linear waveguide is highly localized in the film. Therefore, as the field maximum moves towards and into the cladding with increasing guided wave power, the transmission coefficient is reduced at both linear-nonlinear waveguide boundaries. [Note that there is little inter-mode conversion (e.g. $TE_0 \rightarrow TE_1$) at this boundary due to the inherent difference in their field symmetries.] As the incident TE_1 guided wave power is increased, the fractional power transmitted into the nonlinear waveguide region slowly decreases until it levels off when the maximum in the β-power plot is reached. Since the field extrema are still within the film when the maximum in the β-power plot is attained, this transmission coefficient remains large. As the incident power is decreased from the saturation region until the guided wave power falls below its maximum value, potentially waves on both sides of the peak can be excited. If the cladding is a lossy medium, for example the liquid crystal MBBA, the transmission through the nonlinear waveguide section is now less than for the case when only waves on the low-β side of the curve are excited. Furthermore, since regions characterized by decreasing power with increasing β are probably unstable, waves excited on this branch decay into other modes, which again leads to increased loss. Therefore hysteresis should occur whenever the guided wave power reaches maxima in the β-power plot. This explains completely the behaviour observed experimentally for TE_1.

For TE_0, no hysteresis was observed since the maximum in the β-power curve was not attained in that experiment.

The situation is not as clear-cut for the CS_2 nonlinear cladding medium experiment (Fig 19)[48]. The waveguides were of the ion-exchange type which have different dispersion curves at low powers than the nonlinear thin film waveguide cases analysed to date. Nevertheless, the phenomenon observed for the TE_0 wave was qualitatively the same, although the quantitative details may be different.

III TM Waves Guided by Thin Films

It was shown in the first[10] of this two paper sequence that the analysis of p-polarized (TM) nonlinear guided waves is still a subject of controversy[40]. The problem is that p-polarized waves have two field components, one along the propagation wavevector (E_X), and one orthogonal to the surface(s) (E_Z). It has been argued[40] for nonlinear guiding by thin films that the dominant nonlinearity arises from the E_Z term since $|E_Z|^2 >> |E_X|^2$ for all cases of interest. Hence we assume in the numerical calculations summarized here that $\varepsilon_{XX} = \varepsilon_Y$ and $\varepsilon_{ZZ} = \varepsilon_Y + \alpha|E_Z|^2$ for the pertinent TM components of the dielectric tensor.

(a) TM Nonlinear Waves Guided by Dielectric Films

No extensive calculations of TM nonlinear waves guided by thin dielectric films have been reported using the dominant (E_Z) field component for the uniaxial approximation[36]. It can be argued, however, that the β-power plots for TM waves should resemble closely those already discussed for TE waves. The key point is that the field structure for the E_Z field component is almost identical to that for the E_y field which occurs for the s-polarized case. Therefore, in the absence of detailed calculations, one can presume that the same number of field branches etc. will be obtained for TM waves, although the numerical thresholds will be somewhat different.

(b) TM Polarized Nonlinear Surface Plasmon Polaritons Guided by Metal Films

It is well-known that two surface plasmon polaritons can be supported by thin metal films bounded on both sides by dielectric media. The two surface guided waves take on distinctly different properties when the film thickness is of the order of (or less than) the penetration depth of the fields in the metal. For the long-range mode, so-called because of its enhanced propagation distance in almost symmetrical material geometries, the E_z field remains positive everywhere inside the metal film. The propagation distance is enhanced relative to the semi-infinite dielectric-metal case because, as the film thickness is decreased, a progressively smaller fraction of the guided wave power is carried in the metal, which is the lossy medium. For the short-range mode, the field passes through zero somewhere inside the metal film. There the propagation distance is reduced from that of a surface plasmon polariton guided by a single dielectric-metal interface because reducing the film thickness results in a progressively larger fraction of the guided wave power carried by the metal film. When one, or both of the bounding media exhibit nonlinearities[36,46], one would expect power dependent phenomena similar to those just discussed for thin dielectric films.

(i) $n_{2c}\neq 0$, $n_{2s}=0$

The simplest case is for $n_c=n_s$ with a self-focussing cladding medium. Typical results for the β-power plot are shown in Fig 20 for both the long-range and short-range waves. The material system assumed is described in the figure caption. The behaviour of the long-range wave is very similar to that of TE_1 waves guided by a dielectric film with a self-focussing cladding, see Fig 2. The evolution of the field distributions is also similar, compare Figs 3 and 21. For large values of β, a self-focussed field appears in the cladding medium. On the other hand, the power-dependent changes associated with the short-range mode are relatively uninteresting. The effective index of the short-range solution increases monotonically with guided wave power and the field distributions exhibit little change with increasing power for reasonable optically induced changes in the cladding index.

No significant changes occur for the case $n_c \neq n_s$ with $n_{2c}>0$.

When $n_{2c}<0$, the effective index decreases with increasing guided wave power, as shown in Fig 22. For the same reasons as previously discussed for the dielectric film case with similar nonlinearities, cut-off can occur at either finite ($n_c>n_s$) or diverging ($n_s \geq n_c$) powers.

(ii) $n_{2c} \neq 0$, $n_{2s} \neq 0$

A proliferation of solutions occurs for both bounding media of the self-focussing variety, just as obtained previously for the dielectric film case.

When both bounding media exhibit self-focussing nonlinearities, each of the short-range and long-range solutions split into three branches, see Fig 23. For $n_c=n_s$ and $n_{2c}=n_{2s}$, the fields associated with the branches which have low power limits retain the field symmetry about the center of the respective films. The two sets of degenerate long-range and short-range branches both have power thresholds and exhibit different asymmetric field distributions. For example, the two degenerate long-range branches (uppermost in Fig 23) correspond to self-focussing in either one or the other of the bounding media. The variation in field distribution with β (power) is unusual for this case, as shown in the inset of Fig 23. When the nonlinear symmetry is broken, that is $n_{2c} \neq n_{2s}$, the degeneracies are broken and six separate asymptotic curves occur in the high β (power) limit, as shown in Fig 23.

The solution curves for one self-focussing and one self-defocussing bounding medium are similar to those obtained for the dielectric film case, especially for the long-range mode. As shown in Fig 24, if the self-defocussing nonlinearity is larger in absolute magnitude than the self-focussing nonlinearity, β first decreases with increasing power for the long-range wave. However, eventually the self-focussing nonlinearity dominates and β increases with power until the branch terminates. The long-range wave increases monotonically with power for reasonable changes in β.

For $n_{2c}<0$ and $n_{2s}<0$, the effective index decreases monotonically with power until cut-off is reached for both the short-range and long-range wave solutions. Cut-off is reached at a finite power since the power flow in both bounding media can be described by equations similar to Eq. 3. A sample calculation is shown for the long-range mode in Fig 21 where an unusual effect was found. It has been noted before for both symmetric (identical bounding media with self-focussing nonlinearities) dielectric and metal film cases, that solution branches are possible in which the field distributions are asymmetric above a power threshold. This same effect can be seen in Fig 21 in which the long-range wave acquired two degenerate asymmetric field branches in the β-power plot, but for identical bounding media with self-defocussing nonlinearities. That is, the inherent symmetry of the refractive index of the structure is broken by an asymmetric optical field. To date, this has not been reported for the similar dielectric film case.

IV Summary

If one or more of the media bounding a dielectric or metal film exhibits an intensity dependent refractive index, the guided wave solutions to the resulting dispersion relations take on many interesting properties. The number of wave solutions, the propagation wavevector (and effective index β), the field distributions, the waveguide cut-off and cut-on conditions and the wave attenuation all become power dependent. Tables I and II contain a compilation of the various s- and p-polarized waves which can exist for guiding by a single interface and a film respectively. There are still open questions about the nature of the solutions for the p-polarized case.

In all of the work reported to date it has been assumed that the nonlinearity is of the Kerr type, that is the dielectric constant or refractive index varies with the modulus of the square of the guided wave field. In fact, for semiconductors which exhibit the largest known nonlinearities as well as the largest values of Δn_{sat}, the maximum allowed optically induced change in the refractive index, the index change varies more slowly than quadratically with the guided wave field. However, with plane waves it has been shown that self-focussing and self-defocussing of beams does occur in semiconductor media and hence one would expect the phenomena predicted for Kerr-type nonlinear media in waveguide form to also occur for other forms of the intensity dependent refractive index. How the details of the wave solutions will be affected still remains to be investigated.

Materials in general limit the experimental realization of all of the interesting effects predicted to date. It is clear that the index difference $n_f - n_c$ (and in some cases $n_f - n_s$ also) must be less than Δn_{sat} for interesting power dependent changes in the field distribution to occur. However, for most materials, excluding semiconductors, Δn_{sat} is less than 0.01, and more typically 0.001. This puts severe limitations on the material combinations which can be used for making nonlinear waveguides.

The experimental investigations of nonlinear guided waves are few indeed. However, in both cases, it is clear that features characteristic of nonlinear guided waves have been observed.

The many interesting potential applications of nonlinear guided waves to all-optical computing and signal processing will ensure a growing interest in the field. Of special interest at the moment are operations[50] such as bistability, switching, upper and lower threshold devices. For excitation of nonlinear guided waves through a transverse boundary, the nonlinear fields are established within a few wavelengths of the boundary which can lead to very compact optical devices, in principle. In practise, the incident fields not converted into nonlinear guided waves are probably radiated away, and it will be necessary to extend the nonlinear region along the propagation axis so that these radiation fields can diverge until they do not interfere with the guided waves. However, this should still be possible in tens of wavelengths which would still result in very compact devices. Furthermore, the same type of phenomena should occur in channel waveguide devices which would reduce the waveguide volume and power operating levels considerably. (A channel waveguide is a rectangular region of high index and waves are guided by total internal reflection at all four channel-bounding medium interfaces.)

This research was supported by the National Science Foundation (ECS-8117483 and -8304749), Army Research Office (DAAG-29-85-K-0026), the Joint Services Optics Program (MICOM, BMDSC) and the NSF-Industry Center for Optical Circuitry.).

Table I. Compilation of the various nonlinear s- and p-polarized guided waves that can be supported by a semi-infinite metal boundary, and a thin metal film.

Semi-infinite Metal Boundary

Polarization	Nonlinear Coefficients	Power Independent Solutions	Power Dependent Solutions
p	$n_{2c}>0$	1	1
p	$n_{2c}<0$	1	1
s	$n_{2c}>0$	0	0
s	$n_{2c}<0$	0	0

Metal Film

Polarization	Nonlinear Coefficients	Power Independent Solutions	Power Dependent Solutions
p	$n_{2c}>0$, $n_{2s}=0$	2	2
p	$n_{2c}<0$, $n_{2s}=0$	2	2
p	$n_{2c}>0$, $n_{2s}>0$	2	6
p	$n_{2c}>0$, $n_{2s}<0$	2	2
p	$n_{2c}<0$, $n_{2s}>0$	2	2
p	$n_{2c}<0$, $n_{2s}<0$	2	≥ 3
s	$n_{2c}>0$, $n_{2s}>0$	0	1
s	all other combinations	0	0

Table II. Compilation of the various nonlinear s- and p-polarized guided waves that can be supported by the interface between two semi-infinite dielectrics, and by a thin dielectric film. The p-polarized dielectric film cases are probably the same as for the s-polarized cases.

Two Semi-Infinite Dielectrics

Polarization	Nonlinear Coefficients	Power Independent Solutions	Power Dependent Solution Branches
p, s	$n_{2c}>0$, $n_{2s}=0$	0	1
p, s	$n_{2c}<0$, $n_{2s}=0$	0	0
p, s	$n_{2c}>0$, $n_{2s}>0$	0	2
p, s	$n_{2c}>0$, $n_{2s}<0$	0	1
p, s	$n_{2c}<0$, $n_{2s}>0$	0	1
p, s	$n_{2c}<0$, $n_{2s}<0$	0	0
p, s	$n_{2c}>0$, $n_{2s}>0$	0	1

Dielectric Film

s	$n_{2c}>0$, $n_{2s}=0$	TE_m	TE_m
s	$n_{2c}<0$, $n_{2s}=0$	TE_m	TE_m
s	$n_{2c}>0$, $n_{2s}>0$	TE_m	$2TE_0$, $3TE_m$
s	$n_{2c}>0$, $n_{2s}<0$	TE_m	TE_m
s	$n_{2c}<0$, $n_{2s}>0$	TE_m	TE_m
s	$n_{2c}<0$, $n_{2s}<0$	TE_m	TE_m

References

1. A.E. Kaplan, Sov. Phys. JETP, **45**, 896 (1977)
2. A.E. Kaplan, JETP Lett., **24**, 114 (1976)
3. W.J. Tomlinson, J.P. Gordon, P.W. Smith and A.E. Kaplan, Appl. Opt., **21**, 2041 (1982)
4. D. Marcuse, Appl. Optics, **19**, 3130 (1980)
5. P.W. Smith, J.-P. Hermann, W.J. Tomlinson and P.J. Moloney, Appl. Phys. Lett., **35**, 846 (1979)
6. P.W. Smith, W.J. Tomlinson, P.J. Moloney and J.-P. Hermann, IEEE J. Quant. Electron., **QE-17**, 340 (1981)
7. A.E. Kaplan, IEEE J. Quant. Electron., **QE-17**, 336 (1981)
8. P.W. Smith and W.J. Tomlinson, IEEE J. Quant. Electron., **QE-20**, 30 (1984)
9. P.W. Smith and W.J. Tomlinson, in Optical Bistability, edited by C.M. Bowden, M. Ciftan and H.R. Robl, (Plenum Press. N.Y.), p463 (1981)
10. A.D. Boardman and P. Egan, preceding article in this publication
11. A.G. Litvak and V.A. Mironov, Izv. Vysch. Uch. Zav. - Radiofisika, **11**, 1911 (1968)
12. Yu.R. Alanakyan, Sov. Phys. Tech. Phys., **12**, 587 (1967)
13. W.J. Tomlinson, Opt. Lett., **5**, 323 (1980)
14. A.A. Maradudin, Zeit. Phys. B. **41**, 341 (1981)
15. A.A. Maradudin, in Optical and Acoustic Waves in Solids-Modern Topics, M. Borissov, ed., (World Scientific Publ., Singapore, 1983), p72
16. K.M. Leung, Phys. Rev. A, **31**, 1189 (1985)
17. V.M. Agranovich, V.S. Babichenko and V.Ya. Chernyak, Sov. Phys. JETP Let., **32**, 512 (1981)
18. A.I. Lomtev, Sov. Phys. JETP Lett., **34**, 60 (1981)
19. M.Y. Yu, Phys. Rev. A, **28**, 1855 (1983)
20. K.M. Leung, Phys. Rev. B, in press
21. G.I. Stegeman, C.T. Seaton, J. Ariyasu, R.F. Wallis and A.A. Maradudin, in press, J. Appl. Phys., October, 1985
22. N.N. Akhmediev, Sov. Phys. JETP, **56**, 299 (1982)
23. A.A. Maradudin, in Optical and Acoustic Waves in Solids-Modern Topics, M. Borissov, ed., (World Scientific Publ., Singapore, 1983), p72

24. F. Lederer, U. Langbein and H.-E. Ponath, Appl. Phys. B, **31**, 69 (1983)
25. U. Langbein, F. Lederer, H.-E. Ponath and U. Trutschel, J. Mol. Structure, **115**, 493 (1984)
26. A.D. Boardman and P. Egan, J. Physique Colloq. **C5**, 291 (1984)
27. G.I. Stegeman, C.T. Seaton, J. Chilwell and S.D. Smith, Appl. Phys. Lett., **44**, 830 (1984)
28. D. Mihalache and H. Totia, Revue Roumaine de Phys., **29**, 365 (1984)
29. N.N. Akhmediev, K.O. Boltar and V.M. Eleonskii, Opt. Spektrosk., **53**, 906 and 1097 (1982)
30. A. Boardman and P. Egan, Phil. Trans. Roy. Soc. London, **A313**, 363 (1984)
31. D.J. Robbins, Opt. Comm., **47**, 309 (1983)
32. U. Langbein, F. Lederer and H.-E. Ponath, Opt. Comm., **46**, 167 (1983)
33. F. Fedyanin and D. Mihalache, Zeit. fur Physik B, **47**, 167 (1982)
34. F. Lederer, U. Langbein and H.-E. Ponath, Appl. Phys. B, **31**, 187 (1983)
35. D. Mihalache, R.G. Nazmitdinov and V.K. Fedyanin, Physica Scripta, **29**, 269 (1984)
36. G.I. Stegeman and C.T. Seaton, Optics Letters, **9**, 235 (1984)
37. C.T. Seaton, J.D. Valera, R.L. Shoemaker, G.I. Stegeman, J. Chilwell and S.D. Smith, Appl. Phys. Lett. **45**, 1162 (1984)
38. M.Y. Yu, Phys. Rev. A, **28**, 1855 (1983)
39. L. Wendler, Phys. Stat. Sol., **117**, 241 (1983)
40. C.T. Seaton, J.D. Valera, B. Svenson and G.I. Stegeman, Opt. Lett., **10**, 149 (1985)
41. C.T. Seaton, J.D. Valera, R.L. Shoemaker, G.I. Stegeman, J. Chilwell and S.D. Smith, J. Quant. Electron., **QE-21**, 774 (1985)
42. A.D. Boardman and P. Egan, J. Quant. Electron., in press
43. U. Langbein, F. Lederer and H.-E. Ponath, Optics Comm., **53**, 417 (1985)
44. U. Langbein, F. Lederer, H.-E. Ponath and U. Trutschel, Appl. Physics B, **B36**, 187 (1985)
45. U. Langbein, F. Lederer, H.-E. Ponath and U. Trutschel, Appl. Phys. B, in press
46. J. Ariyasu, C.T. Seaton, G.I. Stegeman, A.A. Maradudin and R.F. Wallis, in press, J. Appl. Phys., October, 1985

47. C.T. Seaton, G.I. Stegeman and H.G. Winful, Opt. Eng., **24**, 593 (1985).
48. H. Vach, C.T. Seaton, G.I. Stegeman and I.C. Khoo, Opt. Lett., **9**, 238 (1984)
49. I. Bennion, M.J. Goodwin and W.J. Stewart, Electron. Lett., **21**, 41 (1985)
50. C.T. Seaton, G.I. Stegeman and H.G. Winful, Opt. Eng., **24**, 593 (1985).
51. G.I. Stegeman, J.D. Valera, C.T. Seaton, J. Sipe and A.A. Maradudin, "Nonlinear s-polarized surface plasmon polaritons", Solid State Commun., **52**, 293 (1984)

Figure Captions

Fig 1 The thin film geometry assumed in this paper.
Fig 2 Field distributions associated with (a) TE_0 nonlinear guided waves and (b) TE_1 nonlinear guided waves for a film thickness of 2μm. The field evolution with increasing β is shown. Here $n_C=n_S=1.55$, $n_f=1.57$, $n_{2C}=10^{-9}m^2/W$.
Fig 3 TE_0 guided wave power versus mode index β for $n_C=1.55$, $n_f=1.57$, $n_S=1.55$ and $n_{2C}=10^{-9}m^2/W$ for three different film thicknesses, h=0.5μm dotted line, h=1.0μm dashed line, h=2.0μm solid line.
Fig 4 TE_1 guided wave power versus mode index β for $n_C=1.55$, $n_f=1.57$, $n_S=1.55$ and $n_C=10^{-9}m^2/W$ for three different film thicknesses, h=1.08μm dotted line, h=1.5μm dashed line, h=2.0μm solid line.
Fig 5 The cut-off power above which the TE_0 wave can be propagated versus film thickness for a nonlinear self-focussing ZnS cladding medium. The inset shows the variation in effective index β with guided wave power for a film thickness of 0.2 μm. Here $n_{2C}=3\times10^{-11}$ m^2/W, $n_C=2.39$) with some composite film characterized by $n_f=2.40$ (for example containing $SrTiO_3$), and substrate $n_S=2.38$. At low powers, the TE_0 mode for this structure is cut-off at a film thickness of ≈0.30 μm.
Fig 6 A method for exciting nonlinear guided waves with threshold characteristics. A linear guided wave incident onto a transverse boundary between a linear and nonlinear waveguide excites a nonlinear wave in the nonlinear waveguide region. This nonlinear wave can excite a

linear wave at the second linear-nonlinear waveguide interface.

Fig 7 The maximum TE_0 guided wave power which can be propagated versus index difference between the film and nonlinear GaAs-$GaAl_xAs_{1-x}$ cladding. Inset is the effective index versus guided wave power for n_f = 3.39 and film thickness of 1.07 μm. Here n_{2c}=-2x10^{-9} m^2/W, n_c=3.385, the substrate is the appropriate bulk $GaAl_xAs_{1-x}$ composition for n_s=3.38, and the film of bulk $GaAl_xAs_{1-x}$ has variable index $n_f(x)$ and thickness.

Fig 8 TE_0 guided wave power versus effective index for three film thicknesses (h=0.5μm dashed line; h=1.25μm dotted line; h=2.0μm solid line). Branches A and B exhibit symmetric and asymmetric field distributions respectively. Here n_s=n_c=1.55, n_f=1.57 and n_{2c}=n_{2s}=10^{-9}m^2/W.

Fig 9 TE_0 guided wave field distributions for (a) branch A and (b) branch B. Here n_c=n_s=1.55, n_f=1.57, h=2.0μm and n_{2c}=n_{2s}=10^{-9}m^2/W.

Fig 10 Real (solid line) and imaginary (dashed line) parts of β versus TE_0 guided wave power for h=2.0 μm, n_f=1.57, n_c=1.55, ε_{Ic}=0.002, n_{2c}=2x10^{-9} m^2/W, n_s=1.55, ε_{Is}=0.001 and n_{2s}=10^{-9} m^2/W. Here $\varepsilon = \varepsilon_R - i\varepsilon_I$.

Fig 11 The effective index versus TE_0 guided wave power for h=2.0 μm, n_f=1.57, n_c=1.56, ε_{Ic}=0.002, n_{2c}=2x10^{-9} m^2/W, n_s=1.55, ε_{Is}=0.001 and n_{2s}=10^{-9} m^2/W.

Fig 12 TE_1 guided wave power versus effective index for three film thicknesses; h=0.8μm dotted line, h=2.0μm solid line, h=4.0μm dashed line. The dotted line curve is for a thickness below mode cut-off. Here n_c=n_s=1.55, n_f=1.57 and n_{2c}=n_{2s}=10^{-9}m^2/W. Branches C, D and E described in the text.

Fig 13 TE_1 guided wave power versus effective index for n_{2c}=2x10^{-9}m^2/W and n_{2s}=10^{-9}m^2/W. The solid line is for n_c=n_s=1.55, n_f=1.57, and h=2.0μm. The dashed line is for n_c=1.56, n_f=1.57, n_s=1.55 and h=3.0μm.

Fig 14 Field distributions associated with TE_1 guided waves with both bounding media nonlinear; n_c=n_s=1.55, n_{2c}=n_{2s}=10^{-9}m^2/W and h=2.0μm. Branches C, D and E correspond to (a), (b) and (c) respectively.

Fig 15 Guided wave power (mW/mm) versus effective index β for n_c=n_s=1.55, ε_m=-10, and n_{2c}=n_{2s}=10^{-9}m^2/W for h=0.001μm (solid line), h=0.005μm (dashed line), h=0.010μm (dotted line) and h=0.015μm (dash-dotted line)

Fig 16 Field distributions with increasing β for n_c=1.55, h=0.005μm, ε_m=-10 and $n_{2c}=n_{2s}=10^{-9}$ m^2/W. (a) n_s=1.55. (b) n_s=1.40. (c) n_s=1.25.

Fig 17 The TE_0 (a) and TE_1 (b) guided wave powers transmitted through a nonlinear waveguide excited in the geometry of Fig 6. The nonlinear waveguide consists of the liquid crystal MBBA on top of a glass waveguide.

Fig 18 The effective index versus guided wave power for TE_0 (dashed line) and TE_1 (solid line) waves. Here h=2.0 μm, n_f=1.57, n_s=1.52, n_c=1.55, ε_{lc}=0.001 and $n_{2c}=10^{-9}$ m^2/W.

Fig 19 Output versus input intensity for a waveguide excited in the geometry of Fig 6. The nonlinear waveguide consists of CS_2 on top of an ion-exchanged glass waveguide.

Fig 20 The real (β, solid line) and imaginary (β_I, dashed line) parts of the effective index as a function of guided wave power for both the long range (lower curves) and short-range (upper curves) surface plasmon waves. Here ε_c = 16 - .0096i, n_s = 4, n_{2c} = 10^{-7} m^2/W, and a metal film with h = 0.05μm and ε_m = -1000 - 160i.

Fig 21 Field distributions associated with (a) long-range and (b) short-range surface plasmon polaritons for increasing values of β. Here n_c = n_s = 4.0, ε_m = -1000, h = 0.05μm and n_{2c} = 10^{-7} m^2/W. The spatial scale inside the metal film has been expanded by ×100.

Fig 22 The real (β) and imaginary (β_I) parts of the effective index versus guided wave power for the long-range solutions. Here ε_c = 16 - 0.0096i, n_{2c} = -10^{-7} m^2/W, h = 0.05 μm and ε_m = -1000 - 160i. n_s = 4 and n_{2s} = 0 for the solid curves, and ε_s = 16 - 0.0096i and n_{2s} = -10^{-7} m^2/W for the dashed lines. The dash-dot line corresponds to the asymmetric long-range wave case. For the the dotted line it is assumed that the TM dielectric tensor components in the nonlinear cladding are $\varepsilon_{xx} = n_c^2 + \alpha|E_x|^2$ and $\varepsilon_{zz} = n_c^2$ for the case n_{2s}=0.

Fig 23 Surface plasmon power versus real part of the effective index for h = 0.05 μm, ε= -1000, n_c = n_s = 4.0. The solid lines are for n_{2c} = n_{2s} = 10^{-7} m^2/W and the dashed line for n_{2c} = 2×10^{-7} m^2/W and n_{2s} = 10^{-7} m^2/W.

Fig 24 The real (β) and imaginary (β_I) parts of the effective index versus guided wave power for the long-range (solid line) and short-range (dashed line) surface plasmon waves. Here $n_c = n_s = 4$, $n_{2c} = 10^{-7}$ m²/W, $n_{2s} = -10^{-7}$ m²/W, h = 0.05 μm and $\epsilon_m = -1000 - 160i$.

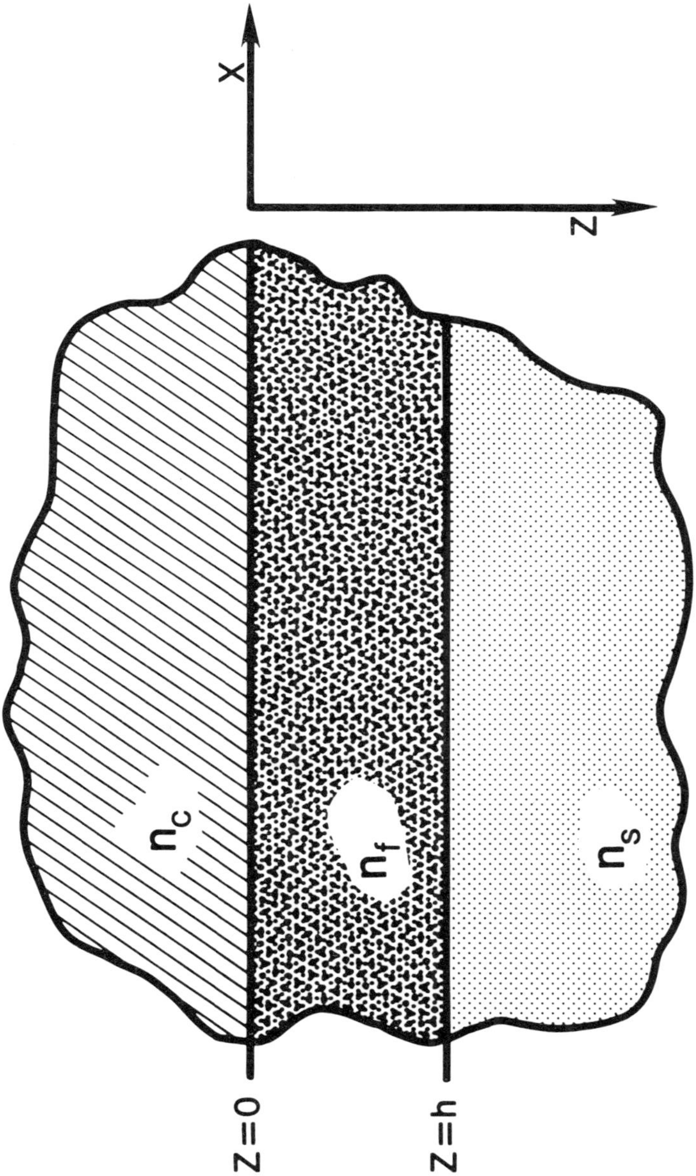

Fig. 1

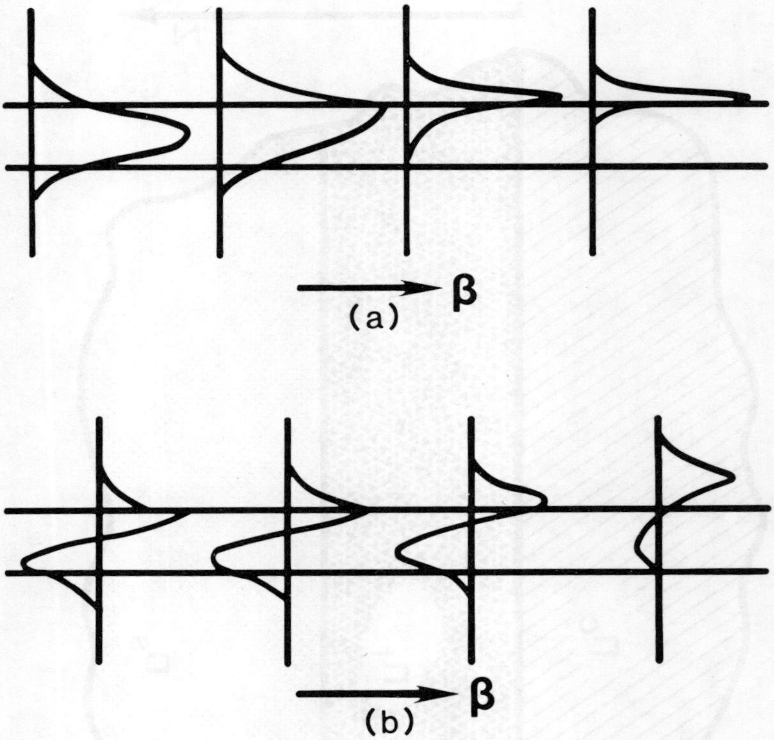

Fig. 2

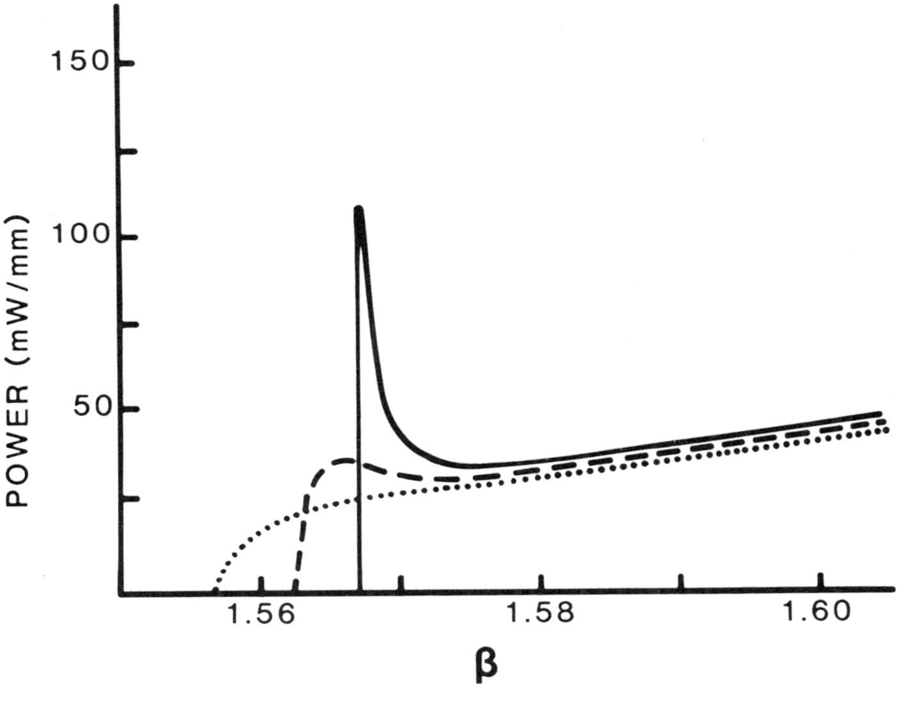

Fig. 3

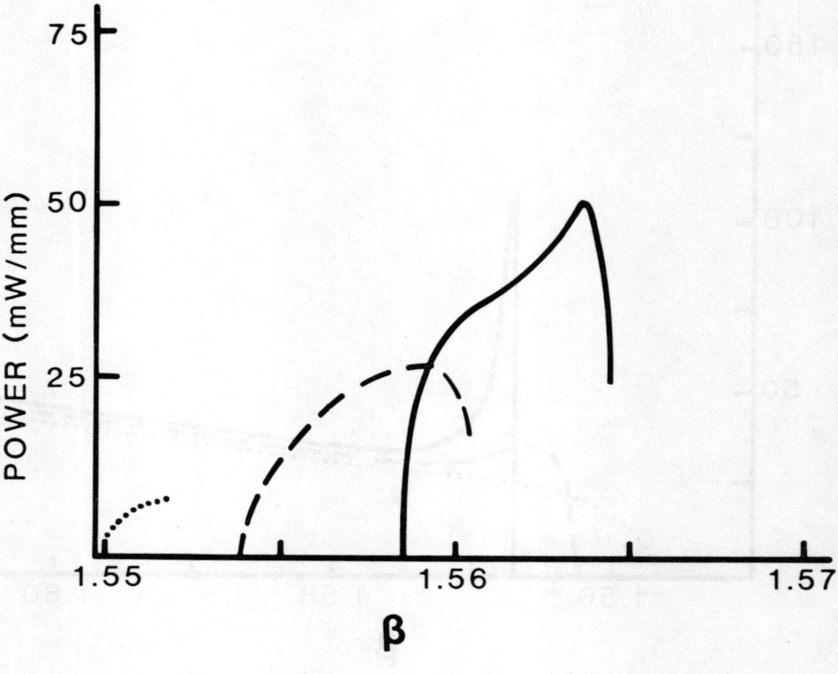

Fig. 4

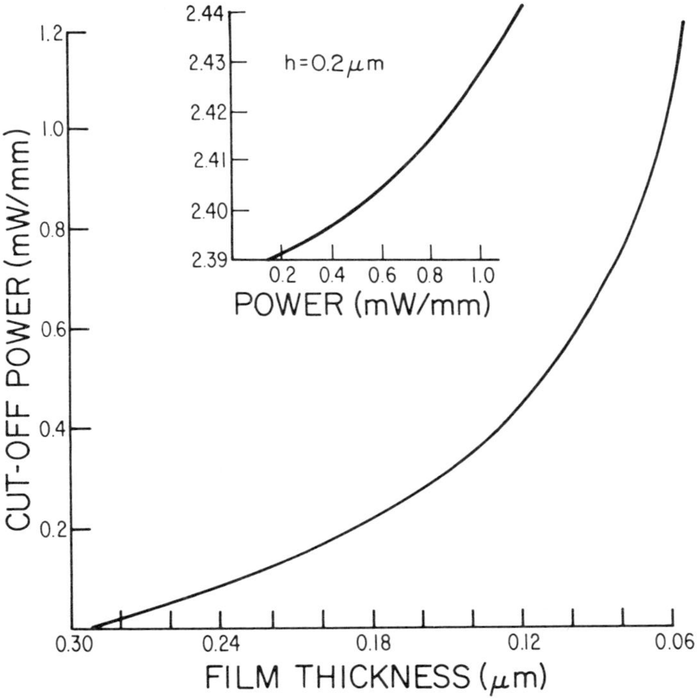
2.44
2.43
2.42
2.41
2.40
2.39
h=0.2 μm
0.2 0.4 0.6 0.8 1.0
POWER (mW/mm)
1.2
1.0
0.8
0.6
0.4
0.2
CUT-OFF POWER (mW/mm)
0.30
0.24
0.18
0.12
0.06
FILM THICKNESS (μm)

Fig. 5

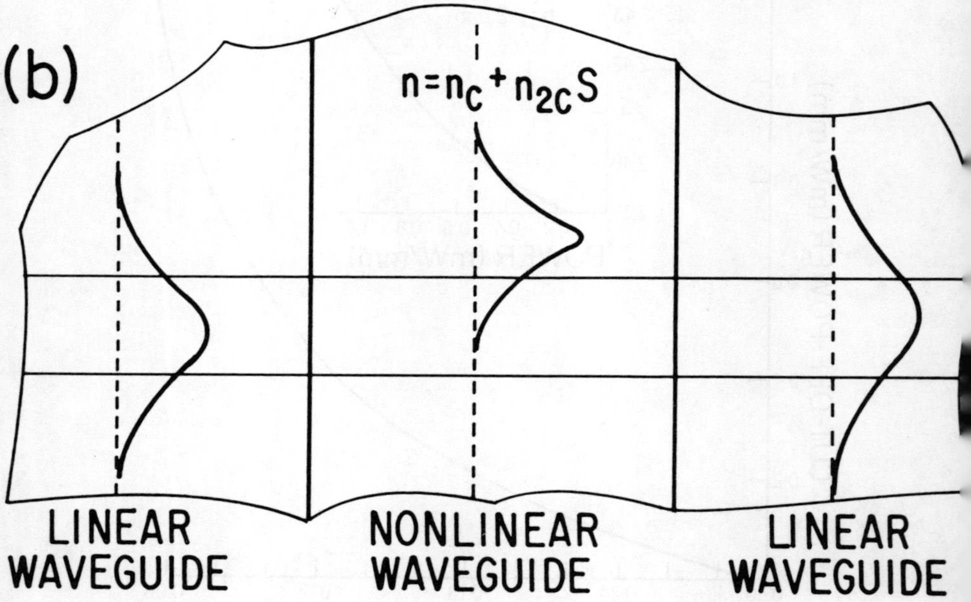
(b)
n=n_c + n_2c S
LINEAR WAVEGUIDE
NONLINEAR WAVEGUIDE
LINEAR WAVEGUIDE

Fig. 6

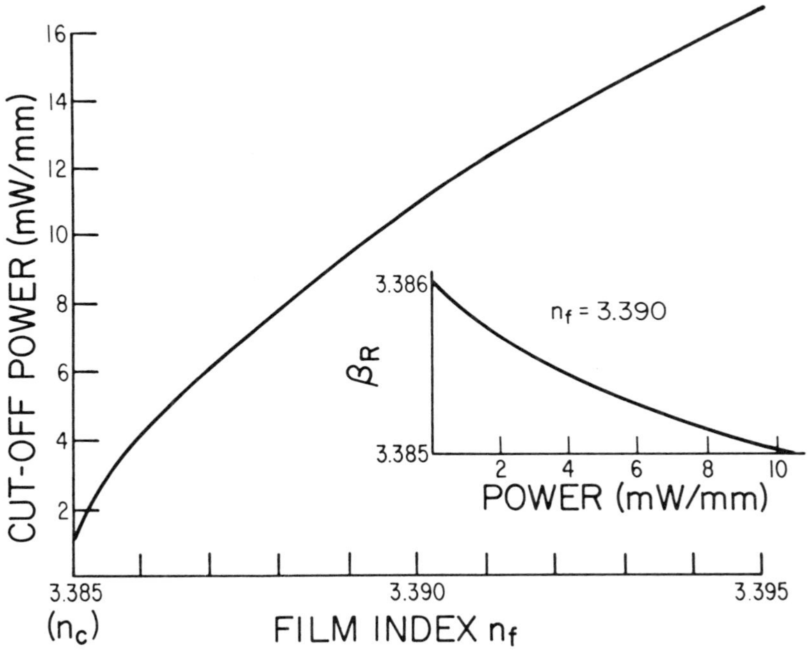
CUT-OFF POWER (mW/mm)
16
14
12
10
8
6
4
2
3.385
(n_c)
3.390
3.395
FILM INDEX n_f
3.386
3.385
β_R
n_f = 3.390
2
4
6
8
10
POWER (mW/mm)

Fig. 7

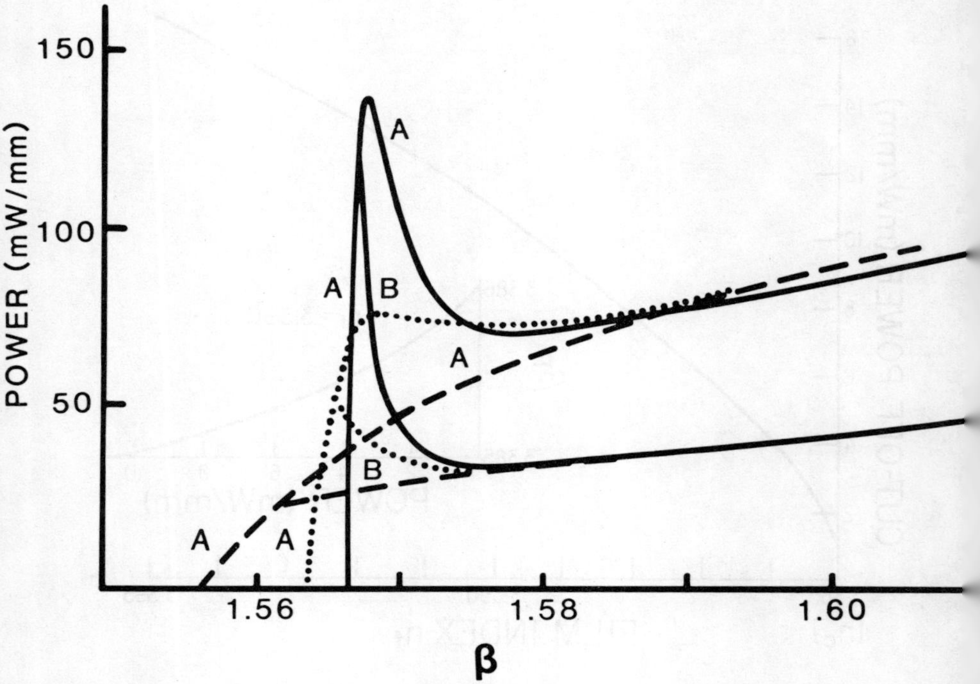

Fig. 8

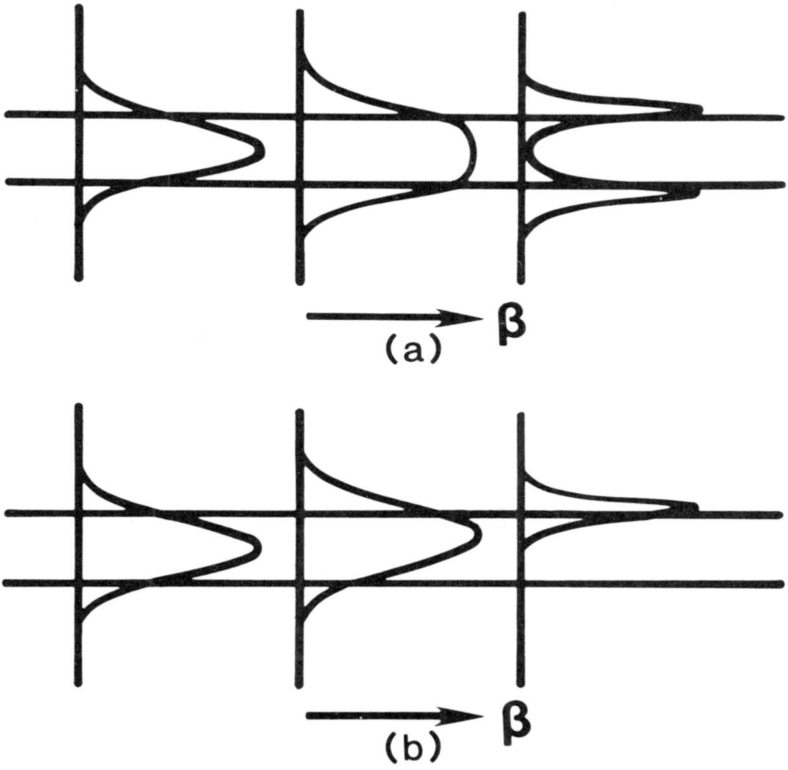

Fig. 9

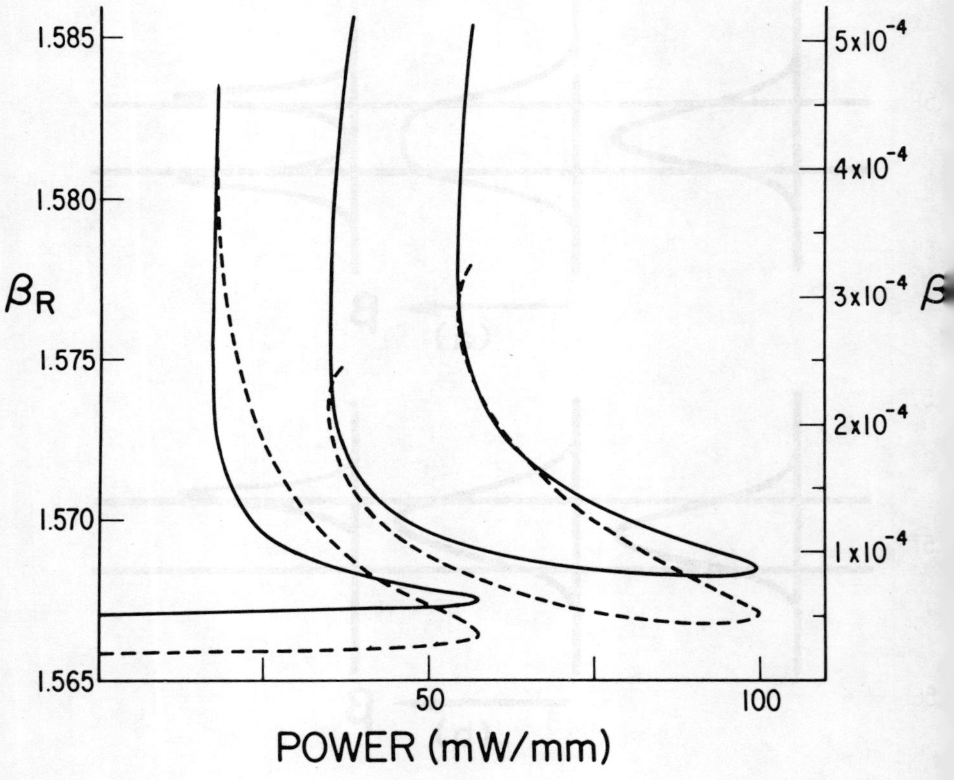
1.585
1.580
1.575
1.570
1.565
β_R
50
100
POWER (mW/mm)
5×10^{-4}
4×10^{-4}
3×10^{-4}
2×10^{-4}
1×10^{-4}

Fig. 10

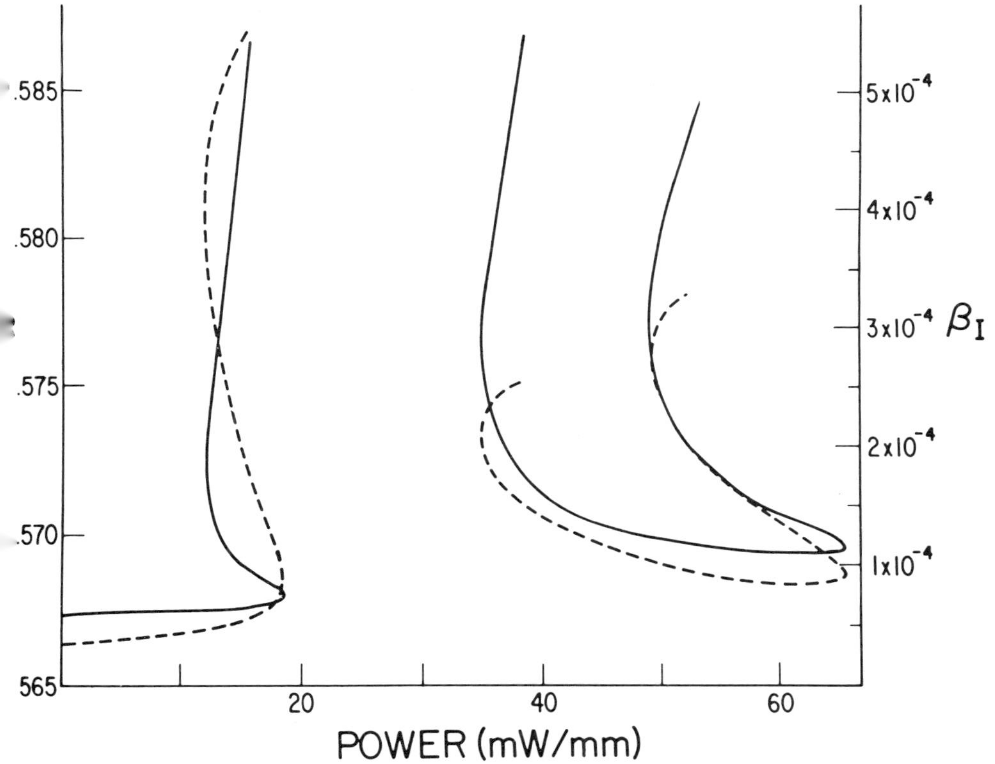
.585
.580
.575
.570
565
20
40
60
POWER (mW/mm)
5x10^-4
4x10^-4
3x10^-4
2x10^-4
1x10^-4
βI

Fig. 11

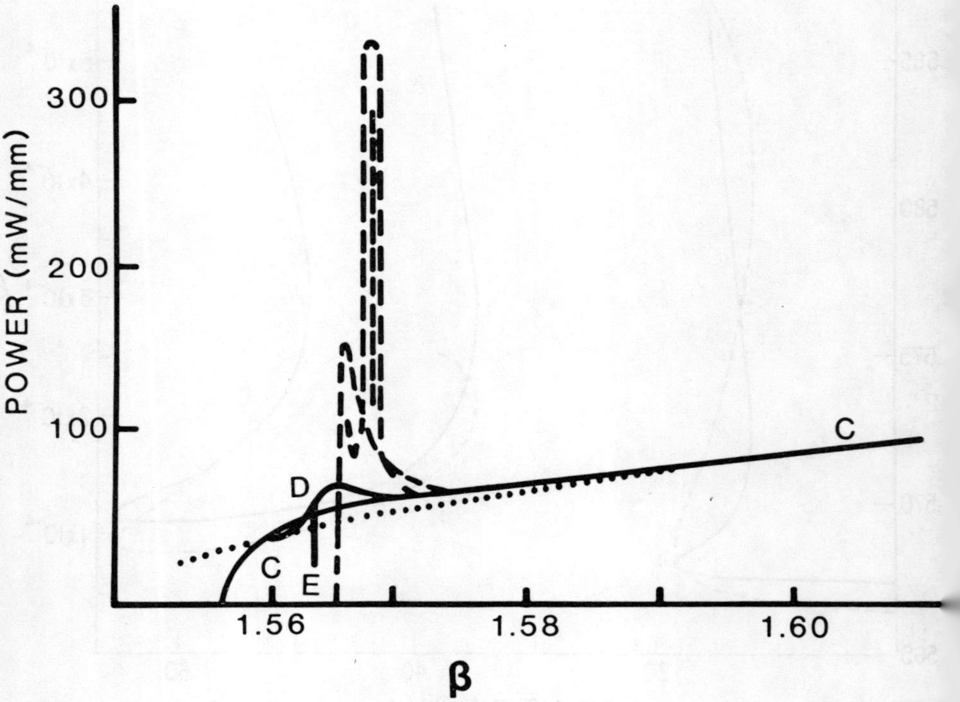

Fig. 12

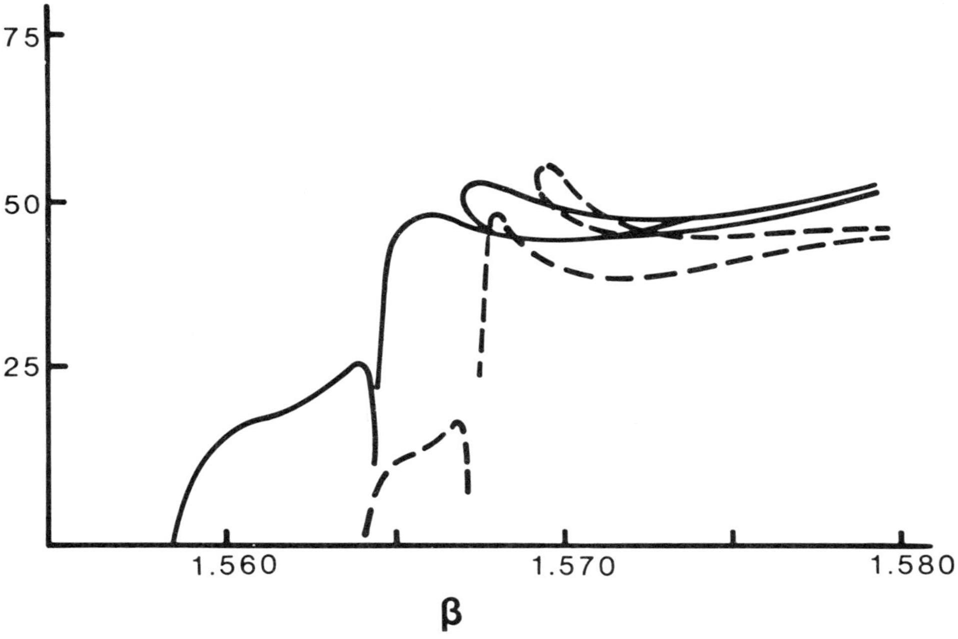

Fig. 13

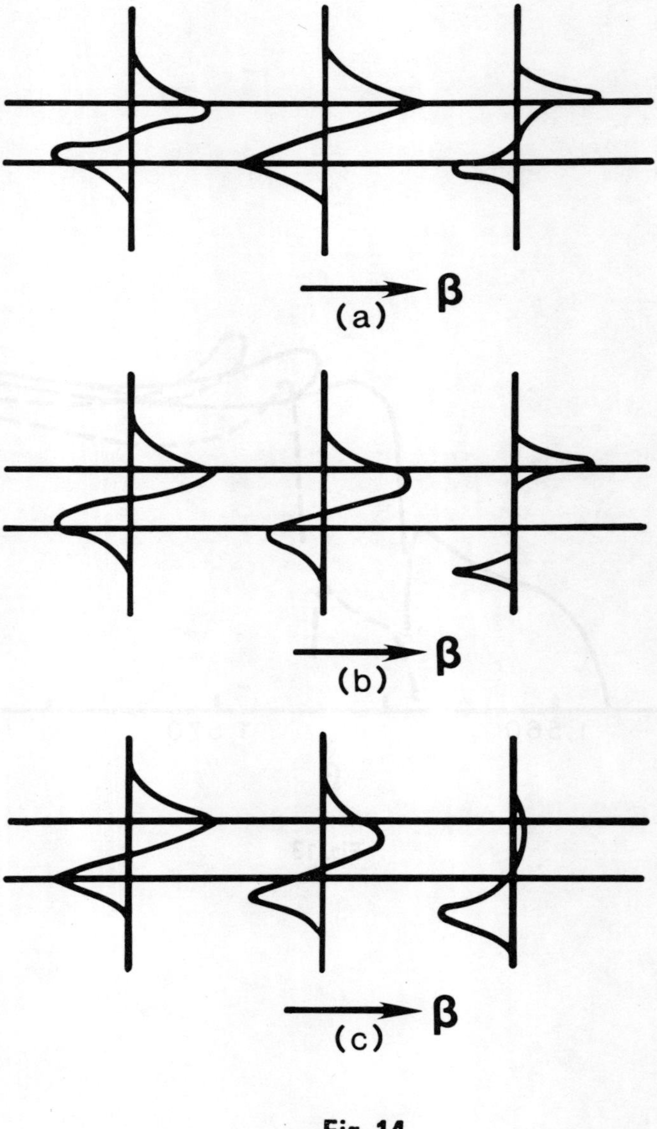

Fig. 14

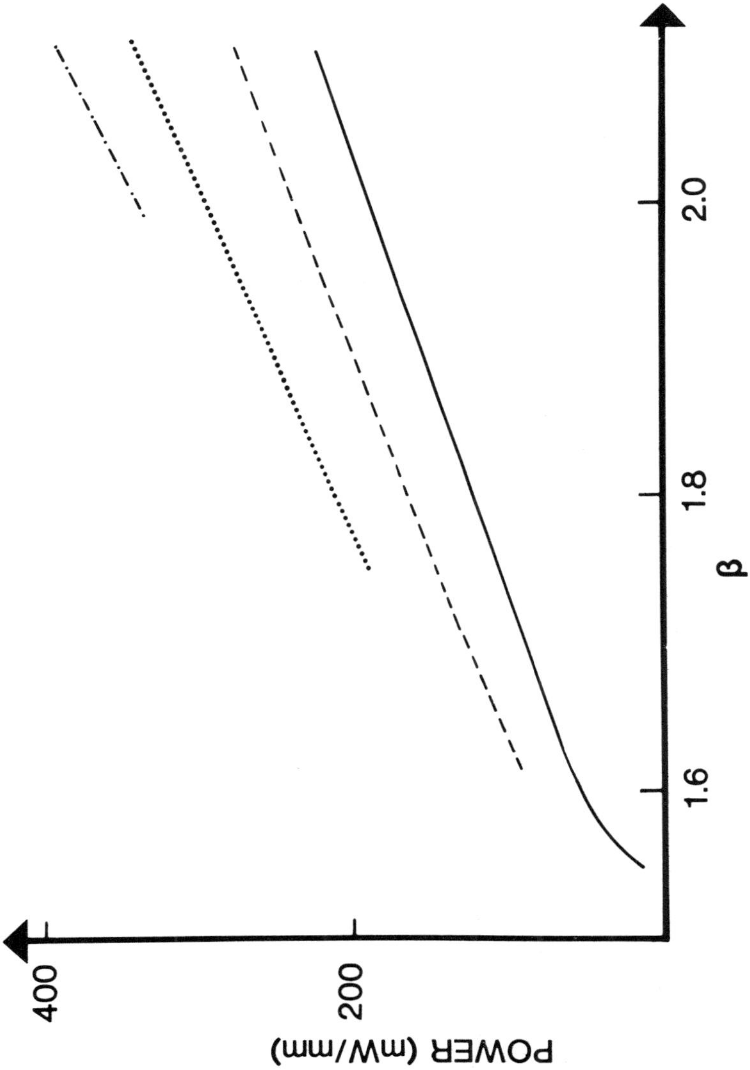

Fig. 15

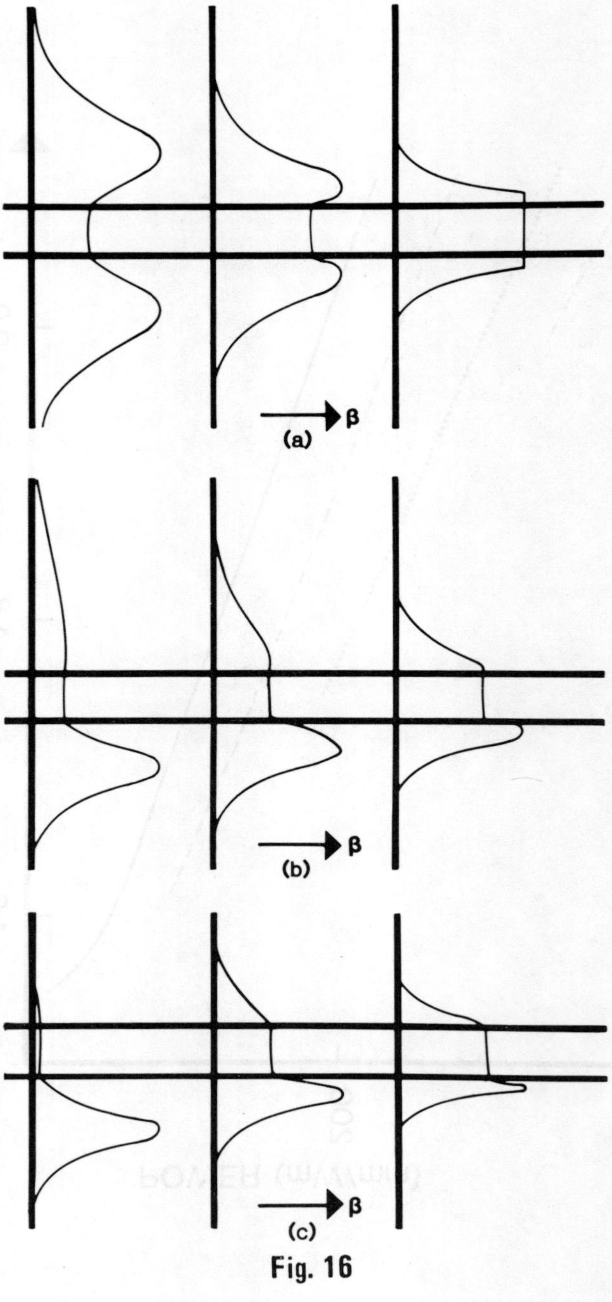

Fig. 16

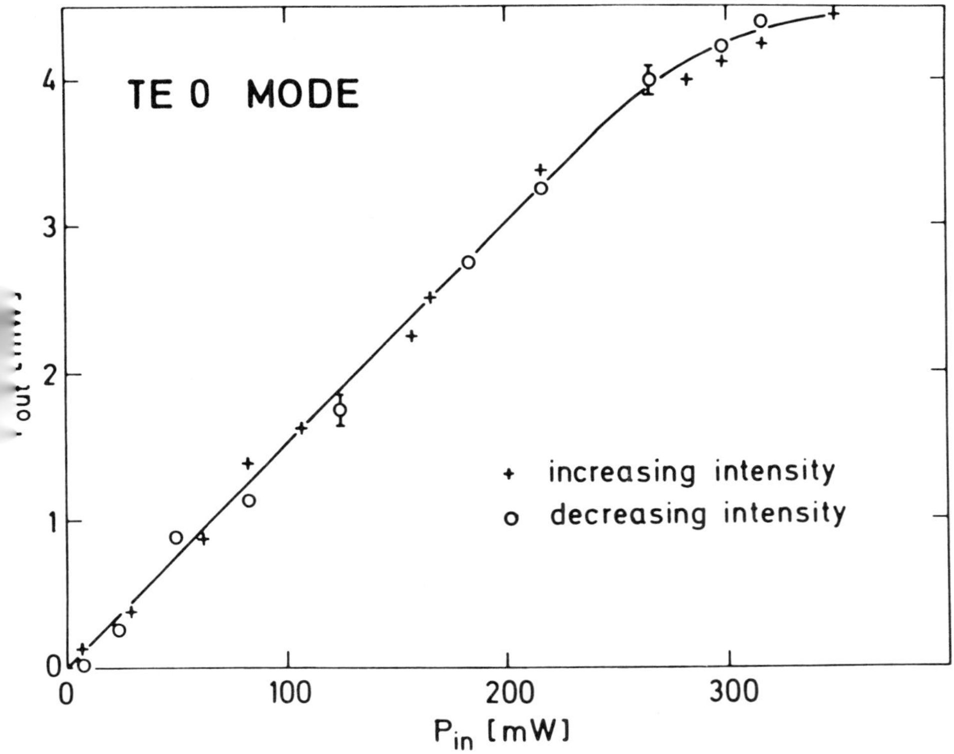
TE 0 MODE
+ increasing intensity
o decreasing intensity
0
1
2
3
4
0
100
200
300
P_in [mW]
P_out [mW]

Fig. 17(a)

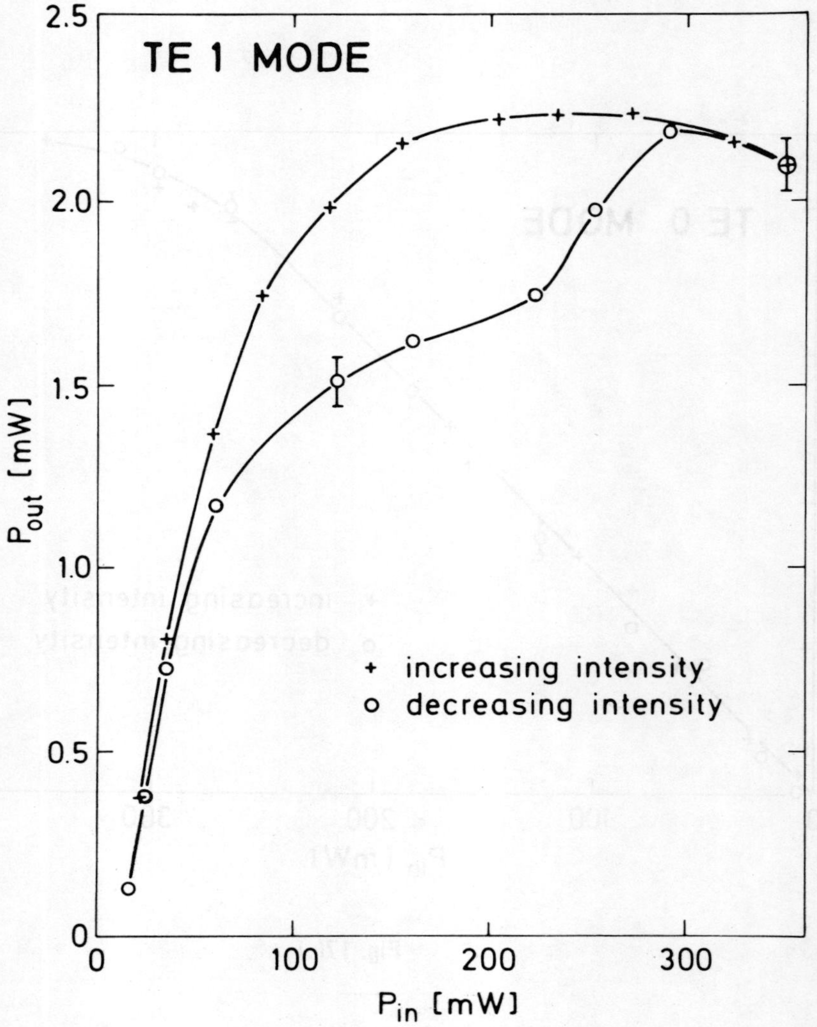
TE 1 MODE
2.5
2.0
1.5
1.0
0.5
0
100
200
300
P_{out} [mW]
P_{in} [mW]
+ increasing intensity
o decreasing intensity

Fig 17(b)

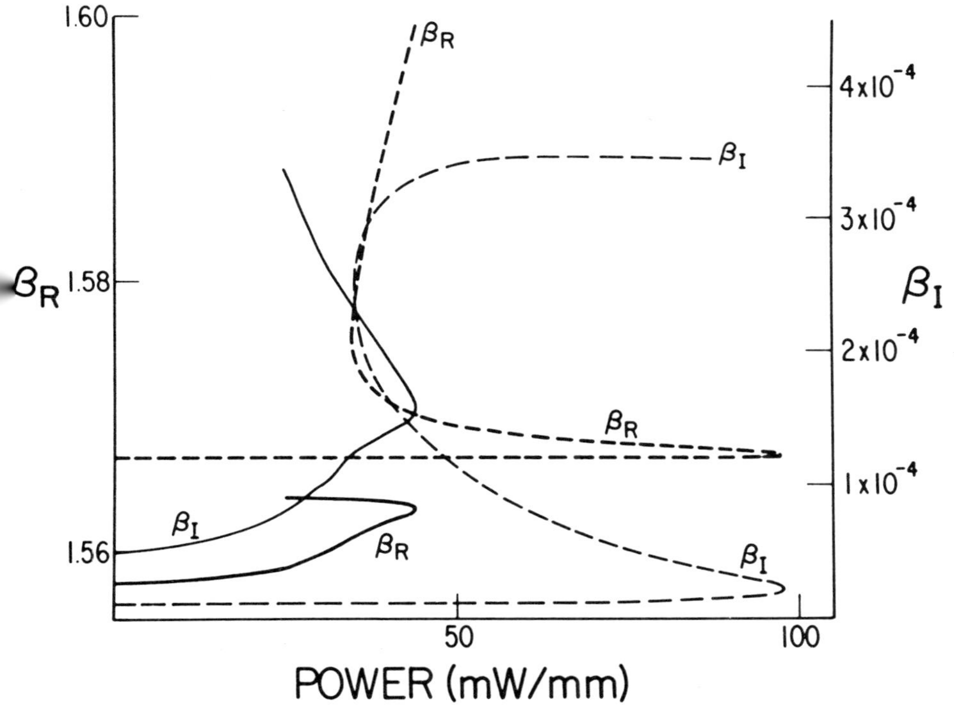
1.60
1.58
1.56
β_R
β_R
β_I
β_R
β_I
β_R
β_I
4×10^{-4}
3×10^{-4}
2×10^{-4}
1×10^{-4}
β_I
50
100
POWER (mW/mm)

Fig. 18

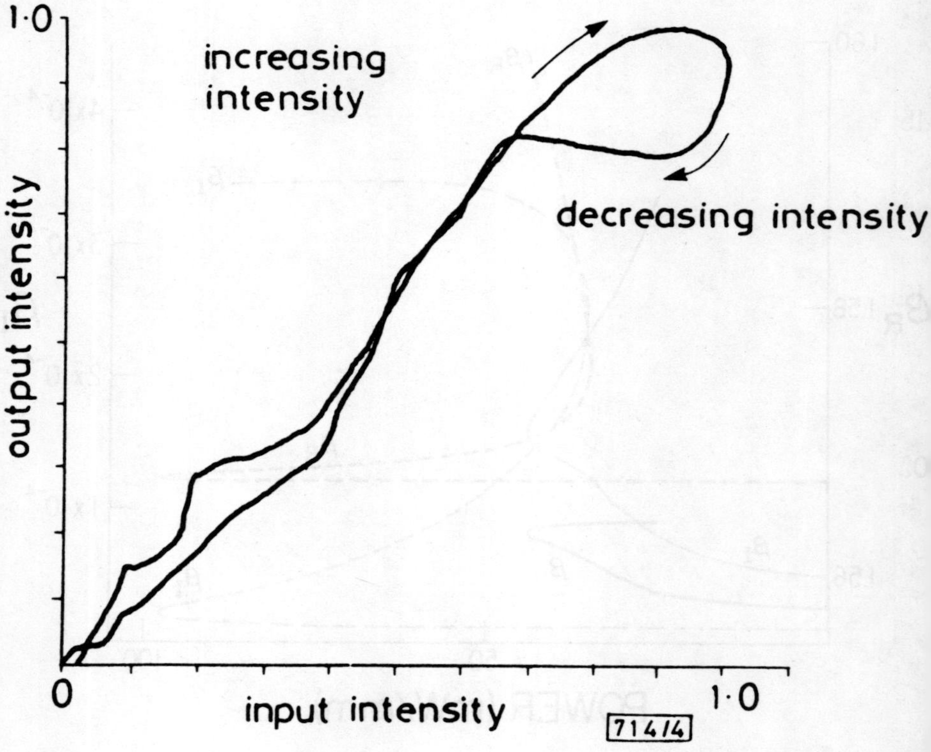
1·0
increasing
intensity
decreasing intensity
output intensity
0
input intensity
1·0
714/4

Fig. 19

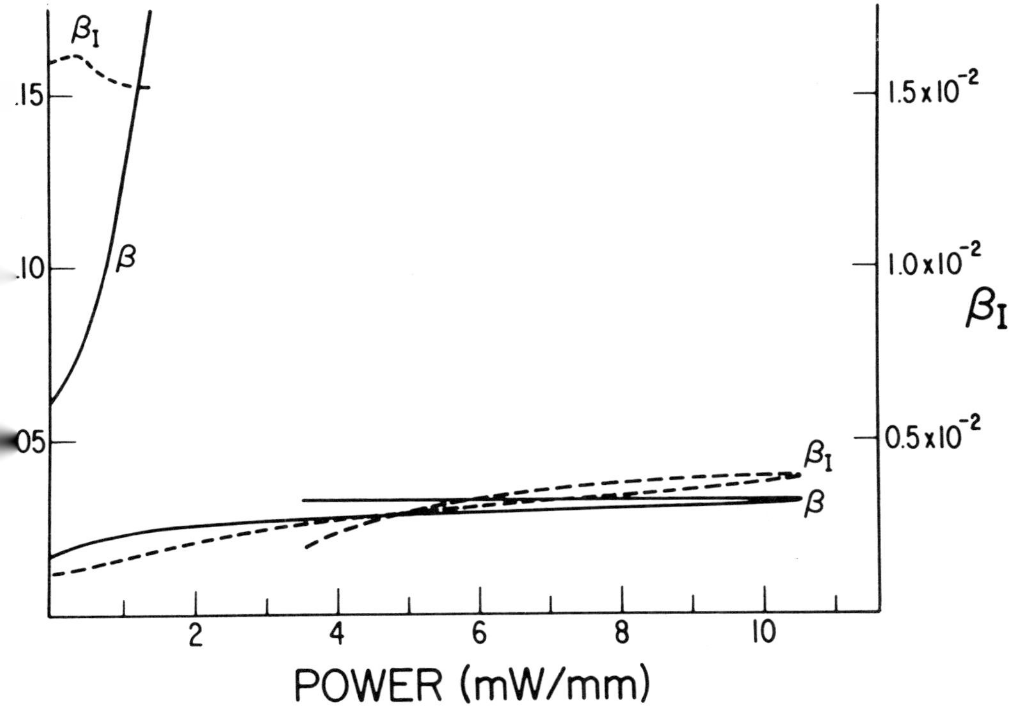

β_I
.15
.10
05
β
1.5 x 10^{-2}
1.0 x 10^{-2}
0.5 x 10^{-2}
β_I
β_I
β
2
4
6
8
10
POWER (mW/mm)

Fig. 20

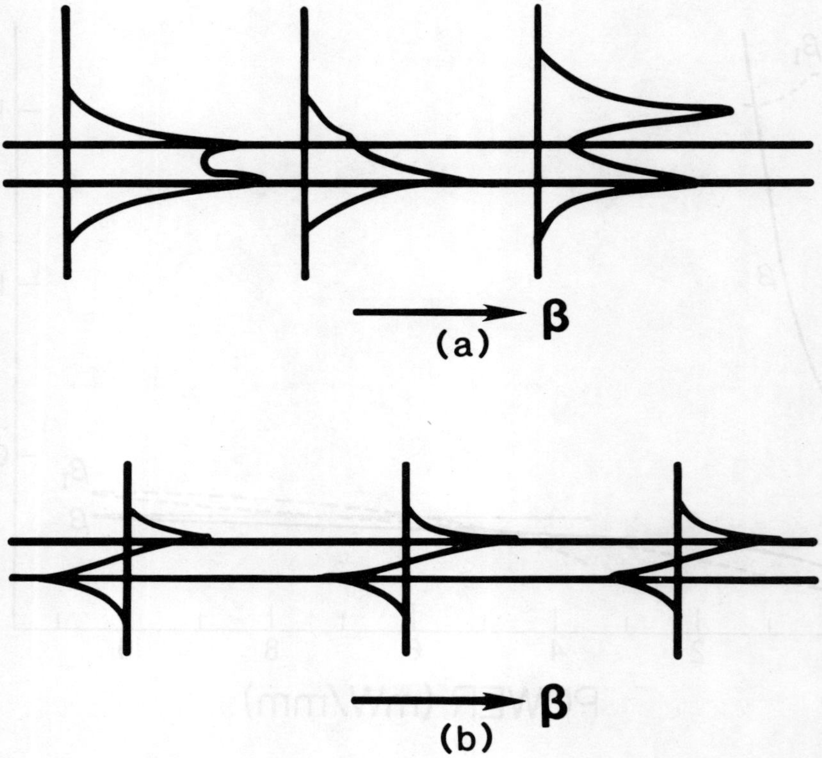

Fig. 21

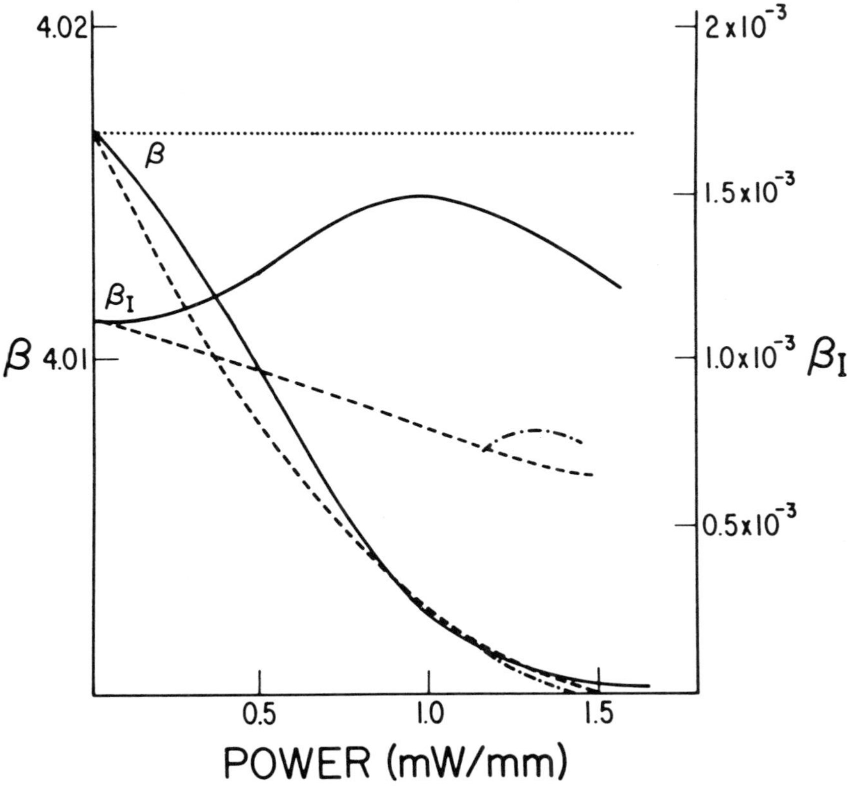
4.02
2x10^-3
β
1.5x10^-3
β_I
β 4.01
1.0x10^-3 β_I
0.5x10^-3
0.5
1.0
1.5
POWER (mW/mm)

Fig. 22

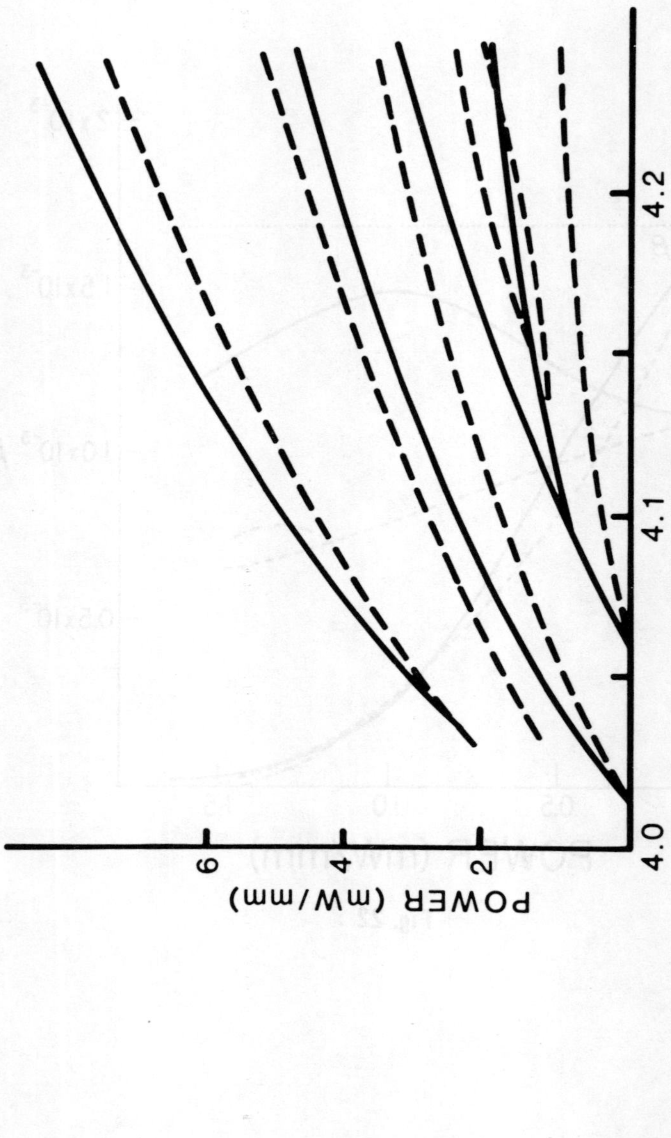

Fig. 23

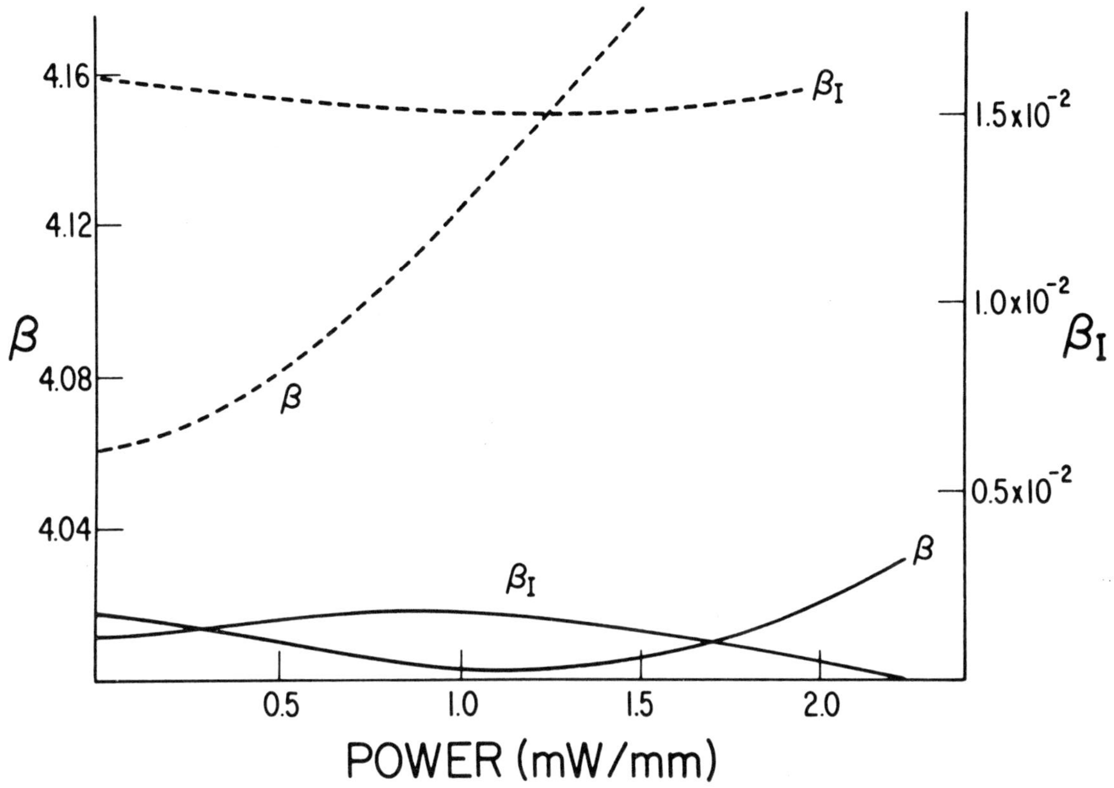

β
4.16
4.12
4.08
4.04
β_I
1.5x10^{-2}
1.0x10^{-2}
0.5x10^{-2}
β
β_I
β
β_I
0.5
1.0
1.5
2.0
POWER (mW/mm)

Fig. 24

HIGH TRANSPARENCY OF OVERDENSE PLASMA LAYERS IN METALLIC FILMS

S. Vukovic

Institute of Physics, P.O.Box 57
11001 Beograd, Yugoslavia

R. Dragila and B. Luther-Davies

Laser Physics Laboratory
Department of Engineering Physics
Research School of Physical Sciences
Australian National University
Canberra, A.C.T. 2601
Australia

I. Introduction

When a beam of electro-magnetic radiation is incident upon a plasma a wide range of physical phenomena can affect the absorption and scattering of the incoming beam. When the plasma is nonuniform the electromagnetic wave penetrates to a maximum density (the critical density, n_c), determined by the condition that the local plasma frequency equals that of the radiation, but this density is reached only when the radiation propagates up the plasma density gradient. For other than this case of normal incidence, the beam is reflected at the turning point density ($n=n_c \cos^2 \theta$), where θ is the angle between the beam direction and the density gradient) beyond which the field decays monotonically. In either case beyond the reflection point an evanescent wave exists. For non-normal incidence, therefore, the field of the electromagnetic wave at the critical density is, in general, much smaller than its in-vacuum value. It is important to note, however, that the evanescent wave can resonantly couple the energy of the incoming radiation to eigen-modes of the system leading to strong absorption. A well-known example of this phenomenon is resonance absorption where a p-polarised incident beam excites longitudinal plasma oscillations (plasmons) leading to up to 50% absorption of the incoming energy. In addition, if the plasma contains a density discontinuity near the critical region, such coupling can also cause the resonant excitation of surface electromagnetic waves which are eigen-modes of that discontinuity[1]. Irreversible damping of the surface waves can result in the incoming energy being totally absorbed, as has been demonstrated in experiments[2] involving the irradiation of a semi-infinite plasma at microwave frequencies.

Here we describe a new phenomenon, also involving surface electromagnetic waves, which can occur when a p-polarised electromagnetic wave irradiates an overdense plasma slab of finite thickness. It involves excitation of a pair of coupled surface waves on either side of the slab by the incoming electromagnetic wave. It is shown that even in the absence of irreversible damping, the energy of the surface waves can be "dissipated" by re-emission of the electromagnetic wave which causes the slab to appear totally transparent. The density profile for which this phenomenon occurs is similar to that which can be expected when a thin foil target is irradiated with an intense laser beam, and hence it may be encountered in laser fusion experiments. It should be emphasized that in contrast to anomalous transparency that has been discussed earlier[3] this effect is purely linear and arises even in a cold collisionless plasma.

Surface waves can not only be excited on density discontinuities within plasmas but can also on interfaces within stratified dielectric media provided a suitable dielectric profile can be constructed. Specifically, a multilayer structure combining materials having dielectric constants with opposite signs is required. This naturally leads to the choice of a metal as one dielectric for which, in general, $\mathrm{Re}\,\varepsilon \ll -1$. Furthermore, in order to be able to couple the incident electromagnetic wave to the surface modes a second restrictive condition must be satisfied, namely that the coating material in contact with the metal should have a positive dielectric constant with a magnitude less than unity (to allow the phase velocity of the surface electromagnetic wave exceeds the speed of light). Normal dielectric materials cannot meet this latter criterion although some special cases such as Quartz in the region of anomalous dispersion around 18 μm and Potassium in the UV can be found. The ATR (attenuated total reflection) technique[4] was developed to overcome this limitation. It involves the addition of a prism coupler thus forming a structure comprising prism-positive dielectric-metal where the incoming electromagnetic wave is normally totally internally reflected at the prism/dielectric boundary. The electromagnetic wave then becomes evanescent within the dielectric layer separating the prism and the metal. Since for total reflection to occur one needs $n_p > n_1$ (n_1 and n_p are the refractive indices of the dielectric and the prism, respectively) then the surface waves with phase velocity $c/n_1 > v_f > c/n_p$ can be excited by an electromagnetic wave which is incident upon the prism-dielectric boundary at some angle θ such that $1 > \sin\theta > n_1/n_p$.

In this paper we study surface wave induced transparency of a stratified dielectric medium and provide expressions which define the conditions in which total transparency of a normally reflecting metal layer can be observed. On the basis of these calculations we have constructed a prototype device which demonstrates the phenomenon and compare its performance with computer predictions. We also suggest some practical applications for the phenomenon.

II. Overdense plasma layers

Let us consider a p-polarized electromagnetic plane wave incident upon a plasma slab with density profile as shown in Fig. 1. Initially, we will consider a dissipationless plasma neglecting both collisional and resonance absorption (d=d'=Ø) of the surface waves. Furthermore, we assume that the profile is symmetric with the characteristic density scalelengths $L=L_1=L_3=n|dn/dx|^{-1} >> \lambda_0$ in the regions 1 and 3 (where λ_0 is the wavelength of the incident wave). The plasma is assumed to be homogeneous in the region 2 ($Ø < x < a$).

The magnetic component of the wave $\tilde{H}=(Ø,Ø,H)$ is a solution of the wellknown wave equation

$$\frac{d}{dx}\left(\frac{1}{\varepsilon}\frac{dH}{dx}\right) + \frac{\omega^2}{c^2}\left(1 - \frac{\sin^2\theta}{\varepsilon}\right) H = 0 \qquad (1)$$

describing propagation of an electromagnetic wave in a cold collisionless plasma. Here $\varepsilon(x)=1-n(x)/n_c$ is the plasma permitivity, n_c, the critical density corresponding to frequency ω of the incident wave, n, the electron density, c, the speed of light and θ, the angle of incidence. Eq. (1) is now solved separately in regions 1 and 3 by using the WKB approximation. Matching the solutions thus obtained at $x=0$ and $x=a$ to the solution corresponding to region 2 by imposing the boundary conditions requiring continuity of H and $(1/\varepsilon)dH/dx$, one obtains a general solution to eq. (1) in terms of the amplitude of the incident wave. Such a solution allows the coefficients of reflection R and transmission T to be determined as follows

$$R = -\frac{\gamma(1+\beta^2) + \beta(1+\gamma^2)\tanh\kappa_2 a}{2\gamma\beta + (\gamma^2+\beta^2)\tanh\kappa_2 a} \tag{2}$$

$$T = \frac{(1-\beta^2)\gamma\cosh^{-1}\kappa_2 a}{2\gamma\beta + (\gamma^2+\beta^2)\tanh\kappa_2 a} \tag{3}$$

where $\gamma = \varepsilon_1\kappa_2/\varepsilon_2\kappa_1$, $\beta_{1,3}=(1-ib_{1,3}^2/2)/(1+ib_{1,3}^2/2)$

$$b_1 = \exp[-\int_{x_r}^{0}\kappa(x)dx]; \quad b_3 = \exp[-\int_{a}^{x_{r_3}}\kappa(x)dx]$$

and $\beta = \beta_1 = \beta_3$, $\varepsilon_1 = \varepsilon(x=0_-) = \varepsilon(x=a_t) =$, $\varepsilon_2 = \varepsilon(0 < x < a)$, $\kappa(x) = (\omega/c)[\sin^2\theta - \varepsilon(x)]^{1/2}$, $\kappa_{1,2} = (\omega/c)(\sin^2\theta - \varepsilon_{1,2})^{1/2}$, $b = b_1 = b_3$, $\kappa(x_{R_1}) = \kappa(x_{R_3}) = 0$.

In certain conditions, which will be elaborated below, eqn. (2) has the solution R=0. This is analogous to the results given in Ref. 5 for the case of a homogeneous plasma occupying the halfspace $x > 0$, where the solution R=0 also arises due to resonant excitation of surface waves propagating along the boundary x=0. This has been demonstrated in microwave experiments[2] and treated selfconsistently, i.e. taking into account profile modification, both numerically[6] and analytically[7]. R=0 occurs because the inflow of energy in the incident wave is exactly balanced by dissipation of the energy in the surface waves it excites and hence $|R|^2=1$ in the limit of a non-dissipative plasma. In our case, however, if a is finite, R can be zero even in the absence of dissipation. This occurs because radiation can leak through the quasiclassical reflection point $x=x_{R_3}$ behind the overdense layer, which, as far as R is concerned, is equivalent to dissipation of the energy of the surface wave. That is the electromagnetic wave is re-emmited at the rear surface. The requirement R=0 ([see eq. (2)]

leads to the following relationship (within the first order approximation with respect to the small parameter b^2)

$$\tanh \kappa_2 a = - \frac{2\gamma}{1 + \gamma^2} \tag{4}$$

which is equivalent to a dispersion equation for two coupled surface waves propagating along the boundaries x=0 and x=a. A similar relationship has been obtained in Ref. 8 for the case of a plasma layer in vacuum, i.e. $\varepsilon_1=\varepsilon_3=1$. However, there is an essential difference between these two cases. The presence of the underdense regions 1 and 3 adjacent to the overdense plasma layer (see Fig. 1) allows the phase velocity of the surface waves to exceed the speed of light. Consequently, these surface waves can be resonantly excited by the incident electromagnetic wave.

Since we consider here a case in which irreversible damping of the surface wave is absent (collisions, resonance absorption, etc.) $|T|^2 + |R|^2$ must equal unity. Thus, the condition R=0 automatically results in $|T|^2=1$ [see ((2) and (3)]. Notice that with these assumptions total transparency can be achieved for any finite value of a, provided that the condition (4) is satisfied, and the divergent behaviour of $\cosh \kappa_2 a$, which apparently leads to vanishing T, is exactly compensated by the denominator in (3).

Obviously, the dispersion relation (4) has a solution only for $\gamma < 0$, i.e. for $\varepsilon_1\varepsilon_2 < 0$. Thus, resonant excitation of the surface waves is possible only when the critical point $\varepsilon = 0$ is localized within the density steps at $x=0$ and $x=a$.

Eq. (4), for unknown $\sin^2\theta$, can be solved analytically if $|\varepsilon_2| >> \sin^2\theta$:

$$\sin^2\theta = \frac{\varepsilon_1\varepsilon_2(\eta^2\varepsilon_2 - \varepsilon_1)}{\eta^2\varepsilon_2^2 - \varepsilon_1^2} \tag{5}$$

where $\eta = -\tanh^{-1}\xi \pm (\tanh^{-2}\xi - 1)^{1/2}$ and $\xi = \omega a|\varepsilon_2|^{1/2}/c$. The requirement $1 > \sin^2\theta > \varepsilon_1$ then implies $|\eta| > -\varepsilon_1(1-\varepsilon_2)^{1/2}/[\varepsilon_2(1-\varepsilon_1)^{1/2}] = \eta_C$. As seen, for a finite value of a there exist two solutions to eq (5), i.e. two surface eigenmodes exist, corresponding to $\eta > -1$ and $\eta < -1$. For large values of a, $\eta \to -1$ and these two solutions degenerate. The surface waves localized at $x=0$ and $x=a$ are then decoupled which leads to $|R|^2=1$ for any θ and b. In this case eq. (5) converts into the standard dispersion relation $1+\gamma=0$, i.e.:

$$\sin^2\theta = \frac{\varepsilon_1\varepsilon_2}{\varepsilon_1 + \varepsilon_2} \tag{6}$$

applicable to the case of a surface wave propagating along the boundary of two semi-infinite plasmas [1].

In the opposite limit, $\xi << 1$, the two solutions of eq. (5) which correspond to $\eta_1 \simeq -2/\xi$ and $\eta_2 \simeq -\xi/2$ are: $\sin^2\theta_1 = \varepsilon_1(1-\varepsilon_1/\eta_1^2\varepsilon_2)$ and $\sin^2\theta_2 = \varepsilon_1(1-\varepsilon_1/\eta_2^2\varepsilon_2)$. Obviously, for $\eta_2 < \eta_c$ the solution θ_2 does not exist. Notice that a large value of η_1 means $\sin^2\theta - \varepsilon_1 << \varepsilon_1$ and thus b need not necessarily be small for a fixed value of L. Then, strictly speaking, the WKB approximation is inapplicable and the asymptotic ($\eta_1 >> 1$) value of θ_1 leading to vanishing R can be viewed only as a crude estimate.

For the intermediate values of $\kappa_2 a \simeq 1$, which are of the most interest, the relationships (2) and (3) were analysed numerically. The illustrative example of the transmission coefficient $|T|^2$ as a function of the angle of incidence θ and a is shown in Fig. 2 corresponding to $L=10\lambda_0$. The analysis shows that for fixed ε_1 and ε_2 the transmission $|T|^2$ has two distinct peaks ($|T|^2=1$) at the optimum angles θ_1, θ_2 when a is small and/or L is large. For a given value of L these two peaks broaden and coalesce as a increases until they degenerate by forming a single peak corresponding to the condition (6). Further increase in a (for fixed L) prevents 100% transmission from being obtained and leads to a reduced value of the peak transmission (see Fig. 2). However, it is worth underlining that for any finite value of a there always exists a value of L such that total transmission can be obtained for any finite $L > L_c$. For example, for $\kappa_2 a = 2.25$ total transmission can still be obtained when $L = 15\lambda_0$ while the value of the classical transmission coefficient for normal incidence is 0.01.

As described earlier, this essentially new effect arises due to perfect coupling of the incident and transmitted waves via coupled surface modes resonantly excited on each side of the overdense plasma layer.

In order to study the role of losses, we will include resonance absorption (i.e. a linear conversion of surface waves into plasma waves) which can be assumed to be the dominant dissipation mechanism consistent with the choice of density profile. Similarly, any other dissipation mechanism (e.g. collisions) can be easily incorporated in this theory. Furthermore, the condition requiring symmetry ($b_1 = b_3$) of the density profile is now relaxed (see Fig. 1). Together with $b_1 \neq b_3$ we consider also $d \neq d'$ but still keep $\varepsilon(x=0) = \varepsilon(x=a+d+d') = \varepsilon_1$ without loss of generality. In contrast to the previous case, we now have to find a solution to eq. (1) in the regions $0 < x < d$ and $a+d < x < a+d+d'$. This can be done by using an iterative procedure with respect to the small parameters $\ell = (\omega/c)d\sin\theta \ll 1$ and $\ell' = (\omega/c)d'\sin\theta \ll 1$. Then, by matching the solutions at the boundaries as before one obtains R and T where, now, $T \propto b_3/b_1$. The presence of resonance absorption implies that $|R|^2+|T|^2 < 1$. Therefore, even when $R=0$, $|T|^2$ cannot reach unity. The analysis shows that the relationship $R=0$ now leads to an extra condition

$$b_1^2 - b_3^2 + \mathrm{Im}\{\frac{\varepsilon_1}{\kappa_1}[\int_0^d \frac{\kappa^2}{\varepsilon} dx + \int_{a+d}^{a+d+d'} \frac{\kappa^2}{\varepsilon} dx]\} = 0 \qquad (7)$$

which has to be fulfilled simultaneously with (4). Here all higher order terms with respect to the small parameters b_1^2, b_3^2, ℓ and ℓ' were ignored. Physically, (7) describes the energy balance, i.e. the energy inflow ($\propto b_1^2$) is fully compensated by resonance absorption [represented by the integral terms in (7)] and the energy outflow ($\propto b_3^2$). This implies $b_1 > b_3$, i.e. $L_1 < L_3$. Although now $|T|^2 < 1$ it can still substantially exceed its classical value.

III. Metallic films

The magnetic component $\underset{\sim}{H} = (0, 0, H)$ of the electromagnetic wave which has its electric field vector, $\underset{\sim}{E}$, in the plane of incidence (x,y) is again a solution to the equation (1) where θ is the angle of incidence measured relative to the normal to the surface of the layered medium at the interface $x=-b_1$ (see Fig. 3). The solution of equation (1) for the case when the spatial distribution of dielectric permittivities is as shown in figure 3, leads to the following expressions for the reflectivity, R, and transmission coefficient, T:

$$R = -(A + B)/(C + D)$$

$$A = \gamma(1-\gamma_0^2)(\beta_1+\beta_2)$$

$$B = [1-\gamma_2\gamma_0^2+(\gamma_2-\gamma_0^2)\beta_1\beta_2+\gamma_0(1-\gamma^2)(\beta_1-\beta_2)]\tanh \kappa_2 a$$

$$C = \gamma[(1+\gamma_0^2)(\beta_1+\beta_2)-2\gamma_0(1+\beta_1\beta_2)]$$

$$D = [1+\gamma^2\gamma_0^2+(\gamma^2+\gamma_0^2)\beta_1\beta_2-\gamma_0(1+\gamma^2)(\beta_1+\beta_2)]\tanh \kappa_2 a$$

$$T = \frac{e^{\kappa_1 b_1}}{e^{\kappa_1 b_1}} \frac{1+e^{-2\kappa_1 b_1}}{1+e^{-2\kappa_1 b_2}} \frac{1+\gamma_0\beta_1+R(1-\gamma_0\beta_1)}{[1-\gamma_0\beta_2+\gamma \quad (\beta_2-\gamma_0)\tanh \kappa_2 a \;]\cosh \kappa_2 a}$$

where

$$\gamma = \frac{\varepsilon_2\kappa_1}{\kappa_2\varepsilon_1}, \quad \kappa_{1,2} = \frac{\omega}{c} \quad (\varepsilon_p \sin^2\theta - \varepsilon_{1,2})^{1/2} \quad ,$$

$$\gamma_0 = \frac{i\,\varepsilon_1}{\varepsilon_p{}^{1/2}\kappa_1} \quad \frac{\omega}{c} \quad (1-\sin^2\theta)^{1/2}, \quad \beta_{1,2} = \frac{1-e^{-2\kappa_1 b_{1,2}}}{1-e^{-2\kappa_1 b_{1,2}}}$$

The maximum transmission, (which is 100% in the case when irreversible dissipation is absent, i.e. when ε_1, ε_2 and ε_p are all real quantities) occurs when R=0. This latter condition is equivalent to a dispersion equation for two coupled surface waves propagating along the boundaries x=0 and x= a [see (4)]. This dispersion relationship has a solution only if $\gamma < 0$, i.e. when the real parts of ε_1 and ε_2 have opposite signs. Since the wave must be evanescent in the regions $-b_1 < x < 0$ and $a < x < b_2 + a$, then for $\varepsilon_p = 1$ (i.e. with the prisms absent), the real part of ε_1 must be within the interval $0 < \mathrm{Re}\ \varepsilon_1 < 1$.

In practice, to increase the range of coating materials that can be used, the prisms are included and the relation $0 < \mathrm{Re}\,\varepsilon_1 < 1$ is then relaced by the condition $n_p > n_1$, where n_p and n_1 are the refractive indices of the prism and material on either side of the central layer respectively. In this case provided $1 > \sin\theta > n_1/n_p$, the wave is evanescent in the regions $-b_1 < x < 0$ and $a < x < b_2 + a$ even when $\mathrm{Re}\ \varepsilon_1 > 1$.

For a typical situation where the middle layer (see figure 3) is a metal foil for which $|\varepsilon_2| \gg \varepsilon_p > 1$, the dispersion relation gives the angular position of the transmission peak as follows:

$$\theta_{max} = \arcsin\left[\frac{\varepsilon_1 \varepsilon_2 (\eta^2 \varepsilon_2 - \varepsilon_1)}{\varepsilon_p (\eta^2 \varepsilon_2^2 - \varepsilon_1)}\right]^{1/2}$$

where

$$\eta = \eta_{\pm} = -(\tanh \xi)^{-1} \pm [(\tanh \xi)^{-2} - 1]^{1/2}$$

and

$$\xi = \frac{\omega}{c}\, a |\varepsilon_2|^{1/2}$$

In general, there are two transmision peaks which correspond respectively to the two cases $\eta = \eta_-$ and $\eta = \eta_+$ (see Ref. 9). Their presence or absence is determined by the obvious extra condition $\theta_{max} < 90°$, which implies that a critical value for η exists given by:

$$\eta_c = \frac{\varepsilon_1(1 - \varepsilon_2/\varepsilon_p)^{\frac{1}{2}}}{\varepsilon_2(1 - \varepsilon_1/\varepsilon_p)^{\frac{1}{2}}}$$

The two peaks are then present if the relevant values of η being either η_- or η_+ are such that $|\eta| > \eta_c$. The η_- peak corresponds to an asymmetric spatial distribution of E_x whilst the η_+ peak is associated with a symmetric E_x distribution. Since the symmetric mode carries more of its energy in the dissipative metallic film it is more heavily damped than the asymmetric mode. While this could be an advantage when high absorption is required, as in the case of experiments demonstrating surface wave induced absorption [4], it is a disadvantage in our case where high transmission is sought and hence the η_+ peak is observed only when using very thin metallic films. The characteristic spatial distribution of the modulus of the field, as well as the Poynting vector, corresponding to a pair of coupled surface waves is shown in Fig. 4 for prism-MgF_2-Ag-MgF_2-prism structure and optimum angle of incidence. Note that the field actually grows in the MgF_2 layer and reaches a value far in excess of that in vacuum at the boundary with the metal. This simply reflects the fact that the surface wave is localised on that boundary but also illustrates graphically the way in which the surface wave facilitates the coupling between the incident and transmitted waves.

A prototype surface wave coupler was constructed by evaporating MgF_2 (n=1.38) and silver films onto the base of a borosilicate crown glass isoceles prism (n=1.515) with an apex angle of 25°. Silver was chosen as the central layer because of it has a small index of refraction ($n \simeq 0.066$) and relatively large extinction coefficient ($k \simeq 4.0$) in the visible region of the spectrum which normally results in it having good reflectivity for the incident wave but low dissipation ($nk<1$) for the surface waves. For the longer wavelength region ($0.5\,\mu m < \lambda < 1.5\,\mu m$) Gold has similar favourable properties whilst Aluminium would also be useful in the UV and blue spectral regions.

The films were deposited using a standard radiant heater evaporation rig onto the cold substrate with the relative thickness being controlled to about 50Å. A simple device using nominally 3000Å MgF_2 layers sandwiching a 600Å Ag layer was constructed. To provide the symmetry required to obtain transmission an identical glass prism was placed in close contact with the films using ethyl salicylate as an index matching fluid ($n \approx 1.52$). The complete structure was placed in a rotating mount and its transmission as a function of angle and polarization for He-Ne laser radiation (λ =6328Å) was determined.

The measured and calculated transmission curves for p-polarized light are shown in Figs. 5 and 6, respectively. Near perfect agreement is obtained if it is assumed that the MgF_2 thickness ws 3250Å rather than 3000Å as determined using the monitoring system. Such a discrepancy is quite feasible since the quartz crystal oscillator film thickness monitor had to be placed some distance from the substrate in a position where slight under reading of the actual film thickness would be expected. No transmission (<0.2%) was obtained using s-polarized light in accordance with expectations. The performance of the device was also measured using other index matching liquids. The transmission fell to one half its maximum value when the refractive index mis-match was only about 5%.

Both the experimental and computed transmission curves show one major transmission peak corresponding to the solution $\eta = \eta_-$. The other solution $\eta = \eta_+$ can just be discerned in the form of the transmission plateau at higher angles.

IV. Conclusion

In conclusion, we have shown that a thick overdense non-dissipative cold plasma slab can be totally transparent due to resonant excitation of coupled surface waves which provide perfect coupling of the incident and transmitted waves in conditions when the density profile is symmetric. An asymmetry and the presence of dissipation result in $|T|^2 < 1$, however, the transmission still remains extremely high when compared to its classical value. It should be noted that the phenomenon described here is purely linear. In the case of metallic films the effect appears to have a number of practical applications. For example, an interface between two dielectrics at which total internal reflection normally occurs can be made highly transparent by the addition of the metal overcoat and further dielectric layers. Such a situation could be encountered using optical fibres where by suitably coating the outside of the fibre it may be possible to couple radiation out through the walls. Furthermore the sensitivity of the transmission characteristic to the properties of the incident radiation suggests the phenonenon may be useful in devices such as polarizers, filters, modulators, and beam splitters.

Acknowledgment:

The authors are grateful to Mr A. Muggleton for preparing the prototype device.

REFERENCES

1 Yu. A. Romanov, Izv. Vyssh. Ucheb. Zeved., Radiofiz. 7, 242 (1964).

2 Yu. Ya. Brodskiy, V. L. Golt'tsman and S. I. Nechuev, Pis'ma v ZhETF 24, 547 (1976); [JETP Lett. 24, 504-508 (1976)].

3 S. S. Moiseev, Fiz. Plazmy 1, 5 (1975); [Sov. J. Plasma Phys. 1, 2-4 (1975)], V. L. Krasovskiy and V. N. Oraevskiy, Dokl. Akad. Nauk SSSR 242, 584 (1978); [Sov. Phys. Doklady 23, 674-675 (1978)], V. L. Krasovskiy, V. V. Lisitchenko and V. N. Oraevskiy, Fiz. Plazmy 5, 1322 (1979); [Sov. Phys. Plasma Phys. 5, 740-742 (1979)].

4 A. Otto, Zeit. Phys. 216, 398 (1968), E. Kretschmann and H. Raether, Zeit. Naturf. 23a, 2135 (1968).

5 Yu. M. Aliev, S. Vukovic, O. M. Gradov, A. Yu. Kyrie and V. M. Cadez, Pis'ma v ZhETF 25, 351 (1977); [JETP Lett. 25, 326-330 (1977)],A. A. Zharov, I. G. Kondratev and M. A. Miller, Fiz. Plazmy 5, 261 (1979); [Sov. J. Plasma Phys. 5, 146-151 (1979)].

6 V. B. Gil'denburg, A. G. Litvak, I. A. Petrova and A. M. Feigin, Fiz. Plazmy 7, 736 (1981); [Sov. J. Plasma Phys. 7, 399-402 (1981)].

7 S. Vukovic, V. M. Cadez, N. Aleksic, Zh. Zh. Kasimov and R. R. Ramazashvili, Phys. Lett 102A, 186-188 (1984).

8 K. N. Stepanov, ZhTF 35, 1002 (1965); [Sov. Phys. JTP 10, 773-780 (1965)].

9 D. Sarid, Phys. Rev. Lett. 47, 1927 (1981).

FIGURE CAPTIONS

Fig. 1 Illustration of a plasma density profile which allows for resonant excitation of coupled surface modes. Total transparency is possible for symmetric profile with d=d'=0.

Fig. 2 Transmission $|T|^2$ as the function of the angle of incidence and of the normalized layer thickness $\kappa_2 a$ for $\varepsilon_1 = 0.2$, $\varepsilon_2 = -2$ and $L = 10\lambda$.

Fig. 3 Schematic showing the spatial distribution of the dielectric permittivity needed to display surface-wave induced transparency.

Fig. 4 The spatial distribution of electromagnetic field associated with a pair of coupled surface waves and the corresponding distribution of the Poynting vector S. The curves correspond to prism-M_gF_2-Ag-M_gF_2-prism structure with $a = 0.1\lambda$, $b_1 = b_3 = 0.5\lambda$, $\varepsilon_p = 2.295$, $\lambda = 6328$ Å and for the optimum angle of incidence $\Theta = \Theta_{max} = 74.7°$. (1)- $|E_y/E_{oy}|$, (2)- $|H/H_o|$, (3)- S_x/S_{ox}. E_{oy}, H_o and S_o correspond to the incident wave.

Fig. 5 The measured transmission characteristics of the prototype prism-MgF_2-Ag-MgF_2-prism device for He-Ne laser radiation. The thickness of the Ag and MgF_2 layers was 600Å and 3000Å respectively. The calculated transmission curves correspond to the thickness of the MgF_2 layer (a) 3000 Å and (b) 3250 Å.

Fig. 6 The calculated transmission characteristics of the prototype device as a function of the coating thickness in the range 1900Å to 5700Å.

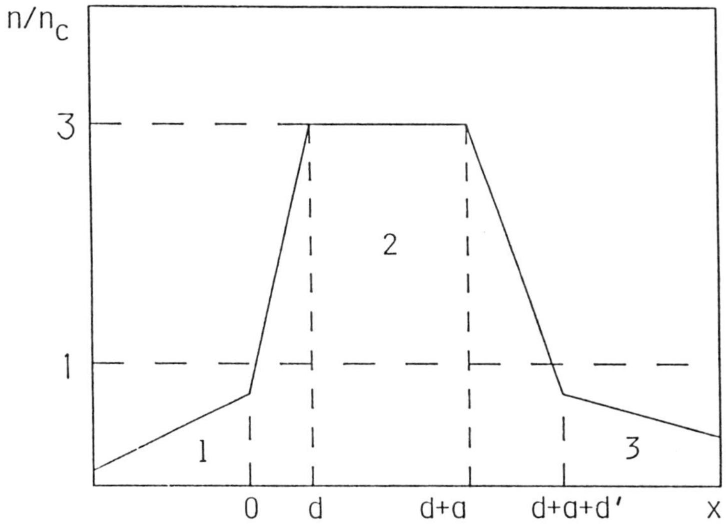

Figure 1

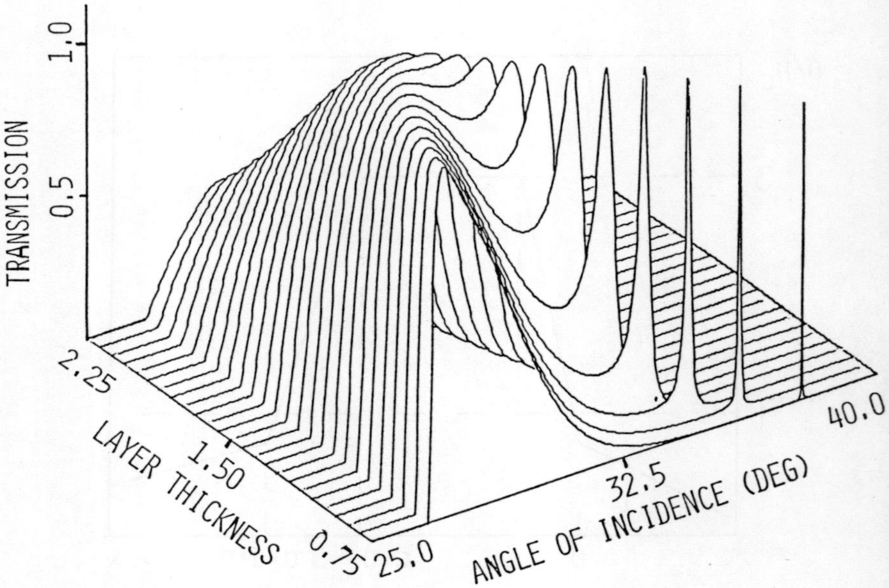
TRANSMISSION
0.5
1.0
2.25
1.50
0.75
LAYER THICKNESS
25.0
32.5
40.0
ANGLE OF INCIDENCE (DEG)

Figure 2

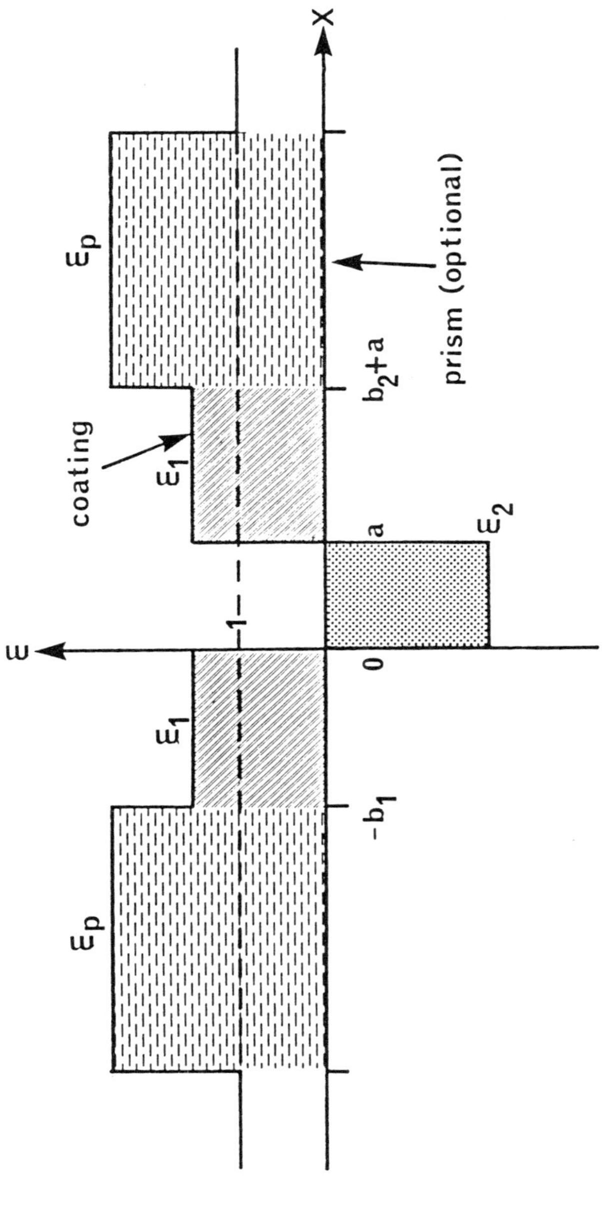

Figure 3

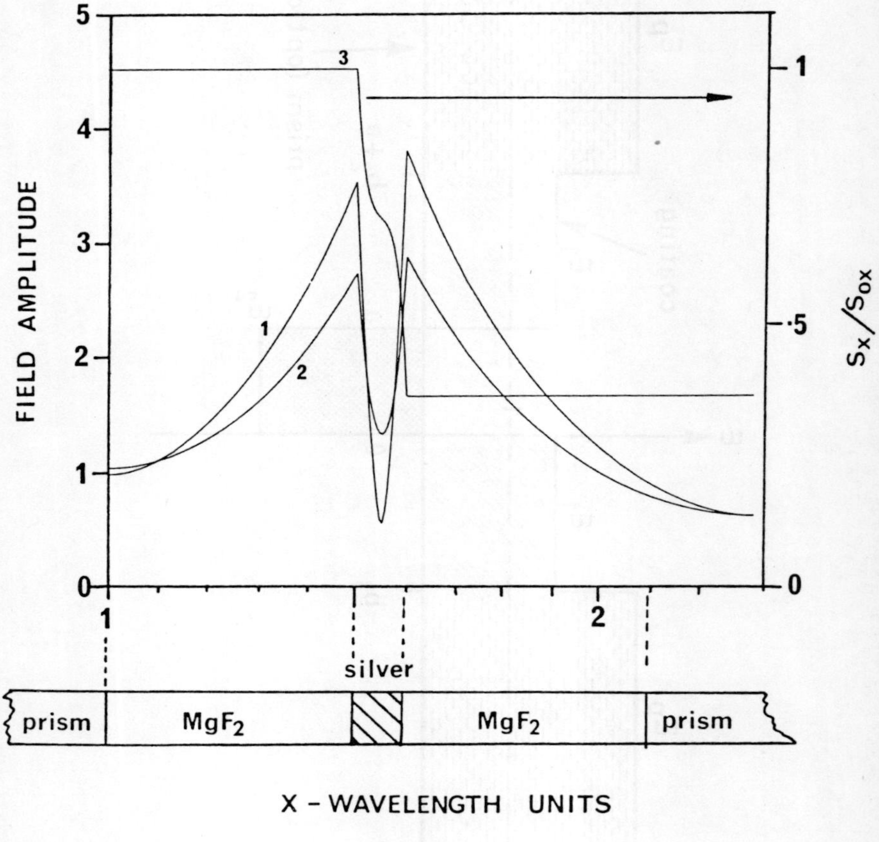
5
4
3
2
1
0
FIELD AMPLITUDE
3
1
2
1
·5
0
S_x/S_{ox}
1
2
silver
prism
MgF_2
MgF_2
prism
X - WAVELENGTH UNITS

Figure 4

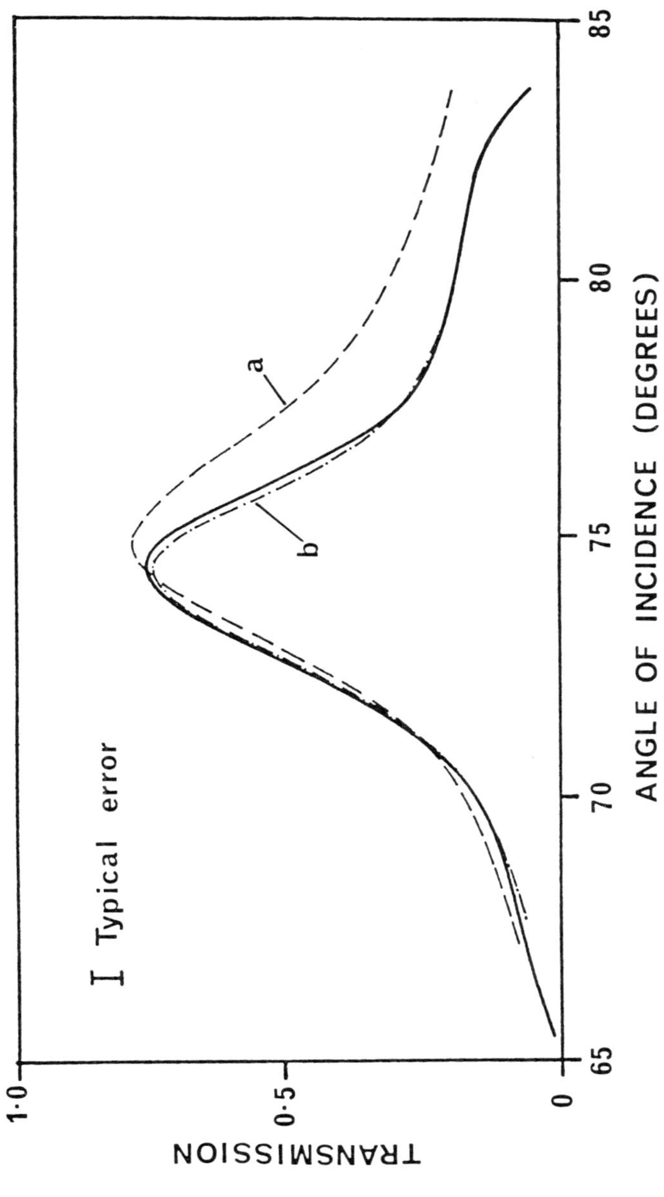
Typical error
a
b
ANGLE OF INCIDENCE (DEGREES)
65
70
75
80
85
TRANSMISSION
0
0·5
1·0

Figure 5

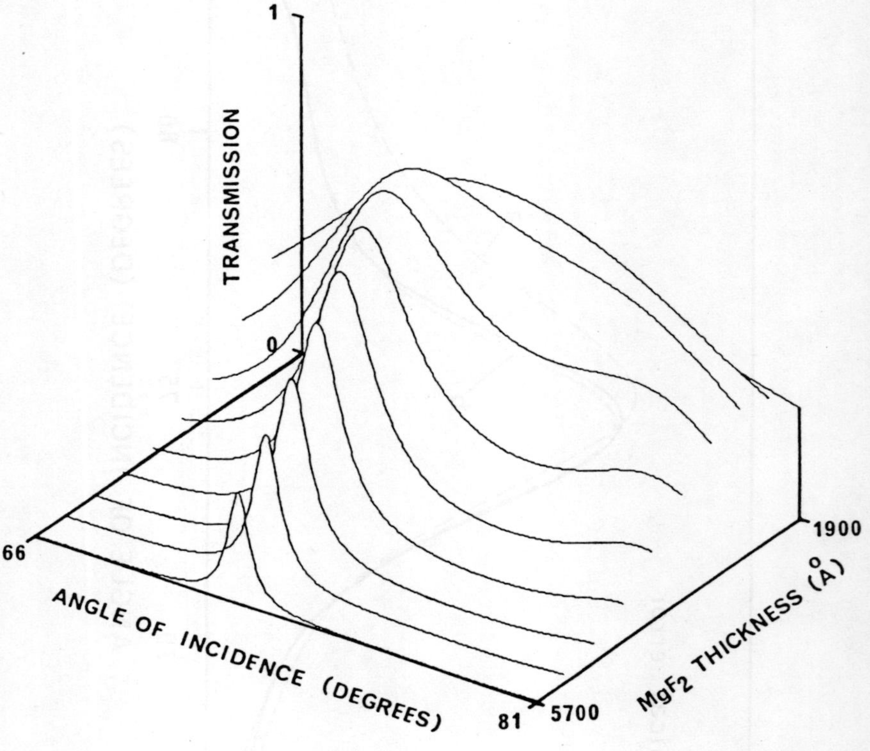
1
TRANSMISSION
0
66
ANGLE OF INCIDENCE (DEGREES)
81
5700
1900
MgF2 THICKNESS (Å)

Figure 6

HYDROMAGNETIC SURFACE WAVES AS POSSIBLE ORIGINATORS OF GEOMAGNETIC MICROPULSATIONS.

by

C. UBEROI

Department of Applied Mathematics, Indian Institute of Science
Bangalore - 560 012, India

1. Introduction

The Alfvén Surface Wave is generally defined to be a wave which can arise due to discontinuity in the Alfvén speed across the interface between two incompressible plasma media along which this wave propagates, but decreased exponentially in amplitude with distance away from the interface. It is to be noted that there is a basic assumption underlying this definition. When the variation in the magnetic field is considered, only the magnitude of the field is allowed to vary but the direction of the field is assumed to be constant. For the study of Alfvén or in general hydromagnetic surface waves along the magnetospheric boundary, bounded by magnetosheath and inner magnetosphere, however, this is not a good assumption. Recent satellite observations have well established that the magnetosheath magnetic fields show variation both in the magnitude and direction. Hence, while studying the surface waves along the magnetospheric boundary the angle between the magnetic field directions on either side of the interface should be considered. The main emphasis in this paper therefore, will be to understand the effect of variation of the magnetic field direction on some of the properties of hydromagnetic surface waves and on their generation mechanism at the magnetopause, with a view to relate these results to the questions which have recently arisen in the literature while understanding the various features of micropulsation phenomenon as a manifestation of the effects of hydromagnetic surface waves.

It has been shown that the angle between the magnetic fields on either side of the interface is a function of the angle between the interplanatery magnetic field (IMF) and the solar wind velocity in front of the bow shock [1]. From theoretical considerations the IMF direction and solar wind velocity are the two key parameters in the understanding of the dependency of pulsation activity on the interplanetary source mechanism [2]. Therefore, the study of the angle effect on the surface waves, considering these as possible originators of micropulsations will give a better understanding of the relationship between pulsation activity and the interplanetary parameters.

2. Alfvén Wave Equation for Magnetospheric Plasmas

The hydromagnetic surface waves along the magnetospheric boundary are a very important source for driving the long period geomagnetic micropulsations of the type pc3 to pc5 with periods ranging approximately between 10 sec to 10 mts. The linear resonance theory of long period micropulsations which has had much success in explaining the gross features of polarization sense and amplitude behaviour of the pulsations is in fact based on the idea of a steady state resonance coupling between a monochromatic surface wave excited at the magnetopause due to Kelvin-Helmholtz instability and a shear Alfvén wave associated with local field line oscillations [3,4]. The detailed analysis of the linear resonance theory in turn refers to the spectral analysis of the Alfvén wave equation in inhomogeneous plasmas [5]. The earlier studies of the Alfvén wave equation for inhomogeneous magnetic fields consider the variation of the magnetic fields only in magnitude assuming the field direction to be constant [6,7]. In order to therefore, understand the role of variation of magnetic field direction at the magnetopause on the micropulsation activity, it is necessary to reconsider the analysis of this equation when magnetic field vary both in magnitude and direction [8].

Consider the variation in density and magnetic fields in the x-direction. The plane in which the field lines lie is taken perpendicular to this direction, and the magnetic field is considered to be

$$\vec{B}_0(x) = [0,\ B_0(x)\ \cos \nu(x),\ B_0(x)\ \sin \nu(x)]$$

The linearized magnetohydrodynamic equations for incomprehensible fluid in inhomogeneous media give rise to the Alfvén wave equation

$$\nabla \{[\frac{\delta^2}{\delta t^2} - \frac{1}{\mu_0 \rho_0} (B_0 , \nabla)^2] \nabla\ v_x\} = 0 \qquad (1)$$

This equation is to be solved subject to the boundary conditions that $v_x \to 0$ as $|x| \to \infty$.

For Fourier analysis of the equation (1) take the time and space dependence as $\exp[-\omega t + \vec{k}\ .\ \vec{r}]$, where the wave vector

$$\vec{k} = (0,\ k \cos\theta,\ k \sin\theta)$$

The equation satisfied by $U_{k\omega}(x,\vec{k},\omega)$, the Fourier transform of v_x, is then

$$\frac{d}{dx} \{ \omega^2 - \omega_A^2(x)]\} \frac{du_{k\omega}}{dx} - k^2 U_{k\omega}[\omega^2 - \omega_A^2(x)] = 0 \qquad (2)$$

where

$$\omega_A^2(x) = (\vec{B}_0\ .\ \vec{k})^2 \ /\ \mu_0\ \delta_0 = k^2 V_A^2(x)\ \cos^2\ (\theta-\nu).$$

$(\theta-\nu)$ is the angle between the direction of the magnetic field and the wave normal vector.

Following the detailed analysis of equation (2) as given by Uberoi[7], the continuous spectrum of the problem is defined as

$$S = \{\omega_1 [\omega_A^2(x)]_{min} < \omega^2 < [\omega_A^2(x)]_{max}\}.$$

Taking x = 0 as the magnetopause and denoting the quantities on either side of the interface x = 0, i.e. in magnetosphere and magnetosheath by suffix 1 and 2, the range of continuous spectrum gives the following inequality to be satisfied by ω/k,

$$[V_{A1}^2 \cos^2\phi,\ V_{A2}^2 \cos^2(\phi-\chi)]_{min} < \omega^2/k^2 < [V_{A1}^2\cos^2\phi,\ V_{A2}^2 \cos^2(\phi-\chi)]_{max}$$

where $\theta - \nu_1 = \phi$, $\theta - \nu_2 = \phi-\chi$ and $\chi = \nu_2 - \nu_1$. It is to be noted that χ is the angle between the magnetic fields on either side of the magnetospheric boundary and is a function of the angle ψ made by IMF with the solar wind direction. The band width of the continuous spectrum $|V_{A1} \cos\phi - V_{A2} \cos(\phi-\chi)|$ is thus seen to depend on χ. Another interesting feature of the variation of the band with with χ is that for a given set of values of V_{A1}, V_{A2} and ϕ, such that $|V_{A1}/V_{A2} \cos\phi| < 1$ there can exist a critical value of $\chi = \chi_c$ for which the range of S can become zero. In this case it becomes a point spectrum. This critical value of χ can be easily calculated from the equation,

$$V_{A2} \cos(\phi-\chi_c) = V_{A1} \cos \phi$$

which gives

$$\chi_c = \phi \pm \cos^{-1}\left(\frac{V_{A1}}{V_{A2}} \cos\phi\right) , \quad 0 < \chi_c < \pi/2 \tag{3}$$

The continuous spectrum of the equation (2) asymptotically corresponds to noncollective oscillations with position dependent frequency and damping proportional to the inverse power of time. There also exist collective oscillations with position independent frequencies with exponential damping [9]. The complex frequency $\omega = \omega_R + i\omega_I$ of the damped surface eigen modes have been calculated in detail in Ref. 9. Following this analysis, to zeroth order in ka, where a is the scale length of variation of density and magnetic field, ω is given as a real value

$$\omega_R^2 = \frac{\rho_1\omega_{A1}^2 + \rho_2\omega_{A2}^2}{\rho_1 + \rho_2}$$

$$= k^2\frac{[B_{01}^2 \cos^2\phi + B_{02}^2 \cos^2\phi-\chi]}{(\rho_1 + \rho_2)} \tag{4}$$

Except for the multiplying factor (ρ_1/ρ_2) the value of ω_R given by equation (4) is the same as the frequency of the surface waves which are excited on the onset of the Kelvin-Helmholtz instability at the magnetopause [10].

When ka $\neq$ 0, the imaginary part ω_I is given by the equation

$$-2\omega_I\omega_R \frac{1}{(\omega_R^2 - \omega_{A1}^2)^2 \rho_1 + (\omega_R^2 - \omega_{A2}^2)^2 \rho_2} + \frac{4ka\pi}{(\rho_2 - \rho_1)\omega_R^2} = 0 \tag{5}$$

using the value of ω_R^2 from (4) and as ω_I is of the order of ka,

$$\frac{\omega_I}{\omega_R} = -2ka\pi \frac{\rho_1^2\rho_2^2}{(\rho_2^2 - \rho_1^2)} \left(\frac{V_{A1}^2 \cos^2\phi - V_{A2}^2 \cos^2(\phi-\chi)}{\rho_2 V_{A2}^2 \cos^2(\phi-\chi) + \rho_1 V_{A1}^2 \cos^2\phi} \right)^2 \tag{6}$$

Considering that Kelvin-Helmholtz instability is the mechanism for excitation of the surface waves at the magnetospheric boundary for the most unstable mode [11], $\theta - \nu_1 \simeq \pi/2$, for which

$$\frac{\omega_I}{\omega_R} = -2ka\pi \frac{\rho_1^2}{\rho_2^2 - \rho_1^2} ,$$

where

$$\omega_R^2 = k^2 V_{A2}^2 \sin^2 \chi/(1 + \rho_1/\rho_2)$$

For magnetopause boundary $\rho_1 < \rho_2$, therefore

$$\omega_I \simeq -2ka\pi(\rho_1/\rho_2)^2 k V_{A2} \sin\chi. \tag{7}$$

The damping time of the collective eigenmodes is therefore proportional to sin χ. From the value of ω_I we can get a rough order of magnitude for the time t_{phase}, after which the collective effects become important and phase-mixing gives rise to resonant interaction of surface modes with standing Alfvén modes along the dipole magnetic lines of force.

$$t_{phase} \sim 2\pi/\omega_I = \frac{1}{ka(\rho_1/\rho_2)^2 \, kV_{A2} \, \sin\chi} \tag{8}$$

Taking $\rho_1/\rho_2 = 1/20$, $a = 0.05\ R_E$

$$k = 1/0.8R_E,\ V_{A2} = 6x10^2\ \text{cm/sec.},$$

it is calculated that

$$t_{phase} = 16/\sin\chi\ \text{hrs.}$$

As χ roughly lies between 30° to 80°, t_{phase} can range between $16 < t_{phase} < 32$ hrs.

For $0 < t < t^*$, consider the solution of the equation (1) for the initial conditions.

$$v_x(\vec{x},0) = V(x) \quad \text{and} \quad \frac{\delta v_x(\vec{x},0)}{\delta t} = W(x)$$

i.e. initially the velocity component is function of x only. In this case eqn. (1) (It can be seen by direct substitution) is satisfied by the well-behaved solution:

$$v_x(\vec{x},t) = V(x)\ \cos[\ \omega_A(x)t] + W(x)\ \frac{\sin[\omega_A(x)t]}{\omega_A(x)} \tag{9}$$

for $(0 < t \leq t^*)$

Solution (9) is not valid for arbitrarily large times as the magnetic fields which require the differentiation of $v_x(\vec{x},t)$ will contain term of order t and, therefore will become very large as $t \to \infty$. The upper limit t* can be approximated to $t < t_{phase}$. Solution (2) represent the undamped oscillations of the field lines with frequency $\omega_A(x) = k(x)V_A(x)\ \cos[\theta - \nu(x)]$. Thus for $t \leq t_{phase}$, which can be of order of tens of hours, the local field line can oscillate at its Alfvén resonance frequency, which consists of sets of harmonics as $k = n\pi/\ell$, where ℓ is the length of the field lines.

The results in this paper have some important bearings on the understanding of geomagnetic micropulsation activity. Firstly, the linear resonance theory of the excitation of micropulsations as developed by Chen and Hasegawa [5] is based on the idea of a resonance coupling between a monochromatic surface wave at the magnetopause and a shear Alfven wave at a local field line in the inner magnetosphere. The range of Alfvén resonance frequencies, at which the resonant coupling occurs, is given by the continuous spectrum of the Alfvén wave equation. The variation of the band width of the continuous spectrum with the interplanetary parameter $\Psi \equiv F(x)$, as discussed in this paper, therefore, not only reinforces the earlier conclusions in the literature that micropulsation activity is sensitive to the angle between the IMF and the solar wind velocity in front of the bow shock, but also gives a quantitative idea of dependence of $S(\omega)$ on χ.

The linear resonance theory have successfully explained a majority of ground and space craft measurements in which the same frequency is observed at different latitudes for a given event with variations in the polarization and amplitude [12].

There are some reports in the literature, however, of ground based observations of pulsations whose period is a function of magnetic latitude for the same event [13]. While the reports of such events are not frequent, their existence cannot readily be explained by the current linear resonant theories. Also, as pointed out by Kivelson and Southwood [14] these theories have no clear way of accounting for the evidence that despite the presence of low-level power in the magnetosphere at all allowed resonant frequencies, observed resonances are often dominated by discrete frequencies.

The second conclusion of this paper is that the well behaved solution of the Alfvén equation for magnetospheric plasmas shows that the local field line can oscillate at the local Alfvén resonance frequency, which consists of set of harmonics, and that this solution can persist for several hours can give a possible explanation for the resonant response of the magnetosphere being dominated by the discrete frequencies. Finally, as the frequency of oscillation of the local field line matches with the resonant frequency, for an L-shell of the resonant field line $\omega_A(L) = k(L)V_A(L)\cos(\theta-\nu(L))$. This shows that the well-behaved solution of the Alfvén wave equation can explain the latitude dependence of the frequency of the magnetic pulsations.

Here, I like to mention that the well-behaved solution of the Alfvén-wave equation for magnetic fields varying only in magnitude has recently been discussed by Hasegawa, King and Assis[15] in the presence of an external wide band source. They used their results to explain the observations of magnetic pulsations with latitude dependent periods. However, the main difficulty with their explanation is that such external wide band sources have not been identified and therefore attributing the observations of pulsations whose period is a function of the magnetic latitude for the same event, to the presence of such sources lack confirmation. From the discussions of the solution of the Alfvén wave equation in this section it appears that for explaining these observations it may not be necessary to involve an external source, but such an explanation is contained in the solutions of the Alfvén wave equation for magnetospheric plasmas.

3. Compressibility Effects

When compressibility is taken into account the plasma-plasma interface can support two types of hydromagnetic surface waves, the slow and fast magnetosonic waves. For large wave numbers normal to the static magnetic field the fast mode represents the compressional surface wave. In a recent paper, Yumoto and Saito [16] have pointed out the difficulties which arise in the interpretation of compressional Pc3 type magnetic pulsations; with respect to the linear resonance theory. They show that though the linear resonance theory is based on the idea of a steady state resonance coupling between a shear Alfvén wave at local field lines in the magnetosphere and a surface wave excited at the magnetopause, the linear coupling oscillations between the standing wave and the compressional propagating wave, with a large value of the component of the wave vector perpendicular to the ambient magnetic field, never occurs in the magnetosphere. Their discussion is based on the fact that the coupling oscillations between a standing wave and a compressional wave can occur only when the following equation is satisfied in the magnetosphere.

$$(\omega_{comp}{}^2 - \omega_{eigen}{}^2) = V_A^2(k_{11}^2 + k_1^2) - (2\pi/T_{eigen})^2 = 0 \qquad (10)$$

where ω_{comp}, k_{11} and k_1 of the compressional wave stand for the frequency and mean wave numbers parallel and normal to the ambient magnetic field, and T_{eigen} is eigenperiod of the standing resonance oscillation at a local field line. To satisfy (10) they take $\omega_{comp} \sim V_A k_1$ and show that the value of $2\pi/\omega_{comp}$ and T_{eigen} can never be matched for large value of k_1 in the magnetosphere.

I like to point out that the value of ω_{comp} as chosen by Yumoto and Saito [16] is based on the dispersion relation of hydromagnetic waves in the bulk media, but since it is the compressional surface mode which resonantly excites the standing Alfven waves in the inner magnetosphere, ω_{comp} should be governed by the hydromagnetic surface wave dispersion equation. I shall now show that a proper choice of ω_{comp} rules out the above criticism against linear resonance theory.

The dispersion relation of hydromagnetic surface waves along the interface x = 0 formed by two compressible conducting media is given as [17,18]

$$\rho_1(k_{11}^2V_{A1}^2 - \omega^2)(m_2^2 + k_1^2)^{\frac{1}{2}} + \rho_2(k_{11}^2V_{A2}^2 - \omega^2)(m_1^2 + k_1^2)^{\frac{1}{2}} = 0 \qquad (11)$$

where ρ_1 and ρ_2 are equilibrium densities in medium $1(x < 0)$ and medium $2(x > 0)$, V_{A1} and V_{A2} are Alfven speeds with $\vec{B}$ being the external magnetic field directed along z direction and

$$m_{12}^2 = \frac{(k_{11}^2 c_{1,2}^2 - \omega^2)(k_{11}^2 V_{A1,2}^2 - \omega^2)}{k_{11}^2 c_{1,2}^2 V_{A1,2}^2 - (c_{1,2}^2 + V_{A1,2}^2)\omega^2}$$

where $c_{1,2} = (\gamma p_{1,2}/\rho_{1,2})^{\frac{1}{2}}$ is the sound speed. Equation (11) has been discussed in detail by Somasundaram and Uberoi [18]. From this discussion it is found that for low-β magnetospheric plasma as $c_1 << V_{A1}$ and for wave propagation angle $\theta > 60°$, only the compressional surface mode exists and is governed by the following approximate dispersion relation:

$$\omega_{comp} \simeq v_{A1} \frac{k_1}{\tan\theta} \left(\frac{1 + B_{02}^2/B_{01}^2}{1 + \rho_{02}/\rho_{01}}\right)^{\frac{1}{2}}$$

As for low-β plasmas the pressure equilibrium conditions implies $B_{01} \simeq B_{02}$

$$\omega_{comp} \simeq (V_{A1}k_1) \frac{1}{\tan\theta} \left(\frac{2}{1 + \rho_{02}/\rho_{01}}\right)^{\frac{1}{2}} \tag{12}$$

Taking the ratio $\rho_{02}/\rho_{01} \sim 20$, the ratio of plasma density in the magnetosheath to outer magnetosphere we get from (12)

$$\frac{2\pi}{\omega comp} \sim \left(\frac{2\pi}{V_A k_1}\right) \tan\theta \sqrt{10}$$

For values considered by Yumoto and Saito [16] near L = 10 and $\theta = 70°$ therefore

$$\frac{2\pi}{\omega_{comp}} \sim 50 \times 2.75 \times \sqrt{10} \sim 434$$

which is close to $T_{eigen} \sim 400$s. Similarly, at L = 5, $2\pi/\omega_{comp} \sim 26$, which is near $T_{eigen} \sim 20$s.

Thus criticism of linear resonance theory based on the calculations of ω_{comp} is not valid. This example points out the fact that compressibility effects are important for understanding the planetary magnetospheres.

Much work has been done to understand the propagation characteristics of the hydromagnetic surface waves [for references, refer the review by Uberoi [19]] for magnetic fields varying in magnitude but constant in direction. The dispersion equation governing these waves, as seen above, is a complicated transcendental equation and will give rise to a polynomial of tenth degree in the phase velocity on squaring. In order to avoid suprious roots and discuss both slow and fast modes of surface wave propagation Somasundaram and Uberoi [18] gave a method based on graphical and analytical analysis of the dispersion equation to find the roots which decay away from the interface. This method was generalized [20] for the case when the magnetic field on either side of the interface varies both in magnitude and direction. The results obtained point out that the existence and nature of the surface mode not only depend on the compressibility factor, the ratio of Alfvén to sound speed, but now also on χ the angle between the directions of the magnetic field on either side of the interface. Fig.1 shows the variation of the phase velocity of the two modes with c_1/V_{A1} for different values of θ and $\chi = 0$. In this figure $\alpha^2 = 0.2$, $\eta = 2$ and $\nu_1 = \nu_2 = 0$. Figures 2, 3, 4 and 5 give the phase velocity for $\phi \equiv (\theta + \nu_1) = 90°, 45°, 25°$ and $0°$ with $\alpha^2 = 0.2$, $\eta > 1.6$. In each figure both the slow and fast modes are shown for $\chi = 0°, 45°, 75°$ and $90°$. On comparing figure 1 and figure 2 the effect of finite value of χ is noted from the fact that the propagation range, the Alfvén surface wave velocity for incompressible case V_{si} and the critical value of the sound speed in medium '1' $C_1 = C_{1c}$, below which slow surface waves cannot propagate, show variation with χ. For smaller values of the angle ϕ, figure 3 shows that due to the band width of the propagating range becoming narrow no surface waves can exist for $\chi = 75°, 90°$. In general, we find that for a given propagation direction and the physical parameters

ρ_{01}/ρ_{02}, B_{01}/B_{02} at the magnetopause there can exist a critical value of the angle $\chi = \chi_c \equiv (\nu_2 - \nu_{1c})$, between the magnetic field directions, for which the width of the propagating band for the surface waves becomes zero. In this case there will be no surface wave propagation. It is interesting to note that χ_c is same as the value of χ for which the band-width of the spectrum S of the Alfvén wave equation becomes zero except due to geometry there is a $\pi/2$ difference in the wave propagation direction . Considering an example, for the magnetospheric conditions $B_{02}/B_{01} = 0.66$, $\rho_{01}/\rho_{02} = 0.1$, $\chi_c = \phi \pm \sin^{-1} (\sin\phi/0.21)$. For $\sin\phi < 0.21$, χ_c will exist. Let $\phi = 10^\circ$, for this value $\chi_c = 65^\circ$ or 45°. Therefore for the magnetic field directions such that $(\nu_2 - \nu_1) = 65^\circ$ or 45° there will be no surface wave excitation at the magnetospheric boundary. If geomagnetic pulsation activity is taken to be consistent with the Kelvin-Helmholtz generation mechanism, then these results once again emphasise the relationship of micropulsation activity with the interplanetary parameters.

4. Hydromagnetic surface waves generated by K-H instability

(a) Phase-Velocity

The Kelvin-Helmholtz (K-H) instability of the magnetopause arising due to shear in solar wind flow velocity is considered to be the major energy source available for hydromagnetic surface wave generation[21]. Taking into account the compressibility and the variation in the direction of the magnetic field on either side of the interface, the hydromagnetic K-H instability was studied by Southwood[11] in detail without making any physical assumptions used by earlier authors. The critical value of the relative streaming velocity for stability was studied by use of the equations for marginal stability. Realizing that the most unstable mode for the dayside magnetopause arises when the wave propagation is nearly perpendicular to the largest magnetic field Southwood[11] gave a simple criterion for the onset of the K-H instability at the magnetopause-magnetosphere boundary which can be written as

$$U > \frac{V_{A_2} \sin(\nu_2 - \nu_1)}{\sin \nu_1} \equiv U_c \qquad (13)$$

where the subscript 1 denotes quantities in the magnetosphere (side 1) and subscript 2 denotes quantities in the magnetosheath (side 2). Here U is solar wind velocity, U_c is the critical flow speed for the instability, V_A is the Alfvén speed and ν is the angle between U and the local magnetic field. Note that $(\nu_2 - \nu_1)$ is the angle between the magnetic fields on the two sides of the magnetopause.

Southwood's criterion has found very useful applications in the understanding of the geomagnetic pulsation activity, especially as it correlates the generation of the hydromagnetic waves at the magnetopause with the solar wind velocity and the angle between the IMF direction and the solar wind velocity in front of the bow shock.

However, due to mathematical complexity of the dispersion equation Southwood's analysis does not give any idea of the frequency of the waves which will be generated at the onset of the K-H instability. Recently it was shown by Uberoi[10] that Southwood's criterion as given by equation (13) can in fact be directly obtained from the marginal instability condition for the pure Alfvén surface waves propagating along the interface between two incompressible media, in the limit $\phi \to \pi/2$. Here ϕ is the angle between the tangential wave vector and the largest magnetic field. The phase velocity of the surface waves excited first at the onset of the instability given by Southwood's criterion therefore will be approximately given by the Alfvén surface wave velocity. From this correspondence between the incompressible and the compressible case in the limit $\phi \to \pi/2$ it was very easy to calculate the phase velocity of the surface waves excited by the K-H instability at the magnetopause. The phase velocity is given as

$$V_{AS} = \left(\frac{\rho_1}{\rho_1 + \rho_2}\right)^{\frac{1}{2}} V_{A2} \sin(\nu_2 - \nu_1) \qquad (14)$$

It is important to note from (14) that the wave oscillation frequency of the magnetopause-magnetosphere boundary depends on the angle $(\nu_2 - \nu_1)$ which in turn is a function of the orientation of the IMF.

(b) Spectral range in structured layers

Recently, there has been a great interest in understanding the K-H instability of structured plasma layers in the magnetosphere[22,23]. This is due to the fact that recent satellite observations of particles and fields have revealed more detailed structure and dynamics of magnetopause regions, such as the presence of the magnetospheric boundary layer and of rapid boundary motions. Most of these studies, however, are based on numerical analysis. The numerical analysis exhibited several characteristics of the real frequency values and growth rate of the boundary waves for fixed sets of values of parameters involved in the dispersion equation. However, in application to the magnetospheric boundary without having a precise knowledge of the parameters involved and to get an understanding of the relationship of K-H instability mechanism to interplanetary parameters it is necessary to establish a general criterion for the onset of K-H instability for finite systems. For this the dispersion equation should be discussed analytically. A simple case where the dispersion equation is amenable to analytical discussions is considered recently by Uberoi[24]. Take the three-layered model as considered by Lee et al[23]. Consisting of three regions of incompressible plasma: magnetosheath (region 1), the boundary layer (region 2), and the magnetosphere (region 3). The dispersion equation gives two modes of instability, the inner and magnetopause mode, corresponding to symmetric and asymmetric perturbations. Assuming that the plasma parameters in region 1 and region 3 are same the symmetric and assymetric perturbations are decoupled and the dispersion equations can be written as product of two factors. Each factor then can be discussed analytically and stability criterion can be established by discussing the marginal stability. This assumption, though unrealistic for the day side magnetosphere, gives an idea of the qualitative feature of the K-H instability criterion for layered system and the spectral range of the surface waves generated at the onset of the instability.

The dispersion equation derived by Lee et al[23] can be written as

$$G_2(G_1 + G_3) + (G_1G_3 + G_2^2)\tanh kL = 0, \tag{15}$$

where

$$G_i = \rho_i[(\omega - \vec{k}.\vec{V}_i)^2 - (\vec{k}.\vec{V}_{Ai})^2], \; i = 1,2,3$$

$$k = (0, k\cos\theta, k\sin\theta)$$

$$\vec{B}_i = (0, B_i\cos\nu_i, B_i\sin\nu_i)$$

$V_i = (0, V_i\cos\phi_i, V_i\sin\phi_i)$, is the streaming velocity.

When $G_2 = G_3$, equation (1) gives two factors

$$(G_1 + G_2\tanh kL/2) = 0 \; : \text{asymmetric modes} \tag{16}$$

$$(G_1 + G_2\coth kL/2) = 0 \; : \text{symmetric modes} \tag{17}$$

When $V_i = 0$, it is easily found that the Alfvén surface waves have phase velocities as

$$\left(\frac{\omega}{k}\right)^2 = \frac{\rho_1 V_{A1}^2\cos^2(\theta - \nu_1) + \rho_2 V_{A2}^2\cos^2(\theta - \nu_2)\left\{\begin{matrix}\tan h\\ \cot h\end{matrix}\right\}\frac{kL}{2}}{\rho_1 + \rho_2\left\{\begin{matrix}\tan h\\ \cot h\end{matrix}\right\}\frac{kL}{2}} \tag{18}$$

for the two modes of propagation when $\theta - \nu_1 = \pi/2$,

$$\frac{\omega}{k} = \frac{V_{A2}\sin(\nu_1 - \nu_2)}{\left[1 + \rho_1/\rho_2\left\{\begin{matrix}\cot h\\ \tan h\end{matrix}\right\}\frac{kL}{2}\right]^{\frac{1}{2}}} \tag{19}$$

Equations (18) and (19) show the angle effect and the effect of finite thickness of the layer on the surface waves propagation velocity.

Writing the equations (16) and (17) as quadratic in ω and following the analysis given in our earlier paper[25], to discuss the marginal stability equation, the results regarding the stability of the system can be summarised as follows:

i) When $U^2 < U_c^2$, both the modes are stable.

Here $U^2 = (V_1 \cos(\theta - \phi_1) - V_2 \cos(\theta - \phi_2)^2$

$$U_c^2 = (V_{A1}^2 \cos^2(\theta - \nu_1) + V_{A2}^2 \cos^2(\theta - \nu_2)$$

ii) When $U_c^2 < U^2 < U^{*2}$,

where $U^{*2} = (\frac{\rho_1 + \rho_2}{\rho_1\rho_2})[\rho_1 V_{A1}^2\cos^2(\theta - \nu_1) + \rho_2 V_{A2}^2\cos^2(\theta - \nu_2)]$,

the asymmetric mode is unstable only for the wave numbers in the range $k_1 < k < k_2$ if $\rho_1 V_{A1} \cos(\theta - \nu_1) < \rho_2 V_{A2} \cos(\theta - \nu_2)$. If $k < k_1$ and $k > k_2$ it is stable.

If $\rho_1 V_{A1}\cos(\theta - \nu_1) < \rho_2 V_{A2}\cos(\theta - \nu_2)$, the symmetric mode is unstable for all the wave-numbers in the range $k_1 < k < k_2$.

iii) When $U^2 > U^{*2}$, both inner and magnetopause modes are unstable for all the wave-numbers $k > k_1$.

$k_{1,2}$ are given by the expressions:

$$\tanh(k_{1,2}L/2) = \frac{(U^2 - U_c^2) \pm [(U^2 - U_c^2)^2 - 4V_{A1}^2 V_{A2}^2\cos^2(\theta - \nu_1)\cos^2(\theta - \nu_2)]^{\frac{1}{2}}}{2(\rho_2/\rho_1)V_{A2}^2\cos^2(\theta - \nu_2)}$$

In the range $U_c < U < U^*$, the wave-number has a minimum value

$$k_m = \frac{2}{L} \tanh^{-1} \frac{\rho_1 V_{A1} \cos(\theta - \nu_1)}{\rho_2 V_{A2} \cos(\theta - \nu_2)} \equiv \frac{2}{L} \tanh^{-1} x$$

when asymmetric mode is unstable, and

$$k_m = \frac{2}{L} \tanh^{-1} (1/x)$$

when symmetric mode is unstable. The corresponding minimum wind speed is

$$U_m^2 = U_c^2 + 2 V_{A1} V_{A2} \cos(\theta - \nu_1) \cos(\theta - \nu_2)$$

The frequency of the surface waves which will be excited when U exceeds this minimum value either due to instability of symmetric or asymmetric mode is

$$\omega_m = k_m \frac{V_1 \cos(\theta - \phi_1) V_{A2} \cos(\theta - \nu_2) + V_2 \cos(\theta - \phi_2) V_{A1} \cos(\theta - \nu_1)}{(V_{A2} \cos(\theta - \nu_2) + V_{A1} \cos(\theta - \nu_1))}$$

Considering a numerical example, let $(\theta - \nu_1) \simeq \pi/2$ such that $x \simeq 0.01$. Then $k_m = \frac{0.02}{L}$, $\omega_m = \frac{0.02 V_1 \cos\nu_1}{L}$, taking $\phi = \pi/2$. If the boundary layer thickness $L = 0.01\ R_E$, R_E is the earth's radius, then $\omega_m = 0.1 \cos \nu_1$ Hz. The frequency therefore is in the hydromagnetic range.

The important qualitative conclusions which can be relevant to the micropulsation theory are that for moderate values of the solar wind velocity lying in the range $U_c < U < U^*$, for given values of V_{A1}, V_{A2}, ρ_1, ρ_2 the instability will set in via the inner or magnetopause modes depending on the angle between the magnetic field directions on either side of the boundaries. For this value of U, the spectral range of the excited surface waves will have upper and lower cut-off frequencies as the wave-numbers for which the instability sets in are in the range $k_1 < k < k_2$. For high values of the solar wind velocity such that $U > U^*$, both the modes are unstable and the surface waves are excited only for frequencies higher than $\omega < \omega_1$, ω_1 corresponds to the wave-number k_1.

Wolfe et al[26] find in their observational study that the dependence of hydromagnetic energy spectra in the magnetosphere on solar wind velocity and interplanetary magnetic field is more effective at higher frequencies. From the results above it appears these observations are consistent with the origin of surface waves by K-H instability of layered systems in the magnetosphere. The finite thickness of the boundary layer as considered above introduces an upper limit to the waves generated by K-H instability mechanism.

5. Conclusion

The discussions in this paper have shown that it is important to consider the variation of magnetic field both in magnitude and direction at the magnetospheric boundary while understanding the geomagnetic micropulsation activity as manifestation of the effects of hydromagnetic surface waves propagating along this boundary.

References

1. L.C. Lee & J.V. Olson, Geophys. Res. Lett., 7, 777 (1980).
2. A. Wolfe, J. Geophys. Res. 85, 5977 (1980).
3. D.J. Southwood, Planetary Space Sci. 22, 483 (1974).
4. L. Chen & A. Hasegawa, J. Geophys. Res., 79, 1024 (1974).
5. L. Chen & A. Hasegawa, J. Geophys. Res., 79, 1033 (1974).
6. C. Uberoi, Indian J. Pure & App. Phys. 2, 133 (1964).
7. C. Uberoi, Phys. Fluids 15, 1673 (1972).
8. C. Uberoi, To be published (1985).
9. A. Hasegawa & C. Uberoi, 'The Alfvén Wave', Technical Information center, U.S. Department of Energy, Oak Ridge, Tenn., 1982.
10. C. Uberoi, J. Geophys. Res. 89, 5652 (1984).
11. D.J. Southwood, Planetary Space Sci. 16, 587 (1968).
12. L.J. Lanzerotti & D.J. Southwood, Solar System Plasma Physics, Ed. L.J. Lanzerotti, C.F. Kennel & E.N. Parker, North Holland, Amsterdam, 3-35 (1979).
13. H. Voelker, Ann. Geophys. 24, 245 (1968); M. Siebet, Planetary Space Sci. 21, 137 (1964); G. Rostoker & J.C. Samson, J. Geophys. Res. 77, 6249 (1972).
14. M.G. Kivelson & D.J. Southwood, Geophys. Res. Lett. 12, 49 (1985).
15. A. Hasegawa, K.H. Tsui & A.S. Assis, Geophys. Res. Lett. 10, 765 (1983).
16. K. Yumoto & T. Saito, J. Geophys. Res., 87 (1982).
17. C. Uberoi, Sol. Phys. 78, 351 (1982).
18. K. Somasundaram & C. Uberoi, Sol. Phys. 81, 19 (1982).
19. C. Uberoi, Contemporary Plasma Physics, Eds. M.S. Sodha, D.P. Tewari & D. Subbarao, Macmillon India Ltd., p.43-58 (1984).
20. C. Uberoi & A. Satya Narayanan, To be published (1985).

21. D.J. Southwood & W.J. Hughes, Space Rev. 35, 301 (1983).

22. R.S.B. Ong & N. Roderick, Planet, Space Sci. 20, 1 (1972); A.D.M. Walker, Planetary Space Sci. 29, 1119 (1981); T. Tamao, J. Geophys. Res. 86, 11,258 (1981); C. Uberoi, (abstract), Proc. Int. Conf. Plasma Phys., Lausanne - Switzerland, Vol.2, 1097 (1984); A. Miura & P.L. Pritchett, J. Geophys. Res. 87, 7431 (1982).

23. L.C. Lee, R.K. Albano & J.R. Kan, J. Geophys. Res. 86, 54 (1981).

24. C. Uberoi, To be published (1985).

25. C. Uberoi & R. Jayakaran Issac, Aust. J. Phys. 21, 931 (1968).

26. A. Wolfe, L.J. Lanzerotti & C.G. Maclennan, J. Geophys. Res. 85, 114 (1980).

Figure Captions

Fig. 1 Variation of the phase velocities of the fast (broken lines) and slow (continuous lines) surface waves with the compressibility factor C_1/V_{A1}, for different angles $\theta = 0^\circ$, 26.6°, 63.4°, 84.3° with $\alpha^2 = 0.2$ and $\eta = 2$.

Fig. 2-5 Variation of the phase velocities of the fast (broken lines) and slow (continuous lines) surface waves with the compressibility factor C_1/V_{A1}, for different angles $\phi = 90^\circ$, 45°, 25° and 0°. For each angle ϕ the variation with $\chi = (\nu_1 - \nu_2)$ is plotted by taking $\chi = 0^\circ$, 45°, 75° and 90°. The maximum value of the propagating range is marked as V_{Amax} and minimum as V_{Amin}. V_{si} varies for every value of the angle and is shown by continuous horizontal lines.

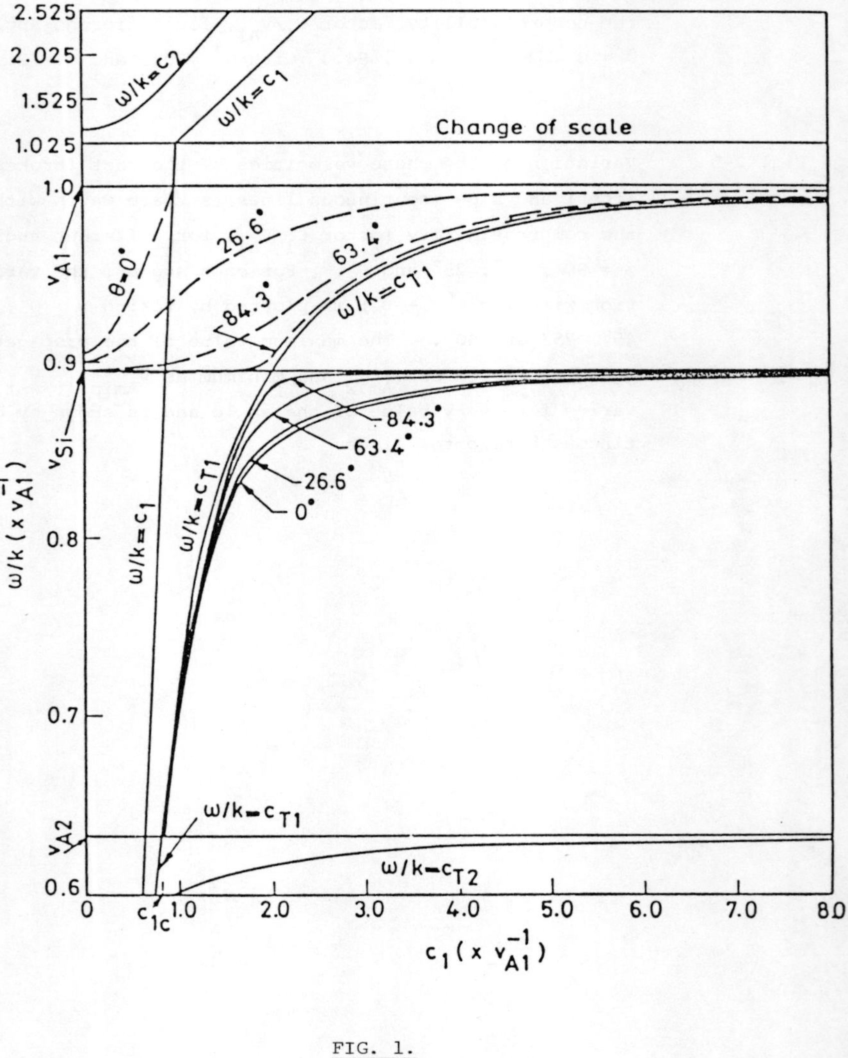
2.525
2.025
1.525
1.025
1.0
0.9
0.8
0.7
0.6
ω/k=c2
ω/k=c1
Change of scale
vA1
θ=0°
26.6°
63.4°
84.3°
ω/k=cT1
vSi
84.3°
63.4°
26.6°
0°
ω/k=c1
ω/k=cT1
ω/k (x vA1^-1)
vA2
ω/k=cT1
ω/k=cT2
0
c1c
1.0
2.0
3.0
4.0
5.0
6.0
7.0
8.0
c1 (x vA1^-1)

FIG. 1.

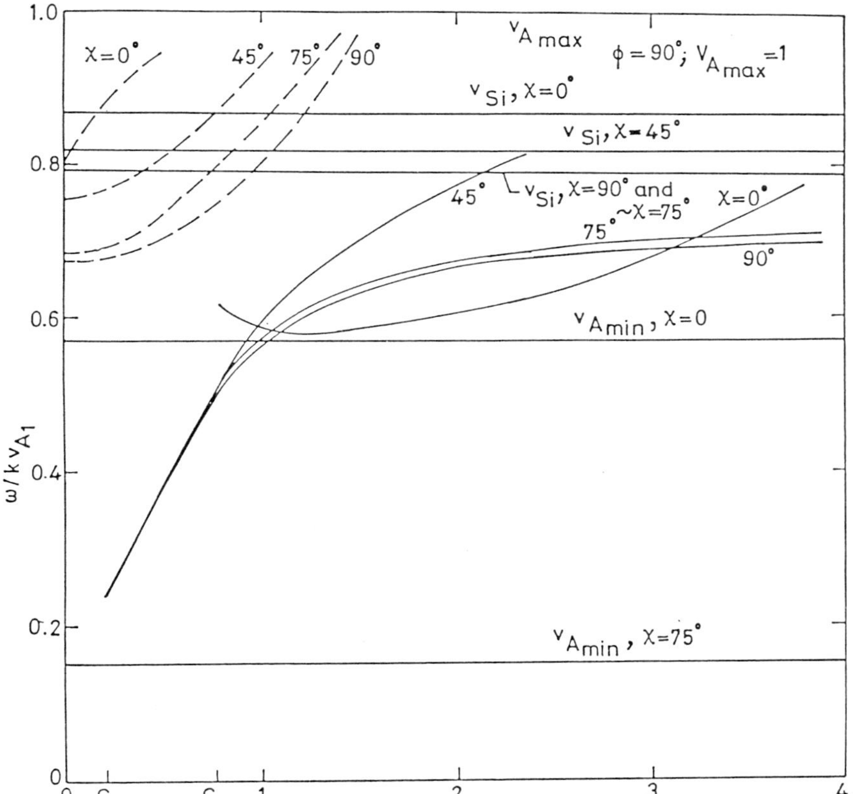

1.0
0.8
0.6
0.4
0.2
0
ω/kv_{A_1}
$\chi=0°$
45°
75°
90°
$v_{A\,max}$
$\phi=90°;\ V_{A_{max}}=1$
$v_{Si},\ \chi=0°$
$v_{Si},\ \chi=45°$
45°
$v_{Si},\ \chi=90°$ and $\sim\chi=75°$
$\chi=0°$
75°
90°
$v_{A_{min}},\ \chi=0$
$v_{A_{min}},\ \chi=75°$
0
C_{1c}
C_{1c}
1
2
3
4
C_1/v_{A_1}

FIG.2.

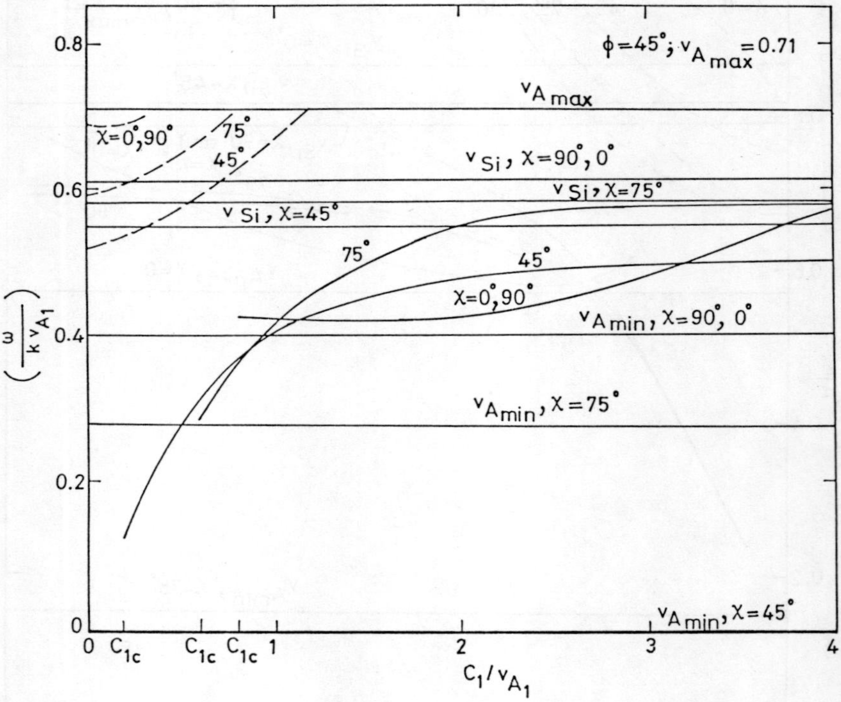

φ=45°; v_A max =0.71
v_A max
χ=0°,90°
75°
45°
v_Si, χ=90°, 0°
v_Si, χ=75°
v_Si, χ=45°
75°
45°
χ=0°,90°
v_A min, χ=90°, 0°
v_A min, χ=75°
v_A min, χ=45°
0.8
0.6
0.4
0.2
0
0
C_1c
C_1c
C_1c
1
2
3
4
(ω / k v_A1)
C_1/v_A1

FIG. 3.

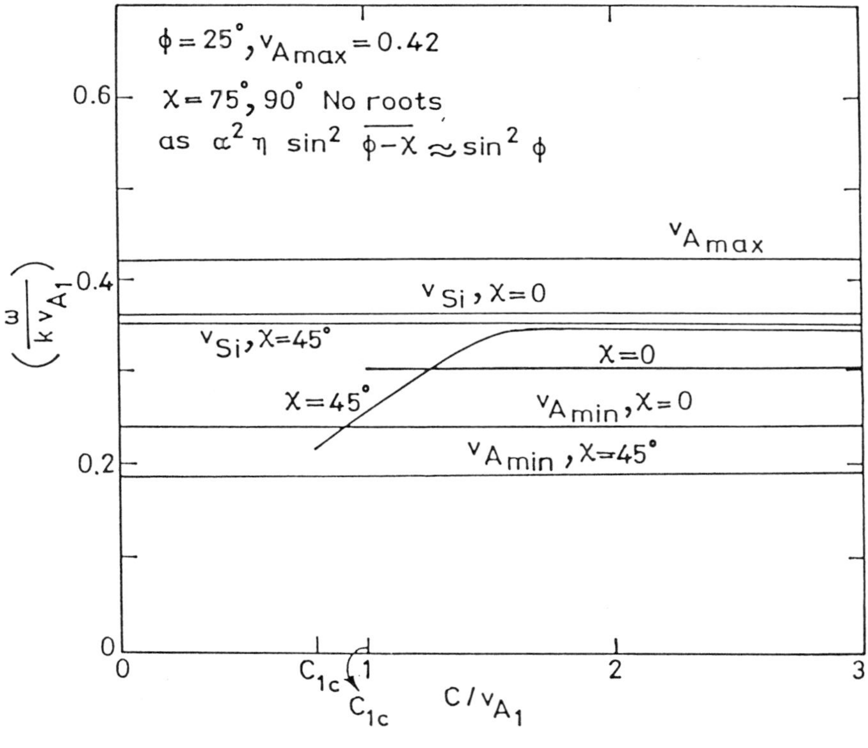

ϕ = 25°, v_Amax = 0.42
χ = 75°, 90° No roots
as α² η sin² ϕ−χ ≈ sin² ϕ
v_Amax
v_Si, χ=0
v_Si, χ=45°
χ=0
χ=45°
v_Amin, χ=0
v_Amin, χ=45°
0.6
0.4
0.2
0
0
C_1c
1
C_1c
2
3
C/v_A1
(ω / k v_A1)

FIG. 4.

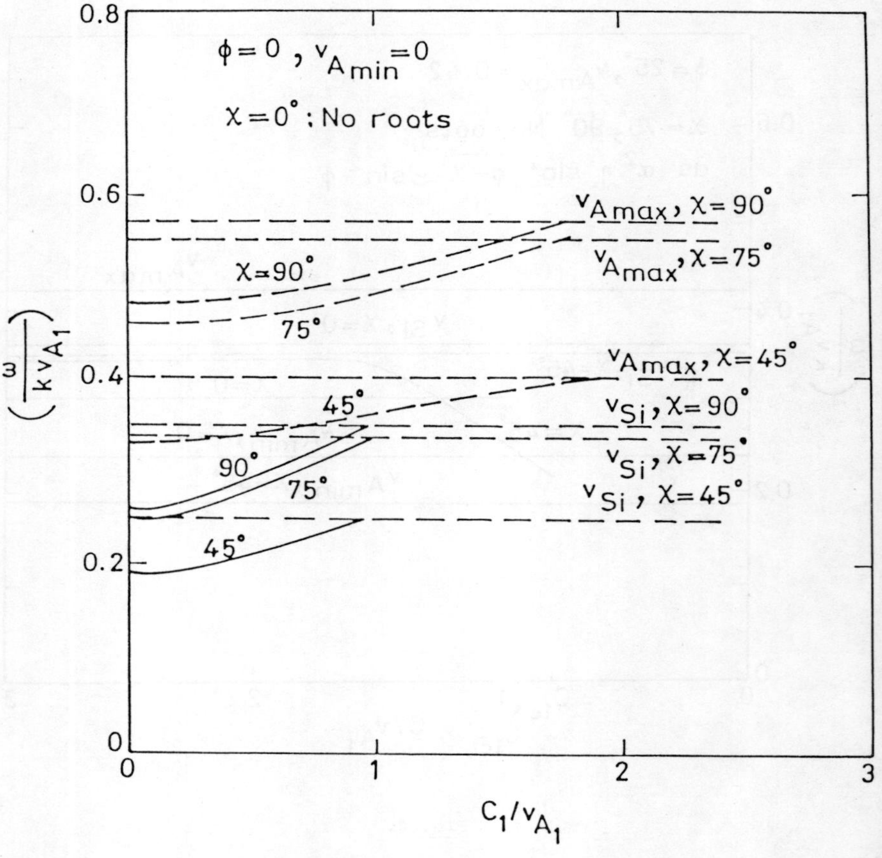

$\phi = 0$, $v_{A_{min}} = 0$
$\chi = 0°$: No roots
$v_{A_{max}}$, $\chi = 90°$
$v_{A_{max}}$, $\chi = 75°$
$\chi = 90°$
75°
$v_{A_{max}}$, $\chi = 45°$
45°
v_{Si}, $\chi = 90°$
v_{Si}, $\chi = 75°$
90°
75°
v_{Si}, $\chi = 45°$
45°
$\left(\frac{\omega}{k v_{A_1}}\right)$
C_1/v_{A_1}
0.8
0.6
0.4
0.2
0
1
2
3

FIG. 5

SURFACE WAVE RELATED PROBLEMS IN FUSION PLASMAS

M. Y. Yu

Institut für Theoretische Physik, Ruhr-Universität Bochum
D-4630 Bochum, Federal Republic of Germany

Introduction All laboratory plasmas are bounded in some manner. Usually, one believes that if the phenomenon under consideration has scalelengths which are short compared with the plasma dimensions, results obtained from infinite-plasma theories should be approximately correct. This, as we know, is not always the case. The existence of a plasma boundary can introduce new phenomena such as emission and absorption of waves, as well as the appearance of surface waves and their related instabilities. By surface waves, we generally mean waves which are localized spatially, for example, near a surface of density discontinuity. Surface waves can also appear near discontinuities of other physical quantities, such as those of the pressure, magnetic field, plasma composition, velocity, etc.

In this article, we shall present a brief survey of situations in fusion plasmas which may allow surface wave propagation. These situations can be quite different from the laboratory experiments for which surface wave problems are traditionally discussed.[1] One finds that in laser fusion as well as magnetically confined plasmas, there exist regions in which surface wave effects may be important. Details of some selected problems shall be given.

In Fig. 1, different types of plasmas are indicated in a density versus temperature diagram.[2] They range from electron plasmas in certain crystals to hot hydrogen plasmas in the interiors of the sun and stars. The region of interest here is the upper right hand quadrant, in which the plasma is well ionized, and both the density and temperature are high. Several

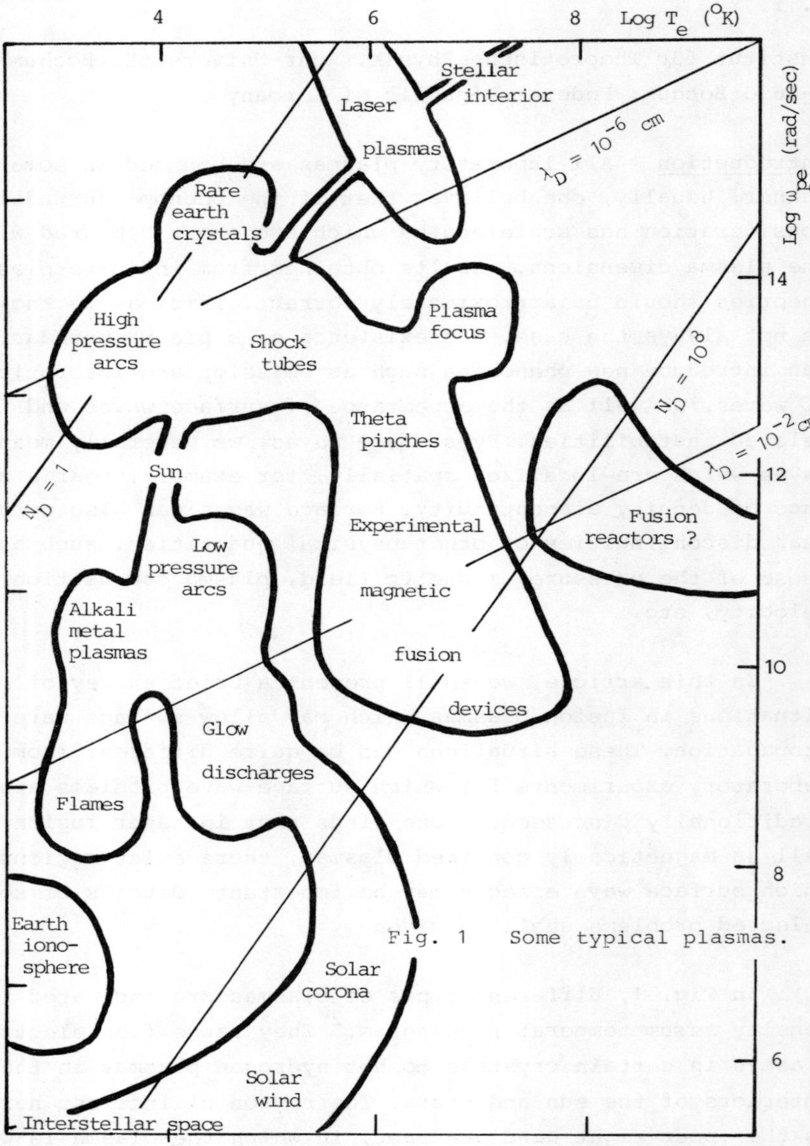

Fig. 1 Some typical plasmas.

sample lines of constant Debye length $\lambda_D = (T/4\pi ne^2)^{1/2}$ and plasma parameter $N_D = (4/3)\pi n\lambda_D^3$ (= the number of electrons in a Debye sphere) are also indicated in Fig.1.

Surface waves in laser plasmas   Fusion by laser-pellet interaction relies on the concept of inertial confinement. The idea is to achieve fusion reaction in the fuel pellet within a time which is short compared to the time for the otherwise unconfined plasma to disperse completely. To do this, short wavelength, high power lasers are focused onto a fuel pellet (frozen deuterium and tritium) of mm size (Fig.2). The light enters the low-density vapor envelope of the pellet and ionizes it. Most of the laser energy is absorbed near the critical surface where the local plasma frequency $\omega_{pe} = (4\pi ne^2/m_e)^{1/2}$ is equal to the light wave frequency. Here, the electromagnetic waves are mode converted into slow electrostatic Langmuir waves. The electromagnetic energy is thus turned into electron kinetic energy. As the Langmuir waves pile up (since their group velocity is much less than the speed of the incoming light waves), collisions and instabilities then convert (dissipate) the electron kinetic energy into electron thermal energy. The ions, due to their heavy mass, are not affected. The hot electrons thus produced diffuse along the negative radial thermal gradient to the interior region of the vapor and heat the pellet surface. The latter then ablates. The inward reaction to the outward explosion of the ablating material drives an implosion, which drastically compresses the fuel pellet. The inward collapse is eventually stopped by the extremely high pressure (10^{12} atm) generated by the process. By drastically reducing the pellet surface area (the final radius is of the order 10 μm) in this manner, the power density is amplified from 10^{15} watt/cm^2 at the laser focus to 10^{19} watt/cm^2 at the compressed pellet surface. The compressional heating as well as the direct heating by the laser photons can cause some of the fuel at the center to react (ignition). Meanwhile, the pellet also starts to expand outward (recoil). The fusion reaction of some of the fuel

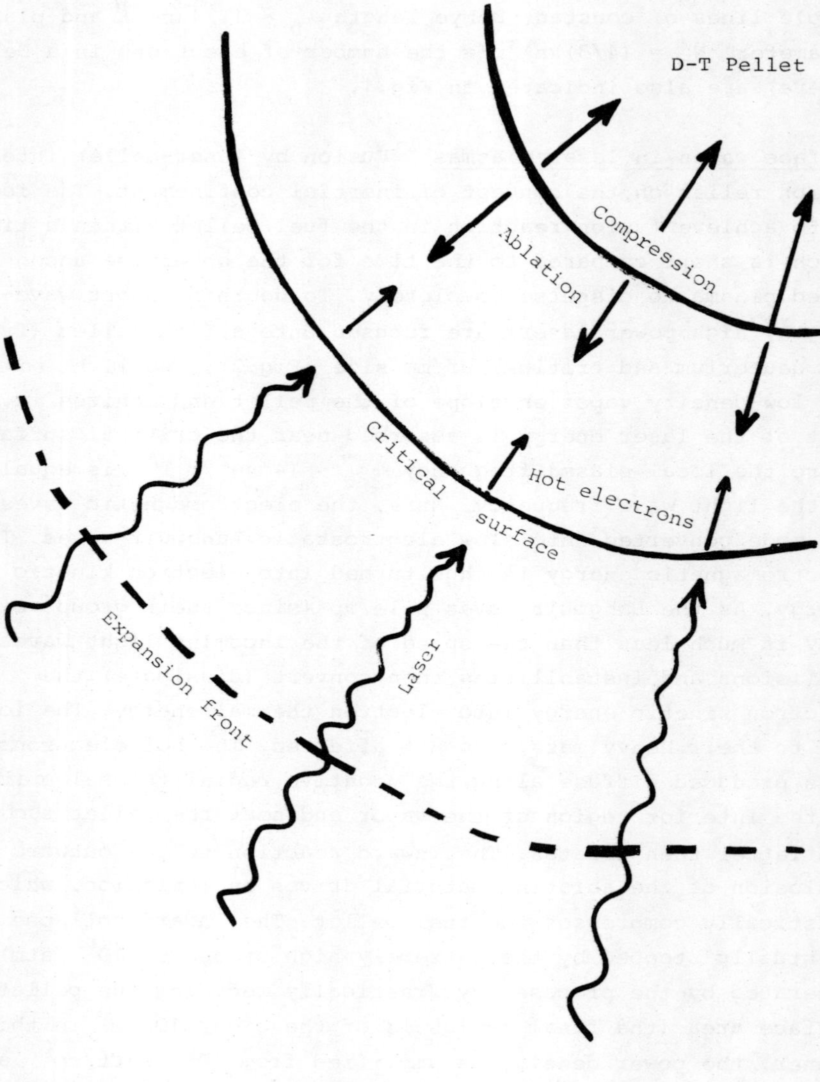

Fig. 2 Laser-pellet interaction.

produces high energy (3 MeV) α particles together with the hot (14 MeV) neutrons. Due to the high temperature and density of the compressed fuel material, collisions cause these α-particles to be re-absorbed before they can leave the pellet. This further reaction results in an outward propagating thermonuclear burn front in the expanding pellet material, effecting more complete burning of the fuel.

In this scenario, there are several regions where surface wave effects can play important roles. The first is at the critical layer. Here, the short wavelength, high speed, electromagnetic waves are mode converted into almost standing electrostatic plasma oscillations. There is a local concentration of wave energy. The resulting wave (or ponderomotive)[3] pressure forces out the plasma, causing the formation of a local density cavity which eventually leads to the appearance of a steep density gradient followed by a almost uniform shelf region. This process is called profile modification. Clearly, the symmetry of the new critical surface,where most of the laser energy is deposited, is important. The stability of this spherical surface is therefore of prime importance. Another problem here is that the intense laser light may cause local nonlinear wave instabilities which generate extremely hot (suprathermal) electrons. The latter, which have very long mean free paths, can penetrate the pellet and heat the interior prematurely (preheat). The process causes thermal expansion of the pellet and can severely reduce the implosion efficiency.

The second region of interest is at the ablation layer. Due to the rapid compression of the pellet, as well as due to the insertion of heavy-element shell(s) in the pellet construction in order to improve compression characteristics, one or more accelerating layers of denser liquid may be present. These layers can be subjected to the Rayleigh-Taylor instability as well as other gravitation (acceleration)-driven instabilities. The latter instabilities can severely distort the pellet symmetry and must be understood. The third surface region is the burn front caused by the reacting fusion α particles in the expanding plasma. The surface is subjected to various instabilities associated with energy addition. Much work has been done in this area in connection with combustion research in gasdynamics. Finally, there is the plasma-vacuum surface at the outermost region. Gurevich and others have[4,5] shown that a shock-like expansion front can be expected because of space-charge effects. However, the physics of this region, due to its possible far-from-equilibrium nature, is still not well understood.

What are the effects of surface waves on the above mentioned surfaces? We shall now give some examples. First, at the steepened profile near the critical surface, experiments, numerical simulations, as well as theories have shown that rippling of the surface can appear. These ripplings are attributed to the parametric generation of surface ion waves.[6] For example, Woo et al.[7] presented a four wave process (filamentation instability) in which the incident laser light excites two sideband (Stokes and anti-stokes) waves together with a low frequency ion wave. The process requires that the incident light to have a s-polarized component (wave electric field parallel to the surface). The existence of these ripples can actually enhance resonant absorption by presenting more angles that can be accessed by the light waves at the surface. They may also help to redistribute the heated electrons across the surface and reduce the hot spot formation near the laser focus. Numerical simulations also confirm that ripple formation enhances light absorption.

The stability of the inward-accelerating ablation layers in thin shell targets are crucial for uniform compression of the pellet. As mentioned, these layers can be subjected to Rayleigh-Taylor type instabilities. Numerical calculations[8] have shown that nonlinear evolution of the latter can lead to the formation of spikes and bubbles. McCrory, et al.[9] found that the spikes tend to saturate. Although the spherical symmetry is thus distorted, the flow exhibits a laminar structure. Another interesting phenomenon at the ablation surface is that magnetic surface waves can be generated. These occur[10] on electron time scales, i.e. much shorter than the Rayleigh-Taylor type instabilities, and may be due to the generation of electron and magnetic vorticities by the baroclinic effect $\nabla n_o \times \nabla T_{e1}$, where T_{e1} is the electron temperature perturbation. These magnetic surface waves are rapidly growing when the gradients are large, but they do not change the shape of the accelerating front. They can nevertheless affect the electron thermal conduction of the plasma there. Inhibition of the heat flux causes local hot spots and thus degrades the heating of the interior region.[11]

Before going to surface problems in magnetically confined plasmas, we shall here present the analysis of a magnetic surface wave instability. As we have seen, discontinuities are intimately associated with several phenomena occuring in laser fusion plasmas. For example, the interaction of the laser radiation and the expanding plasma in the corona region can cause single or multiple shocklike density jumps.[12] Furthermore, during the implosion phase, an ablatively driven medium can compress the material ahead to very high densities within a very short distance. Since the properties of such regions govern the efficiency of laser energy transport, instabilities caused by the discontinuities can play a crucial role in determining the eventual success of inertial fusion.

The main difficulty in the investigation of wave propagation and its stability at the plasma discontinuities is the intrinsic singular nature of the problem. The latter is usually reduced to constructing a suitable model which can be analytically or computationally handled.

Here, adapting wellknown techniques of cold plasma surface wave theory, we present a new dispersion relation applicable to most profiles which might occur in inertial fusion plasmas.[10]

We start with the linearized Maxwell's and nonrelativistic electron equations

$$\nabla \times \vec{E} = -c^{-1}\partial_t \vec{B} \quad ,$$

$$\nabla \times \vec{B} = 4\pi c^{-1} q n_o \vec{v} + c^{-1}\partial_t \vec{E} \quad ,$$

$$\partial_t n + \nabla \cdot (n_o \vec{v}) = 0 \quad , \qquad (1)$$

$$\partial_t \vec{v} = (q/m)\vec{E} - (1/mn_o)\nabla(n T_o + n_o T) + (n/mn_o^2)\nabla(n_o T_o) \, ,$$

$$\partial_t T + \vec{v} \cdot \nabla T_o + (\gamma - 1) T_o \nabla \cdot \vec{v} = 0 \quad ,$$

where $\vec{E}$ and $\vec{B}$ are the electric and magnetic fields, c is the speed of light, q and m are the electron charge and mass, n, $\vec{v}$, and T is the perturbation of the electron density, velocity, and temperature, respectively. The zeroth order electron density and temperature are denoted by $n_o(z)$ and $T_o(z)$, and γ is the adiabatic constant. Furthermore, the electron heat flux has been neglected. The ions form a neutralizing background.

Letting $\partial_t = -i\omega, \nabla = i k_x \hat{x} + \hat{z}\partial_z$, and $\vec{B} = B\hat{y}$, neglecting terms proportional to T_o^2, we can eliminate n, $\vec{v}$, T, and $\vec{E}$ from Eq. (1), and obtain[10]

$$[\partial_z^2 - (\partial_z \ln \varepsilon)\partial_z - k^2]\, B = 0 , \tag{2}$$

where $\varepsilon \equiv 1 - \omega_p^2/\omega^2$, and $k^2 \equiv$

$$k_x^2 + \frac{\omega_p^2 - \omega^2}{c^2} + \frac{k_x^2 T_o}{m\,\omega^2}\,\frac{n_o'}{n_o}\,\frac{[T_o'/T_o - (\gamma - 1)n_o'/n_o - (\omega/\omega_p)^4(n_o'/n_o + T_o'/T_o)]}{(1 - \omega^2/\omega_p^2)^3}.$$

Here, the prime denotes derivative with respect to z. The neglected small terms are of order $T_o^2 k_x^4 B/m^2 c^2 \omega^2$ and higher.

Because of the terms containing the product of the gradients in the expression for k^2, Eq. (2) cannot be integrated across a sharp discontinuity. It is necessary to use a physically more realistic profile in which two regions (say $0 < z < d$ and $d + a < z < \infty$) of different plasma properties are joined by a narrow region ($d < z < d + a$, with $a k_x \ll 1$ and $a \ll d$) of very steep density profile. In the procedure for solution, Eq. (2) is solved by means of the geometrical optics method in the two outer regions, and by successive approximation in the thin transition layer. The dispersion relation is obtained as usual by matching the solutions as the boundaries, assuming no incident waves from the outside towards the lower density side of the profile. Such a model is often used in the theory of cold bounded plasmas, specifically in the leaking eigenmode problem, for which very general results have been elaborately worked out.[13] A straightforward application of the latter to Eq. (2) yields the dispersion relation

$$\frac{k_1}{\varepsilon_1} + \frac{k_2}{\varepsilon_2} \simeq 0 , \tag{3}$$

where the subscripts 1 and 2 denote the two regions of distinct densities and temperatures. Note that the characteristics of the connecting (steep gradient) region do not appear in Eq. (3), although the region is important in applying the boundary conditions. Equation (3) is the main result of interest, it can be solved in general to obtain ω. For simplicity, here we consider the limit $\omega \ll \omega_p$. Equation (3) reduces to

$$\omega^2 = \frac{\omega_1^{*2}/n_{01}^2 - \omega_2^{*2}/n_{02}^2}{(1+\omega_{p1}^2/k_x^2c^2)/n_{01}^2 - (1+\omega_{p2}^2/k_x^2c^2)/n_{02}^2}, \tag{4}$$

where

$$\omega^{*2} \equiv -(T_o/m)(n_o'/n_o)[T_o'/T_o - (\gamma-1)n_o'/n_o].$$

It is clear that, depending on the parameters, oscillating as well as purely growing solutions can exist.

In the limit $n_{01} \ll n_{02}$, Eq. (4) reduces to

$$\omega^2 = \frac{(T_{01}/m)(n_{01}'/n_{01})[(\gamma-1)(n_{01}'/n_{01}) - T_{01}'/T_{01}]}{1+\omega_{p1}^2/k_x^2c^2}, \tag{5}$$

which, except for the T_{01}' terms, is similar to that obtained by Jones.[14] The difference in the detail is due to his somewhat different modelling of the density jump. Thus, our result shows that the profile can be unstable if $T_{01}'/T_{01} > (\gamma-1)(n_{01}'/n_{01})$. The instability is purely magnetic, and relies on a temperature gradient in the same direction as the density gradient. Furthermore, it generates magnetic as well as plasma vorticities which are coplanar with the imhomogeneities, and is associated with the magnetostatic mode.[15] Note also that the instability is purely growing, and may be considered as instability of the Jones mode.

Physically, the instability discussed here is related to the creation of magnetic and electron vorticities by the first order baroclinic ($\nabla n_0 \times \nabla T_{e1}$)effect. Therefore, the inclusion of temperature perturbations is essential for obtaining the present instability. Since the baroclinic effect is independent of the plasma discontinuity, one expects that the instability predicted here should also exist (but locally) in a weakly inhomogeneous plasma. This can in fact be verified by solving the equation k = 0 for ω. However, here, the growth rate, which is the same as (5), but the allowed values of the gradients are severely restricted by the local approximation (n_0'/n_0, $T_0'/T_0 \ll k_x$), is necessarily small in order to be consistent. On the other hand, the inclusion of a heat flux term should reduce ∇T and thus the instability.

It turns out that the magnetic mode discussed here is closely associated with the acoustic gravity mode of a electron fluid. In fact they have almost exactly the same instability criterion.

Surface waves in magnetically confined plasmas. The idea of confining a hot plasma with magnetic fields has been in existence since the beginning of fusion research.[16] Many different types of devices were designed and tested. The most promising are the tokamak and mirror concepts. Unlike laser plasmas, magnetically confined plasmas are relatively uniform and long-lived. Thus, besides the large scale instabilities involving the entire plasma, small scale local instabilities can also be important. The design of present-day tokamaks and mirror machines eliminates most of the classical hydromagnetic instabilities. The latter include the Rayleigh-Taylor, sausage, and kink instabilities. These instabilities and their stabilization are well documented in the literature, and shall not

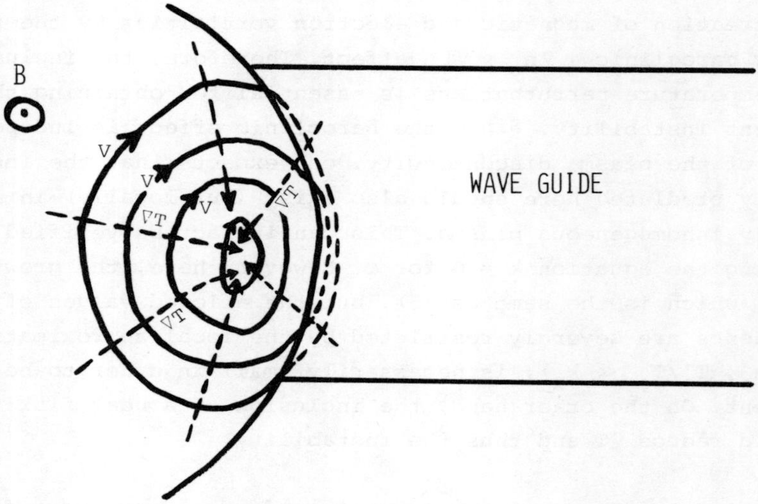

Fig. 3(a) Hot spot in wave-heated plasma.

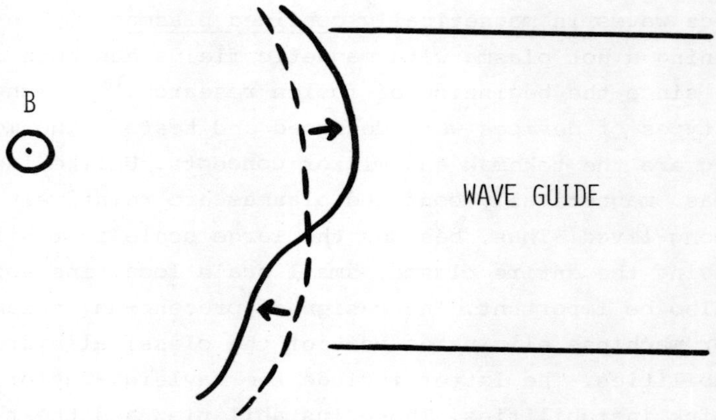

Fig. 3(b) Deformation of plasma surface.

be considered here.

There are also situations in which smaller scaled surface waves might play important roles. For example, in a tokamak with divertors or limiters, which are introduced to remove edge impurities and prevent plasma-wall contact, velocity space anisotropy induced surface waves can alter the orbits of the particle trajectories at the boundary of the plasma and hinder the efficiency of these devices, as well as affect the stability of the edge plasma. Furthermore, in supplementary heating by r.f. waves, there is a possibility of the electromagnetic waves exciting mostly the surface waves at the plasma boundary,[17] without penetrating into the main plasma body to heat the interior. Due to local heating, sometimes a hot spot can appear in the plasma near the wave guide.[18] Radial temperature gradients are established around the hot spot (Fig.3a). Because of the force balance

$$e\vec{E} = -\nabla T_e,$$

assuming nearly constant denstiy, the equipotential (ϕ = constant, where $-\nabla\phi = \vec{E}$) contours coincide with those of the temperarure. The $\vec{E}\times\vec{B}_0$ motion of the plasma is then along the equipotential loops. Since there is acutally small radial density variation of the equi-density contours which clearly do not coincide with the equi-potential contours, the plasma surface can become distorted It was found that an increase of the reflectivity of the plasma occurs. The thermal eddies described here can excite surface waves which are otherwise inhibited by the high temperature and high magnetic field. Note that ponderomotive pressure effects, not included in the discussion above, can also cause surface distortion and excite surface waves.

Besides waves involving physical surfaces, these is another type of wave motion which might be classified as surface waves. These are the tearing modes,[19,20] first predicted theoretically more than 20 years ago.

The tearing mode is of great importance to tokamak plasmas. It can be considered as a surface mode because its eigenfunction is mainly localized near a certain surface, called the rational surface. The latter is defined by $\vec{k} \cdot \vec{B}_o = 0$, where $\vec{k}$ is the wave-vactor of the mode and $\vec{B}_o$ is a sheared but straight magnetic field. The shear is necessary to improve confinement.

Since the tearing mode is rarely discussed as a surface wave, we shall give a brief description of its analysis, using a simplied model.[21]

The sheared magnetic field is given by $\vec{B}_o = B_{zo}\,\hat{z} + B_{yo}(x)\hat{y}$, where B_{zo} is a constant, and $B_{yo}(x)$ is the shear field. The wavevector of the mode shall be taken to be in the y direction. Thus, the rational surface here is the y-z plane (x = 0). In a real tokamak, the rational surfaces are considerably more complicated and resemble concentric twisted toroidal shells.

In the simplied geometry, the eigenfunctions of the tearing mode are $\vec{B} = (B_x, B_y, 0)$, $\vec{v} = (v_x, v_y, 0)$, and the current density $\vec{j} = (j_x, j_y, 0)$. Note that because of the equilibrium shear field, there must be an equilibrium current density layer near x = 0: $\vec{j} = [0, 0, j_{zo}(x)]$, such that $\partial_x B_{yo} = (4\pi/c)\, j_{zo}$. The form of this current layer turns out to be crucial for the instability.

We assume that all the perturbation quantities behave like

$$Q(x) \sim \exp(\, i k_y y + \gamma t\,) + c.\ c.$$

where γ is the growth rate.

The plasma is taken to be quasi-neutral and incompressible. Thus, the MHD equations are

$$\nabla \cdot \vec{j} = 0 ,$$

$$\nabla \cdot \vec{v} = 0 ,$$

$$\eta \vec{j} = \vec{E} + \frac{1}{c} \vec{v} \times \vec{B} , \qquad (6)$$

$$\rho \partial_t \vec{v} = - \nabla p + \frac{1}{c} \vec{j} \times \vec{B} , \qquad (7)$$

where $\eta = m_e \nu_{ei} / n_o e^2$ is the Spitzer resistivity, and the rest of the notation is standard.

Linearizing the equations and introducing the vector potential A_z ($\vec{B} = \nabla \times \vec{A}$, or $B_x = i k_y A_z$), one obtains from (6) and (7)

$$\gamma A_z - B_{yo} v_x = \frac{1}{4\pi} \eta c^2 \nabla_\perp^2 A_z , \qquad (8)$$

$$\gamma \rho \nabla_\perp^2 v_x = - \frac{1}{4\pi} k_y^2 B_{yo} \nabla_\perp^2 A_z + \frac{1}{c} k_y^2 j'_{zo} A_z , \qquad (9)$$

where $j'_{zo} = d j_{zo} / dx$ and $\nabla_\perp^2 = - k_y^2 + \partial_x^2$. Note that we have taken the curl of (7), so that the pressure term is eliminated.

Equation (8) and (9) are a coupled set of second order ordinary differential equations with variable coefficients. They can be solved numerically. Furth, Killeen, and Rosenbluth[20] introduced a novel analytical method of solution, which we shall now follow.

If the resistivity η is small, equation (8) resembles a boundary layer problem: a small parameter is multiplied to the highest order derivative. Thus, resistive effects are only important in a layer where the second derivative is large. Following boundary layer theories in fluid dynamics, we divide the plasma into a narrow inner region, say $|x| < \varepsilon$, and the outer region $|x| \gg \varepsilon$. The outer region is governed mainly by the equilibrium current j_{oz}, one neglects the resistivity η, and since we are interested in a time scale much larger than the Alfvén time ($\gamma \ll k_y v_A$, where $v_A = (B_{zo}/4\pi m_i n_o)^{1/2}$ is the Alfvén speed), we can drop the inertia term in (9). Thus, we have $v_x = 0$, and from (9)

$$[\partial_x^2 - k_y^2 + \frac{4\pi}{cB_{yo}(x)} j'_{zo}(x)] A_z = 0 \; , \qquad (10)$$

for $x \gg \varepsilon$.

Since $B_{yo}(x)$ has a zero at $x = 0$, the solution of (10) should be allowed to have a discontinuity at $x = 0$. For bounded equilibrium current profiles $j_{zo}(x)$, acceptable solutions of (10) may look like

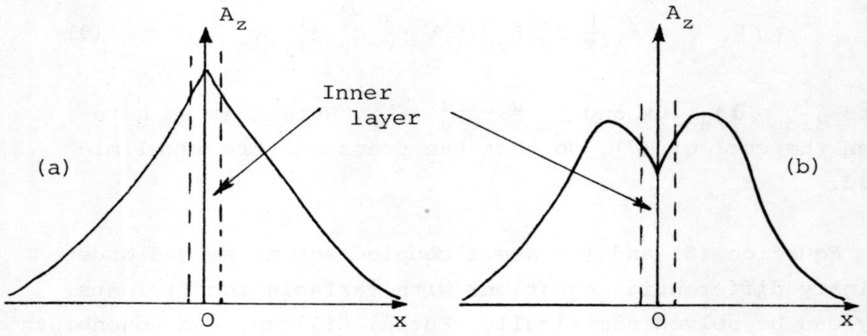

Fig. 4 Possible outer solutions.

Thus, A_z is continuous at $x = 0$, but not $\partial_x A_z$ ($= - B_y$).

One has now to match $\partial_x A_z$ so obtained, which is valid for $|x| \gg \varepsilon$, to an inner solution which is valid near $x = 0$. In the inner region, the plasma motion is dominated by the resistivity and the effect of magnetic field reversal at $x = 0$. One expects $\partial_x^2 \gg k_y^2$ in this layer. We can thus write $B_{yo}(x) \sim B'_y(o)x$, and neglect j'_{zo} in (9).

Combining (8) and (9) for the inner region, one can obtain

$$\gamma \rho \, \partial_x^2 v_x + \frac{k_y^2 B'_{yo} x}{\eta c^2} (\gamma A_z - B'_{yo} x \, v_x) = 0 \; . \tag{11}$$

Equation (11) is incomplete as A_z still appears.

Since v_x is odd and A_z is even with respect to $x = 0$, and that A_z has the (larger) scalelength of the outer region, we can assume $A_z \sim A_z(0)$ = constant in the inner layer. It is instructive to write (11) in the normalized form

$$\partial_\xi^2 V - \xi^2 V = - \xi \; , \tag{12}$$

where $\xi = x/\delta$, $V = v_x/b$, with $\delta = (\gamma\rho\,\eta\, c^2/k_y^2 B'^{2}_{yo})^{1/4}$ and $b = (k_y/B'_{yo}c)^{1/2} (\rho\eta)^{-1/4} \gamma^{3/4} A_z$. Note that the only possible inner layer gradient scale is given by δ.

Equation (12) can now be solved explicitly as an inhomogeneous equation.

We require that the magnetic perturbation $B_y = \partial_x A_z$ should match that from the outer solution. From (8), we get

$$\partial_x A_z = \frac{4\pi}{\eta c}\int (\gamma A_z(o) - B'_{yo}\, x v_x)\, dx ,$$

$$= -\frac{4\pi\gamma\rho}{k_y^2 B'_{yo}}\int \frac{1}{x}\,\partial_x^2 v_x dx ,$$

$$= -\frac{4\pi\rho^{1/4}\gamma^{5/4} A_z(o)}{(k_y B'_{yo} c^3)^{1/2}\eta^{3/4}}\int \frac{1}{\xi}\,\partial_\xi^2 v d\xi , \tag{13}$$

$$\frac{1}{A_z(o)}\frac{dA_z}{dx}\bigg|_{-\varepsilon}^{\varepsilon} = \frac{1}{A_z(o)}\frac{dA_z}{dx}\bigg|_{-\infty}^{\infty} =$$

$$= -\frac{4\pi\rho^{1/4}\gamma^{5/4} A_z(o)}{(k_y B'_{yo} c^3)^{1/2}\eta^{3/4}}\int_{-\infty}^{\infty} \frac{1}{\xi}\, d_\xi^2\, V\, d\xi , \tag{14}$$

where in the integrand the solution of (12) is to be substituted. Since the integral is a pure number (turns out to be of order 1), one obtains

$$\gamma = \left(\frac{I\Delta'}{4\pi}\right)^{4/5} (k_y B'_{yo})^{2/5}\, c^{6/5}\, \eta^{3/5}\, \rho^{-1/5} , \tag{15}$$

where $\Delta' = A_z(o)(d_x A_{zo}/dx)\big|_{x=o}$ is determined by the outer solution, and I is the integral in (14).

Thus, $\Delta' > 0$, corresponding to the outer solution Fig. 4 (b), is necessary for the instability. This usually occurs if the perturbation wavelength (which is in the y direction) is larger than the current layer width (which is in the x direction), as may be inferred from (10) by noting $j'_{zo} = (c/4\pi)\partial_x B_{yo}$.

Physically, the tearing mode appears because a state including magnetic shear can lower its energy by annihilating the magnetic flux due to the shear field. By also generating a B_x field, the tearing mode bends the oppositely directed B_{yo} field to form a chain of magnetic islands. The magnetic field line "reconnection" occurs in a very thin layer whose size ($\simeq \delta$) is considerably smaller than the (magnetic) diffusion skin depth. In this layer, the frozen-field line approximation does not hold, and a part of the magnetic energy is converted into plasma kinetic energy. It turns out that under certain circumstances, the island chain can become nonlinearly unstable, so that longer islands are formed,[22,23] or chaotic behavior in the magnetic field appears.

Conclusion We have shown that in inertial as well as magnetic fusion plasmas, regions of steep gradients in the physical parameters exist and can support surface wave related phenomena. These regions, which may be small in dimension compared with the bulk of the plasma, can nevertheless exert strong influence on the plasma by either altering the local transport properties or by growing into a larger region, thus affecting the entire plasma.

Acknowledgement This work was supported by the Sonderforschungsbereich 162 Plasmaphysik Bochum/Jülich.

References

1. M. Moisan, A. Shivarova, and A.W. Trivelpiece, Plasma Phys. 24, 1331 (1982).
2. W.B. Kunkel and M.N. Rosenbluth, in Plasma Physics in Theory and Application, Ed. W.B. Kunkel (McGraw Hill, New York, 1966) p.8; D.L. Book, NRL Plasma Formulary (Naval Research Laboratory, Washington D.C., 1980) p.41.
3. F. F. Chen, Introduction to Plasma Physics (Plenum, New York, 1974) p.256.
4. A.V. Gurevich and L.P. Pitaevsky, Prog. Aerospace Sci. 16, 227 (1975).
5. Ch. Sack and H. Schamel, Plasma Phys. & Controlled Fusion 27, 717 (1985).
6. K.G. Esterbrook, Phys. Fluids 19, 1733 (1976); K.G. Esterbrook and W.L. Kruer, Phys. Fluids 26, 1888 (1983).
7. W. Woo, J.S. DeGroot and C. Barnes, Phys. Fluids 23, 2291 (1980).
8. R.L. McCrory and R.L. Morse, Phys. Fluids 19, 175 (1976); F.H. Harlow and J.E. Welch, Phys. Fluids 9, 842 (1966).
9. R.L. McCrory, L. Montierth, R.L. Morse, and C.P. Verdon, Phys. Rev. Lett. 46, 336 (1981).
10. M. Y. Yu and L. Stenflo, Phys. Fluids, Dec. (1985).
11. W.M. Manheimer, C.E. Max, and J. Thomson, Phys. Fluids 21, 2009 (1978); M.Y. Yu, P.K. Shukla, and K.H. Spatschek, Phys. Fluids 23,226 (1980).
12. K. Lee, D.W. Forslund, J.M. Kindel, and E.L. Lindman, Phys. Fluids 20,51 (1977); R. Fedosejev, M. D. J. Burgess, G.D. Enright, and M.C. Richardson, Phys. Rev. Lett. 43,1664 (1979).
13. O.M. Gradov and L. Stenflo, Phys. Rep. 94,111 (1983).
14. R.D. Jones, Phys. Rev. Lett. 51, 1269 (1983).
15. C. Chu, M.S. Chu, and T. Ohkawa, Phys. Rev. Lett. 41, 653 (1978).
16. G. Schmidt, Physics of High Temperature Plasmas (Academic, New York, 1979) p.109.

17. F.F. Chen and C. Etievant, Phys. Fluids 13,687 (1970); M. Brambilla, Nucl. Fusion 16, 47 (1976); G.J. Morales, Phys. Fluids 20, 1164 (1977).
18. R.W. Motley, W.M. Hooke, and G. Anania, Phys. Rev. Lett. 43, 1799 (1979).
19. G.S. Murty, Arkiv Fysik 19,499(1961).
20. H.P. Furth, J. Killeen, and M.N. Rosenbluth, Phys. Fluids 6, 459 (1963).
21. See, for example, M.N. Rosenbluth and P.H. Rutherford, in Fusion, Ed. E. Teller (Academic, New York, 1981) p.76.
22. J.M. Finn and P.K. Kaw, Phys. Fluids 20, 72 (1977); D. Biskamp and H. Welter, Phys. Rev. Lett. 44, 1069 (1980).
23. R. Pellat, Fiz. Plazmy 9, 204 (1983) [Sov. J. Plasma Phys. 9, 124 (1983)]; M. Y. Yu and G. Murtaza, Phys. Lett. 111A, 129 (1985).

THE DESIGN OF SURFACE WAVE DISCHARGES TO OBTAIN PLASMA COLUMNS OF SPECIFIED PROPERTIES

Z. Zakrzewski[(a)] and M. Moisan[(b)]

a) Polish Academy of Sciences, IMP-PAN, 80-952 Gdansk, Poland
b) Département de Physique, Université de Montréal, Montréal, Québec, H3C 3J7, Canada

I. Introduction. The fact that an electromagnetic surface wave can propagate along the interface between a plasma column and its surrounding dielectric tube has been known for decades[1]. The results of these investigations have been reviewed in [2] and [3], as concerns, respectively, theory and experiment. More recently, the interest for such plasma waves has shifted to their use as a means of sustaining a plasma column. Recent reviews on the properties and applications of such surface wave produced plasmas can be found in [4-6]. These plasmas have been operated over a large domain of wave frequencies, covering both the RF and the microwave range, and at pressures as low as some 10^{-5} Torr up to a few times the atmospheric pressure. The Surface Wave Discharges (SWD) exhibit many advantageous features, the most spectacular of which might be the possibility of obtaining large volume plasmas by sustaining large length and large diameter columns[7]. The corresponding wave launching structures are simple, compact and assure an efficient energy transfer to the plasma. An extensive modelling work has been achieved on these SWD[7-13]. This modelling reached such an advanced state that a quantitative description of the SWD is now possible in many practical instances. The properties of these plasmas do not depend on the design of the launching structure. They are determined only by the discharge conditions (i.e. nature and pressure of the gas, dimensions of the discharge tube and wall material, wave mode and frequency) and the amount of power absorbed per unit volume of plasma.

The SWD are interesting for various applications. Some applications have already been realized and are well documented, and promising developments can be foreseen in such fields as surface treatment[14,15], lasers[16] and elemental analysis[4-6,17-20]. The range of the discharge conditions and plasma parameters for SWD is broader than for DC discharges. Moreover, the plasma obtained is stable, reproducible and quiescent.

The main features of interest in most applications are: the mean electron energy, the electron density and the column length. The energy is determined by the plasma loss mechanisms and is, because of that, to a first approximation, the same in SWD as in other types of discharges. Thus, it shall not be considered here. As for the electron density, in SWD, it decreases axially away from the launcher. The density value and the axial gradient at the origin (launcher) suffice to determine the plasma column density distribution.

To be able to take full advantage of the exceptional flexibility offered by the application of the SWD principle to plasma generation, the potential user has to be aware of the factors affecting the plasma parameters and of the means of controlling them in practice. Only then, can he address the problem of synthesis (i.e. determining the design and the operational data) of SWD yielding plasmas of specified properties. It is the purpose of this paper to present a general, even if only approximate, description of the influence of various factors on the properties of the plasma column sustained in the SWD. Together with the knowledge of the technical means available, this allows to undertake the synthesis of SWD.

The content of the paper is organized as follows. The next section presents a brief survey of the typical SWD hardware that is now available. Then, a simplified physical model of the SWD is described. Further on, we come to the influence of the discharge conditions and input power on the properties of the SWD sustained in an axially uniform discharge tube. Finally, the possibility of using tapered tubes to influence the properties of SWD sustained plasma column is briefly considered.

II. Surface wave plasma generators (launchers): hardware at the user's disposal. A basic setup with which a long plasma column can be sustained by a surface wave is shown in Fig. 1. It consists, first of all, of a dielectric plasma tube, a source of microwave power and a high frequency structure able to launch the desired surface wave mode. Such a setup ensures a stable operation by using a circulator to reduce the influence of plasma parameter changes on the power generator as well as it provides the possibility of accurately measuring the amount of power delivered to the plasma (power meter and directional couplers, selecting alternatively the incident and the reflected wave). A distinguishing feature of the SWD is that the electric field necessary to sustain the plasma is imposed by a wave propagating along the discharge tube. No external wave-guiding structure is necessary, because the interface between the plasma and the surrounding dielectric constitutes such a guide.

Let us dwell a little more on the launching of the surface wave. Such a problem is not particular to the field of plasma generation and is, in fact, similar to the launching of radio surface waves, for which there exists an ample supply of published informations (see, e.g.,[21]). In plasma physics, this problem has been mainly associated, during the sixties, with the experimental investigation of the (non-ionizing) surface wave propagation along axially uniform plasma columns, usually the positive column plasma of a glow discharge[3].

Any launching device intended to sustain a SWD should essentially perform one single vital task: the efficient launching of the surface wave (i.e. the transformation of most of the electromagnetic energy flowing out from the launcher into surface wave energy) having the required azimuthal mode. For that purpose, it should provide the desired distribution of the electromagnetic field in the wave excitation area and assure the proper impedance matching between the input line and the plasma column. A further requirement may concern the power handling capability of the device.

Various practical realizations of these requirements are possible at the engineering level. In Fig. 2, as an example, a family of plasma generators using the surface wave principle are shown. These have been chosen because of the author's familiarity with them and because they have been thoroughly described in the literature[22-24]. Some design features of these generators and their frequency range of operation are gathered in Table 1. It is worth noting the exceptional wide range of operating frequencies, over which the surface wave principle has been succesfully applied to plasma generation. Let us also point out that these devices can be designed so that they practically assure a complete transfer of the HF power from the generator to the plasma, under a wide range of discharge conditions. Therefore, a potential user has at his disposal all the necessary technical means to assure the necessary flexibility in influencing the properties of the plasma produced.

III. Simplified model of surface wave discharges.

III.1 Outline of the model

A surface wave discharge is a complicated phenomenon. Its model should be based on a set of equations including Maxwell equations and the equations describing the plasma maintenance processes within the discharge. The proper boundary conditions for the electromagnetic field at the media interfaces, as well as those for the charged particles at the surfaces surrounding the plasma, should also be included. Usually, to make this fully self-consistent problem tractable, some simplifying assumptions are sought. In recent years, various attempts of theoretical modelling of SWD have been undertaken[8-13,25,26].

For the practical purposes of the present work, we are looking for a model for the SWD in which both the parameters of the plasma column and the factors affecting these parameters come out in an obvious and readily available way. The relations sought for should be simple and expressed in an analytical form. Such a model is based on the assumption that the formal treatment of the physical processes occurring in a plasma produced by a wave can be divided into two separate parts: one, concerning the plasma maintenance processes within the discharge and the other, concerning the propagation of the wave and the transfer of its power to the plasma.

In this presentation, we consider a steady-state SWD in a cylindrical tube. The geometry of the problem is shown in Fig. 3. The balance of charged particles is readily written. It simply relates the ionization in the electric field of the wave with the particle loss due to ambipolar diffusion to the tube wall. In setting this equation, we neglect the axial electron density inhomogeneity i.e. the conditions for the maintenance of a steady-state discharge at any position along the column, with the exception of the region directly adjacent to the end of the column, are assumed to be the same as if the column was axially uniform (the dart shaped axial inhomogeneity observed at the column end is determined by diffusion and is of the order of a fraction of the tube radius, so that the processes occuring there can be considered to have a negligible influence on the overall behaviour of the discharge).

We shall further assume, in this model, that the physical processes governing the discharge are independent of the way that the electromagnetic field is imposed on the plasma[5]. This means that these processes are determined only by the discharge conditions (nature and pressure of the gas; shape, size and wall material of the vessel; frequency of the electromagnetic field) and the amount of power delivered to the plasma. This assumption can be justified experimentally over a wide range of discharge parameters. The investigation of the discharge processes under these conditions leads to the consideration of the amount of power that is lost on the average per electron, a parameter which proves to be basic and meaningful[5].

As far as the power transfer to the plasma is concerned, it has to be analyzed at two levels, i.e. in terms of the impedance matching between the power generator and the wave launcher as well as in terms of the wave attenuation along the plasma column. The power is carried away from the launcher by the wave and gradually dissipated into the plasma. The propagation of the surface wave along the column it sustains can be studied using standard methods of the electromagnetic field theory (see, e.g.[27]). The influence of the plasma properties on the wave propagation is treated phenomenologically by considering the plasma as a dielectric with a complex conductivity determined by the cross-section averaged electron density and the (effective) collision frequency for momentum transfer. This treatment yields the wave attenuation (e.g. [25]), i.e. the fraction of the wave power that is dissipated into the plasma.

With the above results at hand, the problem of the energy balance in the SWD can be addressed and solved. In the following paragraphs, we shall briefly present this procedure.

III.2 Average power loss per electron in the plasma: Θ

The electrons form the only medium able to effectively absorb the energy from the high frequency field and to transfer it to the heavy particles through collisions of various kinds. The total power loss suffered, under given discharge conditions, by an average electron can be expressed as

$$\Theta \equiv \frac{2m}{M} \langle \nu_m u \rangle + \sum_k \langle \nu_m h_k \rangle eV_k \; ; \quad u = \frac{mv^2}{2} \qquad (3.1)$$

where m and M are the electron and ion mass, v and u are the electron velocity and energy, ν_m is the electron collision frequency for momentum transfer and h_k and V_k are, respectively, the efficiency and the threshold potential for excitation or ionization. The brackets $\langle\ \rangle$ denote the averaging over the electron energy distribution function (EEDF).

In the discharges considered in the present work, the EEDF, and consequently Θ (see Eq. 3.1), are assumed to be independent of the electron density. As indicated further in this text, there exists experimental justification for such an assumption[5]. Also, the dependence of the EEDF and Θ on the electromagnetic field distribution can be, as a first approximation, neglected. Under these assumptions, the value of Θ is a function of the discharge conditions only. Further, in a wide range of discharge conditions, it is also independent of ω. Moreover, it obeys a simple similarity law of the form:

$$\Theta/p = \underset{\sim}{\theta}(p.a). \qquad (3.2)$$

where p is the gas pressure and a is the tube radius.

Figure 4 shows the theoretical[10] and experimental[7] dependence of Θ/p on pa for a SWD in argon. The following conclusions can be drawn. First, the similarity law (3.2) is valid over the full range of the parameters investigated experimentally. Second, the measured values are in good agreement with the available theoretical results. Third, since the electron density and the wave frequency varie considerably under these conditions, it proves that the value of Θ is density and frequency independent.

To date, there are no sufficient experimental data available showing the validity of a similar relation for other gases than argon. However, the fact that this similarity law applies in the case of argon, despite the strongly pronounced Ramsauer effect, sets up an optimistic perspective on the applicability of this similarity law concept to other gases.

In this work, we assume that the value of Θ is given, under known discharge conditions, either from theory or from the existing experimental data (In the case of argon, the user should simply calculate it directly from Fig. 4). When writing the power balance equation (section 3.4), we shall neglect the influence of the axial density gradient. Therefore, the value of Θ that we obtain in this way, if this procedure is correct, should be the same in SWD as in a uniform discharge, either direct current driven or sustained by a high frequency field, under the same operational conditions. This is indeed the case.

III.3 Attenuation of the surface wave

By definition, the attenuation coefficient is a quantity linking the fraction of power lost by the wave, per unit length of the wave path, with the total power flux carried by the wave. Thus, together with Θ, it becomes a key element in the present model of the SWD.

It has been shown, both theoretically and experimentally (e.g., [1,25,29]), that, with a reasonable accuracy, the attenuation coefficient $\alpha(\bar{n})$, at each position z along the plasma column, depends only on the local value of the cross-section averaged electron density $\bar{n}(z)$. The value of the attenuation coefficient can be determined, for given electron density and collision frequency, using standard methods from the electromagnetic field theory. This subject found an ample coverage in the work published over the last decades.

For the purpose of this engineer-oriented work, we would like to have an analytical approximation for $\alpha(\bar{n})$. We found that, in a low pressure plasma ($\nu^2 \ll \omega^2$), this can be satisfactorily achieved with the relation.

$$\alpha(\bar{n}) = \frac{A(\omega,a)\,\nu}{\bar{n} - n_D}, \tag{3.3}$$

where $n_D = (1+\varepsilon_g)n_c$, n_c is the critical electron density and ε_g is the tube dielectric permittivity. This analytical formula directly shows the influence of various factors on the value of $\alpha(\bar{n})$. Therefore, even if less exact than the results from a direct numerical calculation, it better serves our purposes. The coefficient $A(\omega,a)$ has to be determined by fitting Eq. (3.3) onto the results from the numerical calculation of $\alpha(\bar{n})$. Results of such a numerical procedure are shown in Fig. 5 and Fig. 6, for a set of various tube diameters and wave frequencies. In practice, as the conditions corresponding to the electrostatic approximation for the surface wave are often met, things get even simpler. In this case, the value of $A(\omega,a)$ can be obtained analytically and it is proportional to ω/a.

Notice from Eq. (3.3) that the attenuation of the surface wave increases with decreasing electron density and, that within this model, it becomes infinite at $\bar{n}(z) = n_D$.

III.4 Balance of the electron energy and axial structure of the column

In a steady-state SWD, the power loss by electrons in collisions of all kinds has to be compensated by the power acquired by the electrons from the field of the wave sustaining the discharge. The local balance of power can be, in a long SWD, written in the form

$$-2\alpha(\bar{n})\ P(z) + \pi a^2\ \Theta\ \bar{n}(z) = 0 \qquad (3.4)$$

where P(z) is the total flux of wave power going through the z plane. The first and the second terms of (3.4) respectively describe the fraction of the wave power acquired by the electrons and the power loss due to collisions, both per unit length.

The local equation (3.4) can be used to determine the axial structure of the plasma column (i.e. the axial distribution of the electron density $\bar{n}(z)$) sustained by the wave. Simply note that the power flow P(z) decreases along the axis as the wave power is gradually dissipated into the plasma, thus $\bar{n}(z)$ has to follow. The latter quantity is a solution of

$$\frac{d\bar{n}(z)}{dz} = -2\bar{n}(z)\ \alpha(\bar{n})\ \left[1 - \frac{\bar{n}(z)}{\alpha(\bar{n})}\frac{d\alpha(\bar{n})}{d\bar{n}}\right]^{-1}, \qquad (3.5)$$

with the initial condition $\bar{n}(L) = n_D$.

The distribution of the wave power can be found directly from (3.4),

$$P(z) = \tfrac{1}{2}\ \pi a^2\ \Theta\ \bar{n}(z)/\alpha(\bar{n}). \qquad (3.6)$$

Fig. 7 shows a typical example of the axial distribution of the electron density and wave power, as recorded during an experiment*. The broken lines represent the best fit of the results from the present model to the experimental data. There is only one fitting parameter, common to both curves, namely the effective electron collision frequency for momentum transfer ν.

* The measurements were performed with an experimental setup and procedure described in detail in [28].

IV. Properties of surface wave sustained plasma columns: influence of discharge conditions and input power.

IV.1 Electron density and plasma column length

The parameters of the plasma column to be determined are: the electron density n_o at the origin, the corresponding density gradient, and the column length L. We now express these three quantities in a normalized form, respectively:

$$n_o/n_D = \tfrac{1}{2}\left[(8P_o/P_D + 1)^{\frac{1}{2}} + 1\right], \quad (4.1)$$

$$-(\alpha_D n_D)^{-1}\left[dn(z)/dz\right]_o = (1-n_D/2n_o)^{-1}, \quad (4.2)$$

$$\alpha_D L = n_o/n_D - 1 - \tfrac{1}{2}\log_e(n_o/n_D), \quad (4.3)$$

here, we have introduced, besides n_D, two new normalizing parameters:

$$\alpha_D = A(\omega,a)\,\nu/n_D \quad \text{and} \quad P_D = \pi a^2\,\Theta\,n_D/\alpha_D, \quad (4.4)$$

which have the physical meaning of the wave attenuation occurring at the z value for which $n(z) = 2n_D$ and of the corresponding power flux, respectively. As for $(-\alpha_D n_D) = A(\omega,a)\,\nu$, it is the density gradient at the origin, in the high power limit ($n_o \gg n_D$). The convenience for such a choice of the normalizing parameters stems from the fact that n_D, α_D and P_D are constant under given discharge conditions. Thus in equations (4.1) to (4.3), they represent the influence of the discharge conditions, while the influence of the input power enter explicitly through the presence of P_o.

The results of the calculations are shown in Fig. 8 together with experimental points. The procedure for the calculation of $\alpha(\bar{n})$ and for the measurements were described before [25,28]. The experimental conditions were as indicated in the figure. Fig. 8 contains the main outcome of this work in a concise form. Notice that the plasma parameters have been expressed in terms of quantities that are controllable by the user, when designing or operating the SWD. As already pointed out, the impact of changes of P_o is expressed in a straightforward way and the influence of the discharge conditions enters indirectly through the normalizing parameters. To examine the influence of the discharge conditions on these parameters, let us consider their physical meaning (the symbol $\sim$ denoting the proportionality and the arrow $\rightarrow$ referring to the case of the electrostatic approximation):

$$n_D \sim \omega^2(\varepsilon_g+1), \tag{4.5}$$

$$P_D \sim a^2\omega^4(\varepsilon_g+1)\,\underset{\sim}{\theta}(pa)/A(\omega,a) \rightarrow$$

$$(a\omega)^3(\varepsilon_g+1)^2\,\underset{\sim}{\theta}(pa), \tag{4.6}$$

$$\alpha_D n_D \sim A(\omega,a)\;p \rightarrow \omega\;p/a. \tag{4.7}$$

Having at our disposal the above, together with Fig. 8, we can examine the control means that we have over the plasma parameters. Remember that these equations correspond to the low attenuation regime ($\nu^2 \ll \omega^2$).

Case A: the discharge conditions are fixed

Then, the only way to control the plasma parameters is through the variation of P_o. This affects all the parameters considered,

n_o, $[dn/dz]_o$ and L with the electron energy remaining approximately constant. As shown in Fig. 8, both n_o and L increase with P_o and are proportional to $P_o^{1/2}$ at large absorbed power. At low power values, the increase of P_o contributes mainly to that of L, without increasing much the electron density within the column. It occurs that the product n_oL is approximately proportional to P_o in the whole range of the input power. As for the density gradient, it does not depend strongly on P_o for $P_o/P_D > 1$.

Case B: the discharge conditions can be varied

When some freedom of choice for the discharge conditions is granted, the control becomes more effective. Its extent can be readily inferred using the present results. However, we should like to emphasize the following: the changes in p and a also affect the electron energy, while changes in ω and ε_g do not.

Let us point out some particularly interesting results from Eqs. (4.5) to (4.7).

- The electron density n_D, i.e. the density close to the end of the plasma column, increases with the square of the wave frequency (provided $\nu^2 \ll \omega^2$, don't forget). This is definitely a powerful means of increasing the plasma density.

- In the electrostatic approximation, it can be estimated that the power P_D, i.e. the power flow at the plane where $n = n_D$, increases with the third power of ω and, for given (pa) product, with the third power of the radius. This may, in practice, require to a considerable increase in power.

- The density gradient at the origin, for large enough P_o power (i.e., for $n_o/n_D \gg 1$), in the electrostatic approximation, is proportional to $\omega p/a$, a result similar to that of Mateev et al. who found[11]:

$$d\bar{n}/dz = 0.73 \times 10^{-8}\ f\nu/a\ (\mathrm{cm}^{-4}),$$

where the wave frequency f and the collision frequency ν are expressed in Hz and radian per second, respectively.

- For a given plasma column length, because the density gradient increases with ω, increasing the wave frequency will increase the density at the beginning of the column (launcher side) still faster than ω^2.

The control obtained by varying the wave frequency thus comes out as a particularly powerful one and a convenient tool in the hands of the SWD user, seeking a custom-made plasma column.

Recently[29], we have pointed out still another way of controlling, the axial distribution of the electron density in SWD. It consists in sustaining the plasma column in a tapered tube. In that case, for example, an axially uniform density profile can be achieved by a properly decreasing tapering of the tube.

V. Conclusion. The available surface wave plasma sources assure an efficient plasma generation over an exceptionally wide range of frequencies, tube diameters and gas pressures. The main parameters of interest for a plasma column sustained by a surface wave are: the mean electron energy, the electron density and the length of the column. These can be controlled to a large extent by varying the discharge conditions and, for the last two parameters, by varying the wave power.

To a first approximation, the input power only affects the electron density and the column length. Varying the wave frequency gives control over the electron density, on a particularly wide range of values. It also affects the column length and the density gradient. However, the averaged electron energy should remain practically unaffected. Changes in the gas pressure and the tube radius affect all the above mentioned plasma parameters.

References

[1]A.W. Trivelpiece, "Slow-Wave Propagation in Plasma Waveguides", San Francisco Press, (1967).

[2]A. Shivarova and I. Zhelyazkov, Plasma Phys. 20, 1049 (1978).

[3]M. Moisan, A. Shivarova, and A.W. Trivelpiece, Plasma Phys. 24, 1331 (1982).

[4]M. Moisan, C.M. Ferreira, Y. Hajlaoui, D. Henry, J. Hubert, R. Pantel, A. Ricard and Z. Zakrzewski, Revue Phys. Appl. 17, 707, (1982).

[5]M. Moisan and Z. Zakrzewski, "Radiation Processes in Gas Discharge Plasmas", Plenum Press, N.Y., in print.

[6]M. Chaker, M. Moisan, and Z. Zakrzewski, Plasma Chemistry and Plasma Processing, Vol. 5, issue no. 4 (1985).

[7]M. Chaker and M. Moisan, J. Appl. Phys. 57, 91 (1985).

[8]C.M. Ferreira, J. Phys. D: Appl. Phys. 14, 1811 (1981).

[9]Z. Zakrzewski, J. Phys. D: Appl. Phys. 16, 171 (1983).

[10]C.M. Ferreira, J. Phys. D: Appl. Phys. 16, 1673 (1983).

[11]E. Mateev, I. Zhelyazkov, and V. Atanassov, J. Appl. Phys. 54, 3049 (1983).

[12]Yu M. Aliev, A.G. Boev, and A. Shivarova, J. Phys. D: Appl. Phys. 17, 2233 (1984).

[13]I. Zhelyazkov, E. Benova, and V. Atanassov, J. Appl. Phys., in print.

[14]J. Paraszczak, J. Heidenreich, M. Hatzakis, and M. Moisan, Microcircuit engineering 85, to appear in Microelectronic engineering, vol. 3 (Dec. 85).

[15]L. Paquin, D. Masson, M.R. Wertheimer, and M. Moisan, Can. J. Phys. 63, 831-7 (1985).

[16]C. Moutoulas, M. Moisan, L. Bertrand, J. Hubert, J.L. Lachambre, and A. Ricard, Appl. Phys. Lett. 46, 323 (1985).

[17]G. Chevrier, T. Hanai, K.C. Tran, and J. Hubert, Can. J. Chem. 60, 898 (1982).

[18]M.H. Abdallah, S. Coulombe, J.M. Mermet, and J. Hubert, Spectrochim. Acta, 37B, 583 (1982).

[19] G. Loncar, J. Musil, and L. Bardos, Czech. J. Phys. B30, 688 (1980).
[20] J. Musil, Vacuum, in print.
[21] H.M. Barlow and J. Brown, "Radio Surface Waves", Clarendon Press, Oxford, (1962).
[22] M. Moisan, Z. Zakrzewski, and R. Pantel, J. Phys. D: Appl. Phys. 12, 219 (1979).
[23] M. Moisan, Z. Zakrzewski, R. Pantel, and P. Leprince, IEEE Trans. on Plasma Sci: PS-12, 203 (1984).
[24] M. Moisan and Z. Zakrzewski, Canadian Patent Application (1985).
[25] V.M.M. Glaude, M. Moisan, R. Pantel, P. Leprince, and J. Marec, J. Appl. Phys. 51, 5693 (1980).
[26] J. Marec, E. Bloyet, M. Chaker, P. Leprince, and P. Nghiem, in "Electrical Breakdown and discharges in Gases", Pt. B, E.E. Kunhardt and L.H. Luessen, eds., Plenum Press, New York (1983).
[27] W.P. Allis, S.J. Buchsbaum, and A. Bers; "Waves in Anisotropic Plasmas", MIT Press, (1963).
[28] Z. Zakrzewski, M. Moisan, V. Glaude, C. Beaudry, and P. Leprince, Plasma Phys. 19, 77 (1977).
[29] M. Moisan and Z. Zakrzewski, Surface Wave discharges in tapered tubes, this conference, contributed paper.

FIGURES

TABLE 1

DISTINGUISHING AND COMPLEMENTARY FEATURES OF VARIOUS SW PLASMA GENERATORS

NAME	DESIGN	HF CIRCUIT
SURFAGUIDE	INTEGRATED*	WAVEGUIDE
SURFATRON	—"—	COAXIAL
SURFALINE	MODULAR**	COAXIAL OR SYMMETRIC LINE
RO-BOX	—"—	LUMPED LC CIRCUIT

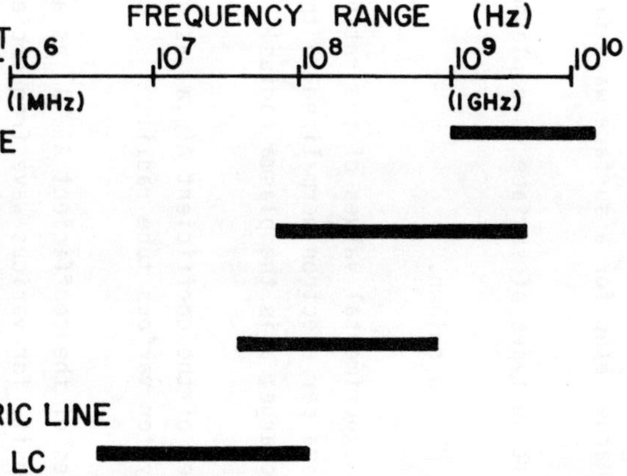

* INTEGRATED DEVICE PERFORMS WAVE LAUNCHING AND IMPEDANCE MATCHING FUNCTIONS

** SEPARATE, INTERCHANGEABLE LAUNCHING AND IMPEDANCE MATCHING MODULES

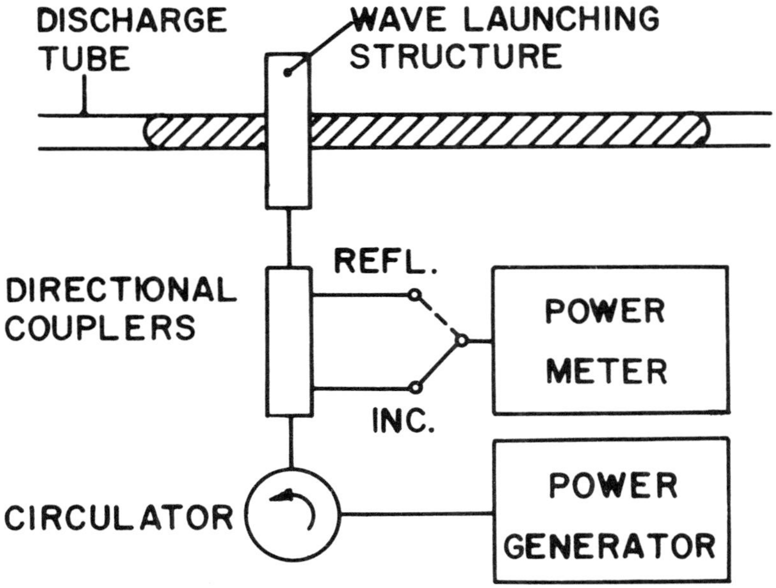
DISCHARGE
TUBE
WAVE LAUNCHING
STRUCTURE
REFL.
DIRECTIONAL
COUPLERS
POWER
METER
INC.
POWER
GENERATOR
CIRCULATOR

Fig. 1

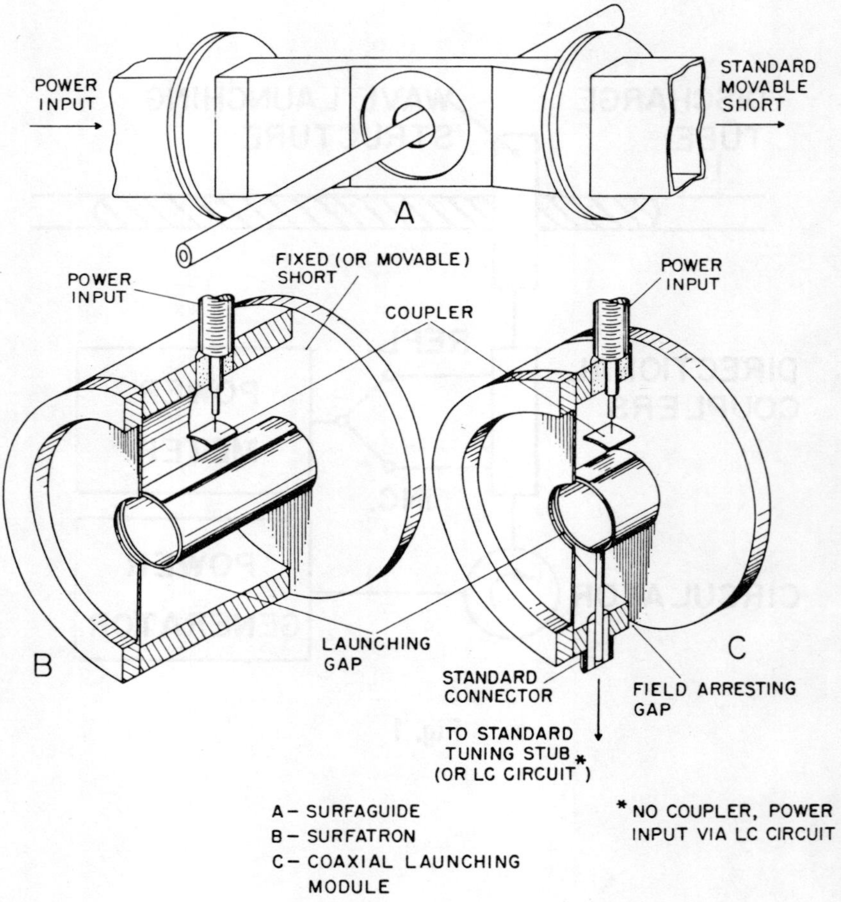
POWER
INPUT
STANDARD
MOVABLE
SHORT
A
POWER
INPUT
FIXED (OR MOVABLE)
SHORT
COUPLER
POWER
INPUT
LAUNCHING
GAP
B
C
STANDARD
CONNECTOR
FIELD ARRESTING
GAP
TO STANDARD
TUNING STUB*
(OR LC CIRCUIT*)
A– SURFAGUIDE
B– SURFATRON
C– COAXIAL LAUNCHING
MODULE
* NO COUPLER, POWER
INPUT VIA LC CIRCUIT

Fig. 2

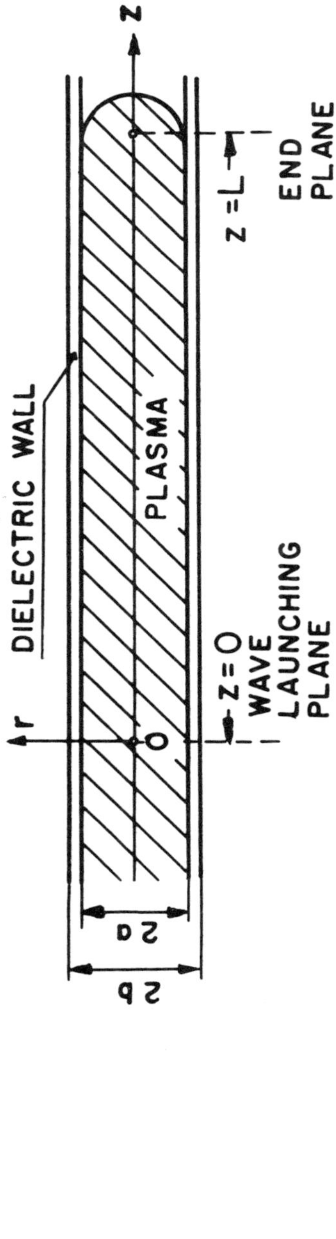
DIELECTRIC WALL
PLASMA
r
O
z
z=O
WAVE
LAUNCHING
PLANE
z=L
END
PLANE
2a
2b

Fig. 3

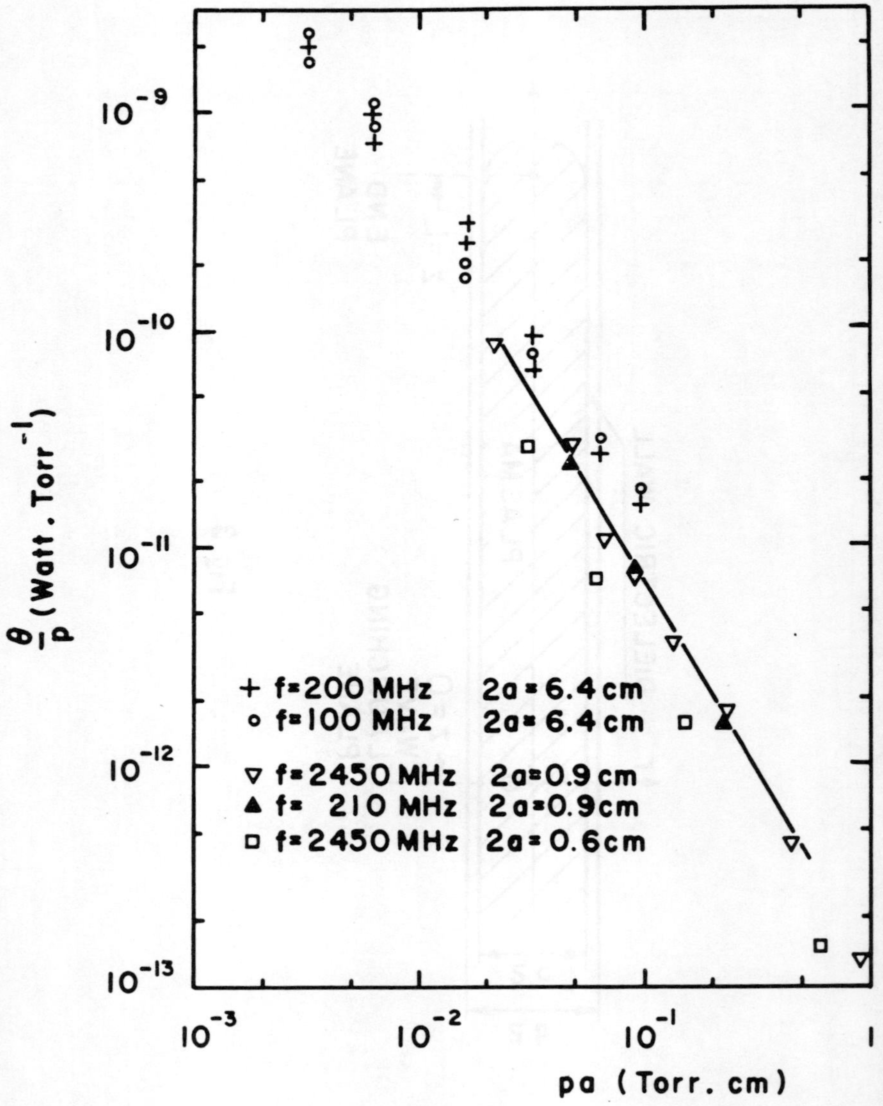
10^{-9}
10^{-10}
10^{-11}
10^{-12}
10^{-13}
$\frac{\theta}{p}$ (Watt . Torr^{-1})
+ f=200 MHz 2a=6.4 cm
o f=100 MHz 2a=6.4 cm
▽ f=2450 MHz 2a=0.9 cm
▲ f= 210 MHz 2a=0.9 cm
□ f=2450 MHz 2a=0.6 cm
10^{-3}
10^{-2}
10^{-1}
1
pa (Torr. cm)

Fig. 4

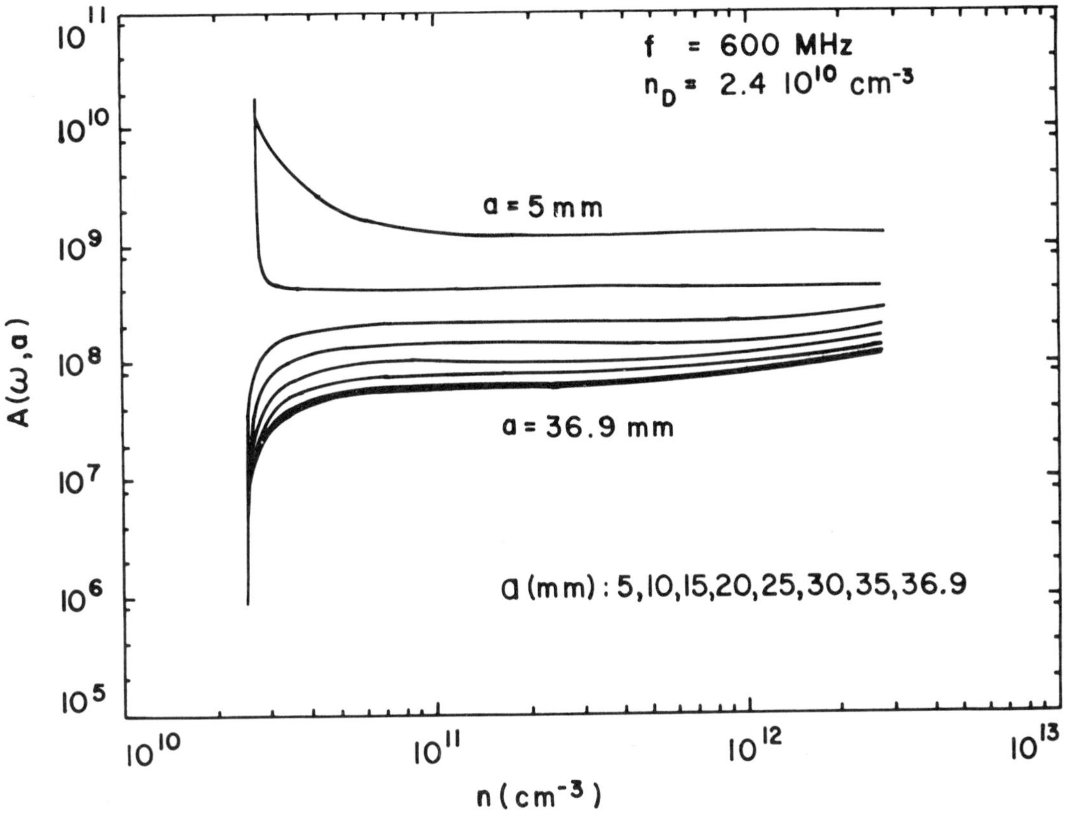
f = 600 MHz
n_D = 2.4 10^10 cm^-3
a = 5 mm
a = 36.9 mm
a (mm) : 5,10,15,20,25,30,35,36.9
A(ω,a)
10^11
10^10
10^9
10^8
10^7
10^6
10^5
10^10
10^11
10^12
10^13
n (cm^-3)

Fig. 5

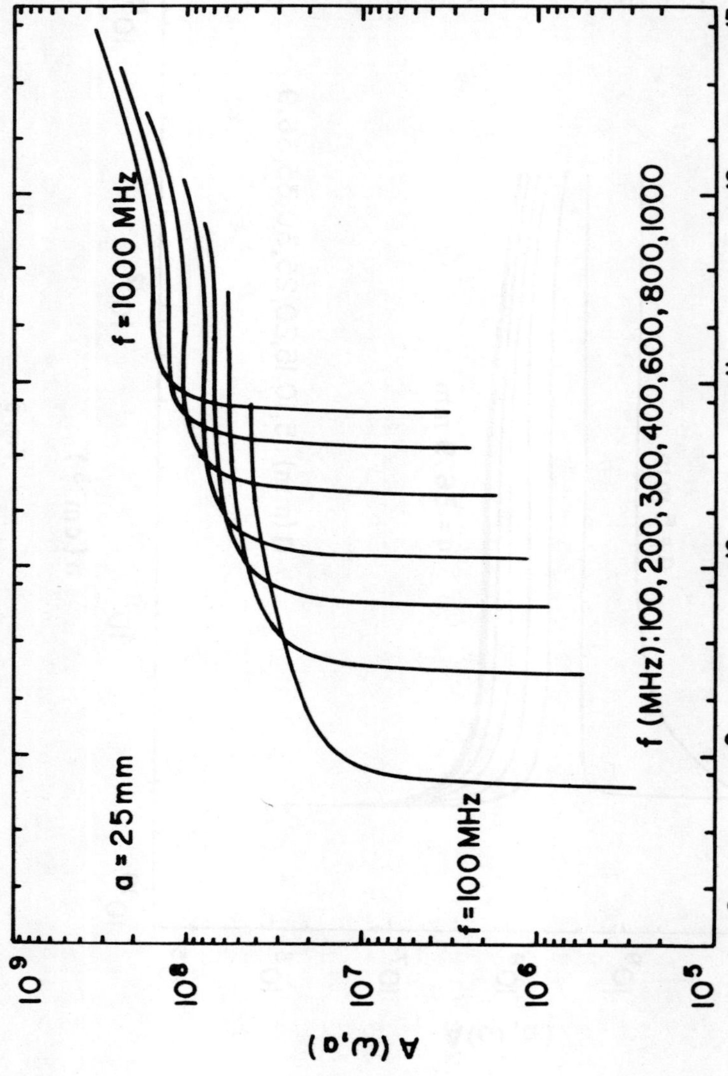
a = 25 mm
f = 1000 MHz
f = 100 MHz
f (MHz): 100, 200, 300, 400, 600, 800, 1000
A (ω, a)
n (cm⁻³)
10^9
10^8
10^7
10^6
10^5
10^8
10^9
10^10
10^11
10^12
10^13

Fig. 6

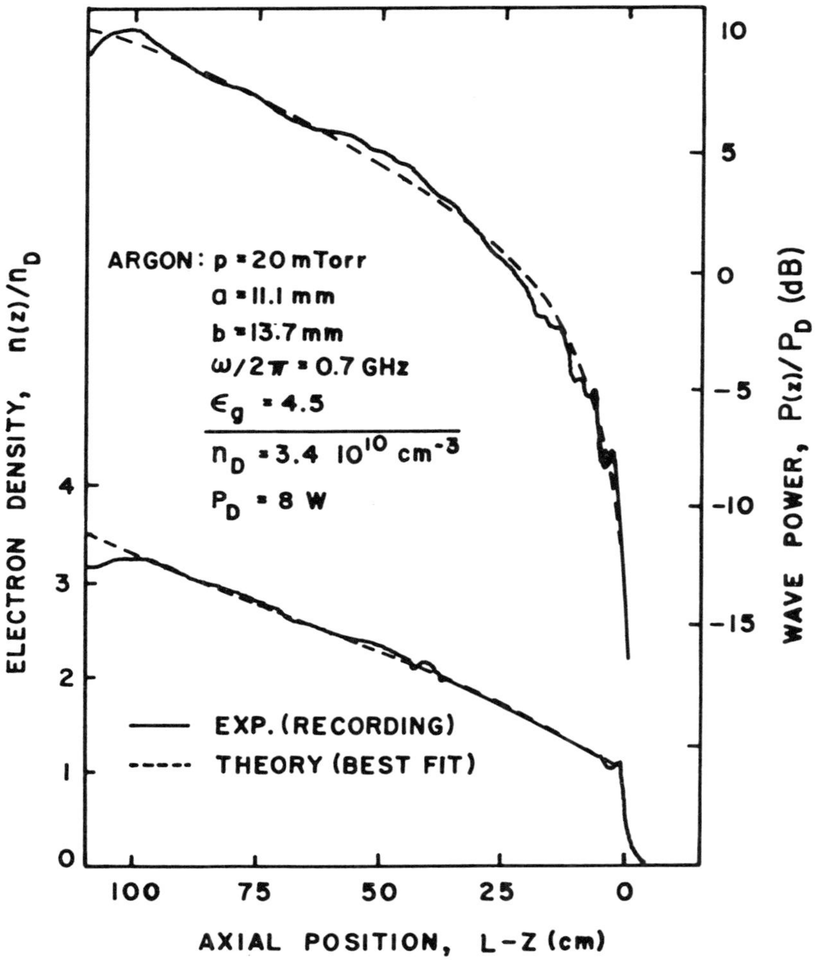
ARGON: p = 20 mTorr
a = 11.1 mm
b = 13.7 mm
ω/2π = 0.7 GHz
ε_g = 4.5
n_D = 3.4 10^10 cm^-3
P_D = 8 W
EXP. (RECORDING)
THEORY (BEST FIT)
ELECTRON DENSITY, n(z)/n_D
WAVE POWER, P(z)/P_D (dB)
AXIAL POSITION, L-Z (cm)
100
75
50
25
0
10
5
0
-5
-10
-15
4
3
2
1
0

Fig. 7

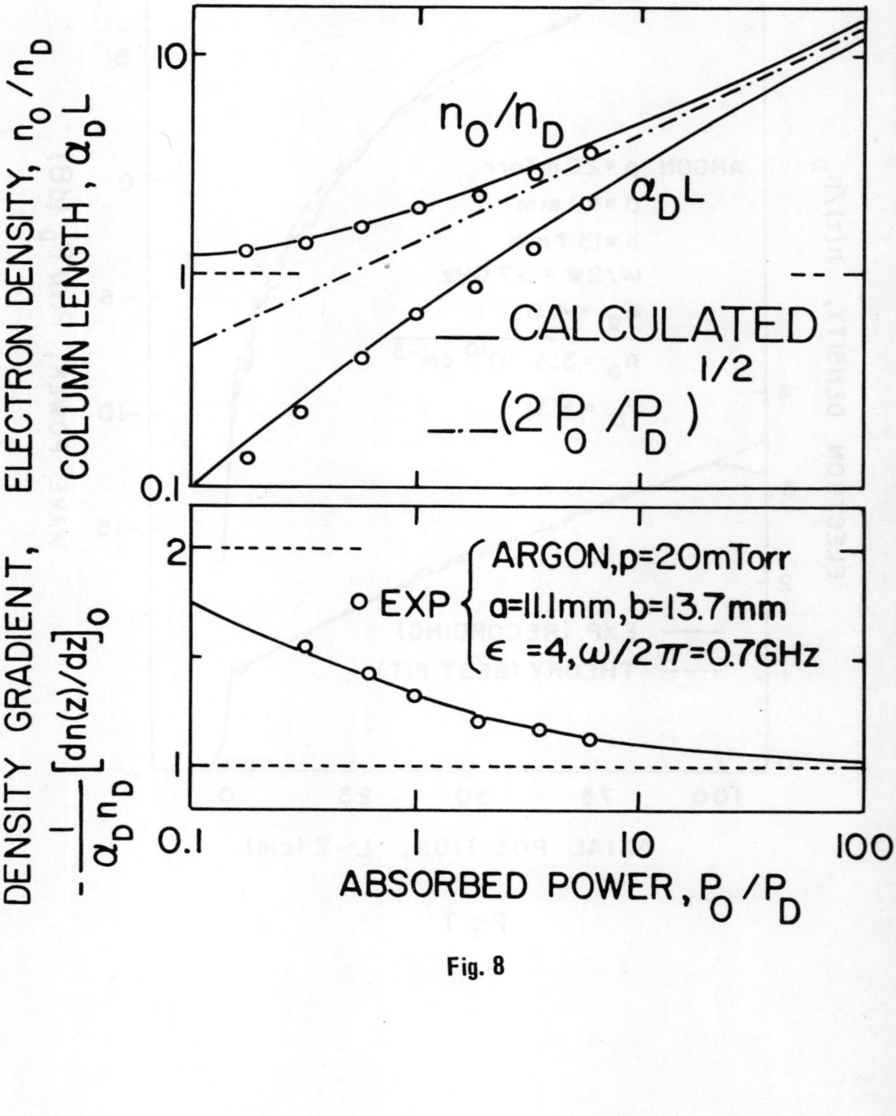
ELECTRON DENSITY, n_0/n_D
COLUMN LENGTH, $\alpha_D L$
10
1
0.1
n_0/n_D
$\alpha_D L$
— CALCULATED
—·— $(2P_0/P_D)^{1/2}$
DENSITY GRADIENT,
$-\frac{1}{\alpha_D n_D}\left[\frac{dn(z)}{dz}\right]_0$
2
1
○ EXP
ARGON, p=20mTorr
a=11.1mm, b=13.7mm
ϵ =4, $\omega/2\pi$=0.7GHz
0.1
1
10
100
ABSORBED POWER, P_0/P_D

Fig. 8

AXIAL STRUCTURE OF A MICROWAVE DISCHARGE SUSTAINED BY A LARGE-AMPLITUDE SURFACE WAVE

I. Zhelyazkov
Faculty of Physics, Sofia University,
BG-1126 Sofia, Bulgaria

V. Atanassov
Institute of Electronics, Bulgarian Academy of Sciences,
BG-1784 Sofia, Bulgaria

E. Benova
Department of Physics, Institute for Foreign Students,
BG-1111 Sofia, Bulgaria

1. Introduction

High frequency gas discharges have long been studied both theoretically and experimentally. In paricular, those that are created and sustained by high frequency surface waves, are the subject of increasing interest in the last years. This is confirmed by the fact that we have not less than 3 invited talks (devoted to different aspects of that problem) at this conference.[1]

A considerable amount of papers which deal with modelling of surface-wave-produced plasma columns appeared in the last four years.[2-17] Our talk considers some general relations associated with such modelling which allow us to obtain "universal" axial profiles of the normalized plasma density, surface wave wavenumber, wave electric field amplitude and wave power depending on a dimensionless axial distance from the end of the column and on two parameters. One of them is associated with the electrodynamic characteristics of the guided wave propagation, $\sigma = \omega R/c$ (ω and R being the wave frequency and tube radius, c denoting the speed of light), while the other, β, is completely determined by the discharge conditions. Generally speaking, two different discharge regimes are possible -- free-fall/diffusion-controlled and recombination-controlled discharges.[18] In contrast to other workers we have studied the axial structure of the surface-wave-produced plasma columns in both discharge regimes. However, the prevailing part of the available experimental data can be classified as typical for free-fall/diffusion-controlled discharges.

2. Basic Assumptions and Governing Equations

The main feature of surface-wave-produced plasmas is that the wave heats the electrons which ionize the medium ensuring in this a way the further wave propagation. We examine the stationary state of a plasma column sustained by a travelling azimuthally-symmetric high frequency surface wave. In studying the surface wave propagation we consider the weakly ionized gas as a collisional cold electron plasma. We presume that:
(i) the plasma is assumed to be a weakly dissipative medium, $\nu \ll \omega$; (ii) the collisional frequency ν is constant both in radial and axial direction; (iii) the plasma density is radially constant, $n \equiv \bar{n} = \frac{2}{R^2}\int_0^R dr\, r\, n(r)$; additionally we ignore (iv) the influence of the thin dielectric tube over the parameters of the plasma created; (v) the wave reflection at the column end and (vi) the electron diffusion in axial direction.

The governing equations are:

(a) local dispersion relation

$$\frac{1-N}{a_p}\frac{I_1(a_p)}{I_0(a_p)} + \frac{1}{a_v}\frac{K_1(a_v)}{K_0(a_v)} = 0 \qquad (1)$$

where

$$N = n/n_{cr}\,, \qquad n_{cr} = m\omega^2/4\pi e^2$$

$$a_p = \left[k^2 - \frac{\omega^2}{c^2}(1-N)\right]^{1/2} R\,, \qquad a_v = \left(k^2 - \frac{\omega^2}{c^2}\right)^{1/2} R\,,$$

$k \equiv |k_z|$ is the surface wave wavenumber, and $I_{0,1}$, $K_{0,1}$ are the modified Bessel functions;

(b) surface-wave energy balance equation

$$\frac{d}{dz} S = - Q \tag{2}$$

where

$$S = \frac{c}{2} \int_0^R dr\, r \langle \vec{E} \times \vec{B} \rangle_z + \frac{c}{2} \int_R^\infty dr\, r \langle \vec{E}^v \times \vec{B}^v \rangle$$
$$= \frac{1}{8} \omega R^3 E^2 \frac{x^2}{a_v^2} \frac{K_1^2(a_v)}{K_0^2(a_v)} \left(\frac{1}{1-N} A - B \right) \operatorname{sgn} k_z \tag{3}$$

is the surface-wave energy flux density (wave power) and

$$Q = 2\pi \int_0^R dr\, r \langle \vec{j} \cdot \vec{E} \rangle = \frac{1}{8} \nu R^2 N E_0^2 \tag{4}$$

with

$$E_0^2 \equiv \frac{2}{R^2} \int_0^R dr\, r |\vec{E}|^2 = E^2 \left(1 - \frac{I_1^2(a_p)}{I_0^2(a_p)} - \frac{x^2}{a_p^2} A \right) \tag{5}$$

is the absorbed surface wave power per unit length. In (3)-(5) x = kR, E is the magnitude of the z-component of the electric field amplitude at the plasma-vacuum interface, $E = |E_z(R)|$,

$$A = 1 - \frac{I_1^2(a_p)}{I_0^2(a_p)} - \frac{2}{a_p} \frac{I_1(a_p)}{I_0(a_p)}$$

$$B = 1 - \frac{K_1^2(a_v)}{K_0^2(a_v)} + \frac{2}{a_v} \frac{K_1(a_v)}{K_0(a_v)}$$

and the angular brackets denote an averaging over the wave period. Equations (1) and (2) have been obtained from Maxwell's equations supplemented with appropriate boundary conditions by assuming that the wave and plasma parameters depend slowly on the axial coordinate z.

The power dissipated by the electrons is expended on ionization, excitation, neutral gas heating and on maintaining the negative potential on the tube wall, i. e.

$$Q = \pi R^2 n \, [\nu_i (\varepsilon_i + \frac{3}{2} \varkappa T_e) + \sum_j \nu_j \varepsilon_j + \nu \frac{3m}{M} \varkappa T_e + 2 \frac{v_s}{R} \varkappa T_e + \nu_i (2 + \ln \frac{M}{m}) \varkappa T_e] \qquad (6)$$

where $v_s = (\kappa T_e/M)^{1/2}$ is the ion sound velocity. Equation (6) has been derived from the internal plasma electrons' energy balance equation by assuming that the electrons are Maxwellian and the temperature T_e is constant both in radial and axial direction. We note that the last term in (6) is associated with the radial electron thermal flux at the boundary and is responsible for maintaining the negative potential at the tube wall. This term has been omitted by some others[3,8] but it should be taken into account in particular at lower gas pressures.[19]

Equation (6) can be rewritten in another form

$$\frac{Q}{\pi R^2 n} \equiv \theta_* = 2 \nu E_0^2 / 16 \pi n_{cr} \qquad (6a)$$

where θ_* is the absorbed wave power per one electron.[3,8]

At low gas pressures the main contribution to the generation of charged particles is the single-step ionization, and the major electron loss process is the diffusion to the wall of the tube. In that case the electron collision frequencies of ionization and excitation do not depend on the number density of charged particles. In such a gas-discharge regime, called diffusion one,[18] from (4) and (6) it follows that

$$E_0^2 = \text{constant} \qquad (7)$$

At higher electron number densities n and gas pressures p (above 1-10 torr for noble gases) the dominant processes are the multi-step ionization and bulk (photo) recombination. The ionization and recombination frequencies are then proportional to n and for so-called recombination gas-discharge regime[18] Eqs. (4) and (6) yield

$$n \propto E_o^2 \tag{8a}$$

A more rigorous analysis shows that at relatively lower electron temperatures[13] another relation takes place, namely

$$n \propto E_o^{2s} \;, \quad \text{where} \quad s > 1 \tag{8b}$$

Dependences of the kind (8a) and (8b) are frequently used in different nonlinear wave problems under appelation "ionization nonlinearity".[20] Equations (7) and (8a,b) can be written in a general form which gives our third governing equation:

(c) power dependence between radially averaged plasma density and squared wave electric field amplitude

$$E_o^2 = E_*^2 N^\beta \;, \quad \text{where} \quad E_* = \text{constant} \tag{9}$$

In Eq. (9) the different values of β correspond to different gas-discharge regimes: free-fall/diffusion-controlled discharge ($\beta = 0$) or recombination-controlled discharge ($0 < \beta \leqslant 1$). In free-fall/diffusion regime E_* is the "discharge electric field" and it is given by[15]

$$E_*^2 = 8\pi n_{cr} U_* \quad , \quad E_*\left[\text{V/cm}\right] = 212\,\frac{\omega}{2\pi}\left[\text{GHz}\right]\sqrt{U_*\left[\text{eV}\right]} \tag{10}$$

where $U_* = \theta_*/\nu$. In this case it is convenient to normalize the wave power S to

$$S_* = \pi R^3 \omega n_{cr} U_* \ , \quad S_*[W] = 39.2\left(\frac{\omega}{2\pi}[GHz]\cdot R[cm]\right)^3 U_*[eV] \tag{11}$$

Further we study the set of governing equations (1), (2) and (9) in order to obtain axial profiles of normalized electron number density $N = n/n_{cr}$, surface-wave wavenumber $x = kR$, wave amplitude $\tilde{E} = |E_z(R)|/E_*$ and wave power $\tilde{S} = S/S_*$ as functions of the dimensionless axial coordinate $\zeta = \frac{\nu}{\omega}\frac{z}{R}$. These profiles depend additionally on two parameters: $\sigma = \frac{\omega R}{c}$ and β, which are determined by the experimental conditions.

3. Approximate Formulae for $N(\zeta)$, $x(\zeta)$, $\tilde{E}(\zeta)$ and $\tilde{S}(\zeta)$

The simplest solutions to Eqs. (1), (2) and (9) are obtained for $\sigma = 0$ (electrostatic approximation). The basic set of equations can be then reorganized to

$$\frac{d\zeta}{dx} = -\frac{1}{2}\frac{d}{dx}\left[\frac{I_0(x)}{I_1(x)} - \frac{K_1(x)}{K_0(x)}\right] + \frac{1}{2}(1+\beta)\left[\frac{I_0(x)}{I_1(x)} - \frac{K_1(x)}{K_0(x)}\right]^2 \tag{12}$$

$$N = 1 + \frac{I_0(x)K_1(x)}{I_1(x)K_0(x)} \tag{13}$$

$$\tilde{E}^2 = \frac{1}{2} x \frac{I_0(x)}{I_1(x)} N^\beta , \qquad \tilde{S} = \frac{1}{2}\left[\frac{I_0(x)}{I_1(x)} - \frac{K_1(x)}{K_0(x)}\right] N^{1+\beta} \tag{14}$$

Equation (12) generalizes Eqs. (13) and (15) derived in Ref. 5 (second paper). We note that in electrostatic approximation it is possible to obtain a solution to Eq. (12), $x = x(\zeta)$, and then to find out $N(\zeta)$, $\tilde{E}^2(\zeta)$ and $\tilde{S}(\zeta)$.

In a thin cylinder limit, x << 1, from (12)-(14) we have

$$\frac{d\zeta}{dx} \approx \frac{1}{x^2 F(\beta, x)}, \quad x(\zeta) \approx x(\zeta_0)\left\{1 - x(\zeta_0) F[\beta, x(\zeta_0)](\zeta - \zeta_0)\right\}^{-1} \tag{15}$$

$$\frac{dN}{d\zeta} \approx -\frac{4}{x\ell}\left(1 - \frac{1}{2\ell}\right) F(\beta, x), \quad N \approx \frac{2}{x^2 \ell} \tag{16}$$

$$\tilde{E}^2 \approx N^\beta, \quad \tilde{S} \approx \frac{1}{x}\left(1 - \frac{1}{2\ell}\right) N^{1+\beta} \tag{17}$$

where

$$\ell = \ln\frac{C'}{x}, \quad C' = 2e^{-C} \approx 1.123, \quad F(\beta, x) = \left[2(1+\beta)\left(1 - \frac{1}{2\ell}\right)^2 + 1 - \frac{1}{2\ell} + \frac{1}{2\ell^2}\right]^{-1}$$

The applicability of these formulae is however restricted by the inequalities $\sigma\sqrt{N} << x << 1$.

In a thick cylinder limit, x >> 1, we get

$$x \approx [2(\zeta_{end} - \zeta)]^{-1/2} \tag{18}$$

$$N \approx 2\left(1 + \sqrt{\zeta_{end} - \zeta}\right) \tag{19}$$

$$\tilde{E}^2 \approx 2^{\beta - 2}(\zeta_{end} - \zeta)^{-1/2}, \quad \tilde{S} \approx 2^{1+\beta}(\zeta_{end} - \zeta) \tag{20}$$

Formulae (18)-(20) describe the asymptotic behaviour of the axial profiles at the end of the column, $\zeta \longrightarrow \zeta_{end}$. This behaviour does not depend on β and it retains itself within the full electrodynamic treatment, provided that $x >> \max(1,\sigma)$. We emphasize that the WKB approximation which has been used to obtain Eqs. (1) and (2) does not break in the limit $\zeta \longrightarrow \zeta_{end}$. It should be also noted that in the experiments the cutoff electron number density is larger than 2 due to the influence of the dielectric tube.

It is easy to see that in thin-cylinder limit, $x \ll 1$, the axial plasma density profile is concave, $d^2N/d\zeta^2 < 0$, while in the opposite limit, $x \gg 1$, it is convex, $d^2N/d\zeta^2 > 0$. Therefore an inflect point at some $x \equiv x_*$ exists, where the slope of $N(\zeta)$ is minimum. The value of x_* is determined[15] as a root of the following equation:

$$V V'' - (V')^2 - V^4 - \beta V^2 (V' + V^2) = 0$$

where $V(x) = \frac{I_0(x)}{I_1(x)} - \frac{K_1(x)}{K_0(x)}$, $V'(x) = dV/dx$ and

$$\left(\frac{dN}{d\zeta}\right)_{min} = \left.\frac{dN}{d\zeta}\right|_{x=x_*} = \frac{2N(x_*)V(x_*)}{V'(x_*) - (1+\beta)V^2(x_*)}$$

The numerical results for x_*, $N(x_*)$ and $\left.\frac{dN}{d\zeta}\right|_{x=x_*}$ depending on β are given in Table I.

Table I

β	0	0.1	0.2	0.5	1
x_*	0.698	0.667	0.642	0.585	0.530
$N(x_*)$	5.83	6.11	6.36	7.02	7.85
$-\left.\frac{dN}{d\zeta}\right\|_{x=x_*}$	3.24	3.11	2.99	2.68	2.26

In the frame of full electrodynamic treatment from governing Eqs. (1), (2) and (9) we obtain approximate solutions for the axial profiles in two limiting cases:

(A) Surface wave field deeply penetrates into both media, $a_v < a_p << 1$; these inequalities imply $\sigma\sqrt{N} << 1$:

$$\frac{d\zeta}{da_v} \approx \frac{x}{a_v^3 F(\sigma,\beta,a_v)} \tag{21}$$

$$\frac{dN}{d\zeta} \approx -\frac{4}{a\ell_v}\left(1-\frac{1}{2\ell_v}\right)F(\sigma,\beta,a_v), \quad N \approx \frac{2}{a_v^2 \ell_v} \tag{22}$$

$$\tilde{E}^2 \approx N^\beta, \quad \tilde{S} \approx \frac{x}{a_v^2}\left(1-\frac{1}{2\ell_v}\right)N^{1+\beta} \tag{23}$$

where

$$\ell_v = \ln\frac{C'}{a_v}, \; F(\sigma,\beta,a_v) = \left[2(1+\beta)\left(1-\frac{1}{2\ell_v}\right)^2 + 2\left(1-\frac{1}{2\ell_v}+\frac{1}{4\ell_v^2}\right) - \frac{a_v^2}{x^2}\left(1-\frac{1}{2\ell_v}\right)\right]^{-1}$$

This approximation (with $\beta = 1$ -- ionization nonlinearity) has been used by Aliev et al.[6] At $\sigma \longrightarrow 0$ formulae (21)-(23) coincide with those obtained in thin cylinder electrostatic approximation (expressions (15)-(17)).

(B) Surface wave field deeply penetrates into the vacuum; the penetration depth in the plasma is relatively small, $a_v << 1 << a_p$. These inequalities require $\sigma\sqrt{N} >> 1$ and

$$\frac{d\zeta}{da_v} \approx \frac{2x}{a_v^3 G(\sigma,\beta,a_v)} \tag{24}$$

$$\frac{dN}{d\zeta} \approx -\frac{2\sigma\sqrt{N}}{x\ell_v}\left(1-\frac{1}{2\ell_v}\right)G(\sigma,\beta,a_v), \quad N \approx \frac{\sigma^2}{a_v^4 \ell_v^2} \tag{25}$$

$$\tilde{E}^2 \approx \sigma N^{\beta+\frac{1}{2}}, \quad \tilde{S} \approx \frac{x^3}{a_v^2 a_p^2}\sigma N^{\beta+\frac{3}{2}} \tag{26}$$

where

$$G(\sigma, \beta, a_N) = \left[4(1+\beta)\left(1 - \frac{1}{2\ell_N}\right)^2 + 2\left(1 - \frac{1}{2\ell_N} + \frac{1}{4\ell_N^2}\right) - \frac{a_N^2}{x^2}\left(1 - \frac{1}{2\ell_N}\right)\right]^{-1}$$

In this case the transition to the electrostatic approximation is in principle impossible.

4. Numerical Results

The results of the numerical solving of Eqs. (1), (2) and (9) are graphically presented on Figs. 1 and 2 for $\beta = 0$ (free-fall/diffusion regime) and $\beta = 1$ (recombination regime) respectively at $\sigma = 0$, 0.05, 0.1, 0.2, 0.5, 1.0 and 2.0 . It follows from the analytical and numerical calculations that the axial plasma density profiles are approximately linear only in a limited range of density variation. This range is relatively large for $\sigma = 0.1 - 0.5$. It is worth noting that the dependence of the axial plasma density gradient $|dN/d\zeta|$ on σ is not monotonous and it passes through a minimum at $\sigma \sim 0.7$. Increasing β decreases the gradient $|dN/d\zeta|$, i. e. the slope of $N(\zeta)$ in recombination regime is less than that for diffusion one. The average slopes (in the range $10 < N < 100$) at $\beta = 0$ and $\beta = 1$ for different σ's are given in Table II.

Table II

$\sigma = \omega R/c$	0	0.05	0.1	0.2	0.5	1.0	2.0
$\lvert dN/d\zeta\rvert$ $\beta = 0$	5.11	4.48	3.64	2.54	1.88	2.11	2.84
$\lvert dN/d\zeta\rvert$ $\beta = 1$	3.20	2.93	2.41	1.71	1.27	1.39	1.88

5. Comparison With the Experimental Data

The "universal" normalized axial profiles $N(\zeta)$, $x(\zeta)$, $\tilde{E}^2(\zeta)$ and $\tilde{S}(\zeta)$ depending on the dimensionless distance $\zeta = \frac{\nu}{\omega}\frac{z}{R}$ and on two parameters, $\sigma = \omega R/c$ and β, allow us to examine large number of experimental data on surface-wave-produced plasmas.

We emphasize that the theoretical plasma density profiles are not exactly linear which has been also shown in Refs. 5, 9, 10, 12. Nonlinear longitudinal electron number density profiles have been measured in the experimental studies of Nghiem et al.[21] and Chaker & Moisan,[22] where large values of $N \sim 10^3$ were achieved.

In view of the absence of marked beginnings and ends of the plasma columns in most of the experiments, it is convenient to compare the slopes of the axial electron number density profiles. The relation between the dimensionless slope $|dN/d\zeta|$ and the actual plasma density gradient is

$$\left|\frac{dn}{dz}\right|\left[cm^{-4}\right] = 1.97|dN/d\zeta|\frac{\omega}{2\pi}\left[GHz\right]\nu\left[s^{-1}\right]/R\left[cm\right] \qquad (27)$$

The results of the comparison between the slopes of the theoretical profiles for the free-fall/diffusion regime and the available experimental density gradients show very good agreement that can be seen in Table III. In calculating $|dN/d\zeta|$ we used slopes of curves[17] (similar to those in Fig. 1) that are averaged within the experimentally examined intervals of plasma density variation. The collision frequency ν has been determined from the gas-discharge theory at a regime of free-fall/diffusion of the charged particles to the walls of the

Table III

Wave frequency $\omega/2\pi$ Tube radius R Kind of the gas & reference number	$\sigma = \frac{\omega R}{c}$	Pressure p [m torr]	Collision frequency ν $[s^{-1}]$	$\left\|\frac{dN}{d\zeta}\right\|$	$\left\|\frac{dn}{dz}\right\|_{theor}$ $[cm^{-4}]$	$\left\|\frac{dn}{dz}\right\|_{exper}$ $[cm^{-4}]$
$\omega/2\pi$ = 27 MHz R = 3.20 cm Argon[22]	0.018	20	1.00×10^{8}	5.69	0.95×10^{7}	1.00×10^{7}
$\omega/2\pi$ = 50 MHz R = 3.20 cm Argon[22]	0.034	20	1.00×10^{8}	5.06	1.56×10^{7}	1.46×10^{7}

$\omega/2\pi$ = 100 MHz						
R = 3.20 cm	0.067	20	1.00×10^{8}	4.18	2.58×10^{7}	2.46×10^{7}
Argon[22]						
$\omega/2\pi$ = 2.45 GHz						
R = 0.15 cm	0.077	500	2.00×10^{9}	4.03	$2.60\ 10^{11}$	2.47×10^{11}
Argon[11]						
		20	1.00×10^{8}	3.37	1.92×10^{8}	2.07×10^{8}
$\omega/2\pi$ = 360 MHz		40	2.15×10^{8}	3.37	4.12×10^{8}	3.59×10^{8}
R = 1.25 cm	0.094	80	3.40×10^{8}	3.52	6.81×10^{8}	7.38×10^{8}
Argon[2]		150	5.30×10^{8}	3.65	1.10×10^{9}	1.09×10^{9}
		300	8.35×10^{8}	3.52	1.67×10^{9}	1.58×10^{9}
$\omega/2\pi$ = 510 MHz						
R = 1.10 cm	0.118	3	6.10×10^{7}	3.35	1.87×10^{8}	1.90×10^{8}
Xenon[15]						

ω/2π = 200 MHz						
R = 3.20 cm	0.134	20	1.00×10^{8}	3.29	4.07×10^{7}	4.68×10^{7}
Argon[22]						
ω/2π = 510 MHz		0.7	1.30×10^{7}	2.47	1.54×10^{7}	1.50×10^{7}
		3	4.70×10^{7}	2.47	5.57×10^{7}	5.40×10^{7}
R = 2.10 cm	0.224	20	2.10×10^{8}	2.47	2.49×10^{8}	2.50×10^{8}
Xenon[15]		50	4.40×10^{8}	2.47	5.22×10^{8}	5.20×10^{8}
ω/2π = 2.45 GHz						
R = 0.45 cm	0.231	200	9.10×10^{8}	2.58	2.53×10^{10}	2.88×10^{10}
Argon[4]						
ω/2π = 2.45 GHz		100	5.60×10^{8}	2.75	1.66×10^{10}	1.92×10^{10}
R = 0.45 cm	0.231	200	9.10×10^{8}	2.58	2.53×10^{10}	2.73×10^{10}
Argon[7]		500	1.66×10^{9}	2.40	4.29×10^{10}	3.91×10^{10}

container.[15,23] The values of ν for argon coincide with those obtained from the theoretical curve on Fig. 5 in Ref. 2 .

Among the large number of available experimental plasma density gradients only that corresponding to the surface-wave-produced plasma in a 12.4-cm i. d. tube[22] does not agree with the theoretical one calculated by expecting a diffusion gas-discharge regime. A resonable agreement can be obtained under the assumption that this case is a recombination-controlled rf discharge. From a theoretical curve $N(\zeta)$, calculated[17] for $\sigma = 0.260$, one finds $|dN/d\zeta| = 1.61$ and with $\nu = 1.04\times10^8\ s^{-1}$ and $\omega/2\pi = 200$ MHz we have $|dn/dz| = 1.07\times10^7\ cm^{-4}$, while the experimental value is $0.83\times10^7\ cm^{-4}$. The discrepancy between the theory and experiment is not larger than 30 %.

6. Conclusions

The theoretical model of surface-wave-produced plasma column presented here and its comparison with the available experimental data allow us to make the following conclusions:

(i) The axial electron number density profile of a surfatron (surfaguide) produced plasma column at low pressures ($\nu \ll \omega$) is completely determined by the wave frequency ω, the electron-neutral collision frequency ν, the tube radius R and the kind of the gas-discharge regime (free-fall/diffusion or recombination). The rf power transferred via the launcher is responsible for the magnitude of the column length and the electron number density at the launcher. In diffusion-controlled discharge the radially averaged square of the wave electric field in the plasma, E_o^2, is constant along the column (the constancy of the wave electric field along almost all the length of the plasma column in diffusion regime is discussed in Refs. 8 and 12, too). In recombination discharge conditions ($\beta = 1$) $E_o^2(\zeta)$ follows the course of the $N(\zeta)$-curve.

(ii) The linearity of the axial density profiles in a more or less wide range of plasma density variations might be used for a quick determination of the electron-neutral collision frequency ν as it was proposed for diffusion discharge regime by Glaude et al.[2] However, for recombination-controlled discharge one should bear in mind the possible weak dependence of ν on the electron number density.

(iii) It is necessary to do carefull experimental studies of surface-wave-produced plasma columns in recombination gas-discharge conditions. One needs more experimental data in order to prove the correctness of the available theories.

References

1. C. M. Ferreira and M. Moisan, "Properties and Modelling of Plasmas Produced by HF and Microwave Discharges"; M. Moisan and Z. Zakrzewski, "Design of SW Discharges to Obtain Columns of Specified Properties", in Proc. ICSW -- Ohrid '85 .

2. V. M. M. Glaude, M. Moisan, R. Pantel, P. Leprince, and J. Marec, J. Appl. Phys. 51, 5693 (1980).

3. C. M. Ferreira, J. Phys. D 14, 1811 (1981).

4. M. Chaker, P. Nghiem, E. Bloyet, Ph. Leprince, and J. Marec, J. Phys.-LETTRES 43, L-71 (1982).

5. V. Atanassov, E. Mateev, and I. Zhelyazkov, in Proc. 1982 ICPP -- Göteborg (Chalmers University of Technology, Göteborg 1982) p. 435;
E. Mateev, I. Zhelyazkov, and V. Atanassov, J. Appl. Phys. 54, 3049 (1983).

6. Yu. M. Aliev, A. G. Boev, and A. P. Shivarova, Phys. Lett. 92A, 235 (1982).

7. M. Chaker, P. Nghiem, E. Bloyet, P. Leprince, and J. Marec, in Proc. CSWP -- Blagoevgrad '81 (Sofia University, Sofia 1983) p. 280.

8. M. Chaker, P. Nghiem, E. Bloyet, P. Leprince, and J. Marec, in Proc. CSWP -- Blagoevgrad '81 (Sofia

University, Sofia 1983) p. 288.

9. Z. Zakrzewski, J. Phys. D 16, 171 (1983).

10. C. M. Ferreira, J. Phys. D 16, 1673 (1983).

11. S. Saada, E. Bloyet, C. Dervisevic, and C. Laporte, Rapport LP 202, Université de Paris-Sud (November 1983).

12. Z. Zakrzewski, Czech. J. Phys. B 34, 105 (1984).

13. Yu. M. Aliev, A. G. Boev, and A. P. Shivarova, J. Phys. D 17, 2233 (1984).

14. J. Wolińska-Szatkowska, Bull. Polish Acad. Sci., Tech. Sci. 32, 583 (1984).

15. V. Atanassov, PhD Thesis, Sofia University (November 1984).

16. Z. Zakrzewski and M. Moisan, in Proc. XVII ICPIG -- Budapest '85 (Organizing Committee ICPIG-XVII, Budapest 1985) Vol. 2, p. 762.

17. I. Zhelyazkov, E. Benova, and V. Atanassov -- to be published.

18. V. E. Golant, Usp. Fiz. Nauk 65, 39 (1958).

19. V. L. Granovskii, Electric Current in Gases (Nauka, Moscow 1971) Ch. 5.

20. A. G. Boev, Zh. Eksp. Teor. Fiz. 77, 92 (1979); Fizika Plazmy 8, 729 (1982).

21. P. Nghiem, M. Chaker, E. Bloyet, Ph. Leprince, and J. Marec, in Proc. CSWP -- Blagoevgrad '81 (Sofia University, Sofia 1983) p. 296.

22. M. Chaker and M. Moisan, J. Appl. Phys. 57, 91 (1985).

23. S. A. Self and H. N. Ewald, Phys. Fluids 9, 2486 (1966).

Figure Captions

Fig. 1 Normalized axial profiles of plasma density $N = n/n_{cr}$, surface-wave electric field amplitude $\tilde{E} = |E_z(R)|/E_*$ and wave power $\tilde{S} = S/S_*$ for free-fall/diffusion-controlled discharge ($\beta = 0$) at $\sigma \equiv \omega R/c = 0$, 0.05, 0.1, 0.2, 0.5, 1.0 and 2.0

Fig. 2 Normalized axial profiles of plasma density $N = n/n_{cr}$, surface-wave electric field amplitude $\tilde{E} = |E_z(R)|/E_*$ and wave power $\tilde{S} = S/S_*$ for recombination-controlled discharge ($\beta = 1$) at $\sigma \equiv \omega R/c = 0$, 0.05, 0.1, 0.2, 0.5, 1.0 and 2.0

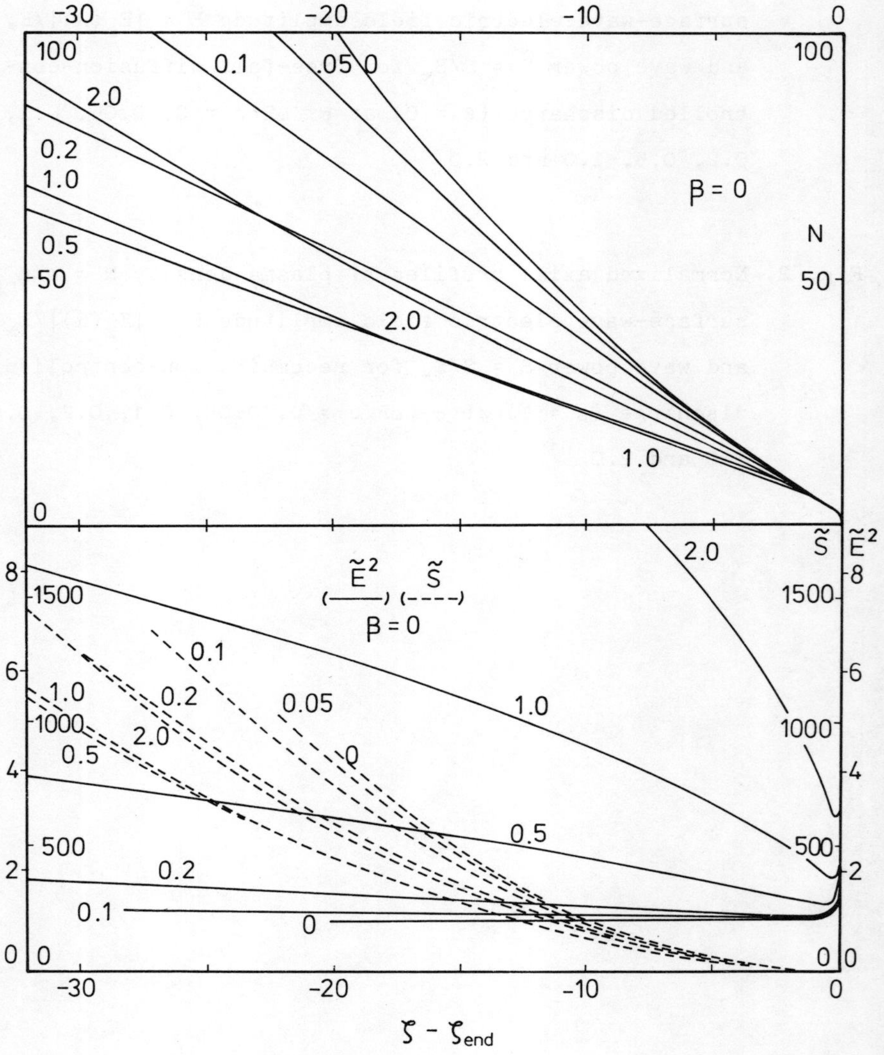

-30
-20
-10
0
100
0.1
.05
0
2.0
0.2
1.0
β = 0
0.5
N
50
2.0
1.0
0
$\tilde{S}$
$\tilde{E}^2$
8
1500
$\tilde{E}^2$ (——)
$\tilde{S}$ (- - - -)
β = 0
0.1
6
1.0
0.2
0.05
1000
2.0
0.5
4
500
0.2
2
0.1
0
$\zeta - \zeta_{end}$

Fig. 1

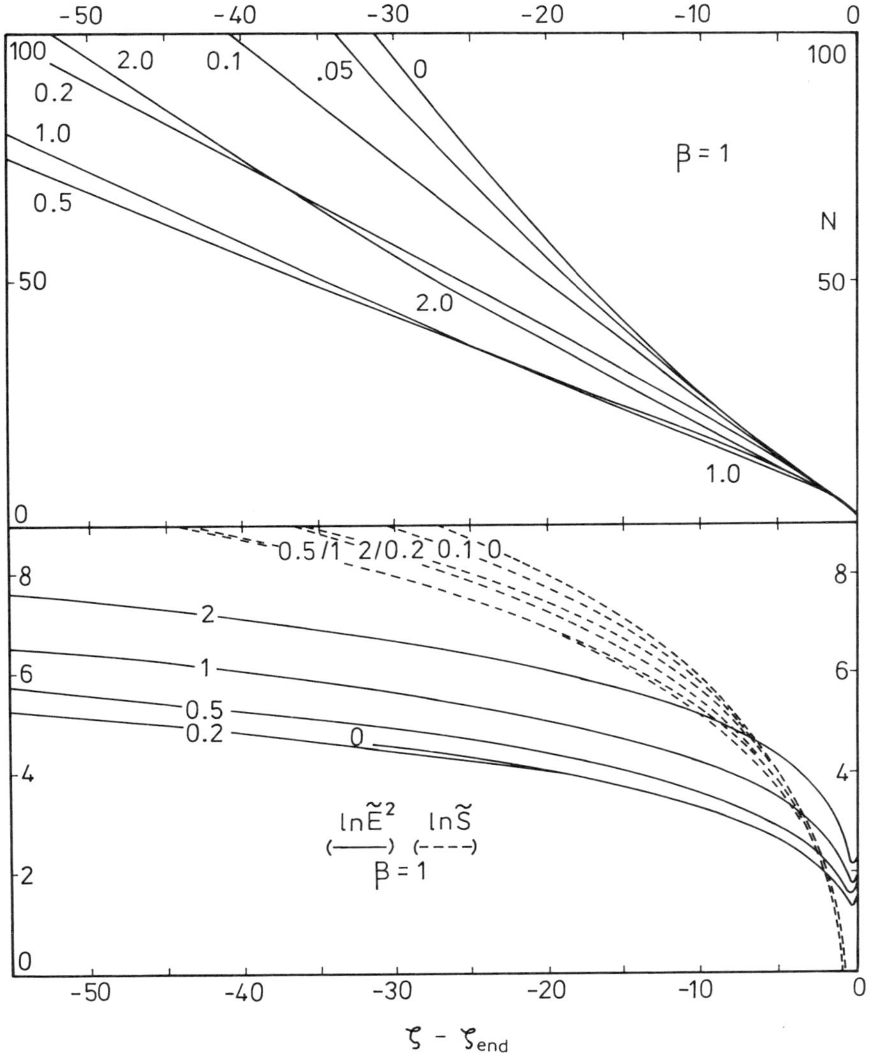
-50
-40
-30
-20
-10
0
100
2.0
0.1
.05
0
0.2
1.0
0.5
β = 1
N
50
2.0
1.0
0
0.5/1
2/0.2
0.1
0
8
2
1
6
0.5
0.2
0
4
$\ln\tilde{E}^2$ (——) $\ln\tilde{S}$ (- - - -)
β = 1
2
0
$\zeta - \zeta_{end}$

Fig. 2

Part II: Contributed Papers

ON THE METHOD OF DETECTION OF WAKE FORCES OSCILLATIONS IN THIN FOILS

E.A.Akopian, L.M.Gorbunov* and G.G.Matevossian
Institute of Radiophysics and Electronics Armenian Academy of Sciences, Ashtarak - 2, Armenian SSR, USSR.

The concept of a wake potential, originating behind a fast moving charged particle and due to polarization of a solid state electron plasma, has been first introduced by N.Bohr[1]. Wake potential expression have been considered in a number of papers [2-5] by means of various models for description of the electron plasma. Wake forces determined by this potential have been detected in numerous experiments, concerning the transmission of ion clusters through thin foils (see surveys [6,7]). However, the small size of clusters in these experiments made impossible the investigation of wake forces action at sufficiently long distances.

We suggest a method for determination of the long-range variable-sign wake forces, which is based on analysis of the fast molecular ion beam transmission through a pair of thin foils separated by a vacuum gap. Molecular ion beam with several Mev energy (same as in [6,7]) is passed through a thin foil of thickness L_0 whose only purpose is to strip the molecular ion of its valence electrones and to form a cluster of correlated ions. For simplicity, we will consider a bicluster of two identical particles. Inside the vacuum gap of thickness L_1 the distance between two ions in the increases because of action of the Coulomb repulsion forces. The value L_1 may be choose like that interaction between the ions inside the second foil of thickness L_2 is determined only by long-range wake forces.

In order to evaluate the features of interaction between the ions in the second foil we consider the dynamics of a single bicluster. The foil thickness L_0 assumed to be sufficiently small so that the distance r_0 between the ions in the

* Physics Institute Academy of Sciences, USSR, Moscow.

cluster and their velocity $u_\circ$ are practically the same as for the initial molecular ion. Bicluster orientation with respect to the velocity vector is determined by an angle α. After the transit time $L_1/u_\circ$ of the ion inside the vacuum gap the particles will become separated by a large enough distance $r=(w_\circ/u_\circ)L_1 \gg r_\circ$ where $w_\circ=\sqrt{4q^2/Mr}$, q and M are the particle charge and mass respectively. At the second foil boundary the particles and their velocitys along the OZ and OX axes will be respectively equal to: $z_\circ=r\mathrm{Cos}\,\alpha$, $x_\circ=r\mathrm{Sin}\,\alpha$, $v_x^{(1,2)}=\pm(w_\circ/2)\mathrm{Sin}\,\alpha$, $v_z^{(1,2)}=u_\circ\pm(w_\circ/2)\mathrm{Cos}\,\alpha$, where 1 and 2 indices refer to the leading and trailling particle in the cluster.

In the second foil interaction forces between the particles are determined by a wake potential. Following the work [2] we consider the electron plasma as a classical gas with the Maxvell distribution function. Then for small angles $\alpha \leqslant 1/\lambda$ (where $\lambda=u_\circ/v_{Te} \gg 1$ and v_{Te} is thermal velocity of plasma electrons) the potential taken at the leading and trailing ion position has the next form:

$$\varphi_1=q\left\{\frac{1}{\sqrt{z^2+x^2}}-\frac{1}{z}\left[1-\exp(-z/\lambda r_{De})\mathrm{ch}(x/r_{De})\right]\right\} \quad (1)$$

$$\varphi_2=\varphi_1-\frac{2q}{\lambda r_{De}}\mathrm{Sin}\left(\frac{z}{\lambda r_{De}}\right)\left[K_\circ\left(\frac{x}{\lambda r_{De}}\right)+\mathrm{ci}\left(\frac{x}{r_{De}}\right)\right] \quad (2)$$

where z and x are the longitudinal and transversal components of interpartial distant respectively, r_{De} is Debye length for electrones, $K_\circ$ is the modified Bessel function of the second kind, and ci is cosine integral.

By obtaining Eqs. (1) and (2), thermal motion of plasma particles has been taken into account in a more consistent way than in the paper [2]. Therefore Eq.(2) does not contain a singularity when x=0 and gives a good agreement with numerical results of the paper [8].

Having in mind the energy spectrum measurments for ions transmitted through the second foil in the incident beam direction, we will limit ourselves by consideration of biclusters with small transversal size ($x < r_{De}$). We also assume that the longitudinal size of cluster has small variation in process of its motion inside the foil L_2. Then we may expand Eqs. (1) and (2) in power series of variables x/r_{De} and $(z-z_o)/z_o$ and obtain simple equations of motion for two ions in the cluster. Upon solving these equations we find the velocity component of trailing ion $v_x^{(2)}$ undergoes the largest charge in process of transmission through the second foil, so that the trailing ion from the foil has the next form:

$$\beta = \frac{v_x^{(2)}}{u_o} = \frac{w_o}{2u_o}\text{Sin}\alpha\left[1 - \frac{w_o r_o L_2}{2r_{De}^2 u_o \lambda}\cdot\frac{w_o L_1}{\lambda r_{De} u_o}\text{Sin}\left(\frac{w_o L_1}{\lambda r_{De} u_o}\right)\right] \qquad (3)$$

As can be seen from Eq.(3) the value of angle β oscillates with variation of the inter-foil distance L_1. This result is due to oscillayions in the OZ direction of the wake potential created by the leading ion. Variation of L_1 leads to change of the cluster size z_o and the trailing ion inside the second foil may be found in the wake potential zones of various sign. The wake forces tend either to bring the trailing ion into the track of the leading ion, or to deflect it from track.

On basis of considerations given above, one may conclude that the energy spectrum of partiles emerging at the initial beam axis should exhibit two peaks (as in the experiments [6,7]). However, the peak height for the faster ions in practically independent of the vacuum gap width L_1, while for the slower ions the peak height oscillates when this distance varies.

As an example, we consider a molecular ion of energy 2Mev (r_o=1Å). Then under plasma frequency $\omega_p=2{,}6.10^{16}\text{sec}^{-1}$, $\lambda=9$ and L_2=200Å the wake forces focus the slower ions (β=0) when L_1=1 μk. On the contrary, one should expect a defocusing when L_1=0,5 μk.

Long-range wake forces have largest manifestation for small values of x_o and consequently for the clusters with small angles $\alpha < \alpha_o = (L_2 w_o r_o/2r_{De}^2\lambda u_o)$. Making use of parameter values assumed above we find $\alpha_o=5.10^{-2}$.

References:

1. Bohr N., Kgl. Danske Videnskab, Nat. Fis. Medd. 18, N 8, 71 (1948).
2. V.N. Neelavathi, R.H. Ritchie, W. Brandt, Phys. Rev. Lett. 33, 302 (1974).
3. M.N. Day, Phys. Rev. B12, 514 (1975).
4. R.H. Ritchie. W Brandt, P.M. Echenique, Phys. Rev. B14, 4808 (1976).
5. Z. Vager, D.S. Gemmel, Phys. Rev. Lett. 37, 1352 (1976).
6. D.S. Gemmel, Nucl. Instr. and Methods, 170, 41 (1980).
7. J. Remillieux, Nucl. Instr. and Methods, 170, 31 (1980).
8. Chong-Lung Wang, G. Joyce, D.R. Nicholson, J.Plasma Phys. 25, 225 (1981).

SURFACE ELECTROMAGNETIC WAVES IN THE WAVEGUIDES FILLED WITH SEMICONDUCTOR PLASMA

E. G. Aleksov and S. T. Ivanov
Faculty of Phisics, Sofia University
BG - 1126 Sofia, Bulgaria

At last years the interest toward the use to waveguide with semiconductor plasma for generation of the electromagnetic waves considerable increased. Achivements at the plasma electronic show /1/ that for succesful examples for generation power microwaves with waveguides filled with semiconductor is necessery a detail picture to dispersion of the waves.

In this paper, the slow electromagnetic surface waves in a metal waveguide of radius b, filled partially with a semiconductor of radius a (waveguide rode) and filled with a bush of semiconductor (waveguide with a bush) is considered. The systems are immersed in an external longitudinal magnetic field B_0.

Linearising Maxwell equations and momentum equation and assuming that all first-order quantities vary as exp(-i (ωt-kz-lφ)) the properties of the semiconductor plasma are expressed a dielectric permittivity tensor $\hat{\varepsilon}$: $\varepsilon_{xx}=\varepsilon_{yy}=\varepsilon_1$, $\varepsilon_{zz}=\varepsilon_3$, $-\varepsilon_{xy}=\varepsilon_{yx}=i\varepsilon_2$

Depending on the external magnetic field the following cases are studied:

a)nonmagnetized plasma, $\omega_c \ll \omega_p$ (where $\omega_c = eB_0/cm$ is the Larmour frequency, $\omega_p = (4\pi e^2 n/m)^{1/2}$ is the plasma frequency). For this case $\varepsilon_1=\varepsilon_3=\varepsilon_p \equiv \varepsilon_L - \omega_p^2/\omega^2$, $\varepsilon_2=0$ and ε_L is the high-frequency constant.

b)magnetized plasma($\omega_c \gg \omega_p$), $\varepsilon_1=\varepsilon_L$, $\varepsilon_2=0$, $\varepsilon_3=\varepsilon_L-\omega_p^2/\omega^2$

c)anysotropical plasma ($\omega_c \sim \omega_p$), $\varepsilon_1=\varepsilon_L-\frac{\omega_p^2}{\omega^2-\omega_c^2}$, $\varepsilon_2=\frac{-\omega_p^2\omega_c}{\omega(\omega^2-\omega_c^2)}$, $\varepsilon_3=\varepsilon_p$

The collision and the termal motion of the cariers in the plasma medium are neglested.

Adding the baundary conditions the Maxwell's equations to and solving them we obtain the dispersion equation:

$$\mathcal{D}(\omega,k_z)=\left\{\Phi-(\beta_1-\beta_2)\frac{\omega}{c}\left[\frac{l}{a}\frac{\omega^2}{c^2}\varepsilon_2\varkappa_v^2+\gamma\varkappa_v f(\varkappa_v a)\right]\right\}\cdot\left\{\Psi 1+(\beta_1-\beta_2)\frac{\omega}{c}\left[\frac{l}{a}k_z^2\varepsilon_2\varkappa_v^2+\gamma\varkappa_v q(\varkappa_v a)\right]\right\}-$$

$$-\left\{\Phi 1+(\beta_1-\beta_2)\frac{\ell}{a}k_z(\varkappa^2\varkappa_v^2-\zeta)\right\}\left\{\Psi-(\beta_1-\beta_2)\frac{\ell}{a}k_z(\varkappa^2\varkappa_v^2-\zeta)\right\}\equiv 0 \qquad (1)$$

there: $\varkappa^2=k_z^2-\varepsilon_1\frac{\omega^2}{c^2}$, $\varkappa_v^2=k_z^2-\frac{\omega^2}{c^2}$, $\zeta=\varkappa^4-\varepsilon_2^2\frac{\omega^4}{c^4}$, $\beta_{1,2}=\frac{\omega}{ck_z\varepsilon_2}(\varepsilon_1\varkappa^2+\varepsilon_3 k_{1,2}^2)$

$$k_{1,2}^2=\left\{-(\varepsilon_1+\varepsilon_3)\varkappa^2-\frac{\omega^2}{c^2}\varepsilon_2^2\pm\sqrt{\left[(\varepsilon_1-\varepsilon_3)\varkappa^2+\frac{\omega^2}{c^2}\varepsilon_2^2\right]^2+4k_z^2\varepsilon_2^2\frac{\omega^2}{c^2}\varepsilon_3}\right\}/2\varepsilon_1 \qquad (2)$$

$$\Phi=\frac{\omega}{c}\varkappa^2\varkappa_v^2\left(\beta_1 k_1 P(k_1a)-\beta_2 k_2 P(k_2a)\right)+k_z\frac{\omega^2}{c^2}\varepsilon_2\varkappa_v^2\left(k_1P(k_1a)-k_2P(k_2a)\right)$$

$$\Psi 1=\frac{\omega}{c}\left(\varkappa^2\varepsilon_1+\varepsilon_2^2\frac{\omega^2}{c^2}\right)\varkappa_v^2\left(\beta_2k_1P(k_1a)-\beta_1k_2P(k_2a)\right)+k_z\frac{\omega^2}{c^2}\varkappa_v^2\varepsilon_2\beta_1\beta_2\left(k_1P(k_1a)-k_2P(k_2a)\right)$$

$$\Phi 1=\frac{\omega}{c}\varkappa^2\varkappa_v^2\beta_1\beta_2\left(k_1P(k_1a)-k_2P(k_2a)\right)+k_z\frac{\omega^2}{c^2}\varepsilon_2\varkappa_v^2\left(\beta_2k_1P(k_1a)-\beta_1k_2P(k_2a)\right)$$

$$\Psi=\frac{\omega}{c}\left(\varkappa^2\varepsilon_1+\varepsilon_2^2\frac{\omega^2}{c^2}\right)\varkappa_v^2\left(k_1P(k_1a)-k_2P(k_2a)\right)+k_z\frac{\omega^2}{c^2}\varepsilon_2\varkappa_v^2\left(\beta_1k_1P(k_1a)-\beta_2k_2P(k_2a)\right)$$

and for waveguide rode:

$$f(\varkappa_v a)=\frac{J'(\varkappa_v a)N'(\varkappa_v b)-J'(\varkappa_v b)N'(\varkappa_v a)}{J(\varkappa_v a)N'(\varkappa_v b)-J'(\varkappa_v b)N(\varkappa_v a)},\quad q(\varkappa_v a)=\frac{J'(\varkappa_v a)N(\varkappa_v b)-J(\varkappa_v b)N'(\varkappa_v a)}{J(\varkappa_v a)N(\varkappa_v b)-J(\varkappa_v b)N(\varkappa_v a)},\quad P(x)=\frac{J'(x)}{J(x)}$$

and for waveguide with a bush:

$$P(x)=\left[J'(x)+\frac{\frac{\ell}{b}\frac{\omega^2}{c^2}\varepsilon_2 J(k_{1,2}b)-\varkappa^2\varepsilon_1J'(k_{1,2}b)}{\varkappa^2k_{1,2}N'(k_{1,2}b)-\frac{\ell}{b}\frac{\omega^2}{c^2}\varepsilon_2N(k_{1,2}b)}\cdot N'(x)\right]\Big/\left[J(x)+\frac{\frac{\ell}{b}\frac{\omega^2}{c^2}\varepsilon_2J(k_{1,2}b)-\varkappa^2\varepsilon_1J'(k_{1,2}b)}{\varkappa^2k_{1,2}N'(k_{1,2}b)-\frac{\ell}{b}\frac{\omega^2}{c^2}\varepsilon_2N(k_{1,2}b)}N(x)\right]$$

$$f(x)=q(x)=J'(x)/J(x)$$

All graph are shown in variables - wavelenght $\Lambda=\frac{\lambda}{b}=\frac{2\pi c}{\omega b}$ and retardation $n=\frac{c}{V_{ph}}=\frac{ck_z}{\omega}$

Let us consider the case of nonmagnetized plasma. There are symetrical E-wave and nonsymetrical EH-waves. The area, where dispersion equation (1) is possible to be solved for surfase modes is shown at fig.1a. The numerical results are displayed on fig.1b. There t=a/b. There are waak dependence at acsial waves number l. The wave, which have biger l is more low-frequency (for waveguide with a bush) or is more high-frequensy (for waveguide rode).

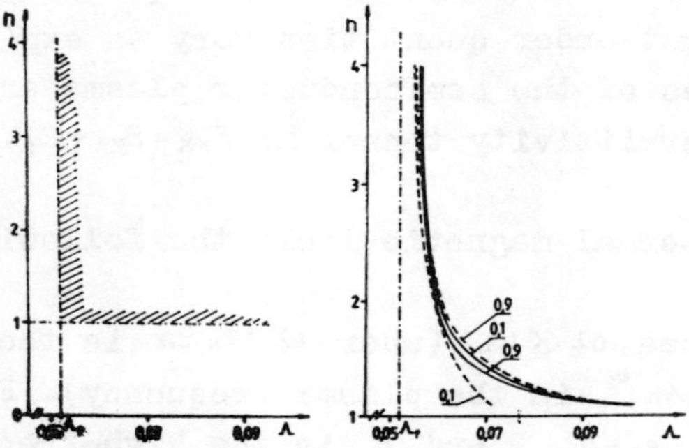

Fig.1. Nonmagnetized plasma.
a) the region where can be exist surface waves.
b) the dispersion curves E_0-wave for waveguide rode(——) and waveguide with a bush(– –).

Let us consider second case - magnetized plasma. There are only symetrical surface wave. In the area, where is propagated E_0-mode (fig.2a), the component B_z of the electromagnetic field have framework spase. Therefore the nonsymetrical waves, which have E_z, B_z and all transverse components, is not surface mode. The dispersion curves are displayed on fig. 2b.

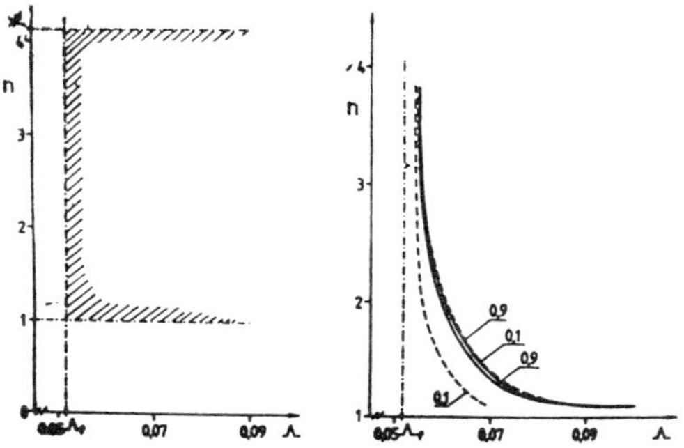

Fig.2. Magnetized plasma.
a) the region where can be exist surface wave.
b) the dispersion curves E_0-wave for waveguide rode (——) and waveguide with a bush (– – –).

Now, let us consider last case -anysotropical plasma. Here all waves, symetrical and nonsymetrical, are EH-modes. The arguments of the Bessel function (2) can be real, imaginary or complex number. When k_1, k_2 and $\varkappa_v$ are imaginary or complex number we have surface waves. On fig.3 is shown thearea, where the dispersion equation may have solutions for surface waves. The same there is shown the area, where k_1 and k_2 have complex value. This area we shall call the region opaqueness. There it may exist radiative modes, i.e. κ_1 or k_2 can be complex numbers.

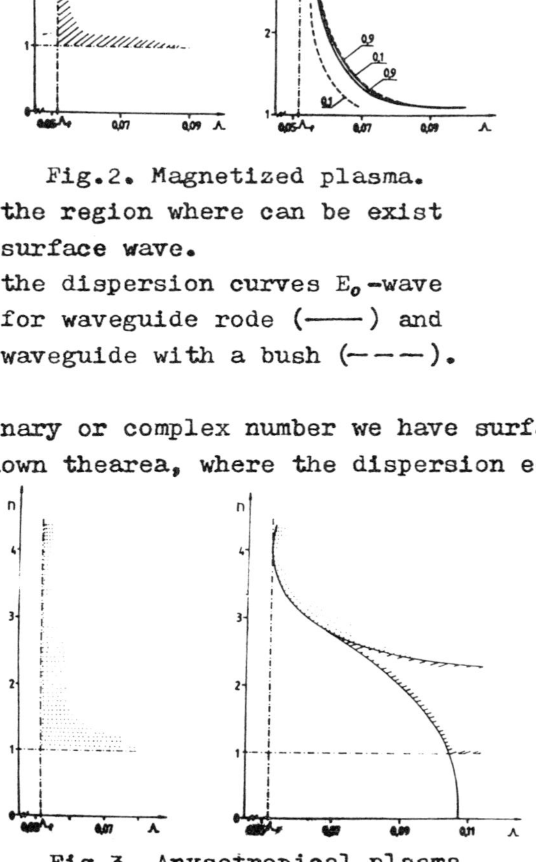

Fig.3. Anysotropical plasma regions where can be exist surface waves. a) case $\omega_c < 2\omega_p/(\varepsilon_L+1)$; $\omega_c = 10^{10} s^{-1}$
b) case $\omega_c > 2\omega_p/(\varepsilon_L+1)$; $\omega_c = 4.10^{12} s^{-1}$
⊠-region opaqueness

Let us continue the analysis to anysotropical waveguide when $\omega_c < 2\omega_p/(\varepsilon_L+1)$. Then the part of the region opaqueness get into the area, where P<1 - fast waves (see /2/). And now it isn't observed a general diferents between waveguide rode and waveguide with a bush. Therefore below the results we shall give only for waveguide rode. The dispersion curves are gived on fig.4a. One can see, that there exist one type modes. In this area the arguments of the Bessel function are complex value. As a consequence from this, the dispersion equation becomes a complex number, too. At the approximation, where we have been work, i.e. when we expand the cylindrical function in order toward the small complex addition, turn out that the imaginary part of dispersion equation is identical zero. In the expression for the real part of equation (1) ω joins in the even powers. It does not present an opening for the determination to type of this veriable. Usually in the literature for this area it consider that there are radiative waves. We assume, that descovered waves are from the same type.

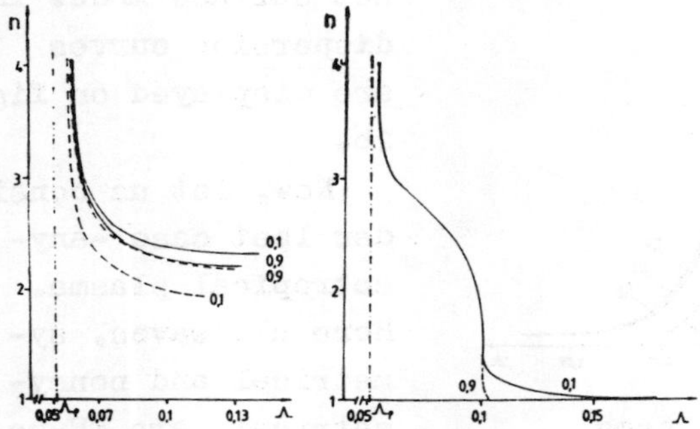

Fig.4. The dispersion curves for waveguide rode, symetrical wave (———) and nonsymetrical wave (— — —). a/ $\omega_c = 10^9 s^{-1}$ b) $\omega_c = 4.10^{12} s^{-1}$

Let us consider the second situation, when $\omega_c > 2\omega_p/(\varepsilon_L+1)$ in this case there are two types waves. Waves which were discovered at $\omega_c < 2\omega_p/(\varepsilon_L+1)$ and waves which lie outside at the region opaqueness .On fig.4б is shown the dispersion for waveguide rode. There are not displayed the dispersion curves for radiative modes because their wavelenght at big B_0 strong increase.

The calculatings are made for: $b=0.015m$, $\omega_p=10^{13}s^{-1}$, $\varepsilon_L=17$

REFERENCES

1.M.V.Kuzelev,F.H.Muhamedzanov,M.S.Rabinovich,A.A.Ruhadze, P.S.Strelkov and A.G.Shkvarunets,ZtETF,4_,83,(1982)

2;E.G.Aleksov,S.T.Ivanov and A.B.Shvachka,Preprint Dubna, P 1184776,(1984)

SURFACE ELECTROMAGNETIC WAVES ON A SEMICONDUCTOR CYLINDER

E. G. Aleksov, S. T. Ivanov and M. R. Nenkov
Faculty of Phisics, Sofia University
BG - 1126 Sofia, Bulgaria

The possibility of excite electromagnetic waves by interaction between a relativistic electron beam and a solid state plasma is of large practical interest. The significant advantages and wide perspectives for the generation of the millimeter microwave radiation have stimulated the study of the dispersion on the electromagnetic wave in a limited systems.

This paper presents a numerical study on the dispersion of a surface wave on a semiconductor cylinder with radius R. The system is immersed in a longitudinal magnetic field B_0 so that $\omega_c = eB_0/cm$ (ω_c is the Larmour frequency). It consider cold collisionless plasma with plasma frequency $\omega_p = (4\pi e^2 n/m)^{1/2}$

The propagate of surface electromagnetic waves on a cylinder from metal /1/ or semiconductor /2/ is a subject of scientific research. In this work we do an attempt for detailed study of the problem for semiconductor cylinder.

The properties of the semiconductor plasma are expressed throught a dielectric permittivity tensor $\hat{\varepsilon}$: $\varepsilon_{xx} = \varepsilon_{yy} = \varepsilon_1$, $\varepsilon_{zz} = \varepsilon_3$, $-\varepsilon_{xy} = \varepsilon_{yx} = i\varepsilon_2$

Depending og the external magnetic field the following cases are investigated:

a)nonmagnetized plasma, $\omega_c \ll \omega_p$, $\varepsilon_1 = \varepsilon_3 = \varepsilon_p \cong \varepsilon_L - \omega_p^2/\omega^2$, $\varepsilon_2 = 0$

Here ε_L is a high-frequency dielectric constant.

b)magnetized plasma, $\omega_c \gg \omega_p$, $\varepsilon_1 = \varepsilon_L$, $\varepsilon_2 = 0$, $\varepsilon_3 = \varepsilon_p$

c)anysotropical plasma, $\omega_c \sim \omega_p$, $\varepsilon_1 = \varepsilon_L - \frac{\omega_p^2}{\omega^2 - \omega_c^2}$, $\varepsilon_2 = \frac{-\omega_p^2 \omega_c}{\omega(\omega^2 - \omega_c^2)}$, $\varepsilon_3 = \varepsilon_p$

Using the Maxwell equations and the boundary continuity conditions for the tangential components of the electromagnetic field, one can easily show that in a linear approach the dispersion relation is of the following form:

$$\mathcal{D}(\omega, k_t) = \{\Phi - (\beta_1 - \beta_2)\frac{\omega}{c}[\frac{\ell}{R}\frac{\omega^2}{c^2}\varepsilon_2 \varkappa_v^2 + \xi \varkappa_v f(\varkappa_v R)]\}\{\Psi 1 + (\beta_1 - \beta_2)\frac{\omega}{c}[\frac{\ell}{R}k_t^2 \varepsilon_2 \varkappa_v^2 + \xi \varkappa_v q(\varkappa_v R)]\} -$$
$$- \{\Phi 1 + (\beta_1 - \beta_2)\frac{\ell}{R}k_t(\varkappa^2 \varkappa_v^2 - \xi)\}\{\Psi - (\beta_1 - \beta_2)\frac{\ell}{R}k_t(\varkappa^2 \varkappa_v^2 - \xi)\} = 0 \qquad (1)$$

there: $\varkappa^2 = k_z^2 - \varepsilon_1 \frac{\omega^2}{c^2}$, $\varkappa_v^2 = k_z^2 - \frac{\omega^2}{c^2}$, $\eta = \varkappa^4 - \varepsilon_2^2 \frac{\omega^4}{c^4}$, $\beta_{1,2} = \frac{\omega}{c k_z \varepsilon_2}(\varepsilon_1 \varkappa^2 + \varepsilon_3 k_{1,2}^2)$

$$k_{1,2}^2 = \left\{ -(\varepsilon_1+\varepsilon_3)\varkappa^2 - \frac{\omega^2}{c^2}\varepsilon_2^2 \pm \sqrt{\left[(\varepsilon_1-\varepsilon_3)\varkappa^2 + \frac{\omega^2}{c^2}\varepsilon_2^2\right]^2 + 4k_z^2\varepsilon_2^2\frac{\omega^2}{c^2}\varepsilon_3} \right\} / 2\varepsilon_1 \qquad (2)$$

$$\Phi = \frac{\omega}{c}\varkappa^2\varkappa_v^2\left[\beta_1 k_1 P(k_1R) - \beta_2 k_2 P(k_2R)\right] + k_z\frac{\omega^2}{c^2}\varepsilon_2\varkappa_v^2\left[k_1 P(k_1R) - k_2 P(k_2R)\right]$$

$$\Psi_1 = \frac{\omega}{c}\left(\varkappa^2\varepsilon_1 + \varepsilon_2^2\frac{\omega^2}{c^2}\right)\varkappa_v^2\left[\beta_2 k_1 P(k_1R) - \beta_1 k_2 P(k_2R)\right] + k_z\frac{\omega^2}{c^2}\varkappa_v^2\varepsilon_2\beta_1\beta_2\left[k_1 P(k_1R) - k_2 P(k_2R)\right]$$

$$\Phi_1 = \frac{\omega}{c}\varkappa^2\varkappa_v^2\beta_1\beta_2\left[k_1 P(k_1R) - k_2 P(k_2R)\right] + k_z\frac{\omega^2}{c^2}\varepsilon_2\varkappa_v^2\left[\beta_2 k_1 P(k_1R) - \beta_2 k_2 P(k_2R)\right]$$

$$\Psi = \frac{\omega}{c}\left(\varkappa^2\varepsilon_1 + \varepsilon_2^2\frac{\omega^2}{c^2}\right)\varkappa_v^2\left[k_1 P(k_1R) - k_2 P(k_2R)\right] + k_z\frac{\omega^2}{c^2}\varepsilon_2\varkappa_v^2\left[\beta_1 k_1 P(k_1R) - \beta_2 k_2 P(k_2R)\right]$$

$$f(\varkappa_v R) \equiv q(\varkappa_v R) = N'(\varkappa_v R)/N(\varkappa_v R), \quad P(x) = J'(x)/J(x)$$

All graph are shown in variables - wavelenght $\Lambda = \frac{\lambda}{R} = \frac{2\pi c}{\omega R}$ and retardation $n = \frac{c}{V_{ph}} = \frac{c k_z}{\omega}$

a) let us consider the case of nonmagnetized plasma. The analysis of the dispersion equation (1) shows that surface waves which exponentially drops with field outside the sample surface, may propagate throught the system at the frequency $\omega < \omega_p/\sqrt{\varepsilon_L}$ ($\Lambda > \Lambda_p = \frac{2\pi c\sqrt{\varepsilon_L}}{\omega_p R}$) fig.1a. All surface waves in this case at big k their frequency acumptotical aspired to $\omega_p/\sqrt{\varepsilon_L}$.

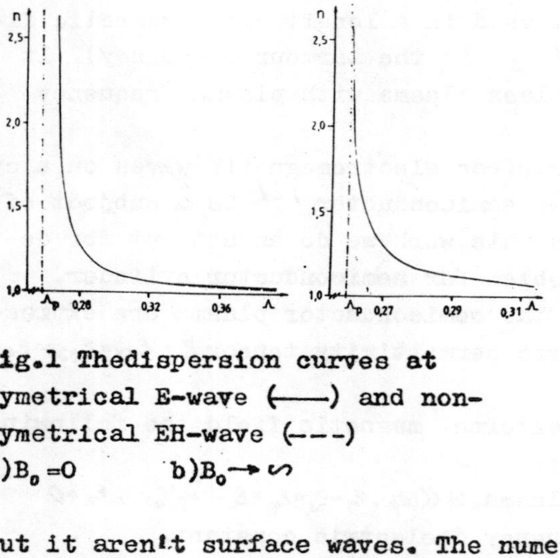

Fig.1 Thedispersion curves at symetrical E-wave (——) and nonsymetrical EH-wave (- - -)
a) $B_0 = 0$ b) $B_0 \to \infty$

b) let us consider the case of magnetized plasma. As it follows from the numerical analysis, here is only one surfase wave E_0. It must be noted, that in this area are propagate many nonsymetrical HE-wave, but it aren't surface waves. The numerical results for E_0-wave are shown on fig.1b. One can easily see, that at big k_z the frequency of the wave asumptotical is aspired to $\omega_p/\sqrt{\varepsilon_L}$. The area, where this wave is surface is limited by the relation: $1 < n < \sqrt{\varepsilon_L}$. When $k_z \to \infty$ the phase velosity asumptotical is aspired to zero and for $V_{ph} < \frac{c}{\sqrt{\varepsilon_L}}$ the wave isn't surface.

c) let us consider the case of anysotropical plasma. Must be define the frontiers of the area, where can be existed surface waves, before begining the numerical analys. It must the arguments k_1 and k_2 be imaginery or complex numbers, in order to propagate the surface wave. Using the equation (2) determine this regions. The area, where k_1 and k_2 are complex number will be cold the region opaqueness. Her asumptotical are described by solution /3/:

$$V_{ph_{1,2}} = \frac{c}{\left[\mp \frac{\omega_p}{\omega_c} + \sqrt{\frac{\omega_p^2}{\omega_c^2} + \varepsilon_L}\right]}$$

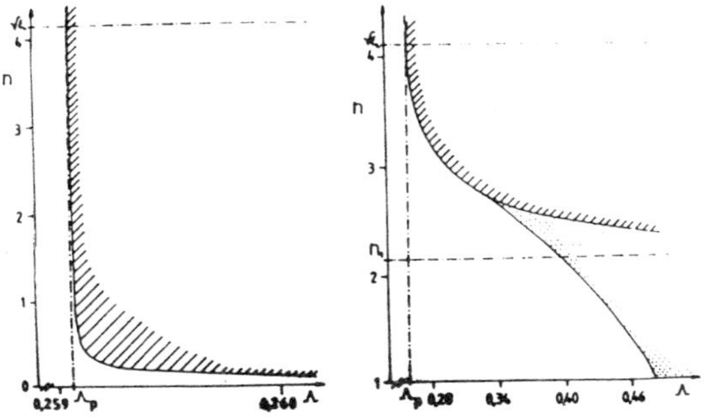

Fig.2.The regions where can be exist surface waves, ▨ region opaqueness
a) $\omega_c = 10^{10}\ s^{-1}$ b) $\omega_c = 3.5 \cdot 10^{12}\ s^{-1}$

On fig.2 are shown the regions for two cases.

Let us continue the numerical analys with case, when $\omega_c < 2\omega_p/(\varepsilon_L+1)$. Then the surface wave can be existed only in region opaqueness. Just that result is gived by numerical analysis (fig.3). This wave appear naturally continuation on the surface wave by a case of nonmagnetized plasma. Here the dispersion equation are gived by complex relation.

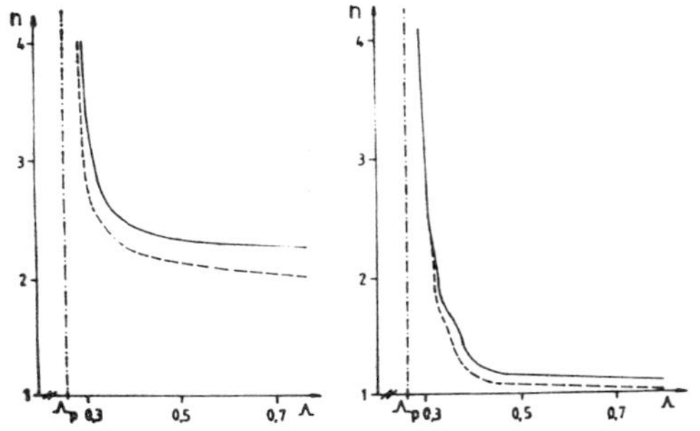

Fig.3.The dispersion curves for symetrical E_0-wave and nonsymetrical EH_1-wave.
a) $\omega_c = 10^{10} s^{-1}$ b) $\omega_c = 10^{8} s^{-1}$

This waves we will called a radiative modes although in our approximation be found only a module to ω. When the external magnetic field B_0 is increased above value $\omega_c > 2\omega_p/(\varepsilon_L+1)$ it appears a new surface wave. Her dispersion curve is outside (for P≈1) of the region opaqueness (fig.4).

Let us studied variation on the dispersion curves of the surface waves when the external magnetic field B alters by $B_0 = 0$ to $B_0 \to \infty$.

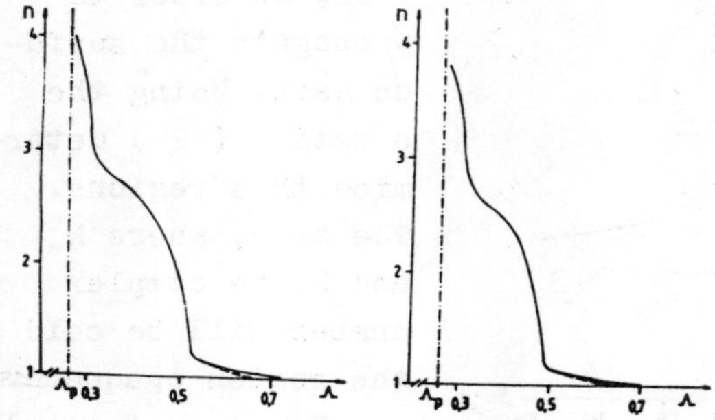

Fig.4. The dispersion curves for symetrical E_0-wave and nonsymetrical EH_1-wave.
a) $\omega_c = 4.10^{12} s^{-1}$ b) $\omega_c = 3.5.10^{12} s^{-1}$

In case of $B = 0$ is shown that one type surface wave exist There we neglest the component ε_2 of the dielectric permittivity tensor. When $B_0 > 0$ and we include the component ε_2 the prinsipal changes for dispersion curves of this type are not observed. But now, in contrast to to nonmagnetized plasma, there the arguments of the Bessel function are complex numbers. At the most phase velosity of this wave are defined by condition:

$$V_{ph} < c \Big/ \left[-\frac{\omega_p}{\omega_c} + \sqrt{\frac{\omega_p^2}{\omega_c^2} + \varepsilon_L} \;\right]$$

When $B_0 \to \infty$, i.e. $\frac{\omega_p}{\omega_c} \to 0$ the region opaqueness are aproached on the line $n = \sqrt{\varepsilon_L}$. Then this surface waves are disappeared. Let us consider the second type surface waves, which appear when $\omega_c > 2\omega_p/(\varepsilon_L + 1)$. Increasing B_0, their dispersion curves become low-frecuency and when $B_0 \to \infty$ they are disappeared, too. Hence at extremaly strong magnetic field, there surface waves aren't. The presence of the surface E_0-mode at $B_0 \to \infty$ be due to approximation for nonlimited strong external magnetic field.

In conclusion we should be noted, that the numerical value for $R = 0.003$m, $\omega_p = 10 s^{-1}$ and $\varepsilon_L = 17$ don't exercise influence on the received results.

R E F E R E N C E S

1. A.D.Boardman, G.C.Aers and B.V.Paranjape, J.Phys.F., 10, 53
2. R.D.Birkhoff, J.C.Ashley, H.H.Hubell and L.C.Emerson, J.Opt. Soc., 67, 564, (1977)
3. E.G.Aleksov, S.T.Ivanov and A.B.Shvachka, Preprint Dubna, P 1184776, (1984)

NONLINEAR INTERACTION OF E.M. RADIATION WITH STRONGLY INHOMOGENEOUS PLASMA

by

Yu. M. ALIEV, A.A. FROLOV
P.N. Lebedev Physical Institute, Moscow, USSR

A.C. SHIKLIN
State University, Kharkov, USSR

Interaction of e.m. radiation with strongly inhomogeneous plasma under condition when exitation of surface waves is essential is investigated. Two mechanisms of the exitation are considered: resonant and parametric. Especially generation of the zeroth and the second harmonic of the pump wave are analysed. Spacial structure of the harmonics under resonance exitation of leaky surface wave has been discovered. In the case of parametric resonance the conditions of emission of the second harmonic and quasi-static e.m. structures are found.

SLOW NONLINEAR E.M. WAVES ALONG THIN PLASMA LAYER

by

Yu.M. ALIEV and S.V. KUSNEZOV

P.N. Lebedev Physical Institute, Academy of Sciences of USSR, Moscow, USSR.

A. P. SHIVAROVA

Faculty of Physics, Sofia University, 5A. Ivanov Blvd., BG-1126 Sofia, Bulgaria

The theory of envelope surface wave solitons in thin plasma layer is presented. The ponderomotive and relativistic nonlinearity are taken into account. The behaviour of near ion-sound solitons is investigated.

ON THE NONLINEAR THEORY OF ABSORPTION OF INTENSE ELECTROMAGNETIC WAVES IN AN INHOMOGENEOUS PLASMA

by

Yu.M. ALIEV, A.A. FROLOV

Lebedev Physical Institute of Academy of Sciences of the USSR,
SU-117924 Moscow USSR

A.A. ZHAROV and I.G. KONDRATJEV

Institute of Applied Physics, Academy of Sciences of the USSR,
46 Uljanov Street, Gorky, USSR

Ponderomotive force-produced step distribution of the permitivity ε (see fig. 1) in the vicinity of plasma resonance [1] may increase the resonance absorption of p-polarized radiation [2]. The numerical calculations [3] indicate that nearly total absorption is achieved for optimal (maximum linear absorption) angle of incidence $\theta = \theta_0$ of a plane wave onto weakly inhomogeneous plasma at moderate values of incident power. The matching effect is explained by synchronized excitation of a leaking quasi-static mode on the density plateau of low-permittivity plasma [4]. Note that our approach and result differ from Ref. 5.

We'll demonstrate that there is another region of strong absorption, which arises from synchronized excitation of a quasi-surface mode and corresponds to stronger deformation of the density profile and high values of the incident wave amplitude [2,3].

Let an intense monochromatic ($\exp i\omega t$) plane wave of p-polarization (B_y, E_x, E_z) be incident onto a plasma with the linear density profile $n_0 = n_c(1+Z/\ell)$ from vacuum ($Z<-\ell$).

The plasma field equations, with striction nonlinearity taken into account, are given by

$$\frac{d^2B_y}{dZ^2} + \frac{1}{\varepsilon}\frac{d\varepsilon}{dZ}\frac{dB_y}{dZ} + \kappa_o{}^2(\varepsilon - \gamma^2)\ B_y = 0$$

$$E_x = -\frac{1}{i\kappa_o\varepsilon}\frac{dB_y}{dZ}\ ,\ E_Z = \frac{\gamma}{\varepsilon}B_y,\ \gamma = \sin\theta,\ \kappa_o = \frac{\omega}{c} \tag{1}$$

$$\varepsilon = 1 - \frac{n}{n_c} - i\nu\frac{n}{n_c}\ ,\ n = n_c\left(1 - \frac{/E_x/^2 + /E_Z/^2}{E_p{}^2}\right),\ E_p^2 = \frac{4m\omega^2 T e}{e^2}\ ,\ \nu = \frac{\nu_{eff}}{\omega}$$

(common factor $\exp(-i\kappa_o\gamma x)$ is omitted).

Equation (1) is solved in a manner similar to that in [3]. The calculation results are displayed in fig. 2,3.

Figure 2 shows lines of constant energy absorption coefficient Q in the parameter plane $\sin^2\theta$, $\ell g(E_o^2/E_p^2)$ for a plasma with the characteristic inhomogeneity scale $\ell = 50\kappa_o^{-1}$, $\nu = 10^{-3}$.

The dashed line, which separates the region of closed curves (separatrix), corresponds to an absorption coefficient slightly exceeding the optimal value $Q_o = 0,5$ in the linear case. Two groups of curves inside the separatrix correspond to radically different types of resonance absorption due to the excitation of a quasi-static mode on the plateau [4] (relatively low power) and of a quasi-surface mode on the density discontinuity (relatively high power). The central points of each group indicate ideally matching (reflectionless) absorption (Q=1). Note that these points correspond to powers which differ by almost three orders of magnitude. The position of the second point of interest associated with another region of consistency can be estimated analytically in the framework of WKP approximation, which holds for rather high density discontinuities. Conditions for total absorption of an incident wave due to synchronized excitation of a quasi-surface mode for model (fig. 1b) permittivity distribution are given by [2,6]

$$\gamma^2 = \sin^2\theta = \frac{\varepsilon_1\varepsilon_2}{\varepsilon_1 + \varepsilon_2} \simeq 2\varepsilon_1$$

$$\nu = 4\varepsilon_1 \exp\{-\frac{4}{3}\kappa_0\ell\,(\gamma^2 - \varepsilon_1)^{3/2}\} \tag{2}$$

where it is taken into account that $\varepsilon_2 \simeq -2\varepsilon_1$, [1,4], $\nu \ll 1$.

The longitudinal component D of the electric field induction on the density discontinuity can be determined from

$$D^2 = \frac{4E_0\varepsilon_1^2}{\nu}\,\gamma^2\,(1 - \gamma^2)\,(\gamma^2 - \varepsilon_1)^{-\frac{1}{2}} \tag{3}$$

where, in a quasi-static limit

$$\varepsilon_1 = 2^{-3/2}(D^2/E_p^2)^{1/3}$$

For comparison, fig. 3 shows dependences of absorption coefficient on incident wave amplitude for optimal angle of incidence $\theta_0 \simeq 0,19$ (curve 1) and $\theta_1 \simeq 0,23$, an angle exceeding the optimal one (curve 2). It is apparent that in the second case the absorption coefficient can exceed 50% in the wide range of incident power.

Thus, it is seen that strong absorption may occur at high levels of radiation power. Such a possibility may help one in search of optimal conditions for energy transfer of powerful electromagnetic field to plasma.

The authors are thankful to Drs. V.B. Gil'denburg and A.V. Khimich for useful discussions.

References

1. Gil'denburg V.B., JETF, 46, 2156 (1964).

2. Zharov A.A., Kondratjev I.G., Miller M.A., Pis'ma v JETP 25, 355 (1977), Fizika Plasmy 5, 261(1979).

3. Gil'denburg V.B., Litvak A.G., Petrova T.A., Feigin A.M., Fizika Plasmy 7, 732 (1981).

4. Sakhorov A.S., Prepring FIAN No. 190, Moscow, 1979; Kotov A.K., Fizika Plazmy 11, 636 (1985).

5. Vuković S., Ramazashvili R.R., Cadeẑ V., Aleksich A., Kasimov Zh.Zh., Phys. Lett. 102A, 186 (1984).

6. Aliev Yu.M., Vuković S., Gradov O.M., Kyrie A.Yu., Cadeẑ V. Phys. Rev. A 17, 2120 (1977).

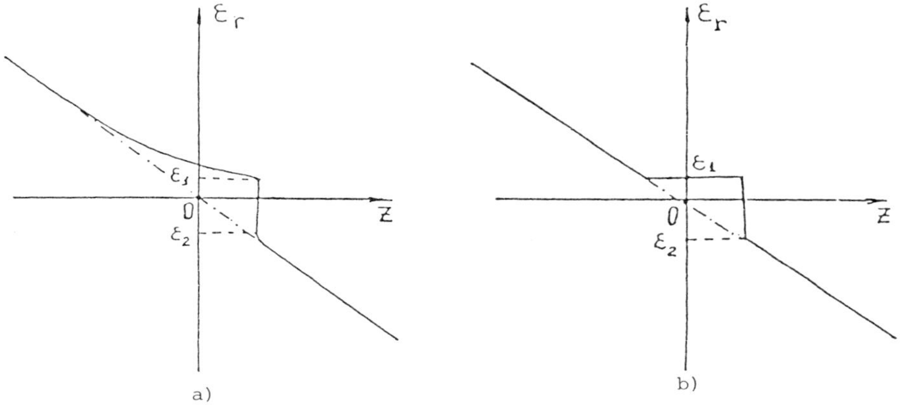

FIG.1.

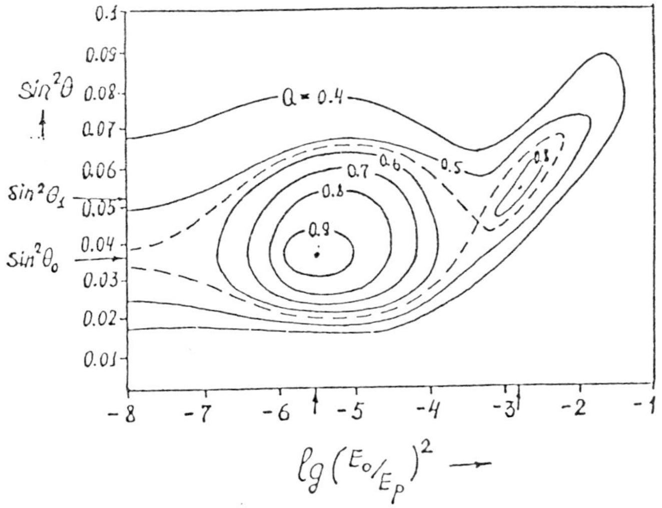

FIG.2.

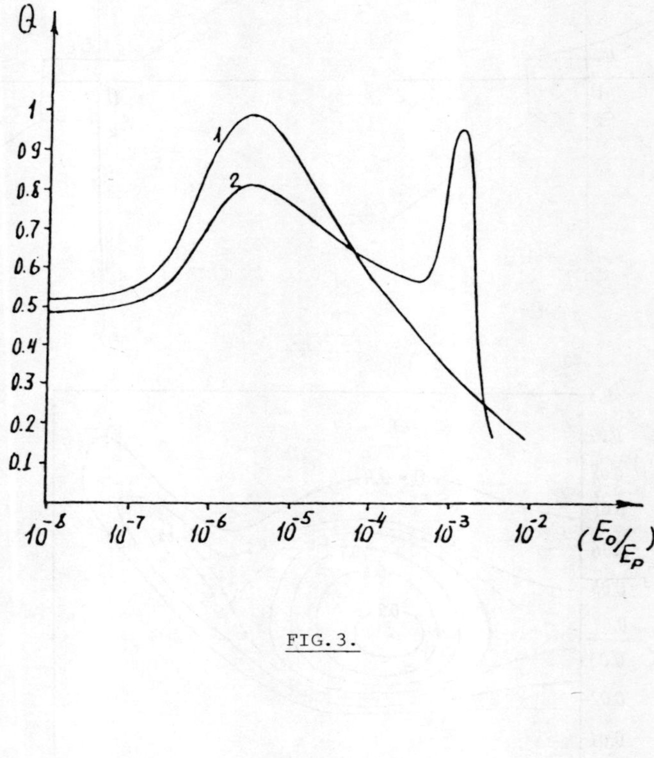

FIG.3.

OBSERVATION OF SELF-EXCITED LOW-FREQUENCY OSCILLATIONS IN A SURFACE WAVE SUSTAINED PLASMA COLUMN

V. Atanassov[+] and E. Mateev[++]

[+]Institute of Electronics, Bulgarian Academy of Sciences
[++]Institute of Solid State Physics, Bulg. Academy of Sciences
BG-1784 Sofia, Bulgaria

During the past decade large interest was shown in studying long plasma columns sustained by a high-frequency (HF) surface wave (SW). The latter is usually launched at the one end of the column by an UHF coaxial structure called the surfatron[1,2]. Although the SW produced plasma is relatively stable and of low noise level[1] unstable behaviour can be observed, in particular, at extremely low or at very high neutral gas pressure.

This report presents results of experimental study on self-excited low-frequency (LF) oscillations in a plasma column sustained by a HF azimuthally-symmetric SW in xenon at low pressure (p_n = 0.1-50 mTorr). Formation of plasma balls and filaments at higher pressure (0.1-1 Torr, in xenon and neon)[3] as well as in argon[4] has been recently reported.

The experimental set up is shown on Fig.1. The HF SW was launched by a surfatron at $\omega/2\pi$ = 510 MHz. The plasma was produced in rasotherm glass tubes of 2R = 4.2 cm i.d./4.6 cm e.d. or 2.2 cm i.d./2.5 cm e.d. The diagnostics includes electron temperature (T_e) measurements (by probe techniques)

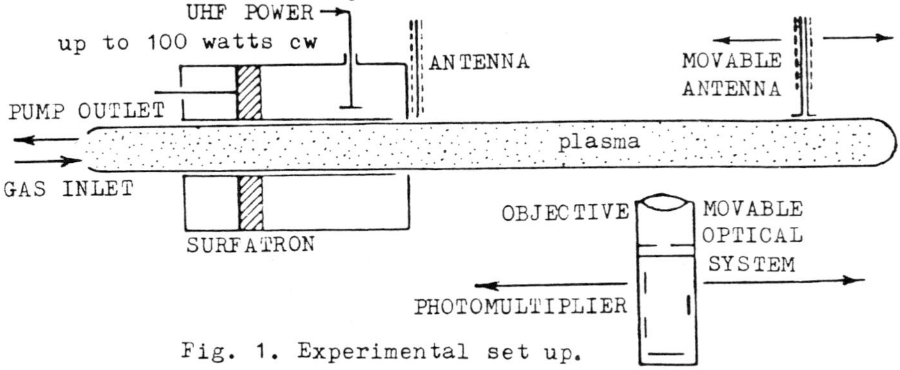

Fig. 1. Experimental set up.

and electron number density (n_e) measurements (by using TM_{010} cavity, double probes or by comparing calculated and measured HF SW dispersion). Axial profiles of plasma light emission intensity I have been obtained by using a movable optical system; its output was calibrated to measure n_e, making use of the fact that light emission intensity depends linearly on plasma density[1], $I = n_e p_n f(T_e(p_n R))$. The measured axial profiles are almost linear, which is well explained by the theory[5,6]. The peak at the end of the column (Fig.3) is possibly caused by a slight increase in electron temperature necessary to compensate increased losses of charged particles.

Fig.2 shows experimental and theoretical $T_e(p_n R)$ curves. Within experimental error of ± 16 per cent T_e does not depend on the absorbed UHF power and, therefore, is constant along the plasma column. The theoretical curve was calculated by the method of Self & Ewald[7] assuming Maxwellian electron velocity

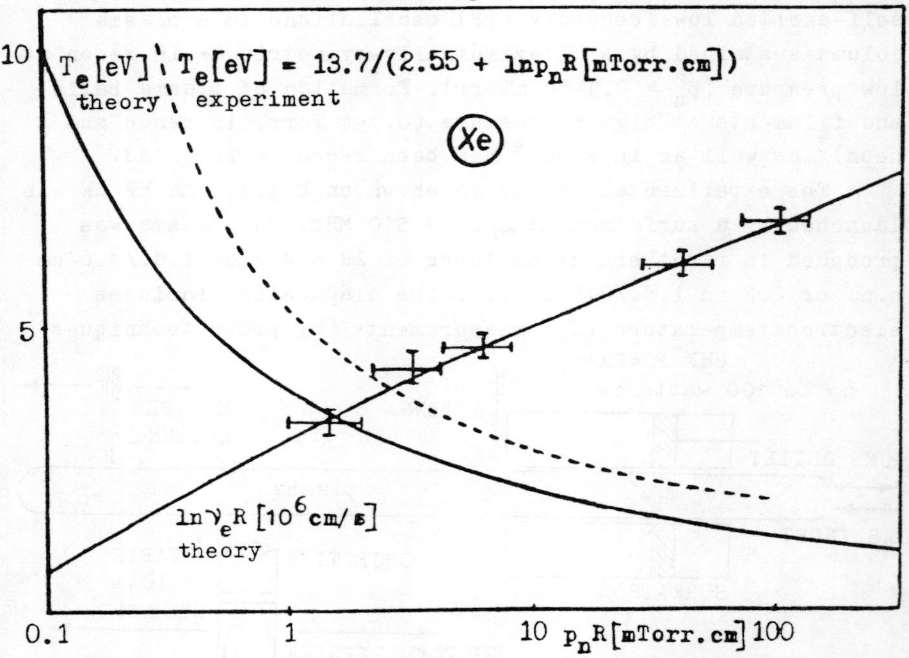

Fig. 2. Electron temperature T_e and electron-neutral collision frequency for momentum transfer ν_e .

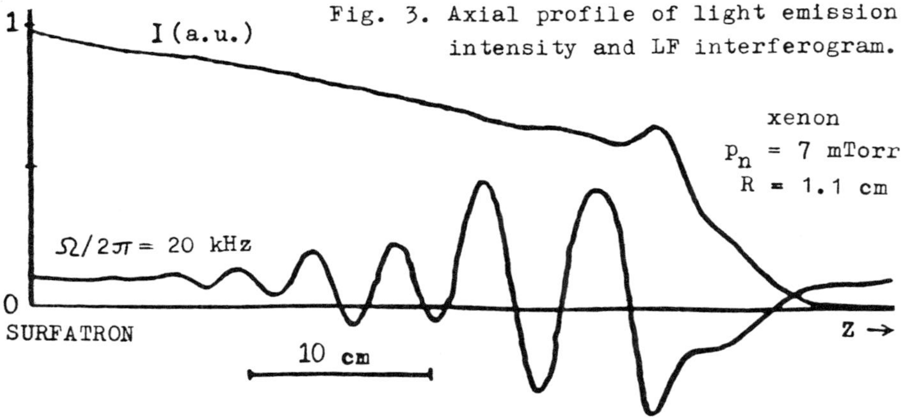

Fig. 3. Axial profile of light emission intensity and LF interferogram.

distribution and by using von Engel's formula for the ionization frequency. Fig.2 presents also the effective electron-neutral collision frequency for momentum transfer ν_e obtained by numerical integration of the properly weighted cross-section, as well as points obtained by fit of theoretical and measured plasma density axial profiles[5,6].

The excitation of LF oscillations was detected by means of an UHF spectrum analyser (sidebands $\omega \pm \Omega$ grow from the noise level) or by means of a scope (output from the movable optical system or detected signal from a movable dipole antenna). The LF oscillations ($\Omega/2\pi$ = 4-70 kHz) arise at pressure lower than some critical value (3 mTorr for the 4.2 cm i.d. tube and 10 mTorr for the 2.2 cm i.d. one). Phase variations have been localized along the column near its end (Fig.3) by using a LF phase detector. Additional phase measurements by means of the scope show that a LF wave propagates toward the surfatron, decreasing in wavelength and amplitude. The results of all observations are presented on a LF dispersion diagram (Fig.4).

It can be seen from Fig.4 that the measured phase velocity is close to the calculated ion-acoustic speed $V_S = (K_B T_e/M)^{1/2}$. However, we cannot conclude that the LF wave results from a HF SW decay instability[6], since the resonant condition for the wave-vectors $\underline{k}_{LF} = 2\underline{k}_{HF}$ is <u>not</u> fulfilled.

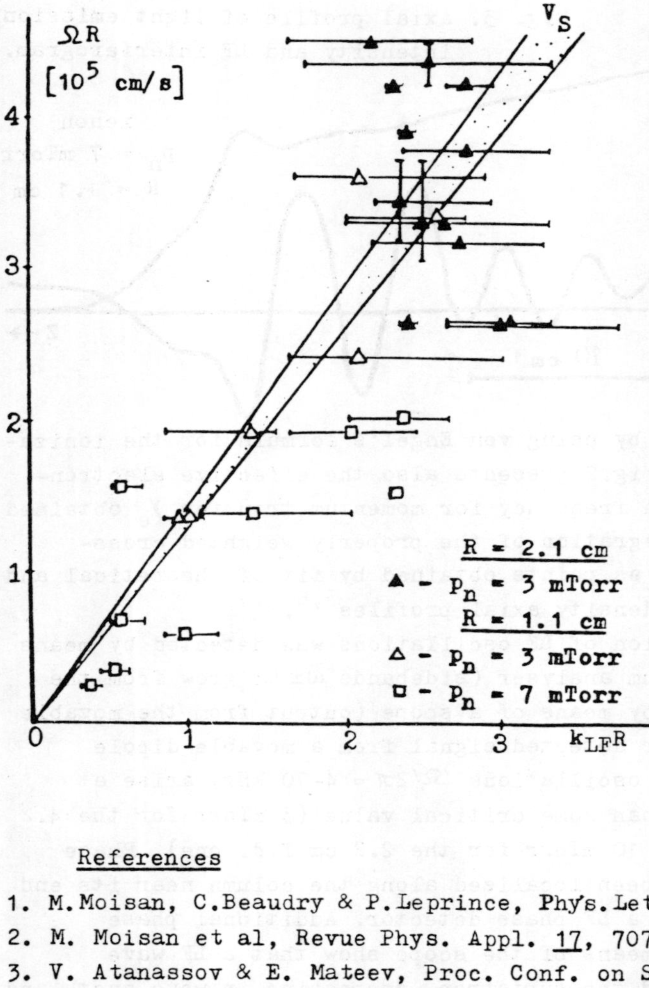

It is quite possible that the excitation of such LF wave is associated with the strong plasma density and electron temperature gradients at the end of the column, rather than with the action of the ponderomotive force.

Fig. 4.
LF dispersion diagram.

References

1. M.Moisan, C.Beaudry & P.Leprince, Phys.Lett. 50A,125(1974).
2. M. Moisan et al, Revue Phys. Appl. 17, 707 (1982).
3. V. Atanassov & E. Mateev, Proc. Conf. on Surface Waves in Plasmas Blagoevgrad 1981, p. 284, Sofia University (1983).
4. M. Chaker & M. Moisan, J. Appl. Phys. 57, 91 (1985).
5. V.M.M.Glaude et al, J. Appl. Phys. 51, 5693 (1980); E.Mateev, I.Zhelyazkov & V.Atanassov, J. Appl. Phys. 54, 3049 (1983); Z.Zakrzewski, J. Phys. D 16, 171 (1983).
6. V. Atanassov, Ph.D. Thesis, Sofia University (1984).
7. S. A. Self & H. N. Ewald, Phys. Fluids 9, 2486 (1966).

ON THE THEORY OF SURFACE MAGNETOPLASMA POLARITONS

by

N.N. BELETSKI, E.A. GASAN, V.M. YAKOVENKO

For the problem of excitation and detection of surface polaritons a principal question is that concerning the resonance frequencies which restrict the existence range of the waves. The resonance frequencies can be obtained from the electrostatic equations (unretarded limit) which take into account the frequency dispersion of dielectric permitivity of the surface-active medium. The resonance frequency of the surface wave in plasma-type media in constant magnetic field $\vec{H}_o$ were studies in /I-4/.

The present paper deals with further development of theory of the surface magnetoplasma polaritons. For the first time the spectrum of resonance frequency of the surface oscillations are obtained on the semiconductor-dielectric boundary for the case of arbitrary wave propagation with respect to $\vec{H}_o$, the latter being directed along the boundary between the media. The conditions have been determined when the solutions of the surface wave-type are absent, i.e. a gap arises in the spectrum of the surface polariton resonance frequencies. The gap is shown to cause the surface wave do not propagate in a definite angle interval. The dependence of both position and width of the gap on H_o and the parameters of semiconductor and dielectric has been obtained.

References

1. V.J. Pakhamov, K.N. Stepanov, Zh. Tekh. Fiz. 37, I393 (1967).
2. N.Z. Abdel-Shahid, V.J. Pakhomov, Plasma Phys. 12, 55 (1970).
3. K.W. Chiu, J.J. Quinn, Phys. Rev. B5, 4707 (1972).
4. S.J. Khankina, V.M. Yakovenko, Fiz, Tverd. Tela 9, 2943 (1967).

SURFACE-WAVES ON A CYLINDRICAL ANTENNA IN AN ISOTROPIC PLASMA

MARIO LE BLANC, MANFRED NACHMAN* AND SYLVAIN PREVOST

Department of Electrical Engineering, Ecole Polytechnique of Montreal, P.O.B. 6079, Station "A", Montreal, Canada.

1. Introduction. An antenna immersed in a plasma is usually surrounded by an electron-depleted layer, the ion-sheath. Under certain conditions, the antenna-sheath-plasma system may support slow surface-waves at frequencies below the plasma frequency[1]. These are the so-called sheath-waves. For a cylindrical monopole of finite length, a standing wave pattern will emerge, leading to a family of resonances in the antenna admittance. The frequencies at which these resonances occur are derived from the condition

$$(2n + 1)\lambda/4 = L ,$$

where λ = wave-length of the sheath-waves and L = length of the monopole. Experimental evidence of the existence of these axial resonances was first reported by J. Marec and G. Mourier[2].

It is the purpose of this paper to present the results of a detailed experimental study of the resonances due to sheath-waves, for the case of a short cylindrical antenna immersed in an isotropic plasma.

2. Experimental. The measurements were performed in a low-pressure, diffusion-type argon plasma (Fig. 1). The antenna consisted of the extension of the inner conductor of a coaxial cable. A specially designed launcher (Fig. 2) served as a transition between the feeding cable and the antenna. DC-biases could be applied independently to the an-

---

*) Author to whom correspondence should be addressed.

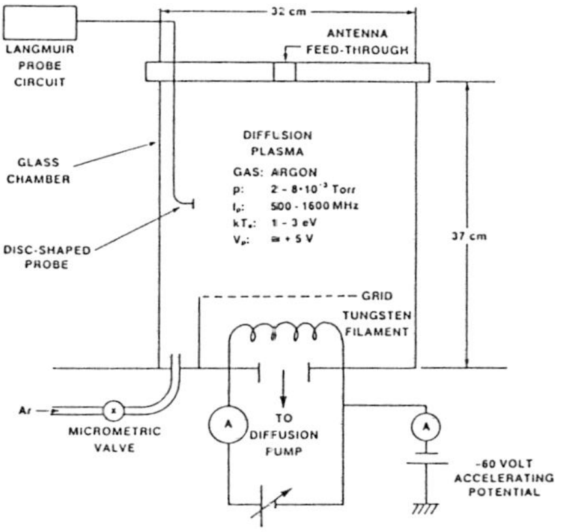

Fig. 1 Plasma chamber

p = neutral gas pressure
f_p = plasma frequency
kT_e = electron temperature
V_p = plasma potential

tenna and the outer conductor. A CW-signal from a sweep oscillator was fed to the antenna and a network analyzer was used for measuring the reflection coefficient (magnitude and phase angle) over the frequency range from 150 to 1000 MHz (Fig. 3). The data were processed by means of a micro-computer which calculated the input admittance (or impedance) of the antenna.

3. Results. Typical plots of the magnitude and phase angle of the complex reflection coefficient, $\Gamma = |\Gamma| e^{j\phi}$,

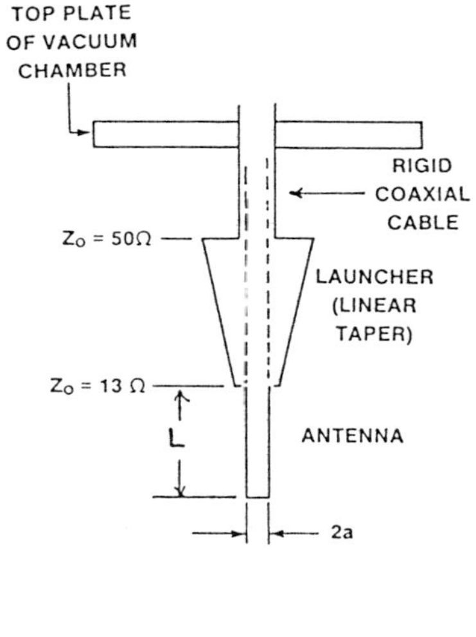

Fig. 2 Antenna and launcher
Z_o = characteristic impedance

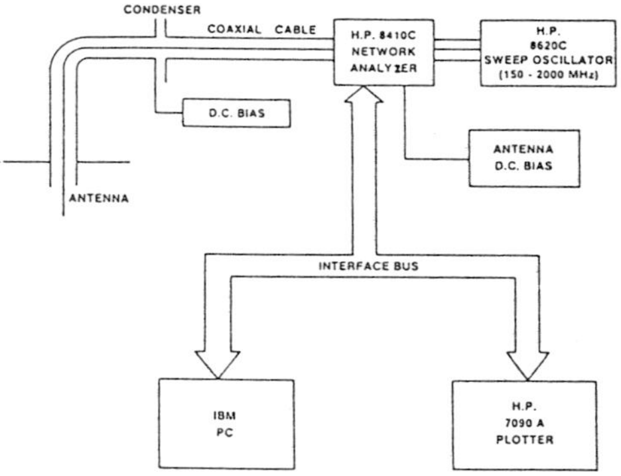

Fig. 3 Measuring circuit

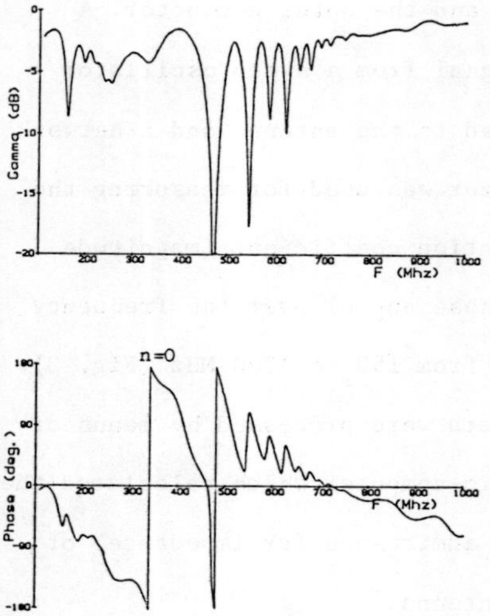

Fig. 4 Frequency dependence of $|\Gamma|$ and ϕ .

versus frequency are shown in Fig. 4 for a cylindrical monopole (L=4.0 cm, 2a = 0.48cm) immersed in a plasma (plasma frequency, f_p = 1.65 GHz, electron temperature, kT_e = 1.5 eV, plasma potential, $V_p \cong + 5V$). These plots were obtained with negative DC-biases (with respect to the ground) applied to the antenna (V_a = -20V) and to the outer conductor ($V_{o.c.}$ = -2V). Two families of resonances may be distinguished: one below, the other above 300 MHz. By decreasing $|V_{o.c.}|$, the resonances below 300 MHz are shifted towards lower frequencies; they disappear completely when $V_{o.c.} \approx V_p$. The 10 resonances at or above 300 MHz are, practically, not affected by the change in $V_{o.c.}$. They are due to surface-waves propagating along the antenna and correspond to n=0 to 9 in the expression $L = (2n+1)\lambda/4$.

The effect on sheath-wave resonances of f_p, V_a and antenna-diameter was investigated. It was found that a shift of the resonances towards higher frequencies was induced by increasing either f_p or the negative DC-bias applied to the antenna. It was also shown that the larger the antenna diameter, the lower the resonance frequencies. A plot of the antenna input admittance versus frequency, as derived from the measured values of the reflection coefficient, is shown in Fig. 5.

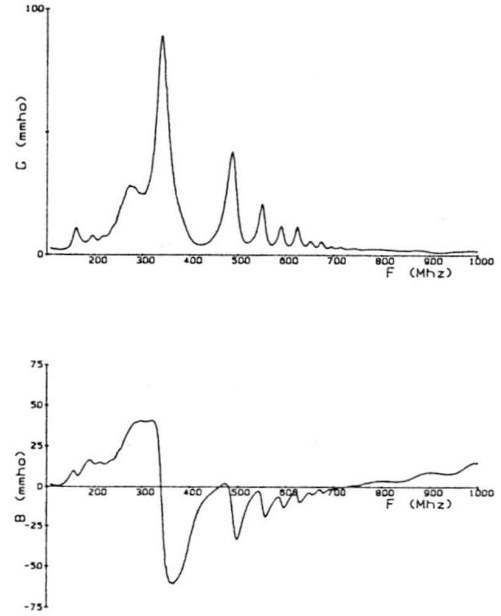

Fig. 5 Antenna input admittance versus frequency
(G = conductance, B = susceptance)

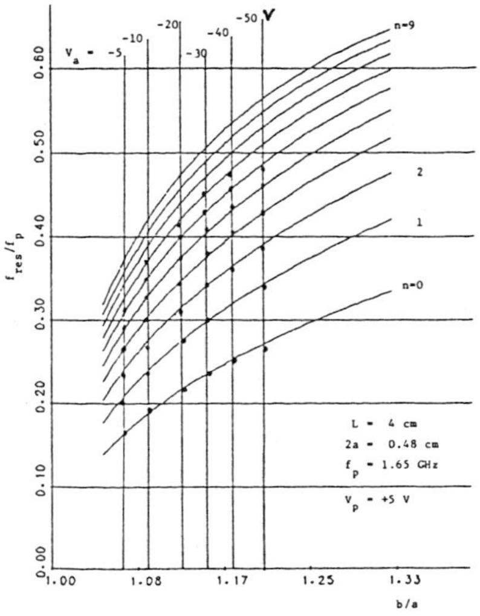

Fig. 6 Resonance frequency, f_{res}, normalized to f_p, versus b/a for the ten resonances n=0 to n=9

4. Discussion. The theoretical dispersion equation for sheath-waves obtained by Marec[3] was used to compute the ratio f_{res}/f_p as a function of b/a (b = a+g, where g = sheath-thickness) for our experimental conditions. The plots thus obtained are shown in Fig. 6 (full lines). Our experimental points, corresponding to resonances for several values of the DC-bias, were placed on the theoretical curves so as to obtain the best fit for the maximum number of points (Fig. 6). It is seen that, except for resonances corresponding to $n \geq 7$, the agreement between theory and experiment is very satisfactory. The results of Fig. 6 may be used to derive the sheath-thickness as a function of antenna-bias and plasma frequency.

References

1. S.R. Seshadri, Proc. IEEE, 112, 877 (1965).
2. J. Marec and G. Mourier, C.R. Acad. Sci. Paris, 271, series B, 367 (1970).
3. J. Marec, D.Sc. Thesis, University of Paris-South, Orsay, No. 1362, 1974.

SURFACE POLARITON SOLITONS

A. D. Boardman*, A. A. Maradudin, and T. P. Shen
Department of Physics, University of California, Irvine, Ca. 92717, U.S.A.

*Permanent Address: Department of Pure and Applied Physics, University of Salford, Salford M5 4WT, United Kingdom

It is of considerable interest to look for surface polariton solitons that have the nature of solitary waves propagating along the planar interface between a linear and a nonlinear dielectric medium.

In a recent search for surface polaritons[1] it was shown that they can exist only if the group dispersion of the linear surface polariton is negative. We show that the condition obtained in Ref. 1 is overly restrictive, and indicate a physical system in which bright surface solitons can exist.

We denote by a(x,t) the (real) electric field of an electromagnetic wave guided in the x-direction by the planar interface between a linear and a nonlinear dielectric medium. The equations determining a(x,t) are[2]: the energy equation

$$\frac{\partial a^2}{\partial t} + \frac{\partial}{\partial x}\left(\frac{\partial \omega}{\partial k} a^2\right) = 0, \tag{1}$$

the dispersion relation,

$$\omega = \omega_o(k) - \frac{1}{2}\frac{\partial^2 \omega_o(k)}{\partial k^2}\frac{a_{xx}}{a} + \omega^2(k)a^2, \tag{2}$$

and the consistency equation

$$k_t + \omega_x = 0. \tag{3}$$

It is convenient to introduce the phase function $\theta(x,t)$ in terms of which ω and k are given by $\theta_t = -\omega$, $\theta_x = k$. We now let $\omega = \omega_o + \bar{\omega}$, $k = k_o + k$, where ω_o and k_o are constants. The phase function θ now becomes $\theta = \theta_o + \bar{\theta}$, where $\theta_o = k_o x - \omega_o t$, $\bar{\theta}_t = -\bar{\omega}$, $\bar{\theta}_x = \bar{k}$. When the preceding results are used in Eqs. (1)-(3) to eliminate $\bar{\omega}$ and $\bar{k}$ in favor of $\bar{\theta}$, we obtain the pair of coupled equations

$$a_t = -\frac{\partial\omega_o}{\partial k_o} a_x - \frac{1}{2}\frac{\partial^2\omega_o}{\partial k_o^2}\left(\bar{\theta}_{xx}a + 2\bar{\theta}_x a_x\right) \tag{4a}$$

$$\bar{\theta}_t = -\frac{\partial\omega_o}{\partial k_o}\bar{\theta}_x - \frac{1}{2}\frac{\partial^2\omega_o}{\partial k_o^2}\left(\bar{\theta}_x^2 - \frac{a_{xx}}{a}\right) - \omega_2(k)a^2. \tag{4b}$$

If we introduce the complex amplitude $A = a\exp(i\bar{\theta})$, we can combine Eqs. (4a) and (4b) to obtain the following nonlinear Schrodinger equation for A:

$$i\left(A_t + \frac{\partial\omega_o}{\partial k_o}A_x\right) + \frac{1}{2}\frac{\partial^2\omega_o}{\partial k_o^2}A_{xx} - \omega_2(k_o)\,|A|^2 A = 0. \tag{5}$$

A solution of Eq. (5) is

$$A(x,t) = \left(-\frac{\partial^2\omega_o/\partial k_o^2}{\omega_2(k_o)}\right)^{1/2}\frac{e^{i\delta t}}{|v_g\Delta|}\,\mathrm{sech}\left(\frac{t-(x/v_g)}{\Delta}\right), \tag{6}$$

where $\delta = (\partial^2\omega_o/\partial k_o^2)/(2v_g^2\Delta^2)$ and v_g is the linear group velocity, $v_g = \partial\omega_o/\partial k_o$.

The bright soliton solution given by Eq. (6) exists only if either of the following conditions holds:

$$\text{a)}\quad \partial^2\omega_o/\partial k_o^2 > 0,\quad \omega_2(k_o) < 0; \tag{7a}$$

$$\text{b)}\ \partial^2\omega_o/\partial k_o^2 < 0, \quad \omega_2(k_o) > 0. \tag{7b}$$

In Ref. 1 only the second of these conditions was considered.

A system in which either the condition (7) is satisfied consists of a linear dielectric medium characterized by dielectric constant ε_1 in the region $z > d$; a linear metal characterized by dielectric constant $\varepsilon(\omega)$ in the region $0 < z < d$; and a nonlinear dielectric medium whose dielectric tensor has the form $\varepsilon_{ij}^{NL}(\omega) = \delta_{ij}\varepsilon_i^{NL}(\omega)$, where

$$\varepsilon_x^{NL}(\omega) = \varepsilon_d+\alpha|E_x(\vec{x};\omega)|^2+\beta\left(|E_y(\vec{x};\omega)|^2+|E_z(\vec{x};\omega)|^2\right) \tag{8a}$$

$$\varepsilon_y^{NL}(\omega) = \varepsilon_d+\alpha|E_y(\vec{x};\omega)|^2+\beta\left(|E_z(\vec{x};\omega)|^2+|E_x(x;\omega)|^2\right) \tag{8b}$$

$$\varepsilon_z^{NL}(\omega) = \varepsilon_d+\alpha|E_z(\vec{x};\omega)|^2+\beta\left(|E_x(\vec{x};\omega)|^2+|E_y(\vec{x};\omega)|^2\right) \tag{8c}$$

The nonlinear dispersion relation for this structure to lowest nonzero order in α and β is given by

$$G(\omega,k) = -\frac{a^2}{4\varepsilon_d}\frac{(\alpha K_3^4+2\beta K_3^2k^2+\alpha k^4)}{K_3^2}, \tag{9a}$$

$$G(\omega,k) = \frac{1+\frac{\varepsilon(\omega)}{K_2}\frac{K_1}{\varepsilon_1}+\left(1-\frac{\varepsilon(\omega)}{K_2}\frac{K_1}{\varepsilon_1}\right)e^{-2K_2d}}{1+\frac{\varepsilon(\omega)}{K_2}\frac{K_1}{\varepsilon_1}-\left(1-\frac{\varepsilon(\omega)}{K_2}\frac{K_1}{\varepsilon_1}\right)e^{-2K_2d}}\frac{\varepsilon(\omega)}{K_2}\frac{K_3}{\varepsilon_d}+1, \tag{9b}$$

where $a \equiv E_x(x,z=0|\omega)$. In Eqs. (9), $K_1 = (k^2-\varepsilon_1\frac{\omega^2}{c^2})^{1/2}$,

$K_2 = (k^2-\varepsilon(\omega)\frac{\omega^2}{c^2})^{1/2}$ and $K_3(k^2-\varepsilon_d\frac{\omega^2}{c^2})^{1/2}$. The effective nonlinearity $\omega_2(k)$ is given by

$$\omega_2(k) = -\frac{1}{4\varepsilon_d}\left[\frac{\alpha K_3^4+2\beta K_3^2k^2+\alpha k^4}{K_3^4}\frac{1}{\frac{\partial G(\omega,k)}{\partial\omega}}\right]_{\omega=\omega_o(k)}, \tag{10}$$

where $\omega_o(k)$ are the solutions of $G(\omega_o,k) = 0$.

We have solved $\omega_o(k)$ and $\omega_2(k)$ numerically in the case that $\alpha=\beta$. We find that the bright soliton exists for $k \gtrsim 1.83\sqrt{\varepsilon_d}\frac{\omega_p}{c}$, $\alpha > 0$ or $k \lesssim 1.83\sqrt{\varepsilon_d}\frac{\omega_p}{c}$, $\alpha < 0$, $\varepsilon_1 = 1.0$, $\varepsilon_d = 1.5$, $\varepsilon(\omega) = 1-\omega_p^2/\omega^2$, and $d = 0.5c/\sqrt{\varepsilon_d}\omega_p$.

Typically solitons of the type considered here have been observed in optical fibers.[3] The similar feature may well be appropriate to the aforementioned planar guides.

This research is supported by ARO under contract DAAG-29-83K-0018 and NSF under contract INT 8312955.

References

1. A. D. Boardman, G. S. Cooper and P. Egan, J. de Phys. C5-197 (1984).
2. H. C. Yuen and B. M. Lake, Phys. of Fluids, v. 18, p. 956 (1975).
3. L. F. Molenauer, R. H. Stolen and J. P. Gordon, Phys. Rev. Lett., v. 45, p. 1095 (1980).

FAST AND SLOW SURFACE GUIDED WAVES ON METALLIC STRUCTURES

Renato G. Bosisio

Electrical Engineering Department

Ecole Polytechnique de Montreal

P.O.B. 6079, Station "A" Montreal, Quebec, Canada H3C 3A7

Introduction Surface guided waves may be simply produced by such simple arrangements as a thin dielectric layer on a conducting surface as suggested by Goubau[1]. In addition metallic periodic structures can also propagate guided waves on a surface defined by the geometry of the structure[2] (e.g. cylindrical or planar). Surface guided waves are of great interest to produce interactions with materials[3] and charged particles[4,5] easily introduced into the propagating path of the surface waves.

Both fast and slow waves have been previously used in microwave devices and reactors to produce low temperature plasmas[6], and to heat industrial products with microwave power. The slow surface guided waves have also been widely used in many microwave electron device interactions[7]. Fast surface guided waves are of interest in industrial applications due to their increased wavelength, increased interaction impedance, and decreased attenuation in the direction transverse to the propagating surface.

This paper shows results on the measurement of both slow and fast space harmonic surface waves propagating on a metallic periodic structure

- the strapped bar structure.

<u>Theory</u> A TM surface wave propagating energy in the positive z direction of a periodic structure can be represented by an axial electric field $E(z)^+$ given by

$$E(z)^+ = \sum_{N=-\infty}^{\infty} E_N \, e^{-i\beta_N z} \qquad (1)$$

where $\beta_N = \beta_o + 2\pi N/L \qquad (2)$

and β_o = fundamental propagation constant for $N = 0$.

L = length of period of mtallic structure.

N is any positive or negative integer.

As a result when both incident and reflected waves are present the total axial electric field $E_T(z)$ is given by $E_T(z)^+ + E_T(z)^-$.

It is found that

$$E_T(z) = \sum_{N=0}^{\infty} 2 \cos \beta_N L z \qquad (3)$$

where $\beta_N = (\beta_o \pm 2\pi N/L)$ and $N = 1,2,3...$ (4)

Re-entrant periodic structures or linear periodic structures terminated by reflecting transverse conductors produce both incident and reflected waves. A Fourier analysis of a measured axial electric field profile $E_T(z)$ can determine the amplitudes E_N of the fast and slow surface waves at a given frequency.

<u>Measurement Apparatus and Results</u> The axial electric field measurements were obtained by using an apparatus similar to one previously described in the literature[8]. An actual recording of the electric field

strength variations is given in Fig. 1 corresponding to the 975 MHz resonant frequency of a circular cavity test structure given in Fig. 2. The complete Brillouin diagram of this periodic structure is shown in Fig. 3 and the amplitudes of the various space harmonics obtained from a Fourier analysis of the field plot (Fig. 1) are presented in Fig. 4.

Conclusion The use of flow surface guided waves has been advantageously used to produce uniform heating of dielectric materials and to produce large volume microwave plasmas. Fast surface guided waves should also proove useful in industrial microwave power applications due to their longer wavelengths, higher interaction impedances and decreased transverse field attenuation.

References

1. G. Goubau, Proc. I.R.E., 39, 619-624, (1954).
2. J.R. Pierce, "Travelling Wave Tubes", Van Nostrand, 1950, Chapter III.
3. Wertheimer, et al, Rubber Chem. & Tech., 47, 473-474 (1973).
4. Harman, W.W., "Fundamentals of Electronic Motion", McGraw-Hill, New York, 1953.
5. Harvey, A.F., "Microwave Tubes - An Introductory Review with Bibliography", Proc. I.E.E., 1960, 107C.
6. Bosisio, R.G. et al, "The Large Volume Microwave Plasma Generator (LMP™): A New Tool for Research and Industrial Processing", Journal of Microwave Power, Vol. 7, No. 4, 1972, pp. 176-197.
7. Slater, J.C., "Microwave Electronics", Van Nostrand, New York, 1950.
8. Ginzton, E., "Microwave Measurements", McGraw Hill.

FIG. 1 Plot of $(E_r)^2$ versus the azimuthal angle θ for the test cavity in Fig. 2 (Perturber situated .225" from the circuit bars).

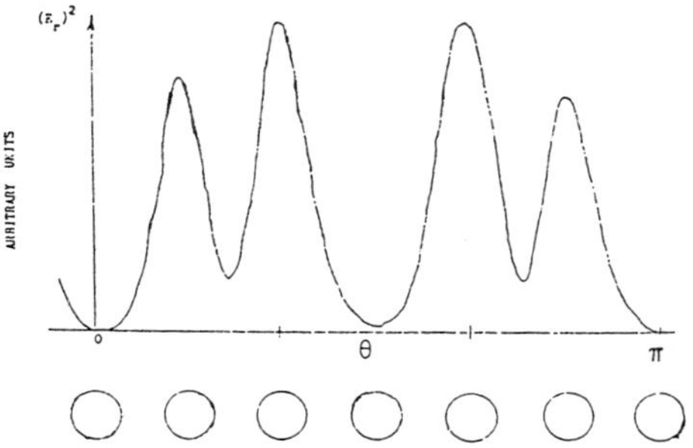

FIG. 2

THE RE-ENTRANT TWELVE BAR AMPLITRON TEST CAVITY

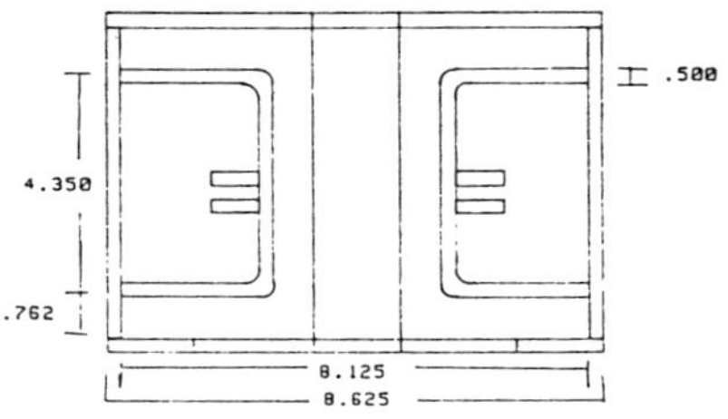

FIG.3 THE BRILLOUIN DIAGRAM OF THE RE-ENTRANT TEST CAVITY SHOWN IN FIG.2

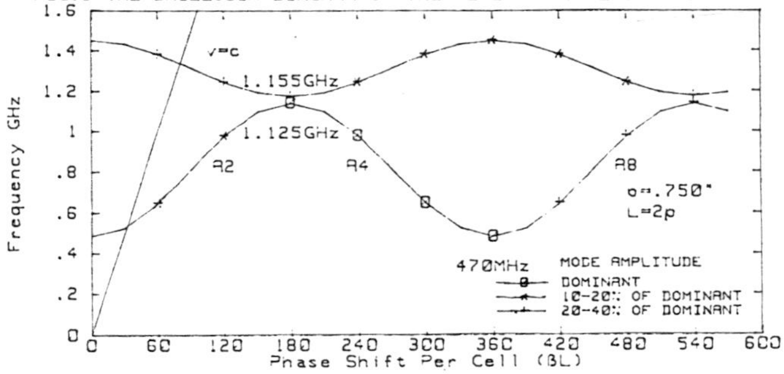

FIG.4 FOURIER ANALYSIS OF SPACE HARMONIC AMPLITUDE MODES AT 975 GHZ

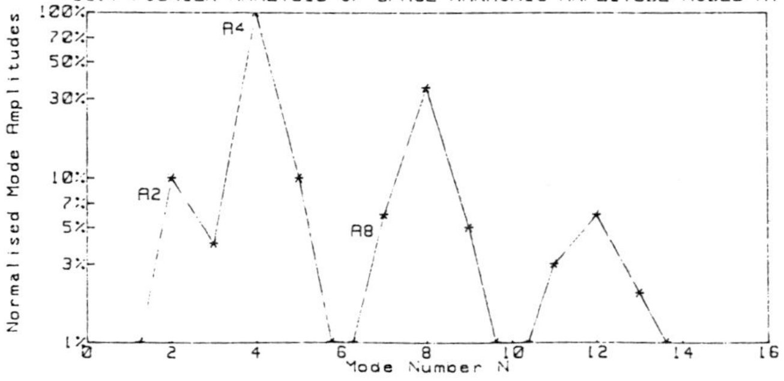

ELECTRO-MAGNETIC SURFACE WAVES ON METALLIC STRUCTURES

by

RENATO G. BOSISIO

Electrical Engineering Department
Ecole Polytechnique
University of Montreal
P.O. Box 6079, Station "A"
Montreal, Que., Canada H3C 3A7.

Periodic metallic structures can be used to propagate both slow and fast surface waves (SSW and FSW) near the surface of the periodic structure.

The characterization of the SSW and FSW for a strapped ladder structure is given from space harmonic analysis of total field strength measurements along a resonant section of the periodic structure. In the case of the present open periodic structure it can be shown that both SSW and FSW can be used to generate large volume microwave plasmas useful in industrial processes related to plasma chemistry (surface treatment, etching, polymerization, etc.). Results obtained on the generation of a large uniform microwave plasmas at 2450 MHz are presented. In addition practical design criteria for generating large volume plasmas are given for both SSW and FSW.

NONLINEAR EFFECTS ON SURFACE WAVE INSTABILITY DUE TO PARTICLE FLOW.

V.M. Čadež
Institute of Physics, P.O.Box 57, YU-11001 Beograd, Yugoslavia,

J.J. Rasmussen
Risø National Laboratory, DK-4000 Roskilde, Denmark.

The effect of nonlinearities in the process of potential surface wave excitation is considered in the case of a narrow plasma-vacuum boundary layer ($x \in (0,a)$) with a particle flow along the boundary. The plasma density and the magnitude of the flow velocity sharply vary across the boundary layer untill they reach their constant values at $x \geq a$.

We shall here restrict ourselves to the heigher frequency branch of potential surface waves for which $\omega \sim \omega_{pe} \gg \omega_{pi}$. This means that the influence of ion motions will not be treated i.e. the electrons only will be considered mobile. Finally, the waves are taken to propagate along the direction of the flow (the y-direction) and their wave numbers $k_{\parallel}$ obey the condition $k_{\parallel} a \ll 1$.

Within the linear theory one obtains the following equation for the electric potential perturbation φ_1 / 1 /:

$$\frac{\partial}{\partial x} \epsilon \frac{\partial \varphi_1}{\partial x} - k_{\parallel}^2 \epsilon \varphi_1 = 0 \tag{1}$$

where $\epsilon(x) = 1 - \dfrac{\omega_p^2(x)}{[\omega - \vec{k}_{\parallel} \cdot \vec{u}_0(x)]^2}$ and $\omega_p \equiv \omega_{pe}$

The relevant dispersion equation that follows from (1) is

$$1 + \epsilon(a) + k_{\parallel} \epsilon(a) \int_0^a \frac{dx}{\epsilon(x)} = 0 \tag{2}$$

which yields the frequency spectrum:

$$\omega - \vec{k}_{\parallel} \cdot \vec{u}_o(a) = \frac{\omega_p(a)}{\sqrt{2}} \tag{3}$$

and the instability growth rate γ /2/:

$$\gamma = sgn\left[\vec{k}_{\parallel} \cdot \vec{u}_o(x_r)\right] \frac{\pi}{8\sqrt{2}} \omega_p^2 \left|\frac{\partial u_o}{\partial x}\right|^{-1}_{x=x_r} \tag{4}$$

The resonant point $x=x_r$, where the instability occurs and the flow excites waves in the considered plasma geometry, is determined from the condition $\epsilon(x_r)=0$.

The purpose of this work is to investigate the influence of nonlinearities on the time evolution of the instability through a slow modification of the initial plasma configuration. Namely, the basic state quantities like $\vec{u}_o(x)$ and $n_o(x)$ are taken stationary before the instability sets in and become slowly time varying functions after that. The corresponding equations describing such time variation follow from perturbation analysis by retaining the zero frequency, nonlinear terms in the expansion. As the first order terms satisfy the linear equations, one finally obtains the following relations for the basic flow vorticity generation and for the plasma density time variation:

- Vorticity generation:

$$\left(\frac{\partial \vec{\xi}_o}{\partial t}\right)_x = \left(\frac{\partial \vec{\xi}_o}{\partial t}\right)_y = 0$$

$$\left(\frac{\partial \vec{\xi}_o}{\partial t}\right)_z \equiv \frac{\partial \xi_o}{\partial t} = -2\frac{\partial}{\partial x} Re\left(v_{1x}^* \xi_{1z}\right) \tag{5}$$

where $\vec{\xi} = \nabla \times \vec{v}$ is the vorticity vector, while the indeces "0" and "1" indicate the basic state and the linear, first order, perturbed quantities respectively.

- Plasma density equation:

$$\frac{\partial n_o}{\partial t} = -2\frac{\partial}{\partial x} Re\left(n_1 v_{1x}^*\right) \tag{6}$$

Equations (5) and (6) can now be expressed in terms

of the linear potential perturbation φ_1 (1) in this form:

$$\frac{\partial \varsigma_0}{\partial t} = 2\,\frac{e^2 k_{\parallel}^2}{m^2 \omega_p^3}\,\frac{\partial}{\partial x}\left(\left|\frac{\partial \varphi_1}{\partial x}\right|^2 \frac{\varsigma_0^2}{\gamma}\right)\Bigg|_{x=x_r} \tag{7}$$

resp.:

$$\frac{1}{n_0}\frac{\partial n_0}{\partial t} = 2\,\frac{e^2 k_{\parallel}}{m^2 \omega_p^3}\,\frac{\partial}{\partial x}\left(\left|\frac{\partial \varphi_1}{\partial x}\right|^2 \frac{\varsigma_0}{\gamma}\right)_{x=x_r} \tag{8}$$

The slow time variation of the basic state quantities ς_0 and n_0 during the instability will affect the initial growth rate γ which is according to (4):

$$\gamma \sim \frac{n_0}{\gamma_0}$$

i.e. slowly time dependent too.

To estimate the initial time evolution of the instability growth one has to express γ in terms of (7) and (8) in the following form:

$$\frac{\partial \gamma}{\partial t} = -2\,\frac{e^2 k_{\parallel}}{m^2 \omega_p^3}\left|\frac{\partial \varphi_1}{\partial x}\right|^2 \tag{9}$$

and

$$\frac{\partial}{\partial t}\left|\frac{\partial \varphi}{\partial x}\right|^2 = 2\gamma\left|\frac{\partial \varphi}{\partial x}\right|^2 \tag{10}$$

Solution of the system (9) and (10) yields an expression for the local growth rate time variation of an unstable surface wave with a given wave number $k_{\parallel}$:

$$\gamma = A\left(1 - \frac{2B}{B + e^{-2At}}\right) \tag{11}$$

where:

$$A = \left[\gamma_0 - 2\,\frac{e^2 k_{\parallel}^2}{m^2 \omega_p^3}\left|E_{\perp}(t=0)\right|^2\right]^{1/2}$$

$$B = \frac{A - \gamma_0}{A + \gamma_0}$$

Here $\gamma_0 \equiv \gamma(t=0)$ and $E_\perp(t=0) \equiv \left.\frac{\partial \varphi_1}{\partial x}\right|_{t=0}$

Expression (11) indicates that the growth rate γ decreases with time due to nonlinearities and the corresponding scaling time T for the considered instability process to be saturated, follows from (11):

$$T = \frac{1}{2\gamma_0} \ln \frac{2\gamma_0^2 m^2 \omega_p^3}{e^2 K_\parallel |E_\perp(t=0)|^2} \tag{12}$$

This relation shows that initially more energetic waves will saturate faster. The same holds for waves with longer wave lengths and, obviously, for those which had larger growth rates at t=0.

References:

1. Stepanov, K.N.: ZhTF 35, 1002, (1965)
2. Vuković, S. and Kyrie, A.Yu.: ZhTF 45, 2315, (1975).

PROPAGATION OF TRANSVERSALLY ELECTRIC SURFACE WAVES IN THE PRESENCE OF DISSIPATIVE PROCESSES.

by

P. K. CIBIN

Institute of Mathematics and Physics,
University "V. Vlahović", Cetinjski put b.b.,
81000 Titograd, Yugoslavia.

It is shown theoretically that the presence of dissipative processes in the ferrites, limits the wavenumber of transversally electric surface waves propagating on the ferrite slabs and semiinfinite systems.

PROPAGATION OF SURFACE WAVES ALONG A PLASMA COLUMN IN THE PRESENCE OF DISSIPATIVE PROCESSES

by

P.K. CIBIN

Institute of Mathematics and Physics
University "V. Vlahović", Cetinjski put b.b., 81000,
Titograd, Yugoslavia

It is shown theoretically that the presence of any kind of dissipative processes in the plasma, limits the wavenumber of surface waves propagating along the cylindrical and annular plasma columns.

This conclusion is valid for all directions and intensities of the external steady state magnetic field and all models of plasma column. Conclusion is also valid for surface waves propagating along corrugated cylinders and ferrite rods.

INFLUENCE OF DISSIPATIVE PROCESSES IN PLANAR PLASMA SLABS AND SEMIINFINITE PLASMAS ON THE WAVENUMBER OF SURFACE WAVES

by

P. K. CIBIN

Institute of Mathematics and Physics,
University "V. Vlahović", Cetinjski put b.b.,
81000 Titograd, Yugoslavia.

It is shown theoretically that the presence of any kind of dissipative processes in the plasma, limits the wavenumber of surface waves propagating in the planar plasma slabs and semiinfinite plasmas.

This conclusion is valid for all directions and intensities of the external steady state magnetic field and all models of plasma, and also valid for all transversally magnetic surface waves (surface waves on corrugated metal surfaces, surface polaritons etc.).

ALFVÉN SURFACE WAVES ON A DIFFUSE LINEAR PINCH

N. F. Cramer and C-M. Yung

School of Physics, University of Sydney, Sydney, N.S.W. 2006, Australia

I. J. Donnelly

Australian Atomic Energy Commission, Private Mailbag, Sutherland, N.S.W. 2232, Australia

Introduction The Bessel function magnetic field, or Lundquist model field, given by $B_z = B_0 J_0(\alpha r)$, $B_\theta = B_0 J_1(\alpha r)$ has often been utilized as a model of a highly twisted force-free magnetic field, as occurs in a reversed field pinch. The unstable modes of a plasma in such a field are well known[1,2]. In this paper we investigate the propagating MHD wave spectrum of such a plasma, and show that the lowest frequency mode of this spectrum is the Alfvén surface wave, which connects to the unstable mode. The wave spectrum of a diffuse pinch is of interest because of the possibility of supplementary heating of such a device by RF waves, as well as astrophysical applications such as wave propagation and energy transport in solar coronal loops.

Analysis A cold collisionless plasma column of radius R is considered, surrounded by an infinite vacuum region. The linearized perturbation equations for arbitrary density and magnetic field profiles may be written[3]

$$\frac{A}{r}\frac{d}{dr}(rQ) = C_1 Q - C_2 P, \qquad A\frac{dP}{dr} = C_3 Q - C_1 P, \tag{1}$$

where $P = \underset{\sim}{b}.\underset{\sim}{B}/B_0^2$ and $Q = -kE_\perp/i\omega B$, with $\underset{\sim}{b}$ the wave magnetic field and

$E_\perp$ the wave electric field perpendicular to $\underset{\sim}{B}$ and $\underset{\sim}{\hat{r}}$. k is the axial wavenumber. The frequency ω is assumed much less than the ion-cyclotron frequency. C_1,C_2 and C_3 are functions of r [3]. The coefficient $A = \omega^2 - v_A^2 k_\parallel^2$, where v_A is the local Alfvén speed, and $k_\parallel = (kB_z + mB_\theta/r)/B$ with m the poloidal mode number. The vanishing of A defines the Alfvén continuum: for given ω and k a singularity in eqns (1) occurs at the radial position where A = 0, and Alfvén resonance absorption occurs there. In order to integrate eqns (1) through the singularity we introduce a small amount of ion-neutral damping. The wave fields at r = R are matched onto the vacuum fields which couple to an antenna positioned in the vacuum. Calculation of the antenna impedance as a function of frequency then yields the (damped) eigenmode frequency and the corresponding resonance damping rate.

<u>Results</u> Fig. 1a (for $\alpha R = 0.05$ and density $\rho = \rho_0$, a constant) and Fig. 2a (for $\alpha R = 2.405$ and $\rho = \rho_0(1-.9(r/R)^2)$) show the dispersion relation for m = 1, in the form of a plot of $f = \omega/\alpha v_{Ao}$ against k/α with v_{Ao} the Alfvén speed at r = 0. The solid lines indicate the Alfvén continuum: for $\alpha R \ll 1$ and for uniform density, as in Fig. 1a, the continuum is of vanishing width. For $\alpha R = 2.405$ and a nonuniform density each line is given by A = 0 for 50 equally spaced radii between r = 0 and r = R. The dashed lines are the dispersion curves for peaks in the calculated antenna resistance. Outside the continuum the mode is an undamped eigenmode of the fast Alfvén wave with typical wave magnetic field profiles shown in Figs. 1b and 2b. We label this mode the Alfvén surface wave: if $\alpha R \cong 0$ the wave field profiles are typical

of the surface wave[3,4], viz constant b_r and b_θ and linear ramp b_z in the plasma. However for $\alpha R = 2.405$ the fields have become more typical of a body wave. Note that the mode connects to an instability, the unstable current-driven external kink mode. The mode also continues into the continuum in Fig. 2a for positive k, where it is Alfvén resonance damped (damping time typically ≅ 3 times the period). A second, lower frequency, peak in the impedance occurs as shown close to the lower edge of the continuum; the resonance damping in this case is greater (damping time ≅ 1.5 times the period). For negative k a heavily damped mode in the continuum connects to the instability. All of these modes may play a role in the RF heating of diffuse linear pinch-type plasmas.

References

1. D. Voslamber and D.K. Callebaut, Phys. Rev., 128 2016 (1962)

2. J.P. Goedbloed and H.J.L. Hagebeuk, Phys. Fluids, 15 1090 (1972)

3. N.F. Cramer and I.J. Donnelly, Plasma Physics and Controlled Fusion, 26 1285 (1984)

4. G.A. Collins, N.F. Cramer and I.J. Donnelly, Plasma Physics and Controlled Fusion, 26 273 (1984)

Figure Captions

Fig. 1a: Dispersion relation of the surface Alfvén wave for $\alpha R = .05$ and $m = 1$. The normalized frequency is continued below the origin by the negative of the normalized growth rate.

Fig. 1b: Wave magnetic fields for $\alpha R = .05$ and $m = 1$. $k/\alpha = .2$ and $f = .43$. b_r ——, b_θ -·-·-·-, b_z ----.

Fig. 2a: As for Fig. 1a, except $\alpha R = 2.405$.

Fig. 2b: As for Fig. 1b, except $\alpha R = 2.405$, $k/\alpha = .4$ and $f = .2$.

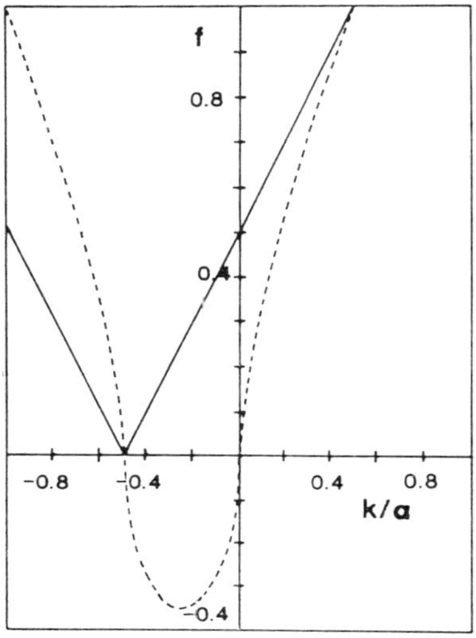

Fig. 1a.

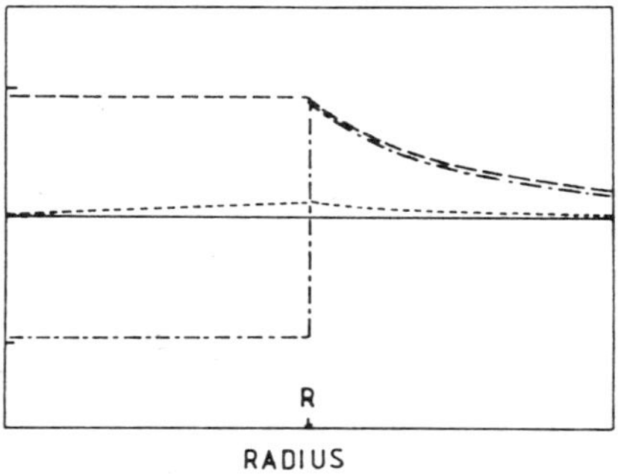

Fig. 1b.

Fig. 2a.

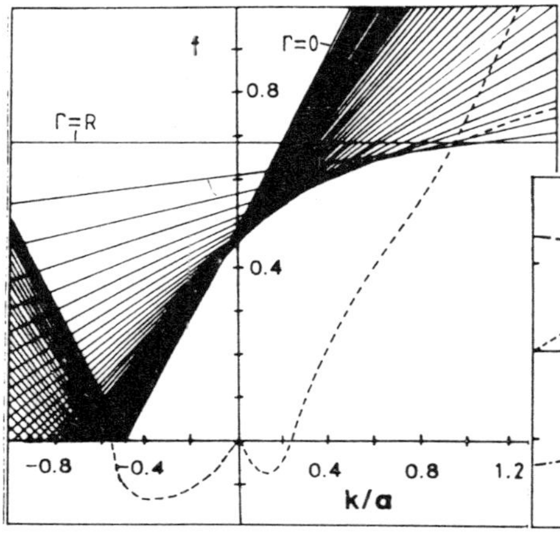

Fig. 2b.

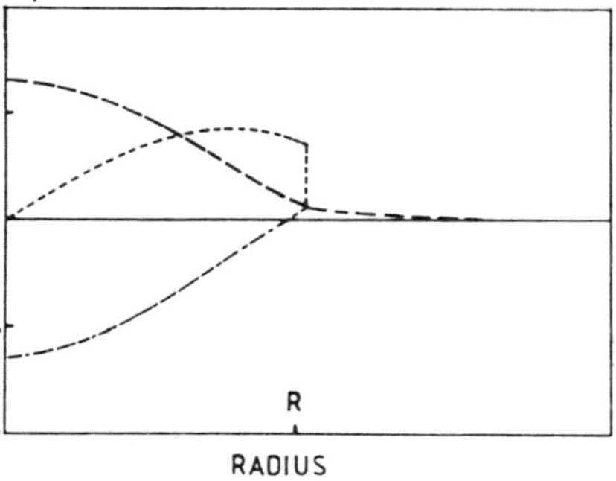

SURFACE WAVES ON THE BOUNDARY OF THE PIEZOSEMICONDUCTOR CRYSTALS

DZAMALOV R.D. AND HABIBULLAEV P.K.
Division of Termal
Physics Academy of Sciences of the Uzbek, SSR.

In this work analities expression was founded for frequency and decreement of connected surface waves in the system of two semi-conductors in the magnetic field. In this work the classic case and the case of quant movement of electrons are inverstigated.

As was shown(I) that in electron-holes pierosemi-conductors with "hot" electrons connected surface waves appear and spectrums of this waves are founded.

It is interesting the question about spectrum of sach vibration in magnetic field. As was shown (I) the condition of unisothermal Te > Th is more comfortable and it realises law helium temprature. By this case in the power- full magnetic field the movement of electrons becoms quant. As to sound waves the law of dispersion determined by the localization of particles (in classic limit it is Larmor radins in the quant case - the " magnetic length" the quant effects appears in the spectrum of surface waves.

Dispersation correlation for the surface electromagnetic waves connected with spreading of volume acoustics waves is formed of the system of equation discribing spreading waves in the piezoelectric media inverroment and equation of potential waves of unhomogemons quant plazma:

$$\Delta \varphi(x) = -\int_{-\infty}^{\infty} dq_x\, \varphi(q_x)\, e^{iq_x x} \sum_{\alpha}\sum_{s'} \Big\{ \Big[1 - \frac{\partial}{\partial x}\Big(\frac{iq_x}{A_{s'}} \times \frac{\partial A_{s'}}{\partial q_\perp^2} - \mathrm{sign}(e_\alpha)\frac{s' q_y}{q_\perp^2}\Big)\Big] A'_{s'} \Big[(\exp \varkappa_{s'})\, Z_{s'}^{-} - (\exp -\varkappa_{s'})\, Z_{s'}^{+}\Big] - \Big[(\exp \varkappa_{s'})\, Z_{s'}^{-} - (\exp \varkappa_{s'})\, Z_{s'}^{+}\Big] \mathrm{sign}(e_\alpha)\frac{\lambda_3^2 q_y}{2}\frac{\partial}{\partial x} A_{s'} \Big\}$$

$$Z_{s'}^{\pm} = J_{+}\Big(\frac{\omega - s'\Omega \pm \beta_\alpha}{|q_{\parallel}|\, V_{T\alpha}}\Big) \Big/ (\omega - s'\Omega \pm \beta_\alpha)$$

the rest marks are standard.

By resieving this equation there was no any suppositon re-latively correlation between typical size of unhomogeneous Lo and the length of waves q-I. That is why boundary conditions on the surface of the semi-conductors are recieving by inter-gration of this eguation on endless thin transitional layer.

The frequency and decreement of the surface waves follow of the dispertion correlation. By this different possible situ-ations when the boundary of the waves appeare on the boundary of the piezoelectrie and semi-conductor or on the boundary of two semi-conductor.

The new branches connected surface waves are diffirent co-mbinations of volume acoustic waves

$$\omega^2 = u_s^2 q_z^2$$

and surface electromagnetic waves

$$\omega^2 = q_z^2 \times \frac{KT_e}{M_h}$$

and as quant waves

$$\omega^2 = q_z^2 \frac{\hbar\Omega_e}{M_h}\left(\frac{q_z^2 \frac{\hbar\Omega_e}{M_h}}{\Omega_h^2} + \frac{4e^{-\frac{\lambda_e^2 q_\perp^2}{2}}}{\lambda^2 q_z^2} \right)$$

The damping condition($\gamma \ll \omega$)of these waves are funded. In this work are also determind the excitement conditions of these waves in the eletric field, wich makes up the unisotermal condition, heating electros.

I.Dzamalov R.D., Habibullaev P.K.

Poverhnostnie svajzannie akustoe- lektromagnitnie volni - Ph . T.P. I98I, vol. I5, p.I575- I578

2.Orlov V. P., Pustovoit V.J.

Svajzannie akusto - pllazmennie volni v piezopolyprovodnicah - Ph.T.P., I968 ,vol. 2,p.I305- I3II

THE SURFACE WAVES EXCITATION BY A CIRCULAR APERTURE ANTENNA

by

E.G. FILONENKO, I.P. SHASHURIN
T.G. Shevchenko Kiev State University
Kiev, USSR.

The surface plasma-waveguide waves in a radial-bounded plasma column is being investigated experimentally. A circular aperture antenna in the butt-end of the column is used as an exciter [1].

The effectiveness of the surface waves excitation is estimated by the antenna input resistance variation [2]. The plasma-waveguide waves amplitude was determined in accordance with the signal along the plasma column.

It's shown, that the surface plasma-waveguide waves excitation leads to the power increase, radiated by the antenna in plasma. The plasma column diameter and the collisional losses influence on the surface plasma-waveguide waves effectiveness has been studied.

References

1. Т.А. Грязнова, Н.Н. Ивануинов, К.С. Карлюк, Е.Г. Филоненко, И.Л. Шашурин. ЖТФ, 1976, 46, № 5, 1081-1085.

2. Т.А. Грянова, К.С. Карялюк, Е.Г. Филоненко, И.Л. Шашурин. у Всес. конф. ло физике низкотемлературной ллазмы; Киев, 1979, стр. 225.

DIPOLE - EXCHANGE WAVES IN UNIAXIAL UNSATURATED THIN FERROMAGNETIC FILMS

O.L.Galkin, Yu.V.Gulyaev, P.E.Zilberman

Institute of Radioelectronics, Moscow, USSR

The long - wave region of the magnetostatic wave (MSW) spectrum of a ferromagnetic film is extremely important for experimental reseach and applications. The reason is that the long waves are easily excited both by the uniform r.f. field and by the r.f. field of the microstrip transduser. The initial region of the spectrum has been previously investigated for the saturated yttrium iron garnet (YIG) films with the magnetization $\vec{M}$ parallel to the magnetic field $\vec{H}$.[1,2] Both perpendicular and parallel resonance has been considered. The dispersion relation, including both dipole-dipole and exchange interaction has been obtained analitically.[1,2] These results permit to use experimental data for the pinned and unpinned boundary conditions to be distinguished.[3]

There is an interesting opportunity to make an external d.c. field perpendicular to the film surface much smaller than the saturation field and still to retain the homogeneous state of the film but with nonparallel $\vec{M}$ and $\vec{H}$.[4] The dispersion relation of the MSW becomes considerably more complicated than that in the saturated case in the dipole approximation.[5] The dipole approximation is unsuitable in the most interesting initial region of the spectrum.[1] Here we would like to consider only the initial region of the MSW spectrum in the homogeneous unsaturated ferromagnetic films and to obtain dispersion relation including both dipole-dipole and exchange interaction. We consider here the ferromagnetic film with a growth-induced tilted uniaxial anisotropy, which is inherent to epitaxial films. Two models of the boundary conditions are considered, namely the pinned and the unpinned models. The calculation consists in solving the magnetostatic equations together with Landau-Lifshitz equation with the effective magnetic field

$$\vec{H}^{eff} = \alpha \Delta \vec{M} + \vec{e}\beta(\vec{M}, \vec{e}) + \vec{H}^{m} + \vec{H}, \qquad (1)$$

where α is the exchange constant, $\beta \ll 4\pi$ is the anisotropy constant, $\vec{e}$ is the unit vector of the anisotropy axis, $\vec{H}^{m}$ is the dipolar field. The usual electrodinamic boundary conditions on the fields and the conditions on the time-varying magnetization and/or its normal derivative must be imposed at the surfaces.

It turned out that the uniform mode of the resonance does not exist for the nonparallel $\vec{M}$ and $\vec{H}$ even in the unpinned case and for the infinite film. The frequencies of the nonuniform modes of the spin-wave resonance (SWR) were proved to be the same in the pinned and unpinned cases

$$\omega_n^2(0) = \left[\omega_b + \omega_m \frac{\alpha}{4\pi}\left(\frac{n\pi}{L}\right)^2\right]\left[\omega_a + \omega_m \sin^2\theta + \omega_m \frac{\alpha}{4\pi}\left(\frac{n\pi}{L}\right)^2\right], \qquad (2)$$

where n = 1,2,3..., $\omega_b = (\gamma H - \omega_m \cos\theta)\cos\theta + \omega_m \frac{\beta}{4\pi}\cos^2(\theta-\theta_u)$,

$\omega_a = \omega_b - \omega_m \frac{\beta}{4\pi}\sin^2(\theta-\theta_u)$, $\omega_m = \gamma 4\pi M_o$, L is the thickness of the film, γ is the gyromagnetic ratio, θ_u is the angle between $\vec{e}$ and normal to the film surface. The frequencies $\omega_n(0)$ strongly depend upon the angle θ between $\vec{M}$ and $\vec{H}$. The distribution of the magnetization in the SWR modes is symmetric or antisymmetric with respect to the film center.

In the unpinned case the frequencies of all the modes alter quadratically with the wave vector. In our case $(\theta \neq 0)$ in the contrast with the case $\theta = 0$[1], the "magnetostatic" mode which in the case $\theta = 0$ tends to the frequency of the uniform resonance does not exist even in the unpinned case and in the infinite film. It means that the absence of the "magnetostatic" branch of the spectrum does not prove that the spins are pinned, as it was in the case $\theta = 0$[1]. The outlined difference between the cases $\theta = 0$ and $\theta \neq 0$ is due to the absence of the uniform resonance in the case $\theta \neq 0$.

In the pinned case the even MSW branches alters quadratically with q, and of the odd ones - linearly.

$$\omega_n(q) = \omega_n(0) + 2qL\left[\frac{\omega_b + \omega_m(\alpha/4\pi)(n\pi/L)^2}{\omega_a + \omega_m \sin^2\theta + \omega_m(\alpha/4\pi)(n\pi/L)^2}\right]^{1/2} \cos 2\theta + O(q^2L^2), \quad (3)$$

where $n = \pm 1, \pm 3, \pm 5 \ldots$. The coefficients before the linear and quadratic members of the dispersion relations depend upon θ, when $\theta = \pi/4$ the sign of the group velocity changes (see equation (3)). The dispersion relation (3) gives the correct behavior provided qL is small enough for the solution not to extend into the crossover region where the repulsion between two adjacent branches is important. It requires

$$|\omega_n(q) - \omega_n(0)| \ll |\omega_{n+1}(0) - \omega_n(0)|. \quad (4)$$

The region of θ where our results are valid is determined by the condition

$$\left[\omega_b(\theta, \theta_H)/\omega_m\right]^3 \sin^2\theta \gg \frac{\alpha q^2}{4\pi} \quad (5)$$

The dependence of the SWR frequencies $\omega_n(0)$ upon the number of the branch n is quadratic for $n < n_1 = 2L[\omega_\ell/(\pi\alpha\omega_m)]^{1/2}$; for $n_1 < n < n_2$ the dependence becomes linear, and for $n > n_2 = 2L \sin\theta/\sqrt{\pi\alpha}$ again quadratic. If the region of the linear dependence is revealed, the numbers n_1, n_2 give an opportunity to estimate θ and θ_u.

The group velocity $V_n(\theta, \theta_u)$ of the waves (3) depends upon θ and hence upon H. The value $V_1(\theta, \theta_u)$ of the first mode was compared with the value of the group velocity in the isotropic saturated film provided that $2L \gg [\pi\alpha\omega_m/\omega_\ell]^{1/2}$. If $H > 4\pi M_0$ than $V_1(\theta, \theta_u)/V_1(0,0) \approx 1$; if $H = (4\pi M_0 - \beta M_0 \cos 2\theta_u)/\sqrt{2}$ than $V_1(\theta, \theta_u) = 0$; if $H \to 0$ than $V_1(\theta, \theta_u)/V_1(0,0) \to -\sqrt{\beta/4\pi} \sin\theta_u$. These equations can also be used to estimate the anisotropy constants and the angles θ and θ_u.

The main terms of the dispersion relation expansion by $qL \ll 1$ which we have revealed in (3) contain only even power of n, the next terms of the expansion also contain odd power of n, that is the degeneracy with respect to the sign of n is removed and each of the branches of the MSW spectrum splits.

REFERENCES

1. R.E. De Wames, T. Wolfram, Phys. Rev. Lett., 26 1445 (1971).

2. T. Wolfram, R.E. De Wames, Solid Stat. Comm., 9 171 (1971).

3. P.E. Wigen, Thin Solid Films, 114 135 (1984).

4. Yu.G. Lebedev, I.G. Tityakov, B.N. Filippov, Fiz. Metallov i Metalloved., 41 1159 (1976).

5. O.L. Galkin, P.E. Zilberman, S.V. Karmazin, Radiotekhnica i Elektronika, 30 735 (1985).

EXCITATION OF SURFACE PLASMONS BY ATR METHOD FOR THIN METAL FILMS.

by

M. GHORANNEVISS, A. AFSHARI, R. ROSTAMI, H. AZODI, M. NARAGHI
and G.H. MIRJALILI.

Atomic Energy Organization of Iran, Nuclear Research Centre,
Plasma Physics Section, P.O. Box 11365-8486 Tehran-Iran

Surface plasma waves are transverse magnetic electromagnetic waves, travelling along the interface of two different media. We will assume one of the media to be a metal and the other to be air.

The waves propagate parallel to the interface (along the x direction) and are exponentially attenuated in the normal direction in both the metal and air. Nonradiative surface plasma waves have been known as solutions of Maxwell's equations. In this paper we relate the amplitude of the surface plasmon mode to the exciting incident radiation.

An electromagnetic wave in a glass prism is incident on a thin metal film at the hypotenuse face of the prism and at an angle of incidence greater than the critical angle for total internal reflection, we assume the electric field vectors are p-polarized and described by monochromatic plane waves.

Our apparatur which produced a vacuum of 10^{-7} torr (Balzer B.A.K. 760), after a few minutes of evaporation the prism was then removed from the evaporator and placed on a suitable rotary platform a He-Ne laser (λ = 6328 A) beam polarized in the plane of incidence by a polaroid sheet was incident on the prism, and the reflected beam could initially be observed on a white piece of paper. At critical angle for total internal reflection, the reflected intensity approximately equalled the incident intensity, but as the external angle of incidence was increased by 2°-3°. The reflectivity was observed first to decrease in the scattered light intensity from the laser spot on the back metal-air surface minima.

When viewed from the hypotenuse face, too thick a film appear while too thin a film will appear semitransparent, however, the apex edge of the prism should be visible through the metal film. If the surface plasmon resonance is not observed, the prism should be recleaned and the evaporation process repeated again.

Since this process requires only a few minutes, new samples of varying thickness may be easily prepared until the desired effect is achieved.

In this paper, the surface plasmons in thin silver, gold, aluminium were excited and studied. By the measurements of plasmon angle at various wavelengths, one can plot the experimental dispersion relation for surface plasmons in conclusion, we have given a theoretical description of the effect of the surface plasmon mode on the reflectivity of thin metal films and have demonstrated how such films may be prepared with a modest apparatus.

RPA CALCULATION OF ELECTROMAGNETIC RESPONSE OF STRONGLY CHARGED JELLIUM SURFACES

P. Gies and R.R. Gerhardts

Institut für Theoretische Physik, Freie Universität Berlin,
D-1000 Berlin 33, Federal Republic of Germany

Optical methods, such as electroreflectance spectroscopy[1] and ellipsometry[2], are important tools for the characterization of metal surfaces in contact with an electrolyte, where strong static electric fields of the order of 1V/Å can be applied. Calculations of optical response properties have been performed using very different microscopic or phenomenological surface models[3-5], and a dramatic dependence on details of the electron density profile has been reported[4-6]. Recently, we have performed[7] the first self-consistent quantum-mechanical calculation of the electron density profile at a jellium surface in a strong static electric field, using the method of Lang and Kohn[8]. Here we present the first calculation of the optical response of a charged metal surface, which is independent of any ad hoc model assumptions about the surface electron distribution. Of course, this is only a first step towards a microscopic theory of the optical response of noble-metal surfaces in strong electric fields, which is largely determined by d-electrons.

As a model for the metal we consider a plane jellium slab situated symmetrically between two infinite potential barriers at distance L. For different values of the surface-charge density σ , the electron density and the effective potential are calculated self-consistently[7]. As shown in Fig. 1, the density profile becomes more diffuse if the surface is negatively charged (curve a), and steeper if the surface charge is positive (curves c-e in order of increasing σ). Simultaneously the mean position (center of mass) z_o of the static induced charge is shifted towards the jellium edge z_b .

From the self-consistent effective potential we obtain the optical surface response from an RPA calculation which is described elsewhere in detail[9]. We solve Maxwell's equations in a mixed Fourier representation[5], using a sine expansion (with $k_m = m\pi/L$),

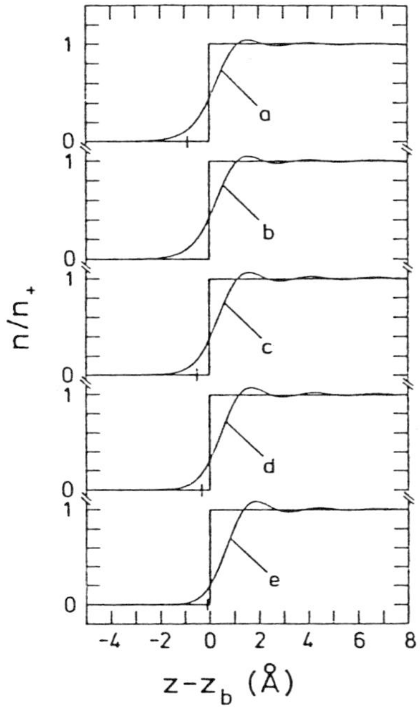

FIG. 1. Self-consistent electron density profile for δ (in 10^{-3} e/Å^2) equal -2.39 (a), 0 (b), 4.77 (c), 11.93 (d), 23.87 (e). The jellium background is indicated, the position of z_o is marked in front of the jellium edge. The density is given by $r_s \equiv (\frac{1}{3}4\pi n_+ a_o^3)^{-1/3} = 3$.

$$E_z(\vec{r},t) = \frac{2i}{L}\sum_{m=1}^{\infty} \mathcal{E}_z(q_x, k_m)\sin(k_m z)e^{i(q_x x-\omega t)} \quad ,(1)$$

for the component of the electric field perpendicular to the slab, and a similar cosine expansion for the parallel component. For the interesting case of p-polarization ($E_y = 0$), longitudinal fields occur in the metal, which decouple approximately from the transverse fields. The material equations read

$$\mathfrak{D}_t(\vec{q}_m) = \epsilon_t \, \mathcal{E}_t(\vec{q}_m) \quad (2)$$

$$\mathfrak{D}_l(\vec{q}_m) = \sum_{n=1}^{\infty} \epsilon_{ll}(k_m, k_n)\, \mathcal{E}_l(\vec{q}_n) \quad , \quad (3)$$

where the subscripts t and l refer to transverse and longitudinal components with respect to $\vec{q}_m = (q_x, 0, k_m)$, and $\vec{\mathfrak{D}}$ and $\vec{\mathcal{E}}$ denote the displacement field and the electric field, respectively. Whereas the transverse fields are determined by the local transverse dielectric function ϵ_t of the bulk metal, the interesting surface effects are determined by the nonlocal longitudinal dielectric function, for which the RPA calculation yields[9]

$$\frac{k_m}{k_n}\epsilon_{ll}(k_m, k_n) = \delta_{mn} - \frac{4L}{\pi^2 a_o}\frac{1}{n^2}\sum_{\nu\nu'} \frac{(\varepsilon_F - \varepsilon_\nu)(\varepsilon_\nu - \varepsilon_{\nu'})}{(\varepsilon_\nu - \varepsilon_{\nu'})^2 - \hbar^2\omega^2} W_m^{\nu\nu'} W_n^{\nu\nu'} \quad , \quad (4)$$

$$W_m^{\nu\nu'} = 2\int_0^L dz \cos(k_m z)\, \varphi_\nu(z)\, \varphi_{\nu'}(z) \quad . \quad (5)$$

Here a_o is the Bohr radius, ε_F the Fermi energy, ε_ν and $\varphi_\nu(z)$ are energy eigenvalues and wavefunctions, respectively. After solving Eq. (3) numerically, Feibelman's surface response function[3]

$$d_\perp = \int_0^{L/2} dz\, z\, \rho_{ind}(z) \Big/ \int_0^{L/2} dz\, \rho_{ind}(z) \tag{6}$$

was calculated, from which reflectance and absorptance can easily be obtained[3]. Disturbing effects from the second surface were suppressed by introducing a finite bulk damping, i.e. by replacing $\omega \to \omega + i\gamma$ in Eq. (4). Numerical calculations were performed on a CRAY-1M computer.

Figure 2 shows the mean position of optically induced charges, $\mathrm{Re}\, d_\perp(\omega)$. The quantity $Y(\omega)$ shown in Fig. 3 reduces in the limit $\gamma \to 0$ to the surface absorptance[3]. Apparently these optical response properties depend strongly on the static surface charge σ. Figs. 2 and 3 clearly reveal a broadened pole structure with a zero of $\mathrm{Re}\, d_\perp(\omega)$ and an accompanying maximum of the corresponding $Y(\omega)$ curve. As the value

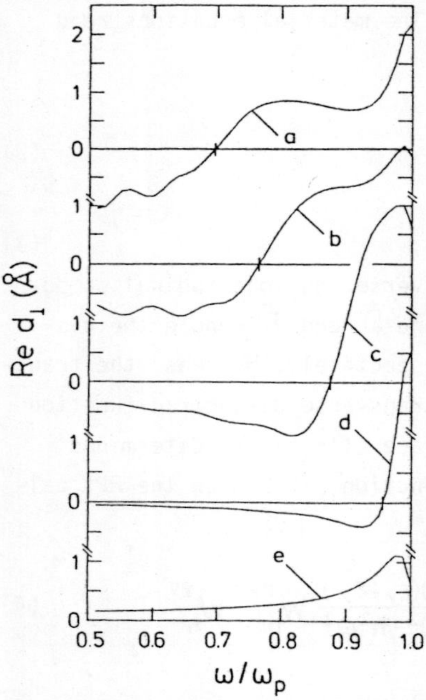

FIG. 2. $\mathrm{Re}\, d_\perp$ for the σ values given in Fig. 1. The model parameters are $L = 64$ Å, jellium thickness 54 Å, $\gamma = 0.025\,\omega_p$, ($\omega_p^2 = 4\pi n_+ e^2/m$).

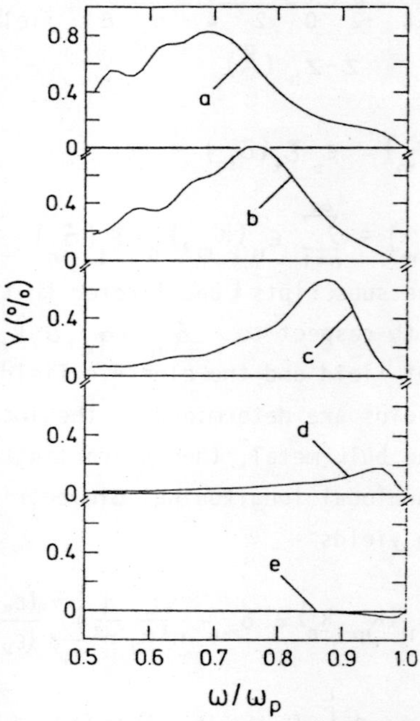

FIG. 3. $Y = \sqrt{8}\,\omega/c\,(\omega^2/\omega_p^2 - 1)\,\mathrm{Im}\, d_\perp$ for the parameters of Fig. 2. Note the finite size effects at low frequencies which depend on the jellium thickness.

of δ increases, the structure shifts towards the plasma frequency ω_p and finally disappears (curve e). These results are in agreement with previous work[3-6] and confirm the interpretation of the pole structure as a multipole surface plasmon which has been shown to exist at aluminum surfaces[6]. The possibility of tuning this resonance by the external electric field is an interesting consequence of the present work.

References

1. R. Kötz and D.M. Kolb, Z. Phys. Chem. (Frankfurt am Main) 112, 69 (1978); F. Forstmann, K. Kempa, and D.M. Kolb, J. Electroanalyt. Chem. 150, 241 (1983)
2. F. Chao and M. Costa, Surf. Sci. 135, 497 (1983)
3. P.J. Feibelman, Prog. Surf. Sci. 12, 287 (1982)
4. for a review see: F. Forstmann and R.R. Gerhardts, in Festkörperprobleme (Advances in Solid State Physics), edited by J. Treusch (Vieweg, Braunschweig, 1982), Vol. XXII, p. 291
5. R.R. Gerhardts and K. Kempa, Phys. Rev. B 30, 5704 (1984)
6. K. Kempa and R.R. Gerhardts, Solid State Commun. 53, 579 (1985)
7. P. Gies and R.R. Gerhardts, Phys. Rev. B 31, 6843 (1985)
8. N.D. Lang, Solid State Commun. 7, 1047 (1969); N.D. Lang and W. Kohn, Phys. Rev. B 7, 3541 (1973)
9. P. Gies, R.R. Gerhardts, and T. Maniv, to be published

Financial support by the Deutsche Forschungsgemeinschaft through Sonderforschungsbereich 6 is gratefully acknowledged.

SOLITARY AND SHOCK WAVES EXCITED BY CHARGED BEAM IN SOLID STATE PLASMAS

by

V.K. GRISHIN

Institute of Nuclear Physics, Moscow State University,
119899 Moscow.

The nonlinear equilibrium of interaction between a charged beam and excited surface wave in collisional solid state plasma is considered. This problem is of particular interest for various experiments where short or shock-front beams and short pulse waves are used. The equilibrium state occurs when beam particles are trapped by the wave and corresponds to the saturation of wave excitation.

The main structural peculiarities of beam and wave state and the analytical relations connecting the parameters of beam, field and plasma, are described self-consistently by a non-linear kinetic equation. The beam-wave equilibrium in the case of collisionally-dominated plasma is investigated. Here the distribution of beam density is described by Korteweg-de Vries-Burgers's equation and the field obtains a profile of classical shock wave. The energy beam spread required to achieve wave field - beam equilibrium is found and field amplitude is evaluated.

TOP EFFICIENCY OF SURFACE WAVE EXCITATION BY CHARGED BEAM IN PLASMA SYSTEM.

by

V.K. GRISHIN
Institute of Nuclear Physics, Moscow State University, 119 899, Moscow, USSR

The top efficiency of surface wave excitation by a charged beam in plasma system in the state of saturation is analysed. The following real scheme is considered. The charged beam is injected into a semi-infinite (along axis Z) waveguide, which excites one of the eigenmodes of the plasma waveguide. The top transfer of beam energy to the electromagnetic field occurs when beam particles are captures by the wave. The state of saturation is evaluated by application of the laws of conservation of energy and momentum in flow form.

Using the conservation of energy and momentum flows in the excited volume and the fact that the velocities of wave and captures beam coincide makes it possible to estimate the energy transfer from the beam to the wave. For a system homogeneous in Z, the maximal coefficient of energy transfer reaches 17% when the transverse distribution of plasma is uniform and about 30% when it is non-uniform. When the plasma density varies along Z, the efficiency can reach 50% for single mode excitation. Evaluations for other surface-wave configurations are also discussed.

RESONANT TRANSMISSION OF PLASMA WAVES THROUGH SUPERCRITICAL DENSITY BARRIERS

E.M.Gromov

Institute of Applied Physics
Academy of Sciences of the USSR
Gorky, USSR

Introduction. The transport of plasma wave energy through density barriers nontransparent for these waves (i.e., supercritical barriers) can be accomplished when plasma waves are transformed into the waves capable of wave energy transfer through such barriers. Thus, the possibility of damped plasma wave transmission due to their transformation into Van-Kampen waves is considered in [I]. However, the transmission coefficient in this case was small as compared to unity.

This paper deals with the transformation of plasma waves into surface electromagnetic (EM) waves at supercritical barriers with monotonic and rectangular density profiles. The possibility of complete transmission of plasma waves through such barriers is shown.

I. Formulation of the Problem. Let us consider the plasma wave incidence onto a plane inhomogeneity in plasma without an external magnetic field. The plasma density gradient is assumed to be directed along the Z-axis, while the plasma wave vector lies in the ZY plane. The equations, describing the propagation of EM and plasma waves with the electric field in the ZY plane take the form [2] :

$$\beta_T^2 \frac{d}{dz}\left\{\frac{N^2}{N_0^2}\;\frac{dE_z}{dz}\right\}+\left[\frac{\omega^2}{c^2}\varepsilon(z)-K_y^2\right]E_z=-iK_y\frac{d}{dz}\left\{\left(1-\beta_T^2\frac{N^2}{N_0^2}\right)E_y\right\}, \quad (I)$$

$$\frac{d^2E_y}{dz^2}+\left[\frac{\omega^2}{c^2}\varepsilon(z)-\beta_T^2K_y^2\frac{N^2}{N_0^2}\right]E_y=-iK_y\left(1-\beta_T^2\frac{N^2}{N_0^2}\right)\frac{dE_z}{dz}, \qquad (2)$$

where $\{E_{z,y}\}=\{\tilde{E}_{z,y}\}\exp(i\omega t-iK_y y)$, $\tilde{E}_z$ и $\tilde{E}_y$ — z и y — are the electric field components of the interacting waves, ω is the frequency of these waves, K_z and K_y are the projections of the plasma wave vector $\vec{K}$ onto the Z and Y axes respectively, $\beta_T^2=3V_T^2/C^2$, $V_T=(T_e/m_e)^{1/2}$ is the thermal velocity of electrons, C is the velocity of light, $\varepsilon(z)=1-\omega_p^2(z)/\omega^2$ is the dielectric permittivity, $\omega_p^2(z)=4\pi e^2N(z)/m_e$, $N(z)$ is the plasma density, N_0 is the unperturbed density in the homogeneous region outside the barrier. When plasma waves are incident upon the plane inhomogeneity at an angle $\theta<\theta_c=\beta_T/(1-\beta_T^2)^{1/2}$ the plasma waves are transformed into bulk EM waves propagating at an angle to the inhomogeneity. At larger angles of incidence θ , i.e., at $\theta>\theta_c$, plasma waves excite the damping surface EM modes propagating along the inhomogeneity. Let us consider, for example, the transformation of plasma waves into EM waves at barriers with monotonic and rectangular density profiles [3] .

2. Monotonic Barriers. We consider as a monotonic barrier the density perturbation produced by an ion-acoustic soliton * $N(z)=N_0[1+\beta_0\,\mathrm{sech}^2(z/\Delta)]$, where $\beta_0=(N_m-N_0)/N_0$, N_m is the maximum plasma density in the soliton, $\Delta=D(6/\beta_0)^{1/2}$ is the width of the soliton. Equations (I)-(2)

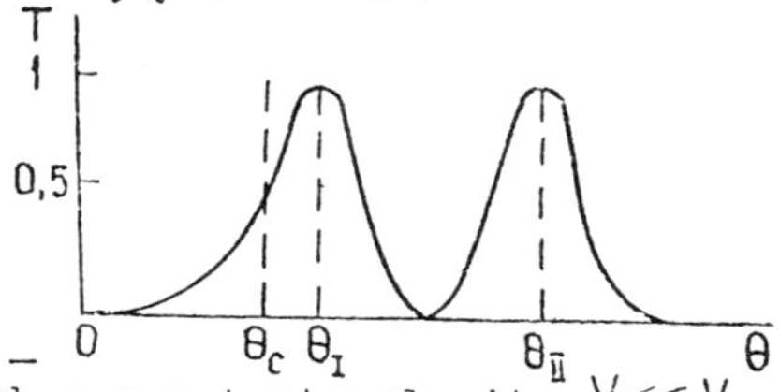

---

* As the inhomogeneity moves with a constant velocity $V \ll V_T$, the scattering of plasma waves in the reference system related to the inhomogeneity, is described by eq. (I)-(2), where $K_z \to K_z - K_*$ ($K_* = V/3V_T D$).

with the given density profile were solved numerically. The transmission coefficient of plasma waves T through the soliton at $\beta_T^2 << K\Delta << 1$ is shown in the figure. For a normal plasma wave incidence onto a soliton, i.e., when there is no interaction with EM, plasma waves reflect at the soliton at $\sigma = K\Delta \lessapprox 1$. When surface EM modes are excited, the transmission coefficient of plasma waves may be close to unity at much smaller wave numbers, namely, at $\sigma \gtrsim \beta_T^2$. Complete transition in this case is accomplished in the vicinity of two values of the angles of incidence θ_I and θ_{II} and corresponds to the resonances with surface EM modes, whose electric field has an even (θ_I) or an odd (θ_{II}) structure.

3. Rectangular Barriers. To define the plasma wave scattering matrix in this case we use the boundary conditions at density barriers. These conditions yielded by equations (I)-(2) have the form:

$$\left[N^2 \frac{dE_z}{dz}\right] = iK_y E_y \left[N^2\right], \quad \left[N^2 E_z\right] = 0, \quad \left[E_y\right] = 0, \quad \left[\frac{dE_y}{dz}\right] = 0, \tag{3}$$

where the brackets [ ] denote the difference in the values of functions at either side of the barrier. Thus, for density barriers of amplitude β_0 and width Δ, which satisfy the relation $\alpha = 3KD^2/\beta_0\Delta << 1$, when the barrier is nontransparent in the absence of plasma wave- EM interaction, the transmission coefficient at the angles of incidence $\theta > \theta_c$ with (3) taken into account is

$$T = \frac{\alpha^2\left[(\theta^2 + p)^2 + p^2\right]}{(\theta^2 + \alpha + p^2)(\theta^4 + p^2)}, \quad p^2 = (1 - \beta_T^2)(\theta^2 - \theta_c^2). \tag{4}$$

The transmission coefficient T is a nonmonotonic function of the angle of incidence θ for a fixed module K and has two maxima close to unity when $\alpha \gtrsim \beta_T^2$. It follows from (4), that the plasma wave angles of incidence θ_I and θ_{II} for which $T = I$, are $\theta_I = (\theta_c^2 + \alpha^2)^{1/2}$ and $\theta_{II} = \theta_c (1+\beta_T^2 K^2 \Delta^2/2)$.

Thus, the resonant interaction of plasma waves and surface EM modes propagating along the supercritical density barriers, may result in complete transmission of plasma waves through such barriers.

References

I. V.V.Lisitchenko, B.N.Orajevsky, Doklady Akad. Nauk SSSR, 1971, 201, No 6, p. 1319.

2. V.L.Ginzburg, Propagation of Electromagnetic Waves in Plasma.- M.: Nauka, 1967.

3. E.M.Gromov, Fizika Plazmy, 1984, 10, No 6, p.1289.

STATIONARY DOUBLE-DIP SOLITONS OF SURFACE PLASMA WAVES

by

D. GROZEV and I. ZHELYAZKOV
Faculty of Physics, Sofia University
BG-1126 Sofia, Bulgaria.

We study the nonlinear propagation of azimuthally-symmetric surface waves in a plasma column at a nonlinearity due to the ponderomotive force action. A coupled set of nonlinear equations for the high frequency field and ion response is solved without using the charge neutrality condition. The evolution equation for the wave field proves to be a cubic nonlinear Schrödinger equation, but the Light-hill's criterion is not fulfilled. Thus the azimuthally-symmetric surface plasma modes are modulationally stable and one may expect the existence of dip-solitons only. The results show that there is a region of Mach numbers where the propagation of double-dip solitons is possible. These solitons, like the Nishikawa double-hump solitons, might be treated as a special case of stable stationary multi-solitons.

SHEAR SURFACE MAGNETOACOUSTIC WAVES IN THE REORIENTATION PHASE TRANSITION REGION

Yu.V. Gulyaev, Yu.A. Kouzavko, I.N. Oleinik, V.G. Shavrov

Institute of Radio Engineering and Electronics, Ac. Sci. USSR, Moscow

Propagation of purely shear surface acoustic waves (SSAW) is possible along certain directions in definite special symmetry crystals where elastic displacements are associated with the existence of electric and magnetic fields [1,2]. SSAW occur due to the fact that the bulk shear wave with polarization lying in plane parallel to the surface is instable and even a small influence (piezoeffect, electrostriction, magnetostriction, piezomagnetism) is enough to transform it into a SSAW. In piezocrystals this SSAW penetrates into the material more deeper in comparison with Raley wave and it is perspective in high-frequency acoustoelectronic devices due to small damping on high frequencies. Similar SSAW can exist also in ferrites as a result of magnetostriction [3]. Besides the magnetostriction in antiferromagnetic crystals there is one more effective mechanism of magnetoelastic coupling, namely a piezomagnetic one. The different symmetry of piezomagnetic tensor and tensors of piezoeffect and magnetostriction lead to the essentially different solutions for SSAW (further SSAW in magnetic materials is referred to as SSMAW)[4,5]. Nevertheless all SSAW are characterized by a large penetration depth into crystals and their velocity is somewhat smaller than that one of a corresponding shear bulk wave.

As it is known the spectrum of one of the transverse acoustic branches varies considerably in the proximity of reorientation phase transition (RPT) region and the sound velocity reduces greatly along a definite direction [6]. If RPT occurs in the magnetic field it is possible to control effectively the sound velocity by changing the field. The velocity of surface Raley waves in magnetics near the RPT becomes much smaller whereas the depth of penetration increases and the polarization approaches to the transversal one [7]. This paper describes SSMAW studied in [4,5] in the proximity of RPT. SSMAW in ferrites is identical to SSAW in piezoelectrics (case I), whereas in antiferromagnetics with anizotropies "easy plane" of NiAs-type (II) and "easy axis" with tetragonal structure (III). SSMAW differ substantially from SSAW and from each other. First of all this concerns the distribution of elastic and magnetostatic oscillations. Long wave SSMAW are studied basing on the elasticity equation with effective elastic modulus and magnetostatic equations. Since the renormalization of the elastic modulus responsible for SSMAW existence is a result of quasi-equilibrious interaction with magnon's subsystem, the study is valid for the following frequency range $\omega \ll \mathcal{E}_{me}/\hbar$ where $\mathcal{E}_{me}$ is the magnetoelastic gap. For the majority of magnetics it is quite a wide frequency range. For example, in α-Fe_2O_3 $\omega_{me} \simeq 10^{10}$ sec^{-1}. The softened elastic dynamic modulus responsible for the existence of SSMAW can be written in the proximity of RPT as

$$\tilde{C} = C(1-\zeta), \tag{1}$$

where $\zeta = \mathcal{E}^2_{me}/\mathcal{E}^2_{1k}$ is the magnetoelastic coupling parameter, $\mathcal{E}^2_{1k} = \Theta^2_N(ak)^2 + \mathcal{E}^2_m + \mathcal{E}^2_{me}$ is the dispersion relation for the low frequency branch of spin waves, $\mathcal{E}_m$ is the gap in the spin wave spectrum which is due to the magnetic interactions and comes to zero in the point of RPT, Θ_N is the Neel temperature, k is the wave vector, and a is the lattice parameter. When approaching to RPT $\zeta \to 1$.

For the I-III considered cases free energies responsible for the piezomagnetism can be written as

$$\begin{aligned} \mathcal{F}_{pm} &= -\gamma(H_x u_{zx} + H_y u_{yz}), \\ \mathcal{F}_{pm} &= -\gamma(H_x u_{zx} - H_y u_{yz}), \\ \mathcal{F}_{pm} &= -\gamma_1(H_x u_{yz} + H_y u_{zx}) - \gamma_2 H_z u_{xy}, \end{aligned} \tag{2}$$

where γ are piezomagnetic constants, u_{ik} is the deformation tensor, and H is the variable magnetic field. The solution for the half-space $y>0$ can be found in the form of a plane wave propagating along the crystal surface, say x-axis (in the case III the geometry is somewhat different: $y \parallel [110]$, $x \parallel [1\bar{1}0]$), with the direction of displacement along the principal axis of a crystal z. Using standard boundary conditions we receive the following results for the elastic displacements u_z, magnetostatic potential ψ ($H=-\mathrm{grad}\,\psi$), for the velocity v and penetration depth Λ of SSMAW in the crystal in the vicinity of RPT.

In the case of ferrit (I) elastic and magnetostatic oscillations can be written in the form

$$u_z = A\exp\left(-\frac{y}{\Lambda}\right), \quad \psi = \frac{4\pi\gamma}{\mu} A\left[\exp\left(-\frac{y}{\Lambda}\right) - \exp(-ky)\right],$$

where A is a constant factor. We shall omit to be brief the space-time factor $\exp[i(kx-\omega t)]$. The SSMAW velocity is

$$v = v_t(1-\zeta+\eta)^{1/2}\left[1 - \frac{\eta^2}{(1+\mu)^2(1-\zeta+\eta)^2}\right]^{1/2} \qquad (3)$$

and penetration depth is

$$\Lambda = \lambda(2+\mu)(1-\zeta+\eta)/(2\pi\eta), \qquad (4)$$

where v_t is the bulk shear wave velocity in the absence of piezomagnetic effect, μ is the magnetic permeability, λ is the wave-length of SSMAW, and $\eta = 4\pi\gamma^2/(C\mu)$ is the constant of magnetomechanical coupling.

For the easy-plane antiferromagnet of the NiAs-type (II) and for the easy-axis antiferromagnet of the CoF_2-type (III) the SSMAW velocity is

$$v = v_t(1-\zeta+\eta)^{1/2}\left[1 - \frac{9\eta^2(1-\zeta)}{4(1+\mu)^2(1-\zeta+\eta)^3}\right]^{1/2}, \quad 1-\zeta \geq \eta, \qquad (5)$$

and penetration depths of the long-ranged ($u_z, \psi \sim e^{-y/\Lambda_2}$ far from RPT $\Lambda_2 \gg \lambda$) and short-ranged ($u_z, \psi \sim e^{-y/\Lambda_1}$, far from RPT $\Lambda_1 \sim \lambda$) components have correspondingly the form

$$\Lambda_2 \simeq \frac{\lambda}{2\pi}\,\frac{(1+\mu)(1-\zeta+\eta)}{\eta}\left[1+(1+\mu)^2\left(\frac{1-\zeta+\eta}{1-\zeta}\right)^2\right]^{1/2},$$
$$\Lambda_1 \simeq \frac{\lambda}{2\pi}\left[1-\frac{\eta}{2(1-\zeta)}\right], \quad 1-\zeta \gtrsim \eta. \tag{6}$$

Formulae (5) and (6) are valid far from RPT as well as near it besides a small neighbourhood of the point of RPT $1-\zeta<\eta$. In all three cases SSMAW velocity diminishes considerably when approaching to RPT, and so do the penetration depths. For the cases II and III in the region of immediate vicinity to RPT $1-\zeta<\eta$ the wave radiation into the bulk appears and oscillations of penetration depth take place.

1. Yu.V. Gulyaev, Pisma v JETP, 9, 63 (1969).
2. J.L. Bleustein, Appl. Phys. Lett., 13, 412 (1968).
3. J.P. Parekh, Electronics Lett., 5, 322 (1969).
4. Yu.V. Gulyaev, Yu.A. Kouzavko, I.N. Oleinik, V.G. Shavrov, JETP, 87, 674 (1984).
5. Yu.A. Kouzavko, I.N. Oleinik, V.G. Shavrov, PTT, 26, 3669 (1984).
6. I.E. Dicshtein, V.V. Tarasenko, V.G. Shavrov, JETP, 67, 816 (1974).
7. S.V. Gerus, V.V. Tarasenko, PTT, 17, 2247 (1975).

FRICTION LOSS OF AN ION TO SURFACE PLASMONS AND ELECTRON-HOLE PAIRS IN HIGH MAGNETIC FIELDS

Godfrey Gumbs
Department of Physics, Dalhousie University, Halifax, Nova Scotia, Canada B3H 3J5

N. J. M. Horing, S. Silverman and H. C. Tso
Department of Physics and Engineering Physics, Stevens Inst. of Tech.
Hoboken, New Jersey 07030, U.S.A.

Abstract: We report here on the effects of a high magnetic field on the frictional energy loss of an ion moving parallel to a solid surface (with the field perpendicular to the surface). Our analysis constitutes a generalization of the corresponding friction parameter calculation of T. L. Ferrell, P. M. Echenique and R. H. Ritchie (Solid State Comm. 32 (1979) 419) to include Landau quantization effects of the ambient magnetic field. We take the semi-infinite solid state plasma response properties to be given by the "diagonal" classical infinite-barrier model with bulk dielectric properties characterized by the RPA (Random-Phase Approximation). In this we consider the quantum strong field limit (all electrons confined to the lowest Landau eigenstate) in our analysis of the part of the frictional energy loss to the excitation of electron-hole pairs in the vicinity of the surface in high magnetic field. The surface magnetoplasmon contribution to the energy loss is determined here using a magnetic field generalization of the hydrodynamic model, which correctly represents the magnetic field dependence of the local surface plasmon. We calculate the frictional energy loss of the ion as a function of its distance from the surface, its velocity and magnetic field strength.

I. INTRODUCTION

This work is directed at generalizing the earlier friction parameter calculation of Ferrell, Echenique and Ritchie[1] (FER) to include the effects of a high magnetic field on the energy loss of an ion passing with velocity v parallel to a solid surface at a distance z_0 from it. The excitations of a quantum plasma behind the surface at z=0 drain energy from the passing ion and are responsible for generating the friction coefficient parameter. Our treatment of the magnetic field (perpendicular to the surface) is undertaken under high field conditions in the quantum strong field limit when all electrons are confined to the lowest Landau eigenstate. In this situation, the electron-hole spectrum[2] is very different from its zero-field counterpart, especially the unquantized hydrodynamic approximation employed by FER in conjunction with electron-hole excitations as well as surface/bulk plasmon modes.

II. FORMULATION OF FRICTION COEFFICIENT IN QUANTUM STRONG FIELD LIMIT

Our analysis proceeds from the friction coefficient formula of FER,

$$\Gamma = \frac{-\hbar}{mv^3}\frac{a_o}{\pi}\int_{-\infty}^{\infty} d\omega \int_{-\infty}^{\infty} dq_y \frac{\omega}{\bar{q}} \exp(-2\bar{q}z_o) Im(\frac{1}{D(\bar{q},\omega)}) , \tag{1}$$

which is valid in the presence and absence of a magnetic field. Γ here is denoted in FER by $M\eta$ where M is the mass of the moving ion. It is defined in terms of the energy loss per unit time $dW/dt = \underset{\sim}{F}\cdot\underset{\sim}{v}$ as $\Gamma \equiv (dW/dt)(e^2/a_0\cdot mv^2/\hbar)^{-1}$ where a_0 is the Bohr radius. Treating InSb, we choose the effective mass of the electron as m =

0.01 m_e and the mobile carrier density in the bulk as $n_B = 10^{17}cm^{-3}$ corresponding to the zero-field Fermi wavenumber $k_F^o = 1.44 \times 10^6\ cm^{-1}$, velocity $v_F^o = 1.66 \times 10^8$ cm/sec and energy $\epsilon_F^o = 0.08$ eV. At a high magnetic field of $H_0 = 150$ kG (quantum strong field limit), the chemical potential $\mu = 0.03$ eV, cyclotron energy $\hbar\omega_c = 0.17$ eV and $k_F \equiv (2m\mu)^{1/2}/\hbar = 8.66 \times 10^5\ cm^{-1}$. In Eq. (1), $\bar{q} = (q_y^2+\omega^2/v^2)^{1/2}$ and $D(\bar{q},\omega)$ is given by

$$D(\bar{q},\omega) = 1 + \frac{\bar{q}}{\pi}\int_{-\infty}^{\infty} dq_z \frac{1}{(q_z^2+\bar{q}^2)\epsilon(q_z;\bar{q},\omega)} \tag{2}$$

and of course $Im(1/D) = -D_I/(D_R^2+D_I^2)$ where the bulk dielectric function $\epsilon(q_z;\bar{q},\omega)$ is given in terms of its real and imaginary parts $\epsilon = \epsilon_R + i\epsilon_I$. For the quantum strong field limit we have[2b]

$$\epsilon_R(q_z;\bar{q},\omega) = 1 - \frac{4\pi e^2}{q^2}\frac{n_B}{2\hbar}\left(\frac{m}{2q_z^2\mu}\right)^{1/2}\exp\left(\frac{-\hbar\bar{q}^2}{2m\omega_c}\right)\sum_{n=0}^{\infty}\frac{1}{n!}\left(\frac{\hbar\bar{q}^2}{2m\omega_c}\right)^n \left\{\ell n\left|\frac{\omega-n\omega_c-\hbar q_z^2/2m + (2q_z^2\mu/m)^{1/2}}{\omega-n\omega_c-\hbar q_z^2/2m - (2q_z^2\mu/m)^{1/2}}\right| + (\omega\to-\omega)\right\} \tag{3}$$

$$\epsilon_I(q_z;\bar{q},\omega) = -\frac{4\pi e^2}{q^2}\frac{\pi\bar{n}_B}{2\hbar}\left(\frac{m}{2q_z^2\mu}\right)^{1/2}\exp\left(\frac{-\hbar\bar{q}^2}{2m\omega_c}\right)\sum_{n=0}^{\infty}\frac{1}{n!}\left(\frac{\hbar\bar{q}^2}{2m\omega_c}\right)^n \{\eta_+(\mu-[\omega-n\omega_c-\frac{\hbar q_z^2}{2m}]^2\frac{m}{2q_z^2}) - (\omega\to-\omega)\}\ , \tag{4}$$

where $(\omega\to-\omega)$ signifies a term which is obtained from the preceding one by replacing the frequency variable ω by $-\omega$. The corresponding high-field relation between the bulk density n_B and chemical potential μ is $n_B = m^{3/2}\omega_c\mu^{1/2}/(2^{1/2}\pi^2\hbar^2)$. The Heaviside unit step function η_+ which appears in $\epsilon_I(g,\omega)$ cuts off the contributing q_z-wavenumber intergation range in D_I with the limits [+(-) for upper (lower) cutoff]

$$q_\pm^2(n) = (k_F^2/\mu)\{2\mu+\hbar(\omega-n\omega_c)\pm[(2\mu+\hbar(\omega-n\omega_c))^2-\hbar^2(\omega-n\omega_c)^2]^{1/2}\} \tag{5}$$

which must be real to contribute.

Our numerical analysis of Γ using Eqs. (1)-(5) is described below (note that in addition to treating the quantum strong field limit, we have also considered the zero-field limit in the RPA in order to provide more detail than the hydrodynamic description of the zero-field case of FER). It should be observed that in our numerical analysis of the quantum strong field limit we have ignored regions where $D_I\to 0$ in calculating the frictional contribution of the electron-hole spectrum as modified by the high magnetic field. We have thus neglected surface plasmon contributions[3,4] corresponding to $-Im(1/D)\pi\delta(D_R)$ in these numerical studies. For high magnetic fields, there are many nonlocal branches of the plasmon spectrum which can contribute, for example, surface Bernstein modes[5,6] as well as the principal local surface magnetoplasmon modes. Generally, such nonlocal plasmon branch contributions are of relatively small amplitude, and we shall neglect them in comparison with the larger amplitude principal mode which we describe in the hydrodynamic model, generalized to include the magnetic field[7], in order to evaluate the surface plasmon contribution to the friction coefficient through $-\mathcal{I}m(1/D)\to\pi\delta(D_R)$. The bulk dielectric function of the hydrodynamic model generalized to include the effect due a magnetic field has been shown to have the form

$$\varepsilon_{Hydno}(\underset{\sim}{q},\omega) = 1 - \frac{\omega_p^2\alpha(\omega,\omega_c)}{q^2[\omega^2-\beta^2\alpha(\omega,\omega_c)]} \tag{6}$$

$$\alpha(\omega,\omega_c) = q_z^2 - \frac{\omega^2\bar{q}^2}{\omega_c^2-\omega^2}, \tag{7}$$

where β is the propagation velocity in the hydrodynamic model. Forming $D(\bar{q},\omega)$ with the use of Eq. (2) $[\omega\to\omega+io^+]$, we have

$$D(\bar{q},\omega) = 1 + \frac{\bar{q}}{\pi}\int_{-\infty}^{\infty} dq_z \frac{q_z^2 - A}{(q_z^2+\beta_+)(q_z^2+\beta_-)}, \tag{8}$$

where

$$\beta_\pm \equiv \frac{1}{2}[B \mp (B^2-4C)^{\frac{1}{2}}] \tag{9a}$$

$$A \equiv \frac{\omega^2\bar{q}^2}{\omega_c^2-\omega^2} + \frac{\omega^2}{\beta^2} \tag{9b}$$

$$B \equiv \left(\frac{\omega_c^2-2\omega^2}{\omega_c^2-\omega^2}\right)\bar{q}^2 + \frac{\omega_p^2-\omega^2}{\beta^2} \tag{9c}$$

$$C \equiv -\left[\frac{\omega^2\bar{q}^2}{\omega_c^2-\omega^2} + \frac{\omega^2(\omega_p^2+\omega_c^2-\omega^2)}{\beta^2(\omega_c^2-\omega^2)}\right]\bar{q}^2 \quad . \tag{9d}$$

The q_z integral in Eq. (8) may be evaluated to yield D_R and D_I for $\omega\to\omega+io^+$ in elementary terms

$$\int_{-\infty}^{\infty} dq_z \frac{1}{q_z^2+\beta_\pm} = \frac{1}{2i\sqrt{\beta_\pm}}\left\{\ell n\left(\frac{q_z-i\sqrt{\beta_\pm}}{q_z+i\sqrt{\beta_\pm}}\right)\right\}. \tag{10}$$

and, writing $\sqrt{\beta\pm} = b_R^{(\pm)}+ib_I^{(\pm)}$ in terms of its real and imaginary parts, we have

$$D(\bar{q},\omega) = 1 + \frac{\bar{q}}{\beta_+-\beta_-}\left\{\frac{\beta_++A}{\sqrt{\beta_+}}\,\mathrm{sgn}(b_R^{(+)}) - \frac{\beta_-+A}{\sqrt{\beta_-}}\,\mathrm{sgn}(b_R^{(-)})\right\} \quad . \tag{11}$$

III. CONCLUSIONS

In discussing our calculations it is useful to denote the electron-hole contribution to the friction coefficient Γ by Γ_{eh} and the surface plasmon contribution by Γ_{sp} so that $\Gamma = \Gamma_{eh}+\Gamma_{sp}$. The numerical calculation of Γ_{eh} has been carried out with the RPA starting first with null magnetic field for ion-surface distances $z=0.1a_o$, $10.0a_o$, $100.0a_o$ and the results are plotted in Fig. 1 showing Γ_{eh} as a function of v/v_F. In the quantum strong field limit, $H_o = 150$ kG, we have also evaluated Γ_{eh} for $z_o = 0.1a_o$, $10.0a_o$ and $100.0a_o$. The results are plotted in Fig. 2 as a function of v/v_F. The value of v where the "maximum" in Γ_{eh} occurs is not sensitive to changes in z_o. In the absence of a magnetic field, the phase velocity of the plasmons which merge with the electron-hole spectrum is about 1.2 v_F, which is much larger than the velocity where the maximum in Γ_{eh} occurs. The numerical results for Γ of FER for H_o=o were obtained by expanding the dielectric function for the hydrodynamic model in powers of $\hbar\omega/\epsilon_F$ and then neglecting terms of order ω^2. In this approximation, the friction parameter Γ (or $M\eta$) of FER is independent of v. These results are in contrast with Fig. 1 where Γ_{eh} is calculated in the RPA with the use of Eqs. (3) and (4) for $\epsilon(q,\omega)$. Numerical results for the contribution Γ_{sp} will be reported elsewhere.

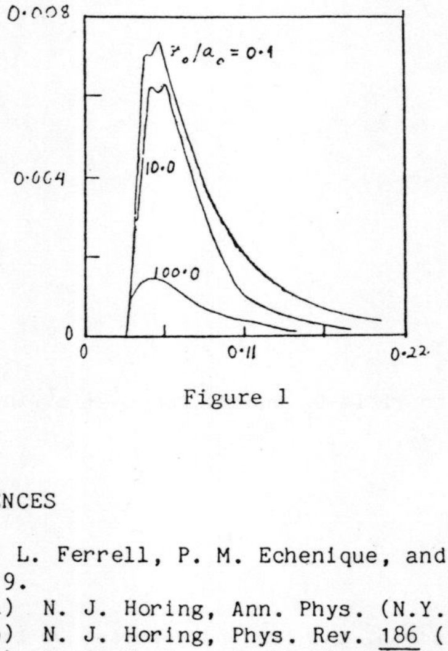

Figure 1

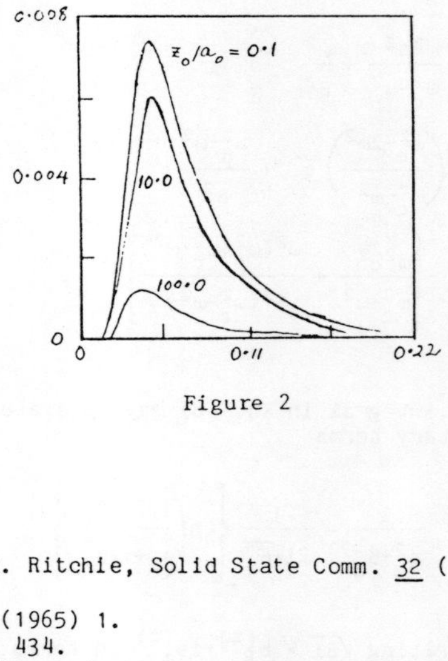

Figure 2

REFERENCES

1. T. L. Ferrell, P. M. Echenique, and R. H. Ritchie, Solid State Comm. 32 (1979) 419.
2. (a) N. J. Horing, Ann. Phys. (N.Y.) 31 (1965) 1.
   (b) N. J. Horing, Phys. Rev. 186 (1969) 434.
   (c) N. J. Horing, in The Many-Body Problem, ed. A. Cruz and T. W. Preist (Plenum, N.Y., 1969) p. 307.
3. N. J. Horing, M. Yildiz, F. Kortel, T. Caglayan, J. Phys. C6 (1973) 2053.
4. G. Gumbs and A. Griffin, J. Phys. F 12 (1982) 1185.
5. N. J. Horing and M. Yildiz, Solid State Comm. 12 (1973) 843.
6. G. Gumbs and N. J. Horing, Phys. Rev. B31 (1985) 4009.
7. S. Silverman, Ph.D. Thesis, Stevens Institute of Technology (1979).

NONLINEAR EVOLUTION OF LOW FREQUENCY SURFACE WAVES

by

Nguyen T. HUNG

Department of Mechanical Engineering
Ecole Polytechnique, P.O. Box 6079, Montreal, H3C 3A7, Canada

The evolution of weakly nonlinear plasma waves has been studies extensively since the mathematical breakthrough in solving such typical equations as the KdV and the nonlinear Schrodinger equations [1]. The basic physical mechanism involved in these equations is a balance between the spreading of the waveform due to dispersion and the steepening of the wave profile due to weak nonlinearities. These (opposing) effects, and consequently the corresponding evolution equations, depend essentially on the type of waves as well as the geometry of the supporting medium. Thus, while the ion acoustic waves in an infinite plasma are governed by the well known KdV equation [2], we will show that the evolution of these waves in a plasma slab is described by a new nonlinear integro-differential equation which admits a closed form algebraic solitary wave solution.

Let us consider a plasma slab made up of cold ions and hot electrons. We are interested in ion acoustic waves propagating along the plasma slab (i.e. the z direction) with typical wave length greater than the thickness d of the plasma. For the low frequency ion waves, a linear analysis of the basic governing equations shows that there exist two modes of propagation, namely, the symmetric and the anti-symmetric "surface modes". In the long wave limit ($kd \ll 1$), we obtain the following dispersion relation for the symmetric mode:

$$\omega = k(1 - |k|/d), \tag{1}$$

where the dispersive term is of second order in k in contrast to the case of a thick plasma slab ($kd \gg 1$) or an infinite plasma where it is of third order in k. This implies that as the thickness of the plasma is reduced, the waves become more dispersive, and, consequently, the wave amplitudes must be larger for solitary waves to form. Moreover, one should expect that the non-analyticity of the wave speed at $k = 0$ will have a decisive effect on the form of the evolution equation.

In order to proceed to the nonlinear regime, we shall use the multiple scaling method and introduce the slow variables

$$\xi = \varepsilon\,(z - t) \text{ and } \tau = \varepsilon^2 t,$$

with ε being a small amplitude parameter.

By expanding the dependent variables in powers of ε and solving successively for the first and second order qualities, we finally obtain the following equation describing the evolution of the ion acoustic waves in a thin plasma slab:

$$\frac{\partial A}{\partial \tau} + A\frac{\partial A}{\partial \xi} + \frac{1}{d}\frac{\partial^2}{\partial \xi^2} H[A] = 0 \tag{2}$$

where

$$H[A] = \frac{1}{\pi}\, p \int_{-\infty}^{\infty} \frac{A\, d\xi'}{\xi' - \xi}$$

It should be noted that this type of equation has been obtained by Benjamin [3] and Ono [4] for internal waves in stratified fluids of great depth, while Ott and Sudan [5] have derived an evolution equation for ion acoustic waves with a somewhat similar integral of the form $(\partial/\partial\xi)H[A]$ corresponding to the Landau damping of ion acoustic waves in an unbounded plasma, in contrast to the present term $(\partial^2/\partial\xi^2)H[A]$ which arises from the wave dispersion in the slab geometry. For stationary solutions of the form $A = A(\xi - c\tau)$, Eq. (2) can be integrated once to yield

$$-cA + \tfrac{1}{2}A^2 + \frac{1}{d}\frac{\partial}{\partial\zeta} H[A] = 0 \tag{3}$$

which admits an exact algebraic solitary wave solution of the form

$$A = \frac{a\Delta^2}{\zeta^2 + \Delta^2} \tag{4}$$

with

$$\zeta = \xi - c\tau, \tag{5a}$$

$$c = \frac{a}{4}, \tag{5b}$$

and

$$\Delta = \frac{2}{ad}. \tag{5c}$$

Here, we note that the width of the solitary wave is inversely proportional to the wave amplitude a, while it is of the order of $a^{-\frac{1}{2}}$ in the case of an unbounded plasma. This implies that for waveforms of the same length scale, the waves must have larger amplitudes, and therefore behave more nonlinearly in a plasma slab than in an infinite plasma. This could have been predicted by noting that for a given wavelength, the dispersive effect in a plasma slab, according to Eq. (1), is $O(k^{-1})$ stronger than that in an unbounded plasma. Consequently, to balance this effect by nonlinearities for permanent waves to form, the wave amplitude must be of the order of k^{-1} larger in a plasma slab than in an unbounded plasma.

In conclusion, we have shown that the nonlinear evolution of ion acoustic waves in a plasma slab is governed by the Benjamin - Ono equation. Recently, this equation has received an increasing interest[6-9], and it has been shown that its solitary wave solutions are in fact solitons in the usual sense. This suggests that the new type of algebraic ion acoustic solitary waves studied here could easily be generated along the boundaries of thin plasma layers.

References

1. A.C. Scott, F.Y.F. Chu and D.W. McLaughlin, Proc. IEEE 61, 1443 (1973).

2. H. Washimi and T. Tanuiti, Phys. Rev. Lett. 17, 996 (1966).

3. T.B. Benjamin, J. Fluid Mech. 29, 559 (1967).

4. H. Ono, J. Phys. Soc. Jpn 39, 1082 (1975).

5. E. Ott and R.N. Sudan, Phys. Fluids 12, 2388 (1969).

6. R.I. Joseph, J. Math. Phys. 18, 2251 (1977).

7. J.D. Meiss and N.R. Pereira, Phys. Fluids 21, 700 (1978).

8. H.H. Chen, Y.C. Lee and N.R. Pereira, Phys. Fluids 22, 187 (1979).

9. Y. Matsuno, J. Phys. Soc. Jpn 48, 1024 (1980).

NONLINEAR INTERACTION OF S AND P SURFACE WAVES AT A NARROW INHOMOGENEOUS LAYER OF THE MAGNETOACTIVE PLASMA

A. M. HUSSEIN
PHYS. DEPT. FACULTY OF SCIENCE, U. A. E. UNIV.

Amplitude of electromagnetic waves with combination frequencies, radiating from a narrow inhomogeneous layer of magmetoactive plasma due to interaction of S and P surface waves, are determined. The method used allow for discontinuity of the tangential components of the wave electric field at the combination frequencies at the plasma boundary.

[1,2] assumed continuity of the tangential components of the wave electric field with combination frequencies at the boundary. Meanwhile, it has been shown by [3] that the tangential components of the electric field of higher harmonic are, in general, discontinuous at the plasma boundary.

It is assumed that the plasma density $n_0(x)$, unperturbed by an electromagnetic field, is zero in the regions $x \leq 0$ and $x \geq a$, and is an arbitrary function of the spatial co-ordinate x in the region $0 \leq x \leq a$. The external magnetic field is in the z direction.

Furthermore an assumption is made that the S and P surface waves with frequencies ω_1 and ω_2 respectively, propagate to meet along the y-axis. Using linear approximation, the electromagnetic field of these waves can be expressed as

$$\{E(r,t); H(r,t)\} = \{E_1(x), H_1(x)\} e^{i(k_1 y - \omega_1 t)} + \{E_2(x), H_2(x)\} e^{i(k_2 y - \omega_2 t)} + C.C. \quad (1)$$

where for S wave

$$H_{z1} = E_{x1} = E_{y1} = 0, \; H_{x1} = n_1 E_{z1}, \; H_{y1} = (ic/\omega_1) \frac{\partial E_{z1}}{\partial x}. \quad (2)$$

and $n_1 = (k_1 c/\omega_1)$

E_{z1} can be drived f[redacted]m the equation

$$\frac{\partial^2 E_{z1}}{\partial x^2} + \varkappa_1^2 E_{z1} = 0 \qquad (3)$$

where $\varkappa_1^2 = (\omega_1^2/c^2)(n_1^2 - \varepsilon_1)$, $\varepsilon_1 = 1 - \frac{\Omega_p^2(x)}{\omega_1^2}$, $\Omega_p^2(x) = \frac{4\pi e^2 n_0(x)}{m}$

e and m are the charge and mass, respectively. c velocity of light.

In the regions $x \leqslant 0$ and $x \geqslant a$, the solution of eq. (3), describin the incident surface waves on the plasma layer, are

$$\left.\begin{aligned} E_{z1} &= E_{10}\, e^{\varkappa_{10} x} & x \leqslant 0 \\ E_{z2} &= E_{1a}\, e^{-\varkappa_{10}(x-a)} & x \geqslant a \end{aligned}\right\} \qquad (4)$$

where
$\varkappa_{10} = (\omega_1/c)(1 - n_1^2)^{1/2}$

Since we are considering large wavelength compared to the width of the plasma layer $|\varkappa_{p}| \ll 1$, the solution of eq. (3) in the region $0 \leqslant x \leqslant a$ will have the form

$$E_{z1} = \varkappa_{10} E_{10}\left[x - \int_0^x dx' \int_0^{x'} dx'' \varkappa_1^2\right] + E_{10}\left[1 - \int_0^x dx' \int_0^{x'} dx'' \varkappa_1^2\right] \qquad (5)$$

For P wave

$$E_{z2} = H_{x2} = H_{y2} = 0,\ E_{x2} = \frac{c}{\omega_2}\cdot\frac{1}{\varepsilon_\perp}\left(\frac{\varepsilon_2}{\varepsilon_1} - k_2 H_{z2}\right),\ E_{y2} = \frac{ic}{\omega_2}\cdot\frac{1}{\varepsilon_\perp}\left(k_2 \frac{\varepsilon_2}{\varepsilon_1} H_{z2} - \frac{\partial H_{z2}}{\partial x}\right) \qquad (6)$$

where $\varepsilon_1 = 1 - \frac{\Omega_p^2}{\omega_2^2 - \omega_c^2}$, $\varepsilon_2 = \frac{-\omega_c \Omega_p^2}{\omega_2(\omega_2^2 - \omega_c^2)}$, $\varepsilon_\perp = \frac{\varepsilon_1^2 - \varepsilon_2^2}{\varepsilon_1}$

H_{z2} can be drived from the equation

$$\frac{\partial}{\partial x}\left(\frac{1}{\varepsilon_\perp}\frac{\partial H_{z2}}{\partial x}\right) - \varkappa_2^2 H_{z2} = 0 \qquad (7)$$

where $\varkappa_2^2 = (\omega_2^2/c^2)\left[\frac{n_2^2}{\varepsilon_\perp} - 1 + n_2^2 \frac{\partial}{\partial x}\left(\frac{\varepsilon_2}{\varepsilon_1 \varepsilon_\perp}\right)\right]$

In the regions $x \leqslant 0$ and $x \geqslant a$, the solution of eq. (7), describing the incident surface waves on the plasma layer, are

$$\left.\begin{aligned} H_{z2} &= H_{z0}\, e^{\varkappa_{20} x} & x \leqslant 0 \\ H_{z2} &= H_{za}\, e^{-\varkappa_{20}(x-a)} & a \leqslant x \end{aligned}\right\} \qquad (8)$$

When $|\varkappa_{2}a| \ll 1$, we have in the region

$$H_{z2} = H_{z0}\left\{1 + \int_0^x dx' \varepsilon_\perp \int_0^{x'} dx'' (\varkappa_2^2/\varepsilon_\perp)\right\} + H_{20}\varkappa_{20}\int_0^x dx' \varepsilon_\perp \qquad (9)$$

In the next approximation of the perturbation theory, we have the following set of equations for the fields at combination frequencies.

The equation for the wave magnetic field with combination frequency ω', can take the form

$$\frac{\partial}{\partial x}\left(\frac{1}{\varepsilon'_\perp}\frac{\partial H'_z}{\partial x}\right)+\varkappa'^2 H'_z = R \qquad (10)$$

where $\varkappa'^2 = (\omega'^2/c^2)-(k'^2/\varepsilon'_\perp)-k'\frac{\partial}{\partial x}(\varepsilon'_2/\varepsilon'_1\varepsilon'_\perp)$.

$$R=\frac{4\pi}{c}\left\{\frac{\partial}{\partial x}\frac{1}{\varepsilon'_\perp\varepsilon'_1}(i\varepsilon'_2 j_{x0}-\varepsilon'_1 j_{y0})-\frac{ik'}{\varepsilon'_\perp\varepsilon'_1}(\varepsilon'_1 j_{x0}+i\varepsilon'_2 j_{y0})\right.$$

$$\frac{4\pi}{c}j_{x0}=\frac{\Omega_p^2}{\alpha mc^2\omega_1}H_{y1}E_{z1}-\frac{i\Omega_p^2 e\omega_c}{\alpha mc^2\omega_1}H_{x1}E_{z1}+\frac{\Omega_p^2 e\omega'}{\alpha mc^2\omega_2\phi}H_{z2}E_{y2}+\frac{i\Omega_p^2\omega_c\omega'}{\alpha mc^2\omega_2\phi}H_{z2}E_{x2}$$
$$+\frac{i\Omega_p^2 e\omega_c}{\alpha mc^2\omega_2\phi}H_{z2}E_{y2}-\frac{\Omega_p^2 e\omega_c^2}{\alpha mc^2\omega_2^2\phi}H_{z2}E_{x2}+\left[\frac{iek_2\omega_c\Omega_p^2}{\alpha mc\omega_2^4\phi}(\omega_2\omega'+\omega_c^2)\right]$$
$$-\left.\frac{i\Omega_p^2 e\omega'(\omega'+\omega_2)}{2\alpha mc\omega_2^3\phi}\frac{\partial}{\partial x}\right]E_{x2}^2+\left\{\frac{-i\Omega_p^2 en_0k'\omega_c}{\alpha\phi^2 mc\omega_2^2}(\omega'+\omega_2)+\frac{i\Omega_p^2 e\omega_c^2[\omega'(\omega'+\omega_2)+\omega_2^2-\omega_c^2]}{2\alpha mc\omega_2^5\phi^2}\right\}E_{y2}^2$$
$$-\frac{\Omega_p^2 e\omega_c\omega'}{2\alpha mc\omega_2^4\phi^2}(2\omega'+\omega_2)E_{y2}\frac{\partial E_{x2}}{\partial x}-\frac{\Omega_p^2 e\omega_c}{2\alpha mc\omega_2^4\phi^2}(\omega_2^2+2\omega_1^2+5\omega_1\omega_2)E_{x2}\frac{\partial E_{y2}}{\partial x}$$
$$+\frac{\Omega_p^2 k_2}{\alpha mc\omega_2^4\phi^2}\left[\omega'(\omega_2^2+\omega_c^2)+\omega_2\omega_c^2\right]E_{x2}E_{y2}.$$

$$\frac{4\pi}{c}j_{y0}=\frac{-i\Omega_p^2 e\omega_c}{\alpha c^2 m\omega_1}H_{y1}E_{z1}+\frac{\Omega_p^2\omega'}{\alpha c^2 m\omega_1}H_{x1}E_{z1}+\frac{i\Omega_p^2 e\omega_c}{\alpha c^2 m\omega_2\phi}E_{y2}H_{z2}-\frac{\Omega_p^2 e\omega_c^2}{\alpha c^2 m\omega_2^2\phi}E_{x2}H_{z2}$$
$$-\frac{\Omega_p^2\omega'}{\alpha c^2 m\omega_2\phi}E_{y2}H_{z2}-\frac{i\Omega_p^2\omega_c\omega'}{\alpha c^2 m\omega_2^2\phi}E_{x2}H_{z2}+\left[\frac{\Omega_p^2 e\omega_c(\omega'+\omega_2)}{2\alpha c\omega_2^2 m\phi^2}\frac{\partial}{\partial x}-\frac{\Omega_p^2 ek_2\omega_c^2}{\alpha mc\omega_3^5\phi^2}\right.$$
$$\left.\cdot(3\omega_2\omega'+\omega_1^2-\omega_c^2)\right]E_{x2}^2+\left[\frac{\Omega_p^2 ek_2}{\alpha\omega_2^3 mc\phi}(2\omega_c^2-\omega_1\omega')-\frac{\Omega_p^2 e\omega_c}{2\alpha mc\omega_2^4\phi^2}(\omega_c^2+\omega_2\omega')\frac{\partial}{\partial x}\right]E_{y2}^2$$
$$-\frac{i\Omega_p^2 e(\omega_2\omega'+\omega_c^2)}{\alpha mc\omega_2^2\phi^2}E_{x2}\frac{\partial E_{y2}}{\partial x}-\frac{i\Omega_p^2 e\omega_c(\omega'+\omega_2)}{\alpha mc\omega_2^4\phi^2}E_{y2}\frac{\partial E_{x2}}{\partial x}+\frac{i\omega_c\Omega_p^2 ek_2(4\omega_2\omega'+\omega_2^2)}{\alpha mc\omega_2^4\phi^2}E_{x2}E_{y2}$$

$$\varepsilon'_1=1-\frac{\Omega_p^2}{\alpha},\quad \varepsilon'_2=\frac{-\Omega_p^2}{\omega'\alpha},\quad \varepsilon'_\perp=\frac{\varepsilon_1'^2-\varepsilon_2'^2}{\varepsilon'_1},\quad \alpha=\omega'^2-\omega_c^2,\quad \phi=1-\frac{\omega_c^2}{\omega_2^2}$$

Outside the plasma(($x\leq 0$ and $x\geq a$), we have R = 0.

The solution of eq. (10) can have the following form:

$$H'_z = H_- e^{-i\varkappa'_0 x} \qquad x\leq 0$$
$$H'_z = H_+ e^{i\varkappa'_0(x-a)} \qquad x\geq a \qquad (11)$$

where H and H the amplitudes of the generated waves at

x=0 and x=a, respectively, are independent of x.

For the region $0 \leq x \leq a$, $|\varkappa' a| \leq 1$, the wave magnetic field with combination frequency can be written in the form

$$H_z' = -\int_0^x dx' \varepsilon_\perp' \int_0^{x'} dx'' \varkappa'^2 \int_0^{x''} dx''' \varepsilon_\perp \int_0^{x'''} dx'''' R + \int_0^x dx \varepsilon_\perp \int_0^{x'} dx'' R + A \int_0^x dx' \varepsilon_\perp' + B. \quad (12)$$

From (11) and (12) and the continuity conditions for $H_z(x)$ and $\partial H_z / \partial x$ at x=0 and x=a, we can find the magnetic field amplitude of the wave at the combination frequency:

$$H_\pm = -\left(\frac{i}{2\varkappa_0'}\right) \int_0^a dx R + i \frac{\varkappa'}{\varkappa_0} \int_0^a dx \frac{\varepsilon_2'}{\varepsilon_1'} \int_0^x dx' R \pm \frac{1}{2} \int_0^a dx \varepsilon_\perp' \int_0^x dx' R. \quad (13)$$

In the last equation we have omitted the terms proportional to a^2.

Therefore the results obtained here differ essentially from the corresponding results in which continuity of the tangential components of the electric field of the high harmonics at the plasma boundary is assumed.

For normal incidence ($k_1 = k_2 = 0$) the waves are emitted; this contrasts with the case of an isotropic layer where the waves for normal incidence are not emitted.

Also, the effect of an external magnetic field leads to a sharp increase in the generated amplitudes near the resonance $\omega' \sim \omega_c$ and $\omega_2 \sim \omega_c$.

REFERENCES

1. R. F. Whitmer & E. B. Barrett, Phys. Rev., 121, 661 (1961)
2. K. S. Karpylyuk & V. N. Oraevsky, Soviet Phys. Tech. Fiz. 13, 1241 (1968).
3. A. R. Barakate, V. V. Dolgopolov & N. M. El-Siragy, Plasma Phys. 17, 89 (1975).

WAVE GENERATION BY NONLINEAR INTERACTION OF INCIDENT RADIATION AND SURFACE WAVE IN MAGNETIZED PLASMA

Sh.M.Khalil, N.M.El-Siragy[(+)], I.A.El-Naggar[(o)] and R.N.El-Sherif

Plasma Phys. Dept., Nuclear Research Centre, Atomic Energy Authority, Cairo/Egypt.

Abstract. The wave generation with combination frequencies due to the nonlinear interaction of incident radiation and a surface wave in plasma layer (both of P-polarization) is studied analytically.

1. Introduction.

Different from other works /1,2/, we shall consider a plasma layer of width a in which the plasma density is an aribtrary differential function of x in the region $0\leqslant x\leqslant a$ and equal to zero in the regions $x\leqslant o$, $x\geqslant a$. This layer is supposed to external static magnetic field directed along z -direction, $\vec{H}_o=\vec{e}_z H_o$. We shall assume that the P-surface wave with frequency ω_1 and P-incident radiation with frequency ω_2 propagate to meet each other along the y -axis. We shall also neglect the thermal velocities, the positive ion current and the pressure gradients.

2. Fundamental Waves.

The interacting waves can be represented linearly by

$$[\vec{E}(\vec{r},t),\vec{H}(\vec{r},t)] = \sum_{n=1,2} [E_n(x),H_n(x)].\exp i(K_n y-\omega t) + CC$$

with wave components: $\vec{E}\equiv(E_x,E_y,0)$, $\vec{H}\equiv(0,0,H_z)$

Using the linearized Maxwell equations and the equation of motion, we can obtain the following expressions for the fundamental electric field components:

$$E_{nx} = \frac{-N_n}{K_n}\frac{1}{\mathcal{E}_{\perp n}}\left(K_n H_{nz}+\frac{\mathcal{E}_{2n}}{\mathcal{E}_{1n}}\frac{\partial H_{nz}}{\partial x}\right), \qquad N_n=\frac{K_n c}{\omega_n},$$

$$E_{ny} = -i\frac{N_n}{K_n}\frac{1}{\mathcal{E}_{\perp n}}\left(K_n\frac{\mathcal{E}_{2n}}{\mathcal{E}_{1n}}H_{nz}+\frac{\partial H_{nz}}{\partial x}\right), \qquad E_{nz}=0, \qquad \dots.(1)$$

where, $\mathcal{E}_{1n}=1-\frac{\omega_p^2}{\omega_n^2-\omega_c^2}$, $\mathcal{E}_{2n}=\frac{\omega_c\omega_p^2}{\omega_n(\omega_n^2-\omega_c^2)}$, $\mathcal{E}_{\perp n}=\frac{\mathcal{E}_{1n}^2-\mathcal{E}_{2n}^2}{\mathcal{E}_{1n}}$, $\omega_c=\frac{eH_o}{mc}$

$n=1,2$ refers to the surface wave (1) and incident radiation (2). The magnetic field components of the fundamental P-polarized surface and incident waves, are subject to the following differential equation:

$$\varepsilon_{\perp n}\frac{\partial}{\partial x}\left(\frac{1}{\varepsilon_{\perp n}}\frac{\partial H_{nz}}{\partial x}\right)+\mathscr{x}_n^2 H_{nz}=0 \quad \text{.......(2)}$$

where, $\mathscr{x}_n^2=(\omega_n^2/c^2)\varepsilon_{\perp n}-K_n^2-K_n\varepsilon_{\perp n}\frac{\partial}{\partial x}\left(\varepsilon_{2n}/\varepsilon_{1n}\varepsilon_{\perp n}\right)$,

and for surface waves $\mathscr{x}_1^2<0$, while for radiation $\mathscr{x}_2^2>0$.

Solutions for equation (2) are:

For surface wave (ω_1 , K_1):

$$H_{1z}=H_{01}\exp(\mathscr{x}_{01}x) \qquad , x\leqslant 0$$

$$H_{1z}=H_{a1}\exp[\mathscr{x}_{01}(x-a)] \qquad , x\geqslant a$$

$$H_{1z}=H_{01}\Big[1+\int_0^x\varepsilon_{\perp 1}dx'\int_0^{x'}(\mathscr{x}_1^2/\varepsilon_{\perp 1})\,dx''\Big]+H_{01}\mathscr{x}_{01}\int_0^x\varepsilon_{\perp 1}dx', \quad 0\leqslant x\leqslant a \quad \text{..(3)}$$

where we considered that the characteristic lengths of the waves under investigation are large compared to the width of the transition layer, i.e., $|\mathscr{x}_i a|\ll 1$ and H_{01} , H_{a1} are the amplitudes of the surface wave at $x=0$, $x=a$ and $\mathscr{x}_{0n}^2=(\omega_n^2/c^2)-K_n^2$ is the decay coefficient in vacuum.

For incident radiation (ω_2 , K_2):

$$H_{2z}=H_{02}\left[\exp(i\mathscr{x}_{02}x)-\Gamma_p\exp(-i\mathscr{x}_{02}x)\right] \qquad , x\leqslant 0$$

$$H_{2z}=H_{a2}\exp[i\mathscr{x}_{02}(x-a)] \qquad , x\geqslant a$$

$$H_{2z}=H_{02}\Big\{(1-\Gamma_p)\Big[1-\int_0^x\varepsilon_{\perp 2}dx'\int_0^{x'}(\mathscr{x}_2^2/\varepsilon_{\perp 2})dx''\Big]+i\mathscr{x}_{02}(1+\Gamma_p)\,\cdot$$

$$\cdot\Big[\int_0^x\varepsilon_{\perp 2}dx'-\int_0^x dx'\varepsilon_{\perp 2}\int_0^{x'}(\mathscr{x}_2^2/\varepsilon_{\perp 2})dx''\int_0^{x''}\varepsilon_{\perp 2}dx'''\Big]\Big\}, \quad 0\leqslant x\leqslant a \quad \text{..(4)}$$

where Γ_p is the reflection coefficient of the plasma layer, and given by:

$$\Gamma_p=\Big[\int_0^a\frac{\mathscr{x}_2^2}{\varepsilon_{\perp 2}}dx-\mathscr{x}_{02}^2\int_0^a\varepsilon_{\perp 2}dx\Big]\Big[2i\mathscr{x}_{02}+\mathscr{x}_{02}^2\int_0^a\varepsilon_{\perp 2}dx+\int_0^a\frac{\mathscr{x}_2^2}{\varepsilon_{\perp 2}}dx\Big]^{-1} \quad \text{...(5)}$$

3. <u>Generation With Frequencies $\omega=\omega_1\pm\omega_2$</u>:

If we allow now for the nonlinear terms in the Maxwell equations and equation of motion, we can derive the differential equation governing the generated magnetic field as:

$$\varepsilon_{\pm\perp}\frac{\partial}{\partial x}\left(\frac{1}{\varepsilon_{\pm\perp}}\frac{\partial H_{\pm z}}{\partial x}\right)+\mathscr{x}_{\pm}^2 H_{\pm z}=R_{\pm}(x) \quad \text{.....(6)}$$

where,

$$\varepsilon_{\pm\perp}=\frac{\varepsilon_{\pm 1}^2-\varepsilon_{\pm 2}^2}{\varepsilon_{\pm 1}},\ \varepsilon_{\pm 1}=1-\frac{\omega_p^2}{\omega_{\pm}^2-\omega_c^2},\ \varepsilon_{\pm 2}=\frac{\omega_c\omega_p^2}{\omega_{\pm}(\omega_{\pm}^2-\omega_c^2)}$$

$$\mathscr{x}_\pm^2 = \frac{\omega_\pm^2}{c^2}\mathcal{E}_{\pm\perp} - K_\pm^2 - K_\pm \mathcal{E}_{\pm\perp}\frac{\partial}{\partial x}\left(\frac{\mathcal{E}_{\pm 2}}{\mathcal{E}_{\pm 1}\mathcal{E}_{\pm\perp}}\right), \quad K_\pm = K_1 \pm K_2$$

$$R_\pm(x) = -\frac{4\pi}{c}\,\mathcal{E}_{\pm\perp}\left\{\left[\text{Curl}\frac{\vec{J}''}{\mathcal{E}_{\pm\perp}}\right]_z + i\left[\text{div}\,\frac{\mathcal{E}_{\pm 2}}{\mathcal{E}_{\pm 1}\,\mathcal{E}_{\pm\perp}}\right]\right\}$$

It is clear from (6) that the generated waves are of P-polarization, and it may be either surface waves ($\mathscr{x}_\pm^2 > 0$) or electromagnetic radiation. J'' are the nonlinear additions and given by:

$$J_x'' = \frac{\omega_p^2}{4\pi(\omega_\pm^2 - \omega_c^2)}\left(\frac{e}{mc}\right)\sum_{\alpha,\beta=1,2}^{\alpha\neq\beta}\frac{\omega_\alpha H_{\beta z}}{(\omega_\alpha^2 - \omega_c^2)}\left[\frac{i\omega_c\omega_\beta}{\omega_\alpha^2 - \omega_c^2}E_{\alpha x} + \right.$$

$$\left. + \left(\omega_\pm + \frac{\omega_c^2}{\omega_\alpha^2}\right)E_{\alpha y}\right] + \frac{1}{4\pi}\left(\frac{\omega_p^2}{n_0}\right)\sum_{\alpha,\beta=1,2}^{\alpha\neq\beta}\frac{\omega_\alpha\,\text{div}(n_0\vec{V}_\beta)}{\omega_\beta(\omega_\alpha^2 - \omega_c^2)}\left(E_{\alpha x} + i\frac{\omega_c}{\omega_\alpha}E_{\alpha y}\right),$$

$$J_y'' = \frac{\omega_p^2}{4\pi(\omega_\pm^2 - \omega_c^2)}\left(\frac{e}{mc}\right)\sum_{\alpha,\beta=1,2}^{\alpha\neq\beta}\frac{\omega_\alpha H_{\beta z}}{(\omega_\alpha^2 - \omega_c^2)}\left[\left(\frac{\omega_c^2}{\omega_\alpha^2} - \omega_\pm\right)E_{\alpha x} - \right.$$

$$\left. - i\omega_c\left(2 + \frac{\omega_\beta}{\omega_\alpha}\right)E_{\alpha y}\right] + \frac{1}{4\pi}\left(\frac{\omega_p^2}{n_0}\right)\sum_{\alpha,\beta=1,2}^{\alpha\neq\beta}\frac{\omega_\alpha\,\text{div}.(n_0\vec{V}_\beta)}{\omega_\beta(\omega_\alpha^2 - \omega_c^2)}\left(E_{\alpha y} - i\frac{\omega_c}{\omega_\alpha}E_{\alpha x}\right) \quad \text{..(7)}$$

If we consider the case of electromagneic waves generation, we can obtain solutions for equation (6) as:

$$H_{\pm z} = H_{\pm 0}\exp.\left(-i\mathscr{x}_{\pm 0}x\right), \quad x \leq 0$$

$$H_{\pm z} = H_{\pm a}\exp.\left[i\mathscr{x}_{\pm 0}(x - a)\right], \quad x \geq a$$

$$H_{\pm z} = H_{\pm 0}\left\{1 - \int_0^x \mathcal{E}_{\pm\perp}dx'\int_0^{x'}\frac{\mathscr{x}_\pm^2}{\mathcal{E}_{\pm\perp}}dx''\right\} - i\mathscr{x}_{\pm 0}H_{\pm 0}\left\{\int_0^x \mathcal{E}_{\pm\perp}dx' - \right.$$

$$\left. - \int_0^x \mathcal{E}_{\pm\perp}dx'\int_0^{x'}\frac{\mathscr{x}_\pm^2}{\mathcal{E}_{\pm\perp}}dx''\int_0^{x''}\mathcal{E}_{\pm\perp}dx'''\right\} + \int_0^x \mathcal{E}_{\pm\perp}dx'\int_0^{x'}\frac{R_\pm(x'')}{\mathcal{E}_{\pm\perp}}dx'', \quad 0 \leq x \leq a \quad \text{...(8)}$$

where, $\mathscr{x}_{\pm 0}^2 = (\omega_\pm^2/c^2) - K_\pm^2$, and $H_{\pm 0}$, $H_{\pm a}$ are the amplitudes of the generated waves in vacuum. To obtain such solution in region $0 \leq x \leq a$, we used the method of successive approximation, taking into consideration that the characteristic lengths of the waves under investigation are large compared to the width of the plasma layer, i.e., $|\mathscr{x}_\pm(x).a| \ll 1$.

From the conditions of continuity of the functions $H_{\pm z}$ and $\partial H_{\pm z}/\partial x$ at points $x = 0$ and $x = a$ we can find $H_{\pm 0}$ and $H_{\pm a}$ as:

$$H_{\pm 0} = \frac{\int_0^a (R_\pm/\mathcal{E}_{\pm\perp})dx - i\mathscr{x}_{\pm 0}\int_0^a \mathcal{E}_{\pm\perp}dx\int_0^x (R_\pm/\mathcal{E}_{\pm\perp})dx'}{2i\mathscr{x}_{\pm 0} + \mathscr{x}_{\pm 0}^2\int_0^a \mathcal{E}_{\pm\perp}dx + \int_0^a (\mathscr{x}_\pm^2/\mathcal{E}_{\pm\perp})dx},$$

$$H_{\pm a} = \left(1 - i\mathscr{X}_{\pm o}\int_0^a \mathcal{E}_{\pm\perp} dx\right) H_{\pm o} + \int_0^a \mathcal{E}_{\pm\perp} dx \int_0^x (R_{\pm} / \mathcal{E}_{\pm\perp})\, dx' \quad \ldots .(9)$$

4. Conclusion.

To get more clear picture and physical meaning of the results, it is necessary find out the values of the integrals $\int_0^a (R_{\pm}/\mathcal{E}_{\pm\perp})\,dx$ and $\int_0^a \mathcal{E}_{\pm\perp} dx \int_0^x (R_{\pm}/\mathcal{E}_{\pm\perp})\,dx'$. These are in fact to long and rather complicated and depends on the density profile which in our case assumed to be arbitrary.

As special case, for low density plasma ($\omega_p \ll \omega_2$), the reflection coefficient r_p can be neglected compared to unity. Besides the quantities $\mathcal{E}_{\pm 1}$, $\mathcal{E}_{\pm\perp}$, $\mathcal{E}_1$ and $\mathcal{E}_{\perp}$ becomes close to unity. In this case, the generated amplitudes are approximately the same and radiated from the plasma layer in both directions, and they given by:

$$H_{\pm o} = H_{\pm a} \cong \left(\frac{2\pi}{c}\right)\left(\frac{K_{\pm}}{\mathscr{X}_{\pm o}}\right)\int_0^a \left(J_x'' - \mathcal{E}_{\pm 2} J_y''\right) dx \quad \ldots .(10)$$

For normal incident of radiation ($K_2 = 0$), it is clear from (10) that $K_{\pm} = K_1$ and the amplitudes tends to decrease.

When we take into account the effect of applied external static magnetic field, this may lead to sharp increase in the generated amplitudes (equation 9), specially near the resonance values, i.e., when $\omega_1 \pm \omega_2 \to \omega_c$ or $\omega_1 \to \omega_c$, $\omega_2 \to \omega_c$.

If $\omega_c \to 0$, the obtained results are reduced to that of Dolgopolov, et. al. /1/.

References.

/1/ V.V.Dolgopolov, A.M.Hussein, I.A.El-Naggar and Sh.M.Khalil, Physica 83 C, 241, 1976.

/2/ A.R.Barakat, V.V.Dolgopolov, N.M.El-Siragy and Y.A.Sayed, Physica 79 C, 419, 1975.

(+) Tanta University, Phys. Dept., Faculty of Science, Tanta, A.R.E.

(o) Alexandria University, Phys. & Chem. Dept., Faculty of Education, Alex., A.R.E.

LEAKING SURFACE WAVES AND NONLINEAR PLASMA TRANSPARENCY

by

A.V. KOCHETOV and A.M. FEIGIN

Institute of Applied Physics, Academy of Sciences
of the USSR, 46 Uljanov Street, Gorky, USSR

The problem on the nonlinear penetration of TM wave into overdense plasma is solved. A stationary one-dimensional model of TM wave interaction with homogeneous layer of overdense plasma is investigated. Nonlinear plasma properties are considered to be determined by the striction. It is shown that TM polarization of the wave plays a leading role for nonlinear penetration associated with the appearance of the regions in plasma, where the plasma permittivity changes the sign. The field structure and plasma density in these regions are determined by the plasma resonance phenomenon. They are a leaking surface wave propagating along the alternating permittivity profile which is formed by this wave. Such field and density distributions are realized if the energy flux passing the layer is not too large ($S < S_m$). Here S_m is the critical flux value depending on the unperturbed plasma permittivity ε_o and the wave angle of incidence on the layer Q. When the flux S approaches the critical value, the characteristic space period of the structure decreases. In accordance with it the dependences of transmission coefficient and reflection coefficient on the incident wave amplitude, which are of pronounced hysteresis character, become strongly curved.

When the energy fluxes ($S > S_m$) are strong, the leaking surface waves are not excited in the layer. In this case, each angle of incidence $Q < \pi/4$ has a characteristic amplitude of the incident wave E_o^*, defined by ε_o and Q, for which a supercritical plasma layer of arbitrary thickness becomes reflectionless. This result is a nonlinear analogue of Brewster effect well-known in the linear theory.

THREE-WAVE INTERACTION OF SURFACE WAVES IN A THIN COLD PLASMA COLUMN

N. Kostov

Institute of Electronics, Bulgarian Academy of Sciences,
BG-1784 Sofia, Bulgaria

I. Zhelyazkov

Faculty of Physics, Sofia University
BG-1126 Sofia, Bulgaria

In a recent paper,[1] we have studied the three-wave interaction of high frequency surface waves in a cold plasma column with sharp movable boundary. The plasma is supposed to be surrounded by vacuum. We have obtained the coupled-mode equations for the nonresonant interaction of plasma surface modes. The coupling coefficients are expressed in explicit forms which allow us to discuss the contributions of both surface charges at the plasma-vacuum interface and movable boundary. The coupled-mode equations for azimuthally-symmetric surface waves travelling along a cold plasma column of radius R surrounded by vacuum are[1]

$$\left(\frac{\partial}{\partial \tau} + v_{g1}\frac{\partial}{\partial \xi}\right)\hat{\phi}_1 = i\lambda_1 \hat{\phi}_2^* \hat{\phi}_3$$

$$\left(\frac{\partial}{\partial \tau} + v_{g2}\frac{\partial}{\partial \xi}\right)\hat{\phi}_2 = i\lambda_2 \hat{\phi}_1^* \hat{\phi}_3 \qquad (1)$$

$$\left(\frac{\partial}{\partial \tau} + v_{g3}\frac{\partial}{\partial \xi}\right)\hat{\phi}_3 = i\lambda_3 \hat{\phi}_1 \hat{\phi}_2$$

where $\hat{\phi}_j I_0(k_j R)$ are surface waves' amplitudes, and λ_j the corresponding coupling coefficients

$$\lambda_1 = -\frac{e}{2m}\frac{\omega_1^2 k_2 k_3}{\omega_2 \omega_3}\left\{\frac{I_1(k_2R)\,I_1(k_3R)}{I_0(k_1R)}\left(\frac{1}{\omega_1} + \frac{1}{\omega_2} - \frac{1}{\omega_3} - \frac{\omega_p^2}{\omega_1\omega_2\omega_3}\right)\right.$$

$$+\frac{1}{\omega_1}\frac{I_0(k_2R)I_0(k_3R)}{I_0(k_1R)}+\frac{1}{\omega_2}\frac{I_0(k_3R)I_1(k_2R)}{I_1(k_1R)}+\frac{1}{\omega_3}\frac{I_0(k_2R)I_1(k_3R)}{I_1(k_1R)}\Big\}$$

$$\lambda_2=-\frac{e}{2m}\frac{\omega_2^2k_1k_3}{\omega_1\omega_3}\Big\{\frac{I_1(k_1R)I_1(k_3R)}{I_0(k_2R)}\left(\frac{1}{\omega_1}+\frac{1}{\omega_2}-\frac{1}{\omega_3}-\frac{\omega_p^2}{\omega_1\omega_2\omega_3}\right)$$

$$+\frac{1}{\omega_1}\frac{I_0(k_3R)I_1(k_1R)}{I_1(k_2R)}+\frac{1}{\omega_2}\frac{I_0(k_1R)I_0(k_3R)}{I_0(k_2R)}+\frac{1}{\omega_3}\frac{I_0(k_1R)I_1(k_3R)}{I_1(k_2R)}\Big\} \tag{2}$$

$$\lambda_3=-\frac{e}{2m}\frac{\omega_3^2k_1k_2}{\omega_1\omega_2}\Big\{\frac{I_1(k_1R)I_1(k_2R)}{I_0(k_3R)}\left(\frac{1}{\omega_1}+\frac{1}{\omega_2}-\frac{1}{\omega_3}-\frac{\omega_p^2}{\omega_1\omega_2\omega_3}\right)$$

$$+\frac{1}{\omega_1}\frac{I_0(k_2R)I_1(k_1R)}{I_1(k_3R)}+\frac{1}{\omega_2}\frac{I_0(k_1R)I_1(k_2R)}{I_1(k_3R)}+\frac{1}{\omega_3}\frac{I_0(k_1R)I_0(k_2R)}{I_0(k_3R)}\Big\}$$

As can be seen from the dispersion characteristics of the surface modes

$$\omega_j=\omega_p\left[1+\frac{I_0(k_jR)K_1(k_jR)}{I_1(k_jR)K_0(k_jR)}\right]^{-1/2}$$

the only way to satisfy the synchronism conditions

$$\begin{aligned}\omega_3&=\omega_1+\omega_2+\Delta\omega\\ k_3&=k_1+k_2\end{aligned} \tag{3}$$

even approximately ($\Delta\omega\to 0$), is by coupling three waves with frequencies small enough to make use of the almost linear portion of the dispersion curve near the origin. This occurs only in a thin cylinder limit ($k_jR \ll 1$), where the coupling coefficients and wave group velocities take the form

$$\lambda_1=-\frac{e}{2m}\frac{\omega_1^2k_2k_3}{\omega_2\omega_3}\left[\frac{1}{\omega_1}+\frac{1}{k_1}\left(\frac{k_2}{\omega_2}+\frac{k_3}{\omega_3}\right)\right]$$

$$\lambda_2 = -\frac{e}{2m}\frac{\omega_2^2 k_1 k_3}{\omega_1\omega_3}\left[\frac{1}{\omega_2} + \frac{1}{k_2}\left(\frac{k_1}{\omega_1} + \frac{k_3}{\omega_3}\right)\right] \tag{4}$$

$$\lambda_3 = -\frac{e}{2m}\frac{\omega_3^2 k_1 k_2}{\omega_1\omega_2}\left[\frac{1}{\omega_3} + \frac{1}{k_3}\left(\frac{k_1}{\omega_1} + \frac{k_2}{\omega_2}\right)\right]$$

$$d\omega_j/dk_j = \frac{1}{2}\frac{\omega_p^2}{\omega_j^2} k_j R^2 \left(1 + 2\left|\ln k_j R\right|\right). \tag{5}$$

Using the renormalizations

$$\hat{\phi}_1 \longrightarrow i(\lambda_2\lambda_3)^{-1/2}F_1,\quad \hat{\phi}_2 \longrightarrow i(\lambda_1\lambda_3)^{-1/2}F_2,\quad \hat{\phi}_3 \longrightarrow i(\lambda_1\lambda_2)^{-1/2}F_3$$

Eqs. (1) become

$$\partial F_1/\partial\xi = F_2^* F_3\ ,\quad \partial F_2/\partial\xi = F_1^* F_3\ ,\quad \partial F_3/\partial\xi = -F_1 F_2 \tag{6}$$

where $\xi = \frac{1}{2}(\tau + \frac{z}{v_g})$. We note that this procedure is applicable only if the group velocities of the interacting waves are equal -- this is approximately true in the case of small wavenumbers, when the resonance conditions are fulfilled. It is well known, that the set of equations (6) admits solutions in the form[2]

$$A_1(\xi) = \sqrt{A_1^2(0) + X(\xi)},\quad A_2(\xi) = \sqrt{A_2^2(0) + X(\xi)},\quad A_3(\xi) = \sqrt{A_3^2(0) - X(\xi)} \tag{7}$$

where $A_j(\xi)$ are the real waves' amplitudes $[F_j(\xi) = A_j(\xi) \times\exp(i\theta_j)]$ and

$$X(\xi) = (X_2 - X_1)\,\mathrm{sn}^2[(X_1 - X_3)^{1/2}\xi + \Psi, k] + X_1 \tag{8}$$

Here $k = \sqrt{(X_1 - X_2)/(X_1 - X_3)}$ is the modulus of the Jacobian elliptic function sn,

$$\Psi = \mathrm{sn}^{-1}\left[\frac{X_1}{(X_1 - X_2)^{1/2}}, k\right]$$

and X_j are the roots of the cubic equation

$$X^3(\xi) - [A_3^2(0) - A_1^2(0) - A_2^2(0)]X^2(\xi) + [A_1^2(0)A_2^2(0) - A_1^2(0)A_3^2(0) - A_2^2(0)A_3^2(0)]X(\xi) + \Gamma^2 - A_1^2(0)A_2^2(0)A_3^2(0) = 0 \tag{9}$$

where $\Gamma = A_1 A_2 A_3 \sin(\theta_3 - \theta_1 - \theta_2)$ is a constant of motion.

If all the initial amplitudes are the same, say A_o, and moreover, $\Gamma = 0$, the modulus of the elliptic function equals 1, i. e. sn function transforms to $\tanh(\xi)$. If we introduce the energies W_j, carried by the interacting waves,[1] we finally get

$$
\begin{aligned}
W_1 &= 8\pi\varepsilon_0 (k_1 R)^2 \frac{\omega_p^2}{\omega_1^2} \frac{A_0^2}{\lambda_2 \lambda_3} \operatorname{sech}^2(\sqrt{2} A_0 \xi) \\
W_2 &= 8\pi\varepsilon_0 (k_2 R)^2 \frac{\omega_p^2}{\omega_2^2} \frac{A_0^2}{\lambda_1 \lambda_3} \operatorname{sech}^2(\sqrt{2} A_0 \xi) \qquad (10) \\
W_3 &= 8\pi\varepsilon_0 (k_3 R)^2 \frac{\omega_p^2}{\omega_3^2} \frac{A_0^2}{\lambda_1 \lambda_2} \tanh^2(\sqrt{2} A_0 \xi)
\end{aligned}
$$

where the values of λ_j should be taken from expressions (4). Solutions (10) can be interpreted in the following way: let initially we have two surface waves at frequencies ω_1 and ω_3 respectively. Due to the nonlinear interaction a new wave at frequency ω_2 appears. We note that in the nonlinear stage of their evolution the two initial waves prove to be bright and dark solitary waves correspondingly, while the born wave is a bright soliton. One may illustrate this possibility by a particular example: let $k_1R = 0.04$, $k_3R = 0.10$. The normalized frequencies of the carrier waves (with respect to ω_p) are $\omega_1/\omega_p = 0.052$ and $\omega_3/\omega_p = 0.107$. The generated wave should have a wave number $k_2R = 0.06$ and frequency $(\omega_3 - \omega_1)/\omega_p \approx \omega_2/\omega_p = 0.071$. The corresponding group velocities of the interacting waves are (at $\omega_p = 1.8\times10^9\ s^{-1}$, $n_o = 10^{15}\ m^{-1}$) $v_{g1} = 5.25\times10^7$ m/s, $v_{g2} = 5.04\times10^7$ m/s and $v_{g3} = 4.72\times10^7$ m/s. We would like to emphasize that on account of the form of the dispersion curve for azimuthally-symmetric waves the three-wave interaction is actually nonresonant even though at small k_jR's. That is why a more realistic treatment should include four-wave interaction and also three-wave interaction in the presence of a cubic nonlinearity.

References

1. T. Lindgren, L. Stenflo, N. Kostov, and I. Zhelyazkov - to be published.
2. J. A. Armstrong, N. Bloembergen, J. Ducuing, and P. S. Pershan, Phys. Rev. 15, 1918 (1962).

LOW FREQUENCY DRIFT INSTABILITIES ON THE PLASMA SURFACE CONFINED BY MAGNETIC FIELD

H. Kozima, K. Yamagiwa, M. Kawaguchi, H. Morishita and H. Shimizu

Department of Physics, Shizuoka University,
836 Oya, Shizuoka 422, Japan

Introduction The leak widths of plasma through point and line cusp fields are interesting problem concerning magnetic traps used widely in laboratory plasma experiments. Recently, it is proposed an explanation[1] of the so-called hybrid width $2d_h = 2(a_i a_e)^{1/2}$ (a_i and a_e is ion and electron gyroradius, respectively) by an influence of fast primary electrons. Our experimental investigations suggest, however, that in the case of quiescent plasma in multidipole type devices the leak width of a line cusp field is closely related with low frequency noises excited around plasma boundaries[2]. On the spectrum of the low frequency noise shown in Fig. 1 for Ar, we found a sharp peak (or peaks) specific to the ion species which is strongest at plasma boundary[3]. We will identify this low frequency noise peak with a drift instability in this paper.

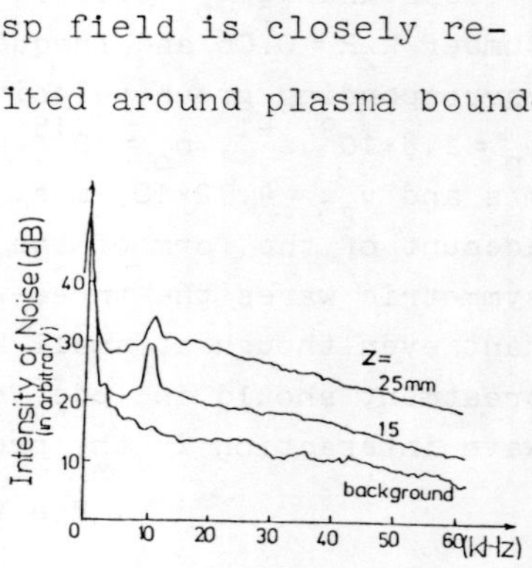

Fig. 1

Experimental Results Figure 2 shows magnetic field lines

around our line cusp fields made by permanent magnet arrays.

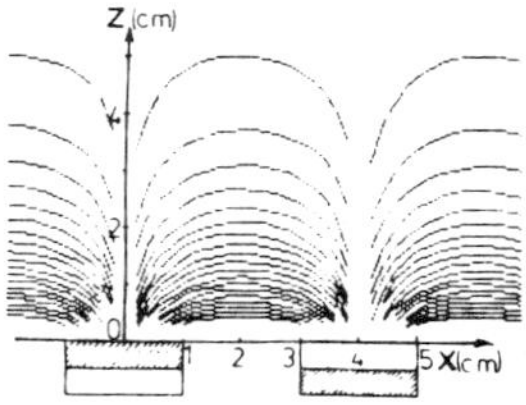

Fig. 2

In Fig. 3, we plotted points where the noise peaks at about 10 (Ar), 15 (Ne) and 30 kHz (He) reach their maximum values for a given height z. These points line up along a plasma boundary as shown in a previous paper[3]. A curve through those points makes an angle θ with field lines in xz-plane. The y-axis is perpendicular to the plane of Fig. 2 . Average values of the angle θ for Ar, Ne and He plasmas are calculated as 17°, 11° and 14° , respectively, from experimental data.

Interpretation with a Drift Wave As shown in Fig. 1 for Ar, the low frequency noise has very broad spectrum extending up to 100 kHz. Main part of the noise might be ion acoustic waves propagating along field lines. The strong peak, however, distributes along plasma boundaries and the distribution profile makes an angle with field lines. Considering those features, we can take up a low frequency drift wave with $\vec{k}\cdot\vec{B} \neq 0$ [4] as a candidate. The dispersion relation of this wave in a low frequency region $\omega \ll \omega_{ci}$ (ω_{ci} ; ion gyrofrequency) is written as follows:

$$\omega = \pm \kappa v_0/2 \pm [(\kappa v_0/2)^2 + K^2 c_s{}^2]^{1/2} \tag{1}$$

where K is a component of the wave vector $\vec{k}$ parallel to the magnetic field, κ is that perpendicular to both magnetic field and density gradient (y component of $\vec{k}$ in our geometry). c_s is ion sound velocity and

$$v_0 = (v_e^2/2\omega_{ce})(1/n)dn/dx.$$

Though we did not succeed to measure k_y, we might try to estimate the dispersion relation of our noise peak using θ determined above for $\tan^{-1}(\kappa/K)$.
For the experimental values of noise peak frequency 10 (Ar), 15 (Ne) and 30 kHz(He), we calculate wave length of the drift wave given by the dispersion relation (1) to obtain follow ing values;

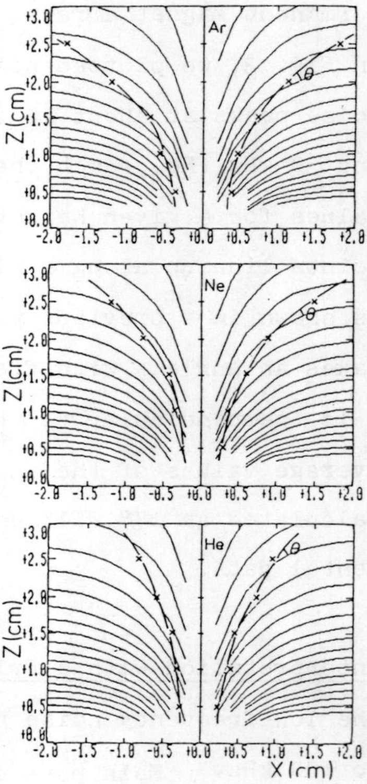

Fig. 3

$$\lambda = 20 \text{ (Ar)}, 38 \text{ (Ne) and } 31 \text{ cm (He)}.$$

The values of wave length thus obtained are permissible ones in our device, because the length of a magnet array is about 40 cm. The condition

$$\omega \ll \omega_{ci}$$

is satisfied only near the magnet surface especially for Ar, and the analysis given above is inevitably qualitative.

In the frequency spectra for He plasma, we can see several peaks around 30 kHz. If we take up a peak at about 25 kHz[3], the wave length becomes 37 cm, a very close value to that in Ne plasma, and is about twice the value in Ar plasma.

The qualitative analysis given above showed that the sharp peaks found in low frequency noise spectra in Ar, Ne and He plasmas excited around line cusp region on the plasma boundary might be the low frequency drift waves with $\vec{k}\cdot\vec{B} \neq 0$. Previous experimental data[2] have shown that the suppression of the noise reduced leak width to a value given by the hybrid gyroradius.

References

1. G. Knorr and R. L. Merlino, Plasma Phys. and Controlled Fusion 26 433 (1984)
2. H. Kozima, K. Yamagiwa, H. Ito and K. Sakurai, Plasma Physics 25 287 (1983)
3. H. Kozima, K. Yamagiwa and M. Kawaguchi, Phys. Letters 106A 252 (1984)
4. N. A. Krall, Drift Waves in "Advances in Plasma Physics" Vol. 1, eds. Simon and Thompson (Interscience, New York 1968)

SURFACE EXCITATIONS IN AN UNDER-DENSE AND OVER-DENSE PLASMA IN THE PRESENCE OF AN EXTERNAL HIGH FREQUENCY FIELD

S. Krishan
Physics Department
Indian Institute of Science
Bangalore 560 012, India.

It is well known that when a high frequency (h.f) dipole field mediates a plasma in the plane boundary half space surface excitations take place provided $\omega_o \simeq \frac{\omega_p}{\sqrt{2}}$, where ω_o is the frequency of the h.f. field and ω_p the electron plasma frequency of the plasma. What is not so well known is the fact that surface excitations can also take place when $\omega_o - \omega_p \simeq \pm \frac{\omega_p}{\sqrt{2}} - \ell\omega_o$, $\ell = 0,1$ for an unmagnetized plasma with zero relative drift between electrons and ions. Defining critical and the mean plasma densities by n_c and n_o, it implies $\frac{n_c}{n_o} \simeq 3$, for the underdense case and $\frac{n_c}{n_o} = 0.72,\ 2 \times 10^{-2},\ 8.5 \times 10^{-2}$ for the over-dense case. Thus one can clearly see the possibility of h.f. field interacting with significantly high over-dense situations. Ion dynamics is essential if a dipole field is to excite instabilities. Thus in terms of lowest order ion contribution and electronic harmonics upto $\omega \pm \omega_o$, the dispersion relation should have terms of the order of $\chi_i^{(o)} J_1^2 (K_\perp \cdot r_e)$ where $\chi_i^{(o)}$ is the free-ion susceptibility at frequency ω_o, $K_\perp$ the surface wave vector r_e the electron excursion length and J_1 the Bessel function. There is also a Landau damping term approximately equal to $\frac{K_\perp}{K_D}$, where K_D is the electron Debye wave-vector. Therefore for instability

$$|\chi_i^{(o)}| J_1^2 > \frac{K_\perp}{K_D}. \tag{1}$$

In terms of irradience I_r, referring to one of the above cited situations i.e. $\omega_p = \pm\omega_p/\sqrt{2} + \omega_o$, (1) yields

$$I_r\ (\mathrm{Watt/cm^2}) > 10^{-9}\, n_c \frac{n_c}{n_o} \frac{M}{m} \frac{K_D}{K_\perp} T_e, \tag{2}$$

where T_e is a pure number standing for electron temperature as number of electron volts and $\frac{M}{m}$ the ratio of ion to electron mass.

For $T_e = 10$, I_r easily exceeds the limit of relativistic threshold. This is also true of other situations mentioned above. Therefore for practical purpose one can completely ignore such instabilities unless the basic state of the plasma is changed. Since the high thresholds are due to high Landau damping, one has to somehow change over to a situation where the prevailing Landau damping turns out to be low. The beam plasma system where the electrons are drifting is the answer to the problem. This has been found to be case for all the above mentioned plasma densities except $\frac{n_c}{n_o} \simeq 2 \times 10^{-2}$.

Let the beam with velocity u be parallel to the plane plasma boundary. The resulting solution to the dispersion relation is found to be

$$\omega - \omega_o = K_\perp u + \beta , \tag{2}$$

where,

$$K_\perp \cdot u = - \omega_p (1+\alpha), \qquad 0 < \alpha << 1, \tag{3}$$

$$\beta = \pm\omega_p \left(\frac{n'}{n_o} (2\alpha - \delta + i\mathcal{I} \right)^{1/2}, \tag{4}$$

$$\delta = 2(-1 \pm i) \frac{m}{M} J_1^2 , \tag{5}$$

$$\mathcal{I} = \sqrt{8/\pi} \frac{K_\perp}{K_D} \left(-1 \pm \left(8 \frac{n}{n'} \alpha^3 \frac{T_B}{T_e}\right)\right), \tag{6}$$

$\frac{T_B}{T_e}$ is the beam to electron temperature ratio, n' the beam density and it has been assumed that

$$\left| \frac{n'}{n} (2\alpha - \delta + i\mathcal{I})^{-3} \right| << 1 . \tag{7}$$

The $\pm$ sings in (6) correspond to the two signs of β in (4). The beam will play a dominant role over the no beam case if

$$| 2\alpha | >> | \delta - i\mathcal{I} | , \tag{8}$$

which gives two values of β. The value of β which gives lower threshold is given by

$$\beta_1 = \omega_p \left(\frac{n'}{2n\alpha} \right)^{1/2} \left(1 - \frac{i}{4\alpha} (\mathcal{I}\Lambda - \delta_i)\right), \tag{9}$$

where

$$\Lambda = (-1 + (8 \frac{n}{n'} \alpha^3 \frac{T_B}{T_e})^{\frac{1}{2}}) , \quad (10)$$

and δ_i the imaginary part of δ . For instability

$$| \delta_i | > | \Lambda | . \quad (11)$$

The threshold is required to be considerably smaller than the no beam case. This necessitates that $| \Lambda | << 1$ or

$$1 \simeq 8 (\frac{n}{n'} \alpha^3 \frac{T_B}{T_e})^{\frac{1}{2}}. \quad (12)$$

Therefore the conditions imposed by (7) and the small spread in the beam can be met by

$$4 \alpha^2 \frac{T_B}{T_e} >> \frac{E_{th}}{mu^2} , \quad (13)$$

where E_{th} is the thermal energy of the beam electrons. For a pulsed source of radiation (12) can satisfied provided the half width is small enough. For a CW source like Gyrotron emitting 28 GHz radiation, the plasma densities are found to be 10^{12} cm^{-3}, 5×10^{12} cm^{-3}, 4.2×10^{13} cm^{-3} at which instabilities will occur. For $\frac{n_c}{n_o} = 0.72$, 8.5×10^{-2}, Te = 100 eV the threshold is found to be 10^6 (Watt/cm^2). A large number of references can be found in Gradov and Stenflo[1] & Moisan et al[2]. The details of the present note can be found elsewhere[3].

REFERENCES

1. O.M. Gradov and L. Stenflo Phys.Rep. 94, 111, 1983.
2. M. Moison, A. Shivarova, A.W. Trivelpiece, Plasma Physics 24, 1331, 1982.
3. S. Krishan, Plasma Physics and Controlled Fusion, in Press (1985).

ABSORPTIVE PROPERTIES AND SURFACE WAVES IN PLANE-STRATIFIED PLASMA-LIKE MEDIUM

I.V.Krivtsun, I.P.Yakimenko, A.G.Zagorodny

Institute for Theoretical Physics, Academy of Sciences of the Ukrainian SSR, 252130, Kiev-130, USSR

Introduction. The study of the influence of an external medium (espesially, the stratified dielectric layers), on the absorption of electromagnetic radiation and the surface waves spectra in a semi-bounded plasma-like medium is the aim of the present work.

Absorption of electromagnetic waves by semi-bounded plasma with stratified dielectric. The absorption under consideration may be calculated by the standard methods of macroscopic electrodynamics, if the solution of the boundary value problem for electromagnetic fields in a plasma is known. Using the results of the kinetic theory of semi-bounded plasma with the specularly-reflecting particles boundary[1] in a case of isotropic medium we obtain the following results for the reflectances R_p, R_s and the absorptances Γ_p , Γ_s for p- and s-polarized waves for the configuration of type stratified dielectric-plasma half-space:

$$R_\nu = \frac{\left|(1+r_\nu^{2,1})(1-r_\nu^{2,3})e^{-ik_{zd}L} - (1-r_\nu^{2,1})(1+r_\nu^{2,3})e^{ik_{zd}L}\right|^2}{\left|(1-r_\nu^{2,1})(1-r_\nu^{2,3})e^{-ik_{zd}L} - (1+r_\nu^{2,1})(1+r_\nu^{2,3})e^{ik_{zd}L}\right|^2}\,, \qquad (1)$$

$$\nu = p, s\,,$$

$$\Gamma_\nu = \frac{16|r_\nu^{2,1}|^2 \mathrm{Re}(r_\nu^{2,3}/r_\nu^{2,1})}{\left|(1-r_\nu^{2,1})(1-r_\nu^{2,3})e^{-ik_{zd}L} - (1+r_\nu^{2,1})(1+r_\nu^{2,3})e^{ik_{zd}L}\right|^2}\,,$$

Here

$$r_p^{2,1} = -\frac{\varepsilon_d \tilde{k}_z}{\tilde{\varepsilon} k_{zd}}, \quad r_p^{2,3} = -\frac{i\varepsilon_d}{\pi k_{zd}} \int_{-\infty}^{+\infty} \frac{dk_z}{k^2} \left[\frac{k_\perp^2}{\varepsilon_L(\vec{k},\omega)} + \frac{k_z^2}{\varepsilon_T(\vec{k},\omega) - k^2 c^2/\omega^2} \right],$$

$$r_s^{2,1} = -\frac{k_{zd}}{\tilde{k}_z}, \quad r_s^{2,3} = -\frac{ic^2 k_{zd}}{\pi \omega^2} \int_{-\infty}^{+\infty} \frac{dk_z}{\varepsilon_T(\vec{k},\omega) - k^2 c^2/\omega^2},$$

$$\tilde{k}_z = \frac{\omega}{c}\sqrt{\tilde{\varepsilon}}\cos\vartheta, \; k_{zd} = \frac{\omega}{c}\sqrt{\varepsilon_d - \tilde{\varepsilon}\sin^2\vartheta}, \; k_\perp = \frac{\omega}{c}\sqrt{\tilde{\varepsilon}}\sin\vartheta,$$

$\tilde{\varepsilon} \equiv \tilde{\varepsilon}(\omega)$ and $\varepsilon_d \equiv \varepsilon_d(\omega)$ are the dielectric constants of the external medium and dielectric layer; L is the thickness of layer; ϑ is the angle of incidence; $\varepsilon_L(\vec{k},\omega), \varepsilon_T(\vec{k},\omega)$ are the longitudinal and transverse dielectric permittivities for an unbounded plasma-like medium[2,3]. The expressions of type (1) for an nonequilibrium anisotropic medium are given in[4]. In the cases $L = 0$ and $\varepsilon_{L,T}(\vec{k},\omega) = \varepsilon(\omega)$ the absorptances (1) are reduced to the results for semi-bounded plasma without cover[1-3] and for the three-layered dielectric structure[5].

The results of the numerical calculations for the case of the degenerate electronic gas in the anomalous skin-effect domain are shown in Figs.1-6. These calculations have been carried out with using the semiclassical relations for $\varepsilon_{L,T}(\vec{k},\omega)$[3]. The given curves correspond to the parameters $v_F/c = 2.83\,10^{-3}$ (metallic potassium), $\gamma = \omega_{pe}\tau = 10^{-4}$-$10^{-5}$, $\tilde{\varepsilon} = 1$ (v_F is the Fermi velocity of electrons, ω_{pe} is the plasma frequency, τ is the relaxation time) for variables $\Omega = \omega/\omega_{pe}$, $\chi = \omega_{pe} L/\pi c$. The influence of wave polarization takes place in the thickness of layer and frequency dependences (Figs.1-3).

Fig.1

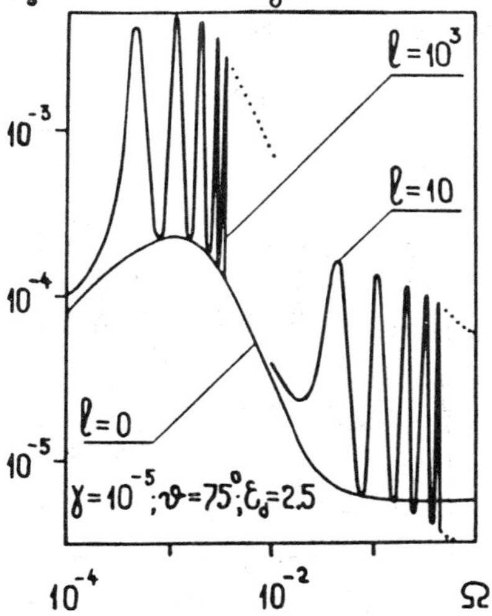

Fig. 2

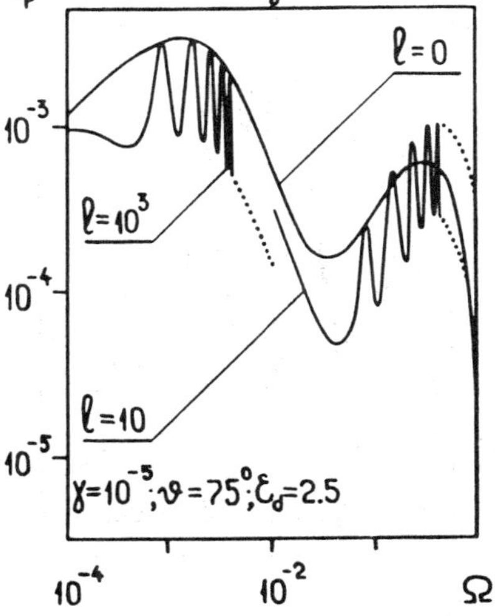

For the non-polarized light it gives the interval between the maxima of order $\Delta L \simeq \pi c/2\omega\sqrt{\varepsilon_d}$ (Fig.4) in contrast with $\Delta L \simeq \pi c/\omega\sqrt{\varepsilon_d}$ for dielectric media.

Fig.3

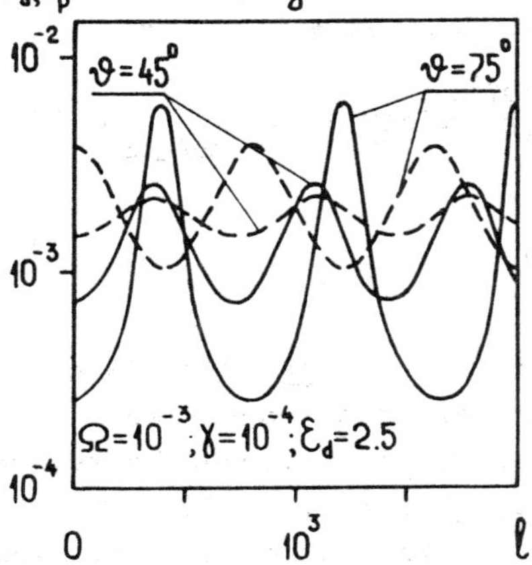

Fig.4

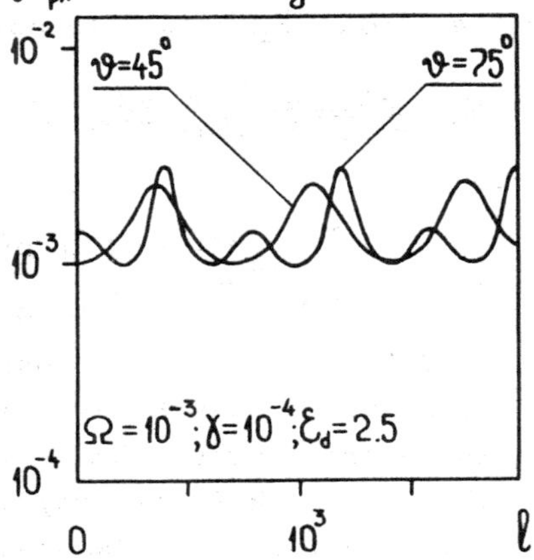

If the layer is absorptive, the global absorptances are decreased (Fig.5). The accounting of the frequency dispersion of the layer (Fig.6) results in the disordering of the equidistant structure of the Γ_p, Γ_s vs Ω maxima.

Fig.5

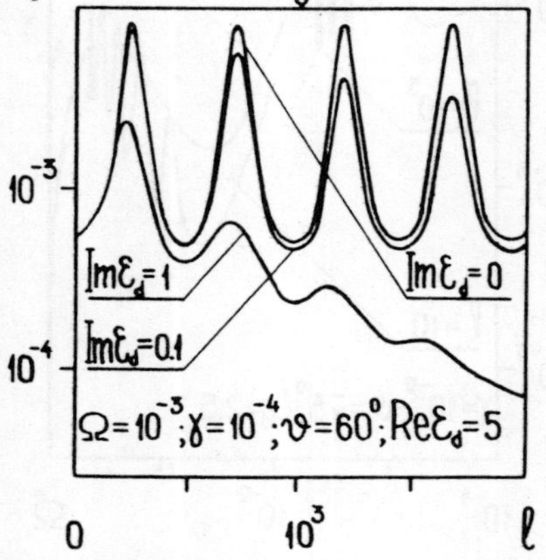

Fig.6

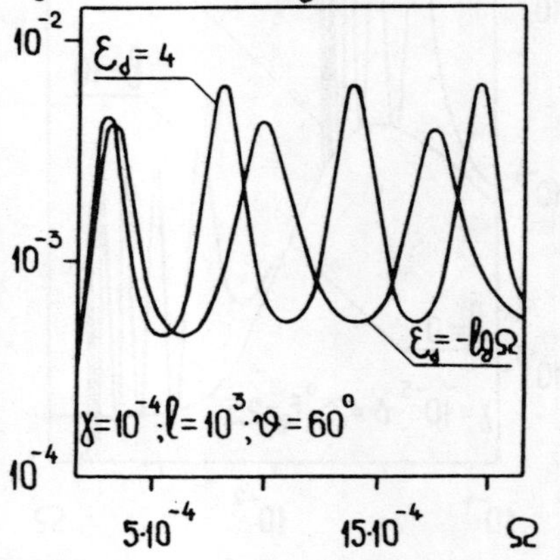

Surface waves in the dielectric layer-plasma medium system.

In the system under consideration surface waves obey the following dispersion equation:

$$(1-r_\nu^{2,1})(1-r_\nu^{2,3})e^{-ik_{zd}L} - (1+r_\nu^{2,1})(1+r_\nu^{2,3})e^{ik_{zd}L} = 0\ . \qquad (3)$$

In the limit $k_\perp c \gg \omega$ its solution can be written as

$$\omega_{\vec k} = \frac{\omega_{pe}}{\sqrt{1+\varepsilon_{eff}}}\left(1+\frac{k_\perp S_e}{2\omega_{pe}}\sqrt{3\varepsilon_{eff}}\right),\ \gamma_{\vec k} = -\left(\frac{2}{\pi}\right)^{1/2} k_\perp S_e \frac{\varepsilon_{eff}}{|\varepsilon_L(\vec k,\omega)|_{k\sim\omega/S_e\sqrt{2}}}, \qquad (4)$$

where

$$\varepsilon_{eff} = \varepsilon_d \frac{\tilde{\varepsilon}(1+e^{-2k_\perp L}) + \varepsilon_d(1-e^{-2k_\perp L})}{\tilde{\varepsilon}(1-e^{-2k_\perp L}) + \varepsilon_d(1+e^{-2k_\perp L})}\ . \qquad (5)$$

1. I.P.Yakimenko, A.G.Zagorodny. Phys.Scr., 14, 242, 1976.
2. V.P.Silin, A.A.Rukhadze. Electromagnetic Properties of Plasma and Plasma-Like Media (in Russian).Moscow,Nauka,1961.
3. K.L.Kliewer, R.Fuchs. Phys.Rev., 172, 607, 1968.
4. V.S.Gvozdetsky, A.G.Zagorodny, I.P.Yakimenko, I.V.Krivtsun. Preprint ITP-85-63R, Kiev, 1985.
5. M.Born, E.Wolf. Principles of Optics. Pergamon, 1968.

STABILITY OF A THIN ANNULAR REB AGAINST SURFACE OSCILLATION MODES

Kukvidze R.R.

General Physics Institute USSR

Academy of Sciences, Moscow

1. Thin annular REB widly usue in high-current relativistic UHF electronics are very sensitive to the various tipe instabilities. In such REB transporting through a cylindrical waveguide there excitate both the volume (Pierce and Buneman) and the surface (diocotron and current-convective) instabilities. The surface instabilities essentially distort the beam geometry leading to the fillamentation and broadening of the beam and it becomes to be unapplicable for the aims of UHF electronics. Such instabilities are very dangerous because they may excitaze in a beam with current much smaller then the threshold currents for excitation of volume instabilities and the vacuum limiting current.

The instabilities of the annular REB such as the Pierce, Buneman and diocotron instabilities are studied sufficiently full, especially, in the linear approximation. But the current-convective instability of such REB up to day was not investigated even in the linear approximation. It is shown bellow that this instability is most dangerous because the threshold current for its excitation is minimal and moreover it excitates under the conditions when the dioctron instability don't excitate.

2. For investigation of the surface instabilities of thin annular REB in a cylindrical waveguide we'l start from the equation [1]:

$$\frac{1}{r}\frac{\partial}{\partial r} r \mathcal{E}_i \frac{\partial \Phi}{\partial r} - \mathcal{E}_i \frac{\ell^2}{r^2}\Phi = -\frac{\ell \Phi \frac{\partial}{\partial r}\omega_b^2}{r\Omega_e \gamma^2(\omega - k_z u - \ell\omega_e)} \quad (1)$$

Here $\mathcal{E}_i = 1 - \frac{\omega_{Li}^2}{\omega^2}$ and $\Phi = \varphi - \frac{u}{c}A_z$ and $\varphi, \vec{A}$ - are the scalar and vector potentials of the oscillations field with $\varphi \sim A_z \gg |\vec{A}_\perp|$. This equation is valied only in the low current limit when the beam current is much less then the threshold currents for excitation of the volume instabilities.

The boundary conditions are

$$\Phi(0) < \infty, \quad \Phi(R) = 0, \quad \{\Phi\}_{r=r_b, R} = 0;$$

$$\left\{\mathcal{E}_i \frac{\partial \Phi}{\partial r}\right\}_{r=r_b} = \frac{2\ell}{r_b}\frac{\omega_d \Phi(r_b)}{\omega - k_z u}; \quad \left\{\mathcal{E}_i \frac{\partial \Phi}{\partial r}\right\}_{r=R_b} = \frac{2\ell}{R_b}\frac{\omega_d \Phi(R_b)}{\omega - k_z u - \ell\omega_e(R_b)}, \quad (2)$$

where $\omega_b = \sqrt{\frac{4\pi e^2 n_b}{m}}$, $\omega_{Li} = \sqrt{\frac{4\pi e^2 n_i}{M}}$; Ω_e, Ω_i - are the Lengmiur and Ω_e , Ω_i - the Larmour frequencies of electrons and ions correspondely, ω , k_z and ℓ - are the oscillation frequency, longitudinal and azimuthal wave-numbers, R - is the cylinderical waveguide radius, r_b and R_b - are the internal and external radiusses of REB, u - is the longitudinal velocity of beam electrons, $\gamma = (1 - u^2/c^2)^{-1/2}$, $\omega_e = -\omega_d\left(1 - \frac{r^2}{r_b^2}\right)(1 - \gamma^2 f)$ - is the angular velocity of electrons with account of the beam space charge and magnetic fields, $\gamma^2\omega_e^2 r_b^2 \ll 1$, $\omega_d = \frac{\omega_b^2}{2\gamma^2\Omega_e}$ and finally $f = \frac{n_i}{n_b}$ = const - is the degree of beam charge neutralization.

The eigenvalue problem (1)-(2) was solved in the assumption that the beam electrons are magnetized and ions-nonmagnetized. This means, that

$$\omega^2 \gg \Omega_i^2, \quad k_z u \gg \omega_{Li}, \quad |\omega - k_z u - \ell\omega_e|^2, \quad \frac{\omega_b^2}{\gamma^3} \ll \frac{\Omega_e^2}{\gamma^2} \quad (3)$$

Moreover, as it was mentioned above, the beam current is assumed to be much less then the threshold current of Pierce instability and therefore

$$\varkappa = \frac{\omega_b^2 \Delta R_b}{\gamma^3 u^2} \ll 1 , \quad \alpha = \frac{\ell L u \gamma}{R_b^2 \Omega_e} \sim \frac{1}{\varkappa} , \tag{4}$$

where $\Delta = R_b - r_b \ll R_b$, Δ – is the beam thickness and L - is the length of the waveguide.

3. It follows from the solution of the formulated eigenvalue problem that in the noncompensated beams , $f = 0$, when $\delta = R - R_b \gg \Delta$ the diocotron instability is dominative. The maximal growth rate and the threshold current (which determinated from the condition $\operatorname{Im}\omega = u/L$ because this instability is convective) are equal

$$\operatorname{Im}\omega = 0{,}45\,\omega_d , \quad I_{th} = 40(\gamma^2 - 1)\frac{\Omega_e R_b^2}{c L}\cdot\frac{\Delta}{R_b}\ \kappa A, \tag{5}$$

and $y = \ell \frac{\Delta}{R_b} = 0{,}8$.

If the beam is partially compensated $f \neq 0$ then it is possible of exciting the current-convective instability which is result of ion inertion (i.e. $\varepsilon_i \neq 1$). This instability is absolute and its growth rate and threshold current are equal

$$\operatorname{Im}\omega = \frac{\sqrt{3}}{2} k_z u \left(\frac{\Omega_i^2 y^{-1}}{\omega_{Li}^2}\right)^{1/3} , \quad I_{th} = 17\frac{(\gamma^2-1)^{3/2}}{\gamma}\cdot\frac{M}{m}\ \frac{\Omega_i^2 k_z^2 \Delta R_b}{\omega_{Li}^2 y}\ \kappa A, \tag{6}$$

if $y \ll 1$, and equal

$$\operatorname{Im}\omega = \frac{\sqrt{3}}{2}\left(\frac{k_z u}{2}\omega_{Li}^2\right)^{1/3} , \quad I_{th} = 17(\gamma^2 - 1)\frac{\Omega_e R_b}{c} k_z R_b \frac{\Delta}{R_b}\ \kappa A, \tag{7}$$

if $y \gtrsim 1$.

It is easely seen that if $y \ll 1$ then the threshold current and growth rate of the current-convective instability decreases when y increases and if $y \gtrsim 1$ then they pracfically don't depend on ℓ (i.e. y). On the other hand these quantities depend on $k_z = \frac{\pi n}{L}$ and increase when the number $n = 1, 2 \ldots$ increases. Therefore in the real system it will be excited that mode of instability the threshold current of which is nearer to the beam current.

Note that when $\gamma^2 f$ increases the threshold current of the diocotron instability increases also, but its growth rate decreases. In the same time for the current-convective instability these quantities both increase. As a result it must be easely shown that if $\gamma^2 f > 1/2$ then the current-convective instability is dominative in the system.

4. If the split δ between REB and the walls of the waveguide is small, $\delta \ll \Delta$, then with decreasing δ the growth rate of dioctron instability tends to zero and the threshold current becomes infinity. Therefore if $\delta = 0$ the system occurs stable against the diocotron instability. But as regard to the current-convective instability then the system remains unstable in this case. Moreover if $y \ll 1$ then the threshold current and the growth rate increase and become equal

$$I_{th} = 17\left(\gamma^2 - 1\right) \frac{R_e R}{c} \frac{k_z \Delta}{2y} kA, \quad \Im m\,\omega = \frac{\sqrt{3}}{2}\,\omega_{Li} \left(y \frac{\omega_{Li}}{\Omega_i}\right)^{1/3} \tag{8}$$

But if $y \gtrsim 1$ then they remain the same as are given by expressions (7).

Finally it must be noticed that in the real condition allways will excited the modes with maximal ℓ (when $\frac{\Delta}{R}$ is given) because the threshold current for excitation of these modes is minimal and the growth rate is maximal.

REFERENCES

1. Karbushev N.I., Rukhadze A.A., Udovichenko S.Iu., Sov. Phys. 1984, v.10, p.268.
2. Karbushev N.I., Rukhadze A.A., Udovichenko S.Iu., Sov. Phys. Kratkie Soobchenia Physiky 1983, v.7, p.50.

SURFACE POLARITON DIFFRACTION

by

T.A. LESKOVA

Institute of Spectroscopy, USSR Academy of Sciences
Troitsk, Moscow r-n, 142092, USSR

In this report we present the results of an analytical theory of the refraction of a surface polariton at an interface:

a) The refraction of a surface polariton by the edge of a thin dielectric film deposited on a metal surface is studies and its description by means of the surface analogues of the Fresnal relations is presented. The resonance effects in the diffraction spectra are discussed.

b) The theory of a Fabry-Perot type interferometer for surface polaritons is developed. The resonance interference phenomena are investigated.

c) Wave-guide surface polaritons in a dielectric strip deposited on a metal substrate are investigated.

SURFACE WAVE DISCHARGES IN TAPERED TUBES

M. Moisan and Z. Zakrzewski*

Université de Montréal, Département de Physique, Montréal
(Québec) H3C 3J7

Introduction Surface wave discharges (SWD) in axially uniform tubes have been investigated both theoretically and experimentally over the last ten years[1,2]. Recently[3], we have devised and tested a SWD setup with a plasma tube diameter varying along the axis. In this note, we concentrate on the axial distribution of the electron density within a conical tube. We shall show that such a setup may be of practical interest.

Discharge model The geometry of the problem to be considered is presented as an insert in Fig. 1. The waves sustaining the SWD are excited in the z_1 plane and travel in both directions along the z axis. The waves travelling in the z and -z direction shall be called the "upward" wave and the "downward" wave, respectively. Due to the axial nonuniformity of the tube and to both axial and radial nonuniformity of the plasma, a self-consistent description of the discharge would result in an intractable problem. In this work, we are interested mainly in finding the general influence of various factors on the axial density profile. Thus, we have chosen to introduce rather severe, although physically justified, approximations in the discharge model. This allows us to treat the problem analytically. The main assumptions are:

1. the cone angle $\emptyset$ is small enough so the balance processes take place locally in each cross-section of the plasma column. For the same reason, the local value of the wave attenuation coefficient $\alpha(z)$ at any axial position z is the same as in a uniform tube with a radius equal to

* Permanent address: Polish Academy of Sciences, IMP-PAN, 80-952 Gdansk, Poland.

a(z). 2. The power Θ lost per one electron*, in collisions of all kinds, is independent of the electron density and it obeys a simple similarity law $\Theta/p = f(pa)$, where p is the gas pressure. 3. The electron collision frequency is constant within the plasma volume.

With the above assumptions, the equation for the local balance of power is

$$2\alpha(z)\, P(z) = \pi a^2.\Theta.n(z), \qquad (1)$$

where n(z) is the cross-section averaged electron density in the z plane and P(z) is the total power flux of the wave through that plane. Th left hand side and the right hand side of Eq. (1) represent, respectively, the wave power absorbed into the plasma and the power lost by electrons, both per unit column length. Further, we assume that the dependence of $\alpha(z)$ on the tube radius (i.e. position) and on the electron density can be separated:

$$\alpha(z) = A.f_z(z).f_n(n). \qquad (2)$$

Let us also recall that

$$P(z) = P_1 \exp\left[\mp 2 \int_{z_1}^{z-z_1} \alpha(z)dz\right];\ P_1 = P(z_1), \qquad (3)$$

---

* The total power loss, suffered by an average electron in collisions of all kinds, can be expressed as

$$\Theta \equiv \frac{2m}{M} \langle \nu_m\, \varepsilon \rangle + \sum_k \langle h_k \nu_m \rangle\, eV_k ,$$

where m and M are the electron and ion mass, ε is the electron energy and ν_m, the collision frequency for momentum transfer, and h_k and V_k are, respectively, the efficiency and the threshold potential for excitation by collisions leading to the k-th level. The brackets $\langle\ \rangle$ denote the averaging over the electron energy distribution.

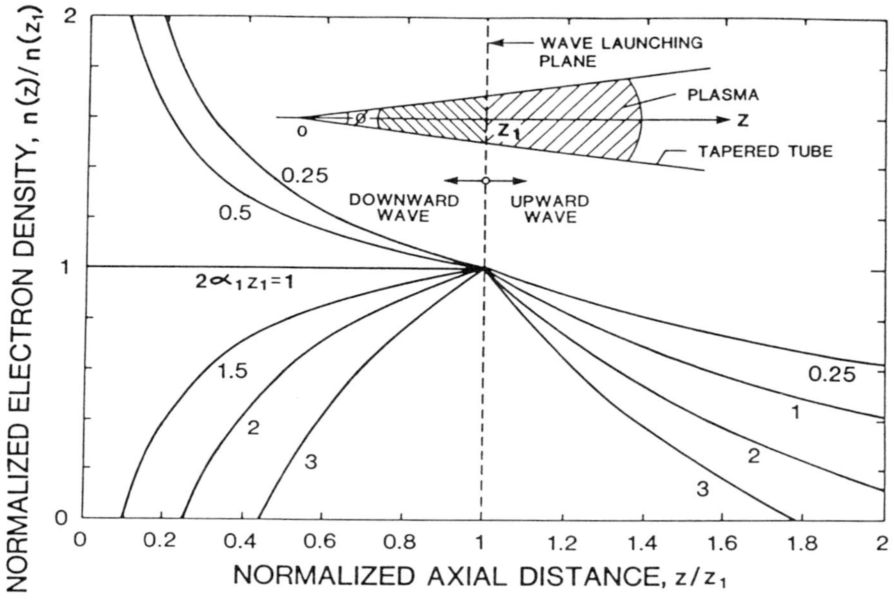

Fig. 1.

where the upper and the lower signs correspond to the upward and the downward waves, respectively. Also, we shall limit ourselves to the case of argon, for which the above mentioned similarity law is known ($\Theta . a^2 \cong$ const., see Fig. 7 in [4]). Substituting (2) and (3) into (1) and taking the logarithmic derivative on both sides of the resulting equation, gives

$$\frac{dn}{dz} = \frac{\frac{n(z)}{f_z(z)} \frac{df_z(z)}{dz} \mp 2\,\alpha(z)\, n(z)}{1 - \frac{n(z)}{f_n(n)} \frac{df_n(n)}{dn}}. \qquad (4)$$

The solution to this equation with respect to n(z) and with the initial condition $n(z) = n_1$, leads to the axial distribution of the electron density. The following approximation may be chosen[5] for the attenuation coefficient (Eq. 2) of the surface wave: A = const., $f_z(z) = z^{-1}$, $f_n = n^{-1}$. Then, the solution to Eq. (4) is

$$n(z)/n_1 = (1 \mp 2\alpha_1 z_1)\,(z/z_1)^{-\frac{1}{2}} \pm 2\,\alpha_1 z_1. \qquad (5)$$

Results and discussion A normalized plot of the axial density profiles for different values of $2\alpha_1 z_1$ is shown in Fig. 1. To bring into the open the physical meaning of this parameter, let us notice that $2\alpha_1 z_1 = 2\,\alpha_1\, a(z_1)\, \mathrm{ctg}\, \emptyset/2$ and indicate that α_1 is approximately proportional to the gas pressure and inversely proportional to the electron density. The upward wave always sustains a column with an axially decreasing electron density. This case is of practical interest, as it allows[3] to sustain large diameter plasmas using a launcher with a small aperture. The electron density in a column sustained by the downward wave decreases, increases or remains constant with an increasing distance from the wave launching plane, depending on the value of $2\alpha_1 z_1$. Thus, conditions (Ø, gas pressure, electron density) may be sought, for which the density is axially uniform. This feature can be of interest for some applications.

Acknowledgment The authors would like to thank M. Chaker and G. Sauvé for discussions.

References

1. C.M. Ferreira, M. Moisan, invited lecture, this conference.
2. M. Moisan, C.M. Ferreira, Y. Hajlaoui, D. Henry, J. Hubert, R. Pantel, A. Ricard, and Z. Zakrzewski, Rev. Phys. Appl. 17, 707 (1982).
3. M. Moisan, Z. Zakrzewski, Proc. XVIIth Int. Conf. on Phenomena in Ionized Gases (in print), Budapest 1985.
4. M. Chaker and M. Moisan, J. Appl. Phys. 57, 91 (1985).
5. Z. Zakrzewski, M. Moisan, Proc. XVIIth Int. Conf. on Phenomena in Ionized Gases (in print), Budapest 1985.

OPTICAL GAIN OF A SURFACE WAVE SUSTAINED He-Ne PLASMA COLUMN

C. Moutoulas, M. Moisan, and Z. Zakrzewski,
Université de Montréal, Département de Physique, Montréal, (Québec)
H3C 3J7

Introduction An external mirror He-Ne laser (632,8 nm) has been realized[1] using a surface wave (SW) produced plasma as the active medium. It is the purpose of this paper to report the results of experimental investigations on this laser. The attention is focused on the dependence of the optical gain coefficient of the medium on the SW frequency and electron density.

Experimental apparatus and procedure The plasma is produced within a 5 mm i.d. T-shaped quartz tube with a SW launcher (surfatron) located at the base of the T (Fig. 1). The wave emerges from the launcher toward the junction of the T, where it divides into two waves of the same power flow, propagating in opposite directions in the arms of the T. The plasma column obtained is symmetric with respect to the T junction and the length of the active region increases with increasing HF power to the launcher. The dependence of the optical gain $g_0(\bar{n})$ on the cross-section averaged electron density $\bar{n}$ has been determined from simultaneous measurements of the total optical gain $G_0(z')$ of the column and of

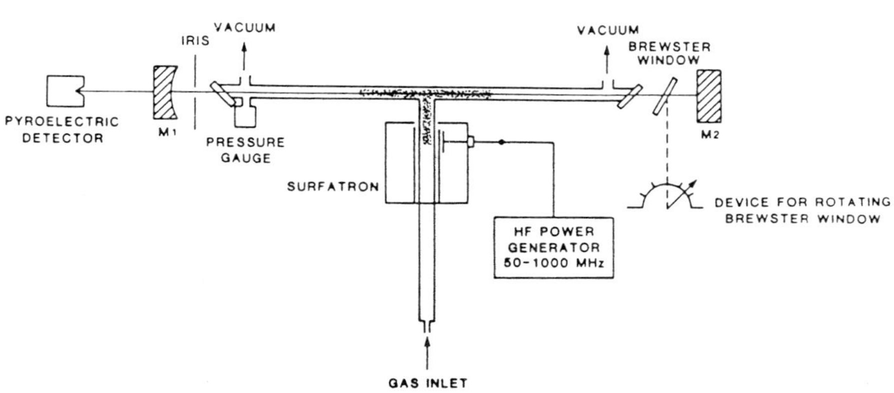

Fig. 1

the axial density profile $\bar{n}(z)$. The above quantities are related as follows.

$$G_o(z') = \int_0^{z'} g_o(z)dz = \frac{1}{2}\int_{-z'}^{z'} g_o(z)dz, \quad z' = \ell/2, \qquad (1)$$

where z is a linear coordinate coinciding with the tube axis (z = 0 at the junction), $g_o(z)$ is the local value of the gain coefficient in the z cross-section, and ℓ is the total length of the active region. A set of $G_o(z')$ values, measured[1] for various lengths of the column is obtained. The column length is controlled by changing the microwave power to the launcher. A third-order polynominal is fitted on these data. Such choice of the polynominal order is justified a priori by a semi-empirical model developed[2] for $G_o(z')$ and a posteriori, by the quality of the fit. The derivative of $G_o(z')$ gives the value of $g_o(z)$. On the other hand, the axial profile of the electron density $\bar{n}(z)$ is obtained by using a TM_{010} cavity. With both $\bar{n}(z)$ and $g_o(z)$ known, $g_o(\bar{n})$ can be readily calculated.

The results of measurements of the total gain carried out at a gas pressure of 0.6 Torr in a 7:1 He-Ne gas composition, are reported in Fig. 2, in the frequency range 100 to 915 MHz, for various plasma column

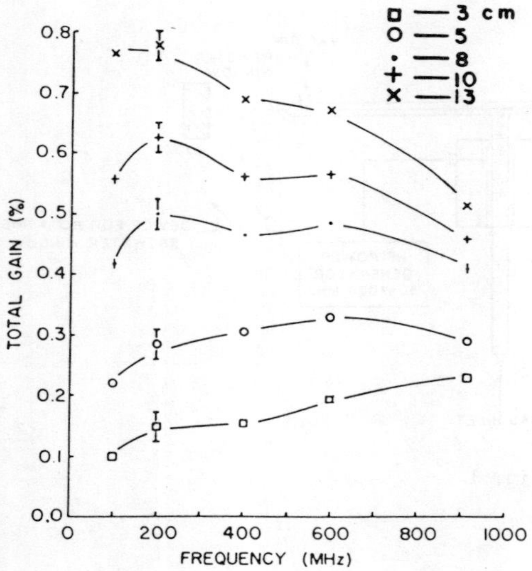

Fig. 2

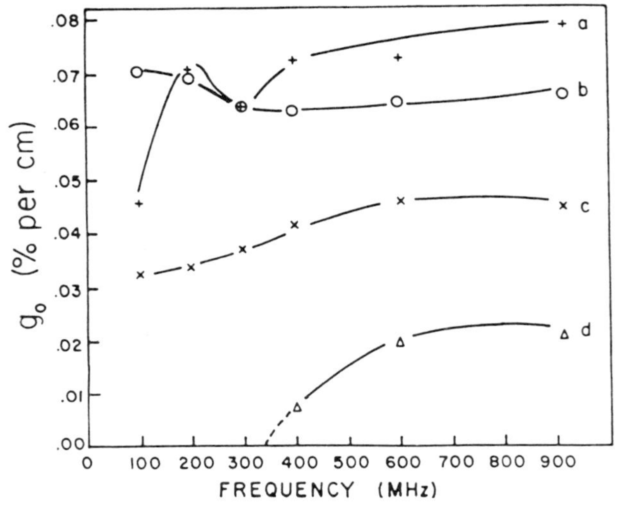

Fig. 3
$\bar{n}(cm^{-3})$:
a - 0.3 10^{11}
b - 1.0 10^{11}
c - 2.3 10^{11}
d - 4.0 10^{11}

lengths. We observe a tendency for G_0 to increase with the wave frequency, at small ($z' \leqslant 5$ cm) active length while, for larger column lengths, it decreases with frequency. Looking only at the results for $z' = 13$ cm or $z' = 3$ cm, one would conclude differently about the influence of frequency on the laser efficiency. This difference is in fact due to the axial inhomogeneity of the plasma column as it shall become obvious from the results of the measurements of the gain coefficient $g_0(\bar{n})$. These results are presented in Fig. 3 as a function of frequency, for given density values. It indicates that gain values in excess of 7% per meter are easily achieved at frequencies of 400 MHz and up, as compared to 6%/m for DC pumped lasers[4] under the same discharge conditions. The physical picture obtained from g_0 as a function of frequency is quite different from that of G_0.

Conclusion We have observed that there exists an influence of the applied frequency on the gain coefficient in a SW pumped He-Ne laser. This coefficient increases with the wave frequency. The same conclusion had been arrived at with a completely different type of HF produced plasma[3].

We note that using a SW pumped He-Ne laser at GHz frequencies to obtain large beam intensity is not very efficient. This is due to a large spread in electron density values along the plasma column as a result of a strong axial density gradient. However, this pumping scheme could be used efficiently to operate short length lasers ($\leqslant$ 5 cm), which could be supplied with low power transistors ($\sim$10 watts) at GHz frequencies.

Various methods can be used to level the density gradient (stacked T-tubes, one surfatron at each end of a cylindrical tube) but only those yielding a low density value along the columns are of interest (e.g. a decreasing diameter tapered tube)[5].

Finally, we note that the SW pumping technique allows to determine the gain coefficient, for a given density values as a function of the applied frequency, over both the RF and microwave range of frequencies. This could be useful in optimising the operation of HF pumped lasers in general.

Acknowledgements The authors are grateful to Messrs. F. Roy, R. Lemay, R. Martel, and J. Mathieu for their expert technical assistance.

References

1. C. Moutoulas, M. Moisan, L. Bertrand, J. Hubert, J.L. Lachambre, A. Ricard, Appl. Phys. Lett. 44, 323 (1985).
2. C. Moutoulas, M.Sc. Thesis, Université de Montréal, Département de Physique, Canada (1985).
3. Yu. N. Muller, V.A. Krustalev, Sov. J. Quantum. Electron. 11, 401 (1981).
4. C.S. Willet, Introduction to gas lasers; Population inversion mechanisms, (International series of monographs in natural philosophy) V.67 Pergammon Press, Oxford (1974).
5. M. Moisan, Z. Zakrzewski, contributed paper, this conference.

NEW TYPE SURFATRON FOR SURFACE WAVE PRODUCED PLASMAS

by

Shigehiko NONAKA

Toyota Technological Institute,
2-12 Hisakata, Tempaku, Nagoya 468, Japan

It is well known that a re-entrant type cavity with a narrow gap can sustain the plasma which is produced in a dielectric tube by an azimuthal symmetrical surface wave. The produced plasma is quite quiescent, and the plasma density is determined by the input r.f. power and the size of the tube.

This paper proposes a new type surfatron with a resonant circuit consisted of a helix and a condenser. The characteristics of such a helix surfatron are considered to be easily constructed and to make possible an available r.f. frequency wide by matching the length of helix to the excited surface wavelength. The first step experiment has clarified that the helix can excite the surface wave and can produce the plasma.

HELIX SURFATRON FOR SURFACE WAVE PRODUCED PLASMAS

Shigehiko NONAKA
Toyota Technological Institute
Hisakata 2-12, Tempaku, Nagoya, 468, Japan

Michel MOISAN, Z. ZAKREZEWSKI
Unversitate de Montreal, Physics Department,
Montreal, Quevec H3C-3J7, Canada

Edward J. POWERS, Bob ROGERS
ENS, The University of Texas at Austin,
P.O.Box7728, Austin, Texas 78712, USA

1. INTRODUCTION

It is well known that a re-entrant type cavity with a narrow gap can produce the plasma which is sustained without a DC magnetic field in a dielectric tube by an azimuthally symmetrical surface wave excited by high power r.f. signals (1). Such a exciter is called a surfatron. The produced plasma is quite quiescent.

This paper proposes a new type surfatron with a helix and a condenser as a resonant circuit. The advantages of the helix surfatron are to be relatively easy-to-construct and to be available for relatively wide frequency ranges by adjusting the helix length. The first step experiment demonstrated that the helix could excite the surface wave and produce the plasma.

2. Principle of Helix Surfatron

Electromagnetic wave propagating along a helix wire is a slow and forward wave along the axis as well as an azimuthally symmetrical surface wave on a plasma column. The electric field structure in the tube cross-section also has a resemblance to that of surface wave. These waves are ,therefore, considered to make a strong coupling if a wire is wounded on a dielectric tube surrounding the plasma and if the pitch is correctly adjusted to match the wavelength to surface wave.

From such a point of view we constructed two kinds of surfatron with a helix. The prototype is useful for both a comparatively high frequency r.f. signal and a small radius glass tube, which is shown in Fig. 1(a). A feeding point should be positioned at the loop of current standing wave on a helix.

For frequencies above 500 MHz a coupling condenser may be removed for the sake of stray capacitances between wires, but for intermediate frequencies between 500 MHz and 100 MHz it would be needed to decrease the reflection of an input r.f. power. The second one is for both a comparatively low frequency signal and a large radius glass tube. In this case a coupling condenser also serves as a resonant circuit together with an inductance of helix, as shown in the figure(b). This type surfatron seems to be available in the wide frequency ranges from 20 MHz to 500 MHz.

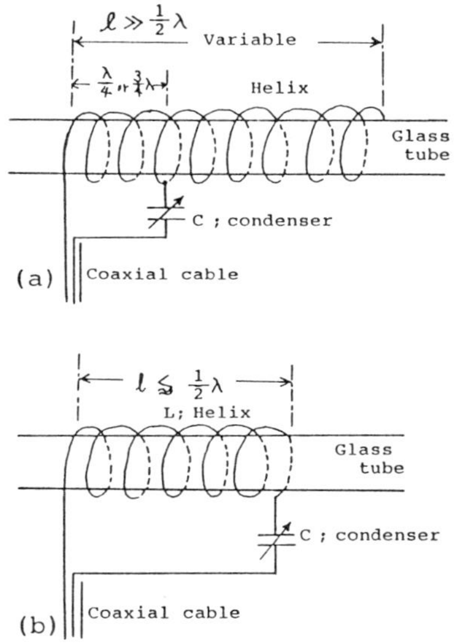

Fig. 1; Two types of Helix-surfatrons constructed. (a) is for high frequency r.f. signals, and (b) for low frequency ones, respectively. ℓ is a wirelength, λ the wavelength of E.M. wave along a wire, respectively.

3. EXPERIMENTS

Experiments were performed using an Argon gas in the range of gas pressure from 10mtorr to 300mtorr. Figure 2(a) shows the experimental view obtained using the high frequecy surfatron, in which two portions of white glows are plasmas sustained by surface waves.

In this experiment the pyrex glass tube, inner and outer diameters 14mm ϕ and 16mm ϕ, was used. The helix, diameter 32mm ϕ and length 22cm in maximum extension, was used. The whole of the helix was covered by a hollow brass tube of the cut-off size for electromagnetic cavity modes inside the brass tube to prevent the radiations. Signal frequency was varied in the range from 200 MHz to 900 MHz. In the experiment of the photo, a gas pressure was too high, and the reflection of r.f. power was so high as 60 percents at the portion connected with surfatron that the absorbed power was about 40 Watts into plasma, so that the plasma length is relatively short as seen in the photo. But it became long when we appropriately adjusted a helix pitch and a gas pressure. Figure 2(b) shows one of the interferometer patterns obtained at the plasma column in Fig. 2(a). It is shown clearly that surface wave propagates producing the plasma inside the glass tube.

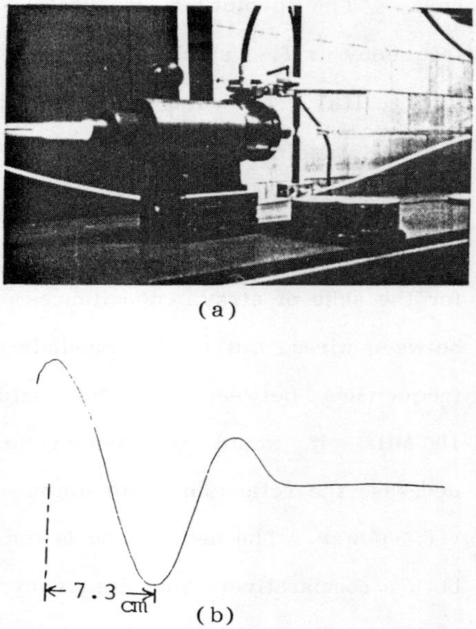

Fig. 2; Experimental results obtained by the high frequency surfatron. (a):experimental view. (b):typical interferometer pattern. Gas pressure=300mtorr. r.f.=670MHz.

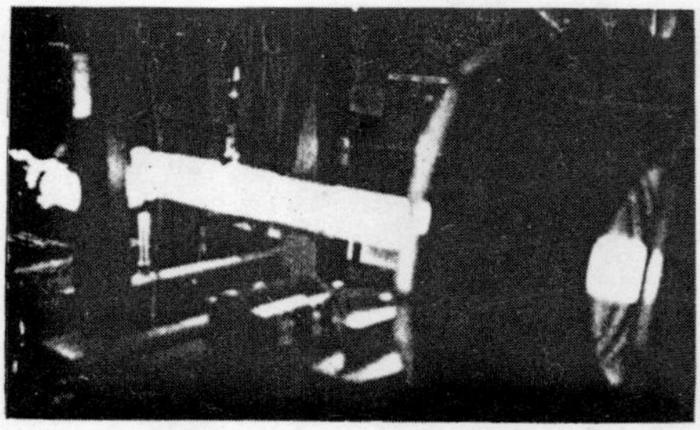

Fig. 3; Experimental view when the low frequency surfatron was used without a DC magnetic field applied. Gas pressure=50mtorr. r.f.=30MHz.

It also proves that a helix can excite the surface wave that would be symmetrical mode and sustain the plasma.

Figure 3 shows the case when using the low frequency surfatron shown in the Figure 1(b). Diameter of the glass tube used is 80mmϕ. The helix has 10 turns and 15cm in length. Signal frequency used is 30 MHz. The r.f. power reflection is so low that the total input power is 100 watts into plasma. The sustained plasma column attained to 200cm in length without owing to an external DC magnetic field as shown in the photo. Electromagnetic radiations from the helix which is not surrounded by a hollow brass tube were not observed in this experiment.

4. Discussions

From experiments above it was demonstrated that a helix can excite the surface wave which sustains the plasma column. Though it was not checked whether a helix may excites an azimuthally higher modes including a dipole mode (2-4), it seems to be impossible in our surface wave experiments because those are all backward waves in almost all regions of wavenumber(5-6).

ACKNOWLEDGMENT ; Experiments have been done at the Physics Department in The University of Montreal in CANADA and also at the Electrical Engineering Department in The University of Texas at Austin in USA within the studying abroard of one (S.N.) of the authors. One (S.N.) of the authors would express his gratitude to Prof. M. Moisan in Montreal and to Prof. E.J. Powers in USA for their acceptance for abroad, and also to the directors of T.T.I. in Nagoya JAPAN for their permission for abroard.

REFERENCES

1) M. Moisan et.al, Revue Phys. Appl. 17(1982)707-727.
2) Aubert, Messiaen and Vandenplas, Appl. Phys. Lett.,18(1971)63-65.
3) B. Kampmann, Z. Naturforsch. 34a(1979)414-422.
4) G. Lisitano et.al, Appl. Phys. Lett.,16(1970)122.
5) A. Shivarova and I. Zhelyazkov in: Electromagnetic Surface Modes, edited by A. D. Boardman (John Wiley 1982) pp.465-520.
6) S. Nonaka and Y. Akao, J. Appl. Phys.,54(1983)3798-3806.

SURFACE PLASMONS IN INHOMOGENEOUS PLASMA OF CONDUCTION ELECTRONS AT THE BOUNDARY OF A METAL

V.I.OKULOV

Institute of Metal Physics, Ural Scientific Centre
Academy of Sciences of the USSR, Sverdlovsk USSR

The spectrum of potential surface plasmons in a metal depends on the distribution of electron density at the boundary. For the waves the frequency of which at small wave vectors q is described by the formula

$$\omega = (\omega_0/\sqrt{2})(1 + \alpha q/2) \tag{1}$$

where ω_0 is the plasma electron frequency, the properties of surface layer influence essentially on the magnitude of the coefficient α. But apart from the waves with the frequency (1) because of the inhomogeneity of electron density the new modes of surface oscillations can arise. The present report is devoted to the consideration of those oscillations the frequencies of which differ from (1).

The effects due to the distribution of electron density at the surface of a metal can be described by introducing the suitable nonlocal complex dielectric tensor $\varepsilon(\omega, q, z, z')$; we believe that the z-axis is directed along the normal to the plane boundary of a metal and the wave propagates along the y-axis. The dispersion equation for surface plasmons in simple metal at $qd \ll 1$, $cq/\omega \gg 1$ (d is the effective thickness of nonuniform surface layer, c is the velocity of light) can be written in the form [1]:

$$1 + \varepsilon^0(\omega) + q\int_{-\infty}^{\infty} dz\,[\varepsilon^0(\omega)\Phi(z) - \varepsilon^0(\omega) - 1 + \int_{-\infty}^{\infty} dz'\,\varepsilon_{yy}(\omega, 0, z, z')] = 0 \tag{2}$$

where the function $\Phi(z)$ can be found from the equation:

$$\int_{-\infty}^{\infty} dz'\,\varepsilon_{zz}(\omega, 0, z, z')\,\Phi(z') = 1 \tag{3}$$

with the boundary conditions $\Phi(\infty) = 1$, $\Phi(-\infty) = 1/\varepsilon^0(\omega)$; $\varepsilon^0(\omega)$ is the bulk high-frequency dielectric function.

The formula (2) is valid in linear approximation in qd and describes those surface plasmons for which, in particular, the formula (1) is justified. On the other hand, the frequencies of normal oscillations modes of electron density in surface layer are determined from the existence condition of nontrivial solution of the equation

$$\int_{-\infty}^{\infty} dz' \, \varepsilon_{zz}(\omega, 0, z, z') \, \varphi(z') = 0 \qquad (4)$$

with the boundary conditions $\varphi(\pm\infty)=0$. From the equation (4) in simple model the results of the paper [2] in which the new surface oscillation modes have been predicted are obtained. The modes of another type have been investigated by us in frameworks of models of smooth extensive surface layer and very high potential barrier for electrons at the boundary of a metal.

In model of smooth change of electron density at the surface it is assumed that the complex dielectric tensor is local:

$$\varepsilon_{yy}(\omega, 0, z, z') = \varepsilon_{zz}(\omega, 0, z, z') = \left(1 - \frac{\omega_L^2(z)}{\omega^2}\right) \delta(z - z') \qquad (5)$$

here $\omega_L^2(-\infty) = \omega_0^2$, $\omega_L(\infty) = 0$. This model can be applied to the metal surface distorted by deformations and the layers of impurities and defects.

If the function $\omega_L^2(z)$ is monotone, the equation (2) with approximation (5) gives only the known formula of the form (1). However, in the case when the function $\omega_L(z)$ has an extremum the oscillations with extremal frequencies can exist. Let us believe that near the point $z = z_m$ in region $|z - z_m| \lesssim \mathcal{D}$ the expansion $\omega_L^2(z) = \omega_m^2 \pm [(z - z_m)/\Delta]^2$, is valid, where $\omega_m^2 = \omega_L^2(z_m)$ and the signs $\pm$ correspond with maximum and minimum. Then in the limit of small q the equation (2) has besides (1) the following solution for quadrate of the frequency of new oscillation modes:

$$\omega^2 = \omega_m^2 \mp \left[\frac{\pi}{2} \, \frac{\omega_m^2(\omega_0^2 - \omega_m^2)}{\omega_0^2/2 - \omega_m^2}\right]^2 (q\Delta)^2 \qquad (6)$$

The solutions of this type exist in the case of minimum (the upper sign) at $\omega_m^2 < \omega_0^2/2$ or at $\omega_m^2 > \omega_0^2$ and in the case of maximum (the lower sign) at $\omega_0^2/2<\omega_m^2<1$. If near the extremum the function $\omega_L^2(z)$ varies very slowly, i.e. $\Delta \gg D$, then at $q\Delta \gg D/\Delta$ the new oscillation mode exists always and its frequency is defined as follows:

$$\omega^2 = \omega_m^2 + \omega_m^2 \frac{\omega_0^2 - \omega_m^2}{\omega_m^2 - \omega_0^2/2} qD \qquad (7)$$

The surface oscillations described by the formulas (6) and (7) are of local dipol oscillations at the extrema of electron surface density.

For describing the electron oscillations at ideal metal surface in simple approximation the quantum-mechanical model of infinite high potential barrier of the boundary can be applied. Substituting the dielectric tensor $\varepsilon_{zz}(\omega,0,z,z')$ in this model into equation (4) at $\beta \equiv 4\varepsilon_F/\hbar\omega_0 \gg 1$ (ε_F is the Fermi energy) one can obtain the following integral equation:

$$\varepsilon_{\parallel}^{cl}(\tilde{\omega},t)\psi(t) = \frac{3}{2}\int_{-1}^{1} dt' \frac{t'^2\psi(t')}{\tilde{\omega}^2 - \beta^2 t^2 t'^2} \qquad (8)$$

where $\tilde{\omega}=\omega/\omega_0$, $t=q/2K_F$, K_F is the Fermi momentum, $\varepsilon_{\parallel}^{cl}(\tilde{\omega},t)$ is the classical longitudinal dielectric function. The function $\psi(t)$ is simply related to the Fourier transform of the function $\varphi(z)$. From the equation (8) the asymptotic expansion in β^{-1} of the integral $\int_{-1}^{1} dz\, z^4 \psi(z) = I$ is obtained

$$I = \frac{3}{2} I \int_{-1}^{1} dt \frac{t^2}{\varepsilon_{\parallel}^{cl}(\tilde{\omega},t)} \cdot \frac{1}{1-\tilde{\omega}^2} + \cdots \qquad (9)$$

Each term of this expansion is expressed by I and the condition $I \neq 0$ gives the dispersion equation for the oscillations of corresponding type. In the first approximation the frequency of oscillations is written in the form

$$\omega = \omega_0 \frac{3}{\beta} = \frac{\hbar q_D^2}{2m} \qquad (10)$$

where q_D is Debye wave vector.

The condition $\beta \gg 1$ corresponds the high electron density. In the case of low electron density from the equation (4) one can obtain the differential equation the eigenvalues of which are found by WKB method and have the form

$$\omega_n = \omega_0 C \beta^{1/3} n^{1/3} (n + 1/2)^{2/3} \tag{11}$$

where n is the integer, C is the constant of order of unity. The results obtained show that the structure of surface layer of electron density which is formed near high potential barrier of the boundary admits the existence of normal oscillation modes.

Thus the potential surface modes with the frequencies different from (1) can exist at the very various types of electron structure of metal surface.

References

1 ZYRYANOVA N.P., OKULOV V.I., Fiz. Met. Metalloved. 50, 496 (1980)

2 BENNET A.J. , Phys. Rev. B1, 203 (1970)

NONLINEAR INTERACTION OF VOLUME AND SURFACE WAVES IN A RELATIVISTIC PLASMA

V.N. Pavlenko and V.P. Zakharov
Institute of Physics, Ukrainian Academy of Sciences,
Kiev, U.S.S.R.
and
E. Näslund and L. Stenflo
Department of Plasma Physics, Umeå University,
S-90187 Umeå, Sweden.

In the theory and experiments on relativistic plasma electronics [1] and free-electron-lasers [e.g. 2] one has in detail investigated the nonlinear excitation of volume beam modes in bounded plasmas. In this paper we shall complement previous descriptions by considering an additional decay channel for the pump wave, namely the nonlinear excitation of surface beam modes.

With regard to the experimental situations it would, of course, be natural to consider wave propagation in cylindrical geometry. If the surface wave electric fields are nonzero only within a distance from the boundary that is much smaller than the cylinder radius, it is, however, possible to simplify the theory by considering a semi-infinite plasma, which we here describe by the dielectric tensor $\varepsilon_p(\omega)\delta_{ij}$. We will, in addition, assume that a relativistic electron beam, with constant density n_0 and velocity $\underset{\sim}{v}_0 = v_0 \hat{\underset{\sim}{x}}$, is confined to the region $z>0$. Charge neutralization is provided by a background of immobile ions. The halfspace $z<0$ is supposed to be a dielectric, with relative dielectric constant ε_r.

To consider the nonlinear interaction of volume and surface waves it is now convenient to adopt, and, in order to include relativistic effects, slightly generalize, the theory which Sitenko et al. [3] previously developed from a Vlasov description with particles reflected specularly from the boundary. Then, neglecting the nonlinear terms, we first deduce the wellknown eigenmodes of our plasma. We thus find the transverse volume mode

$$\omega^2 \varepsilon_p(\omega) \approx k^2 c^2 + \omega_{pb}^2/\gamma \tag{1}$$

and, for sufficiently large beam velocity, the longitudinal volume modes

$$\omega \approx \underset{\sim}{k} \cdot \underset{\sim}{v}_0 \pm \omega_{pb}/\gamma^{3/2} \tag{2}$$

where the beam plasma frequency ω_{pb} is equal to $(n_0 q^2/\varepsilon_0 m)^{\frac{1}{2}}$, and where $\gamma = (1-v_0^2/c^2)^{-\frac{1}{2}}$ is the relativistic factor. The upper and lower signs in (2) represent the fast and slow beam modes, respectively.

Due to previous theory [e.g. 3] there are also two kinds of surface waves, that are s- and p-polarized. As s-polarized waves are nonpropagating we will here only consider p-polarized waves. Generalizing equations (16)-(18) of

Ref. [3] and considering electrostatic oscillations, one then obtains

$$\omega \approx \underset{\sim}{k}\cdot\underset{\sim}{v}_0 \pm \omega_{pb}/\gamma^{3/2}(1+\varepsilon_r)^{\frac{1}{2}} \tag{3}$$

We next study the decay of a transverse volume wave $(\omega_3, \underset{\sim}{k}_3)$ described by (1), into a fast quasi-electrostatic surface wave $(\omega_2, \underset{\sim}{k}_2)$, described by (3), and another transverse volume wave $(\omega_1, \underset{\sim}{k}_1)$, supposing that the matching conditions $\omega_3 = \omega_1+\omega_2$, $k_{3x} = k_{1x}+k_{2x}$ and $k_{3z} = k_{1z}$ are satisfied. We also assume that $k_{1,2,3y} = 0$, $\omega_3 \gg \omega_{pb}$, and that the volume waves propagate almost parallel to the surface, i.e. $k_{1,3x} \gg k_{1,3z}$. From (1) and (3) we then obtain

$$\omega_1/\omega_3 \approx [1+(v_0/c)(\varepsilon_p(\omega_3))^{\frac{1}{2}}]/[1-(v_0/c)(\varepsilon_p(\omega_1))^{\frac{1}{2}}] \tag{4}$$

As we are interested in values of $\omega_1/\omega_3 \gg 1$, it is here necessary that $v_0 \approx c$.

With $\underset{\sim}{E}_{1,3} \approx \hat{\underset{\sim}{z}}E_{1,3}(x)$ and $\underset{\sim}{E}_2 = \hat{\underset{\sim}{x}}E_2(x)$, we can then, for the stationary case, generalize the theory of Ref. [3] to derive three coupled equations for the wave amplitudes. For simplicity we then consider the case where the pump wave amplitude E_3 is much larger than E_1 and E_2, and regard E_3 as constant. The threshold electric field turns then out to be

$$|E_3| = 16(m/q)(k_3/k_2)(k_{1x}v_0/\omega_{pb})^{\frac{1}{2}}c^2(1+\varepsilon_r)^{-1/4}\gamma^{9/4}(\mathrm{Im}k_1)^{\frac{1}{2}}(\mathrm{Im}k_2)^{\frac{1}{2}} \tag{5}$$

where $\mathrm{Im}k_{1,2}$ stands for the imaginary part of $k_{1,2}$.

The fields defined by (5) might be exceeded in experiments (e.g. [4]). As compared to Refs. [3]-[5], the result above is of particular interest because the frequency of the generated transverse wave now can be much larger than the pump wave frequency due to the presence of the relativistic beam. This could be of interest for free-electron-laser applications.

References

1. Kuzelev, M.V., Rukhadze, A.A., Strelkov, P.S. and Schkvarunetz, A.G. in Int. Conf. on Plasma Physics, Lausanne, Vol. II, (Ed. by M.Q. Tran and R.J. Verbeek) pp. 701 (1984).
2. Davidson, R.C. and Yin, Y.Z., Phys. Rev. A, 30, 3078 (1984).
3. Sitenko, A.G., Pavlenko, V.N. and Revenchuk, S.M., Sov. J. Plasma Phys. 7, 499 (1981).
4. Shivarova, A. and Zhelyazkov, I. in Electromagnetic Surface Modes (ed. by A.D. Boardman) pp. 463 Wiley (1982).
5. Ray, R.N., Atre, M.V., Gopalswamy, N. and Krishan, S., Plasma Phys. 24, 319 (1982).

RADIATION AND COLLISIONAL DAMPING OF FAST MAGNETOSONIC SURFACE WAVES IN A PLASMA COLUMN

W. Sahyouni, Zh. Kiss'ovski and I. Zhelyazkov
Faculty of Physics, Sofia University
BG-1126 Sofia, Bulgaria

The mechanisms which convert the kinetic energy of the solar photosphere and convection zone into the thermal energy of the corona have been under study for a long time. One of the possible ways of energy transfer is that due to MHD surface waves.[1] All investigations however are devoted to the processes in the corona without examining the wave propagation in the photosphere. In this paper, we show that a part of the solar energy can be transferred from the convection zone to the corona through fast magnetosonic surface waves travelling along the photospheric flux tubes.

The dissipation of surface waves in the photosphere is studied under the symplifying assumptions: (i) gravity and stratification are ignored; (ii) plasma is fully ionized and isothermal; (iii) surface wave properties are calculated via small-amplitude linearized theory. The MHD waves propagate on a homogeneous cylinder of radius a directed along the $\hat{z}$-axis and embedded in a magnetic environment. The magnetic field $B_o\hat{z}$ is much larger than the background one $B_e\hat{z}$. The gas pressure and density within the cylinder are p_o and ρ_o, outside p_e and ρ_e.

The basic equations are

$$\frac{\partial \rho}{\partial t} + \nabla\cdot(\rho\vec{v}) = 0$$

$$\rho\left[\frac{\partial \vec{v}}{\partial t} + (\vec{v}\cdot\nabla)\vec{v}\right] = -\nabla\left(p + \frac{B^2}{4\pi}\right) + \frac{1}{4\pi}(\vec{B}\cdot\nabla)\vec{B} \tag{1}$$

$$\frac{\partial \vec{B}}{\partial t} + (\vec{v}\cdot\nabla)\vec{B} = (\vec{B}\cdot\nabla)\vec{v} - \vec{B}(\nabla\cdot\vec{v})$$

$$\frac{\partial p}{\partial t} + \vec{v}\cdot\nabla p + \gamma p(\nabla\cdot\vec{v}) = 0, \quad \gamma = c_p/c_v$$

Following a standard procedure[2] we obtain the dispersion relation governing the propagation of fast magnetosonic surface waves $\propto \exp[i(-\omega t + n\theta + kz)]$

$$\rho_e (k^2 v_{Ae}^2 - \omega^2) m_0 \frac{I_n'(m_0 a)}{I_n(m_0 a)} = \rho_0 (k^2 v_{A0}^2 - \omega^2) m_e \frac{K_n'(m_e a)}{K_n(m_e a)} \qquad (2)$$

where I_n, K_n are the modified Bessel functions,

$$m_0 = \frac{(k^2 c_0^2 - \omega^2)^{1/2} (k^2 v_{A0}^2 - \omega^2)^{1/2}}{(c_0^2 + v_{A0}^2)^{1/2} (k^2 c_T^2 - \omega^2)^{1/2}} > 0 \qquad v_{A0,e} = \frac{B_{0,e}}{\sqrt{4\pi \rho_{0,e}}}$$

$$m_e = \frac{(k^2 c_e^2 - \omega^2)^{1/2} (k^2 v_{Ae}^2 - \omega^2)^{1/2}}{(c_e^2 + v_{Ae}^2)^{1/2} (k^2 c_{Te}^2 - \omega^2)^{1/2}} > 0, \qquad c_{0,e} = \left(\gamma \frac{p_{0,e}}{\rho_{0,e}}\right)^{1/2}$$

$$c_T = \frac{c_0 v_{A0}}{(c_0^2 + v_{A0}^2)^{1/2}}, \qquad c_{Te} = \frac{c_e v_{Ae}}{(c_e^2 + v_{Ae}^2)^{1/2}}$$

In a thin cylinder limit, $ka \ll 1$, for $n = 1$ (kink mode) Eq. (2) becomes

$$\frac{\omega}{k} = c_k \left\{1 - \frac{1}{2} \frac{\rho_0 \rho_e (v_{Ae}^2 - v_{A0}^2) m_e^2 a^2 K_0(m_e a)}{(\rho_0 + \rho_e)^2 c_k^2}\right\} \qquad (3)$$

where

$$c_k = \left(\frac{\rho_0 v_{A0}^2 + \rho_e v_{Ae}^2}{\rho_0 + \rho_e}\right)^{1/2}$$

Further we consider that the wave dissipation is due to the following mechanisms: ion viscosity, electron heat conduction and radiation. The volumetric heating rate associated with the viscosity, Q_{vis}, is[3]

$$Q_{vis} = \frac{\eta_0}{3} (\nabla . \vec{v})^2, \quad \text{where } \eta_0 = 1.69 \times 10^{-16}\, T^{5/2} \qquad (4)$$

The heat conduction flux is carried mainly by the electrons. If the electrons are magnetized, the volumetric heating rate due to the heat conduction is[1]

$$Q_{ther} = \kappa_{\parallel e} \left(\frac{k}{\omega}\right)^2 T_0 (\gamma - 1)^2 (\nabla . \vec{v})^2 \qquad (5)$$

where $\kappa_{\parallel e} = 14.2\times10^{-7}\,T^{5/2}$. The volumetric rate at which radiation extractes energy from the waves is given by[4]

$$Q_{rad} = \Lambda\left(\frac{n}{\omega}\right)^2 (\nabla.\vec{v})^2 \tag{6}$$

where[5] $\Lambda = 10^{-24}$ ergs.cm^3.s^{-1}, and n is the electron number density.

The damping length of the waves is calculated by equating the volumetric heatings and radiation rates to the negative of the Poynting flux $\vec{S}$. If we denote the e-folding distance for the Poynting flux by L_z, we have

$$L_z = \frac{\int_0^\infty r\langle \bar{S}_z\rangle dr}{\int_0^\infty r\langle \bar{Q}_{dis} + \bar{Q}_{ther} + \bar{Q}_{rad}\rangle dr} \tag{7}$$

where the overbar denotes the average over time and the angle brackets denote the average over θ. The calculation of integrals in (7) yields

$$L_z = \frac{\alpha^2 A_1 B_1 + A_2 B_2}{\alpha^2 C_1 D_1 + C_2 D_2}$$

where

$$\alpha = \frac{\rho_e}{\rho_0}\frac{c_e^2}{c_0^2}\frac{K_1(m_e a)}{I_1(m_0 a)}$$

$$A_1 = \frac{B_0^2}{4\pi}\left(\frac{c_0^2}{k^2 v_{A0}^2 - \omega^2}\right)^2 \frac{k}{\omega}, \quad B_1 = \left(\frac{m_0^2 a^2}{2} - 1\right) I_0^2(m_0 a) - \frac{m_0^2 a^2}{2} I_1^2(m_0 a) + 1$$

$$A_2 = \frac{B_e^2}{4\pi}\left(\frac{c_e^2}{k^2 v_{A0}^2 - \omega^2}\right)^2 \frac{k}{\omega}, \quad B_2 = \left(1 - \frac{m_e^2 a^2}{2}\right) K_0^2(m_e a) + \frac{m_e^2 a^2}{2} K_1^2(m_e a)$$

$$C_1 = \frac{\eta_0}{3} + \varkappa_{\parallel e}\left(\frac{k}{\omega}\right)^2 T_0 (\gamma - 1)^2 + \Lambda\left(\frac{n_0}{\omega}\right)^2$$

$$D_1 = \frac{a^2}{2}\left[I_1^2(m_0 a) - I_0^2(m_0 a)\right] + \frac{a}{m_0} I_0(m_0 a) I_1(m_0 a)$$

$$C_2 = \frac{\eta_0}{3} + \varkappa_{\parallel e}\left(\frac{k}{\omega}\right)^2 T_0 (\gamma - 1)^2 + \Lambda\left(\frac{n_e}{\omega}\right)^2$$

$$D_2 = \frac{a^2}{2}\left[K_0^2(m_e a) - K_1^2(m_e a)\right] + \frac{a}{m_e} K_0(m_e a) K_1(m_e a)$$

Typical parameters of the photosphere are:[6] T = 6035 K, $\rho_e = 2.89\times10^{-7}$ g.cm^{-3}, $p_e = 1.12\times10^5$ g.cm^{-1}.s^{-2}. We have taken $\rho_c/\rho_e = 0.64$ $p_c = B_e^2/8\pi + p_e - B_0^2/8\pi$, $a = 10^7$ cm, $B_0 = 1000$ G, $B_e = 10$ G. In calculating the dissipation associated with Q_{rad} we have assumed $n_0/n_e = 0.64$, where $n_e = 10^{13}$ cm^{-3}. The value of L_z, corresponding to these conditions, is 5.77×10^8 cm. Thus, the waves' dissipation mainly due to the radiation and the electron heat conduction might be a mechanism for energy transfer if the wave periods are longer than 200 s, when the thin cylinder approximation is valid.

References

1. B. E. Gordon and J. V. Hollweg, Astrophys. J. 266, 373 (1983).
2. P. M. Edwin and B. Roberts, Solar Phys. 88, 179 (1983).
3. S. I. Braginskii, in Voprosy Teorii Plazmy, ed. M. A. Leontovich (Gosatomizdat, Moscow 1963), Vol. 1, p. 195.
4. D. P. Cox and W. H. Tucker, Astrophys. J. 157, 1157 (1969).
5. R. W. P. McWhirper, P. C. Thonemann, and R. Wilson, Astron. & Astrophys. 40, 63 (1975).
6. O. Gingerich, R. W. Noyes, and W. Kalkofen, Solar Phys. 18, 347 (1971).

SURFACE RESONANCES IN RADIATION SPECTRUM OF BOUNDED PLASMA-MOLECULAR SYSTEM

A.Yu.Shevchenko, I.P.Yakimenko, A.G.Zagorodny

Institute for Theoretical Physics, Academy of Sciences of the Ukrainian SSR, 252130, Kiev-130, USSR

Introduction. In the present work, the radiation intensities of the plasma-molecular column have been calculated with taking into account the molecular subsystem dynamics and the electron thermal motion. The influence of the molecules on the resonance spectrum has been investigated in detail.

Thermal radiation of plasma-molecular column. The spectrum of thermal radiation can be calculated on the basis of Langevin approach. The final result for the intensity of radiation $\dot{I}_\omega$ is[1]

$$\dot{I}_\omega = \frac{2 I_{0\omega}}{\pi^2 \tilde{k} a} \sum_m \frac{\tilde{k}^2}{\tilde{q}\,|\Delta H_m^{(1)}(\tilde{q}a)|^2}\, \mathrm{Im}\Big\{(\alpha_1-\alpha)(|\Delta_1|^2+|\delta_1|^2) + \quad (1)$$

$$+(\alpha_2-\alpha)(|\Delta_2|^2+|\delta_2|^2)-\beta(\Delta_1^*\delta_2+\delta_1^*\Delta_2)-\Delta^*(\Delta_1+\Delta_2)\Big\}$$

where

$$\Delta=\Delta_1\Delta_2-\delta_1\delta_2\,,\quad \Delta_1=h-\frac{\tilde{q}^2 k_1^2}{q_1^2\tilde{k}^2}\gamma-\gamma\gamma_p h\frac{k^2}{q_1^2}\,,$$

$$\Delta_2=h-\frac{\tilde{q}^2}{q_1^2}\gamma-\gamma_p\frac{m^2\tilde{k}^2}{a^2q_1^2}\,,\quad \delta_1=\Big(1-\frac{\tilde{q}^2}{q_1^2}\Big)\frac{mk}{a\tilde{k}}-\gamma\gamma_p\frac{mk\tilde{k}^2}{a\tilde{k}q_1^2}\,, \quad (2)$$

$$\delta_2=\Big(1-\frac{\tilde{q}^2}{q_1^2}\Big)\frac{mk}{a\tilde{k}}-\gamma_p h\frac{mk\tilde{k}^2}{a\tilde{k}q_1^2}\,,\quad h=\frac{H_m^{(1)\prime}(\tilde{q}a)}{H_m^{(1)}(\tilde{q}a)}\,,\quad \gamma=\frac{J_m'(q_1a)}{J_m(q_1a)}\,,$$

$$\gamma_p=-\frac{\varepsilon_e J_m(q_pa)}{\varepsilon_0 J_m'(q_pa)}\,,\quad \alpha=h\,,\quad \alpha_1=\gamma_p\delta^*\frac{\tilde{k}^2mk}{q_1^2a\tilde{k}}\,,$$

$$\alpha_2=\gamma_p\gamma\frac{k^2}{q_1^2}\left(h-\frac{\tilde{q}^2k_1^{*2}}{q_1^{*2}\tilde{k}^2}\right),\ \beta=\gamma_p\gamma\delta^*\frac{k^2}{q_1^2}-\gamma_p^*\frac{\tilde{k}^2mk}{q_1^{*2}a\tilde{k}}\left(h^*-\frac{\tilde{q}^2k_1^2}{q_1^2\tilde{k}^2}\gamma\right),$$

$$\delta=\left(1-\frac{\tilde{q}^2}{q_1^2}\right)\frac{mk}{a\tilde{k}},\ \tilde{k}^2=\frac{\omega^2}{c^2}\tilde{\varepsilon},\ k_1^2=\frac{\omega^2}{c^2}\varepsilon,\ \varepsilon=1+\chi_e+\chi_i+\chi_m,$$

$$k_p^2=\frac{\varepsilon_0(\omega+i\nu_e)\omega-\omega_{pe}^2}{\varepsilon_0 S_e^2},\ \varepsilon_0=1+\chi_i+\chi_m,\ \chi_\sigma=-\frac{\omega_{p\sigma}^2}{\omega(\omega+i\nu_\sigma)}\ (\sigma=e,i),$$

$$\chi_m=-\frac{\omega_{pm}^2}{\omega^2-\omega_r^2+2i\omega\gamma_m},\ \tilde{q}^2=\tilde{k}^2-k_\perp^2,\ q_{1,p}^2=k_{1,p}^2-k_\perp^2,\ k_\perp^2=\frac{\omega^2}{c^2}\tilde{\varepsilon}\sin^2\theta,$$

$$\omega_{p\sigma}^2=\frac{4\pi e_\sigma^2 n_\sigma}{m_\sigma},\ \omega_{pm}^2=\frac{4\pi e^2}{m_e}(N_1-N_2)f_m,$$

$$\omega_r=\frac{E_2-E_1}{\hbar},\quad \dot{I}_{o\omega}=\frac{\omega^2\tilde{\varepsilon}}{4\pi^3c^2},$$

$\tilde{\varepsilon}$ is the dielectric constant of the external medium; e_σ, n_σ, m_σ and ν_σ are the charge, the number density, the mass and the collision frequency for the particles of σ-species; N_1 and N_2 are the molecular level populations with E_1 and E_2 energies; γ_m is a dissipative constant for the molecular polarization; a is the column radius; S_e is the thermal velocity of the electrons. Eq(1) can be considered as the generalization of the well-known Rytov's formula[2] to the case of plasma with space dispersion. It follow's from Eq(1) that the resonance frequences in intensity spectrum satisfy the equation $\Delta = 0$. In the case of small column radius it takes the simple form

$$1+\frac{\tilde{\varepsilon}}{\varepsilon}=-\gamma_p\frac{m}{a}, \tag{3}$$

giving at $S_e = 0$ the resonances $\omega_1=\omega_{pe}/(1+\tilde{\varepsilon})$, and $\omega_2=\omega_r$, which correspond to the main plasma and molecular resonances with the molecular band-width equal to the natural one.

Increase of the column radius gives rise to the dipole resonance spliting and the molecular line broadening (Fig.1)

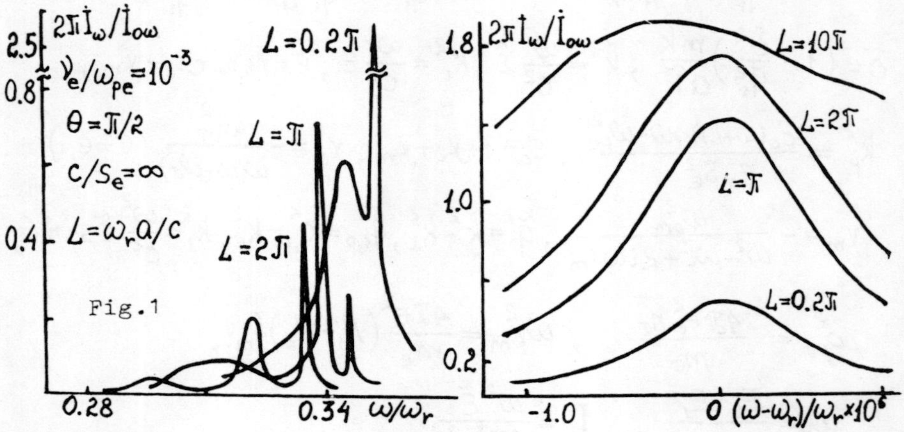

Fig.1

The curves on this and others figures correspond to the parameters $\gamma_m / \omega_r = 10^{-6}$, $\omega_{pm} / \omega_r = 5\ 10^{-4}$, $\omega_{pe} / \omega_r = 0.5$, $\tilde{\varepsilon} = 1$. These effects dependend on the dissipative parameters. Particularly, for $\nu_e / \omega_{pe} = 10^{-3}$ the splitting of resonance is observed for $\omega_r a / c = 0.2\pi$, yet for $\nu_e / \omega_{pe} = 10^{-2}$ there is an unique broadened peak (dotted line in Fig.2). Thermal motion of electrons causes the ion-acoustical peaks

$$\omega_{mn}^2 = \omega_{pi}^2 / \left(1 + \frac{k_D^2}{k_{mn}^2}\right), \quad k_D^2 = \omega_{pe}^2 / S_e^2, \quad k_{mn} = X_{mn}/a \qquad (4)$$

where X_{mn} satisfy the equations

$$1 - \frac{\omega_{pe}^2 a^2}{2c^2} \sin^2\theta \ln\left[\frac{\omega_{pi} a \sin\theta}{c(1+k_D^2 a^2/X^2)^{1/2}}\right] + \frac{S_e^2}{2c^2} X \cos^2\theta \frac{J_0(X)}{J_0'(X)} = 0, \quad m=0; \qquad (5)$$

$$J_m'(X) = \frac{mX}{k_D^2 a^2} J_m(X), \quad m \neq 0.$$

Besides that the Tonks-Dattner's resonances are observed

$$\omega_{mn}^2 = \omega_{pe}^2 + k_{mn}^2 S_e^2 (\omega_{mn} \neq \omega_r), \quad k_{mn} = \frac{X_{mn}}{a}; \quad \left(1 + \frac{2X^2}{k_D^2 a^2}\right) J_m'(X) = \frac{m}{X} J_m(X). \qquad (6)$$

Moreover, the imaginary solutions of equation (7) define the spectrum in the vicinity of main resonance

$$\omega_m^2 = \frac{\omega_{pe}^2}{2}\left(1 + \frac{m\sqrt{2}}{k_D a}\right) \tag{8}$$

in full agreement with the numerical analysis (Fig.2). The coincidence of the frequency ω_r with one of the resonances (6) does not cause the simple overlaping of the spectra. As is clear from Eq.(3), for $\omega_r \approx \omega_m$ the Tonks-Dattner's resonance splitting takes place according to $\omega_{MN}^2 \to \omega_{MN(1,2)}^2 = \omega_r^2 \pm \omega_{pm}\omega_{pe}$ (Fig.3). The resonance spectrum due to the nonlinear interaction of waves scattered by the plasma sphere has been investigated in[4].

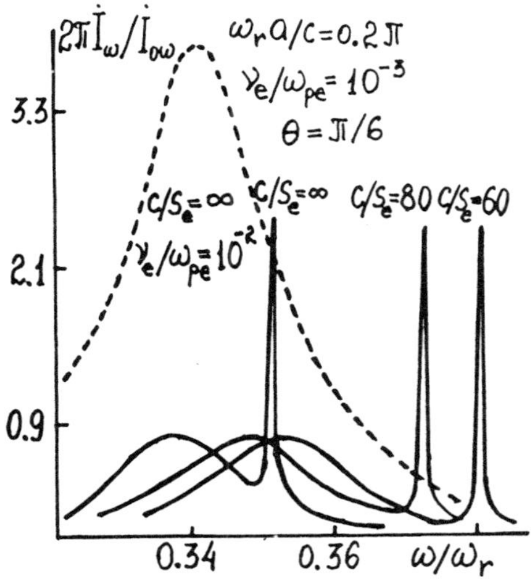

Fig.2

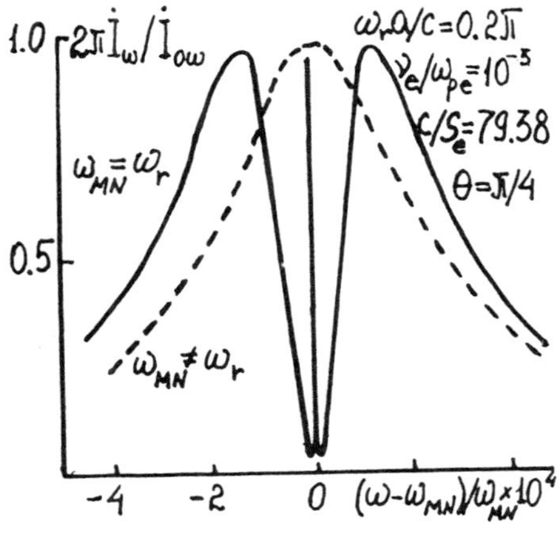

Fig.3

1. A.G.Zagorodny, A.Yu.Shevchenko, I.P.Yakimenko. Preprint ITP-84-169R, Kiev, 1984.
2. S.M.Rytov. Theory of Electric Fluctuations and Thermal Radiation. Moscow, Izd. AN SSSR, 1953.
3. N.A.Krall, A.W.Trivelpiece. Principles of plasma Physics. Mc Graw Hill, N.-Y, 1973.
4. A.G.Zagorodny, A.Yu.Shevchenko, I.P.Yakimenko. Preprint ITP-85-63R, Kiev, 1985.

P-POLARIZED NONLINEAR ELECTROMAGNETIC WAVES IN CYLINDRICAL LAYERED STRUCTURES

A. Shivarova*& and M. Y. Yu§
*Institut für Experimentalphysik II
§Institut für Theoretische Physik I
Ruhr-Universität Bochum, D-4630 Bochum 1, F. R. Germany

Introduction

The problem for the existence and the propagation of nonlinear surface waves in bounded media has recently received a great deal of attention. Plane geometry (layered structures) have mainly been considered so far in the investigations of p-polarized nonlinear waves. Because of the two-dimensional character of the surface wave field, usually various simplifying assumptions have to be made. For gaseous plasmas the wave field of the nonlinear modes has been obtained for thin slab only.[1,2] In the case of dielectric structures the permittivity of the nonlinear media is simplified to a uniaxial form.[3-5] When the problem is to determine only the dispersion relation of the nonlinear waves, some of these simplifications can be avoided.[6,7]

Here we investigate p-polarized nonlinear waves in systems consisting of a dielectric rod (linear medium) coated with a nonlinear medium (dielectric with arbitrary but diagonal permittivity tensor), and surrounded by another linear dielectric. The solutions for the nonlinear wave fields are presented for the case when the nonlinear medium is gaseous plasma. A method giving directly the dispersion relation of the wave is also presented.

Derivation of the equations for the wave field

We consider the propagation of TM surface wave (E_r, B_θ, $E_z \neq 0$) in z-direction along a cylindrical structure consisting of three media: a linear medium I - dielectric rod ($r < R_1$) with dielectric constant ε_1, nonlinear thin layer II ($R_1 < r < R_2$) with dielectric tensor of general form $\varepsilon_{ij} = \delta_{ij}\, \varepsilon_{ij}(\omega, |E|^2)$ and a linear medium III ($r > R_2$) with dielectric constant ε_3. We limit our investigations to the case of weak nonlinearity and take into account only the longitudinal electric field component in the dielectric tensor:

$$\varepsilon_{rr} = \varepsilon_{or} + \alpha_r |E_z|^2; \quad \varepsilon_{zz} = \varepsilon_{oz} + \alpha_z |E_z|^2 \tag{1}$$

with $\alpha_{r,z}|E_z|^2 << \varepsilon_{or,z}$.

From the Maxwell equations with wave field variation $\propto \exp(-i\omega t + ikz)$ we obtain the following equation describing the E_z-field component:

$$\frac{1}{r}\frac{\partial}{\partial r}\{r \frac{\varepsilon_{rr}}{k^2[1 - (\varepsilon_{rr}/\eta)]} \frac{\partial E_z}{\partial r}\} = \varepsilon_{zz}E_z \tag{2}$$

where $\eta = (kc/\omega)^2$.

Considering thin coating ($\Delta R = R_2 - R_1 << R_1$), in the first order approximation with respect to the radius, we reduce Eq. (2) for the nonlinear medium to the corresponding one for a plane layer ($r = R_1$).

The first integral of Eq. (2) is:

$$(\frac{dE_{z2}}{dr})^2 = k^2 \frac{\eta - \varepsilon_{or}}{\eta} \frac{\varepsilon_{oz}}{\varepsilon_{or}} E_{z2}^2 - \frac{k^2}{2} \frac{\varepsilon_{oz}}{\varepsilon_{or}}(\frac{3\alpha_r}{\varepsilon_{or}} - \frac{\alpha_z}{\varepsilon_{oz}} \frac{\eta - \varepsilon_{or}}{\eta}) E_{z2}^4$$

$$+ [1 - \frac{2\alpha_r}{\varepsilon_{or}} \frac{\eta}{\eta - \varepsilon_{or}} E_{z2}^2] C_2' \tag{3}$$

where $C_2' = [(\eta - \varepsilon_{or})/\eta\varepsilon_{or}]^2 C_2$ and the inequality $[\alpha_r/(\eta - \varepsilon_{or})]E_z^2 << 1$ is used. For the constant C_2' determined by the boundary conditions at $r = R_1$:

$$\{E_z\}|_{r = R_1} = 0; \quad \{B_\theta\}|_{r = R_1} = 0 \tag{4}$$

and the solutions in the first region I ($E_{z1}(r) = C_1 I_o(k_1 r)/I_o(k_1 R_1)$) we have:

$$C_2' = k_2^2\{\frac{\varepsilon_1^2}{\varepsilon_{or}^2} \frac{k_2^2}{k_1^2} \frac{I_1^2(k_1R_1)}{I_o^2(k_1R_1)} - \frac{\varepsilon_{oz}}{\varepsilon_{or}}\}C_1^2 - \frac{\varepsilon_{oz}}{2\varepsilon_{or}}\{k^2 \frac{\alpha_r}{\varepsilon_{or}} + k_2^2 \frac{\alpha_z}{\varepsilon_{oz}}\}C_1^4 \tag{5}$$

where $k_n^2 = k^2 - (\omega/c)^2\varepsilon_n$, $\varepsilon_{n=1,3} = \varepsilon_{1,3}$, $\varepsilon_{n=2} = \varepsilon_{or}$. Thus, the first integral (3) can be written in the form:

$$(\frac{dE_{z2}}{dr})^2 + a\, E_{z2}^2 + b\, E_{z2}^4 = C_2' \tag{6}$$

where $a = -[(k_2^2\varepsilon_{oz}/\varepsilon_{or}) - (2k^2\alpha_r C_2'/k_2^2\varepsilon_{or})]$,

$b = \{(3k^2\alpha_r\varepsilon_{oz}/\varepsilon_{or}^2) - (k_2^2\alpha_z/\varepsilon_{or})\}/2$.

Depending on the type and the properties of the nonlinear medium a, b and C_2' have different signs, and solutions for different types of nonlinear waves can be obtained.

Solutions for structures with gaseous plasma

In the case when the second region is occupied by gaseous plasma, the nonlinear medium is isotropic but its permittivity (with nonlinearity due to the ponderomotive force action) includes the two electric field components. By considering the case of weak nonlinearity ($|E_o|^2 = |E|^2/E_p^2 << 1$, where E_p is the plasma field) and having in mind that $|E_r|^2 < |E_z|^2$ we reduce the expression for the plasma permittivity to (1) with $\varepsilon_{or} = \varepsilon_{oz} = \varepsilon_o$, $\alpha_r = \alpha_z = \alpha/E_p^2$ where $\varepsilon_o = 1 - \alpha$, $\alpha = \omega_{po}^2/\omega^2$. We have $C_2' > 0$, $a < 0$ but for b the two cases ($b > 0$ as well as $b < 0$) are possible.

In the case of $b > 0$ the solution for the field in the nonlinear medium is obtained in terms of Jacobi elliptic functions:

$$E_{z2} = \delta_1 \, \mathrm{cn}[\gamma_1(r - r_o)|m_1] \tag{7}$$

where $\delta_1 = [(|a| + D_1)/2b]^{1/2}$, $\gamma_1 = \sqrt{D_1}$, $m_1 = (|a| + D_1)/2D_1$ with $D_1 = (|a|^2 + 4C_2'b)^{1/2}$. Because of our assumption $|E_r|^2 << |E_z|^2$ we do not allow in the second region zeros of the Jacobi function.

For $b < 0$ the solution is:

$$E_{z2} = \delta_2 \, \mathrm{cs}[\gamma_2(r - r_o)|m_2] \tag{8}$$

where $\mathrm{cs}(u|m) = \mathrm{sn}(u|m)/\mathrm{cn}(u|m)$, $\delta_2 = [2C_2'/(|a| - D_2)]^{1/2}$, $\gamma_2 = [(|a| + D_2)/2]^{1/2}$, $m_2 = 2D_2/(|a| + D_2)$ with $D_2 = (|a|^2 - 4C_2'|b|)^{1/2}$. The result (8) shows that the field in the nonlinear coating has a singularity at $r = r_o$ because losses are not included in the dielectric tensor.

Derivation of the dispersion relation

The dispersion relation of the surface polaritons can be obtained directly[6], without first obtaining the solution for the wave field.

Eq. (2) can be formally integrated to yield:

$$\left[r(A/\varepsilon_{zz}) \frac{dE_z}{dr}\right]^2 = k^2 \int r^2 A \, dE_z^2 + C \tag{9}$$

where $A = \varepsilon_{zz}\varepsilon_{rr}/[1 - (\varepsilon_{rr}/\eta)]$. Using the relation between

dE_z/dr and B_θ and the fact that $\Delta R \ll R_1$ we obtain for region II:

$$B_{\theta 2} = (-i/\sqrt{\eta}) \left(\int^{E_{z2}^2} A_2 \, dE_z^2 + C_2''\right)^{1/2} \quad (10)$$

where $C_2'' = C_2/k^2$. With the boundary conditions (4) applied both at R_1 and R_2 one has:

$$B_{\theta 1}^2(R_1) = \eta^{-1} \left(\int^{E_{z1}^2(R_1)} A_2 \, dE_z^2 + C_2''\right) \quad (11)$$

$$B_{\theta 3}^2(R_2) = \eta^{-1} \left(\int^{E_{z3}^2(R_2)} A_2 \, dE_z^2 + C_2''\right) \quad (12)$$

where E_{z1}, $B_{\theta 1}$ and E_{z3}, $B_{\theta 3}$ are the linear solutions in regions I and III. Since $\Delta R \ll R_1$ the right hand side of (12) can be somewhat simplified. Using the Taylor expansion, one obtains:

$$B_{\theta 3}^2(R_2) = \eta^{-1} \left(\int^{E_{z3}^2(R_1)} A_2 \, dE_z^2 + C_2'' + (\Delta R) \, A_2(R_1) \frac{\partial E_z^2(R_1)}{\partial r}\right) \quad (13)$$

The dispersion relation is obtained by substituting the linear solutions $E_{z1}(R_1)$ and $E_{z3}(R_1) = C_3 K_o(k_3 R_1)/K_o(k_3 R_2)$ at the boundaries, together with their magnetic field counterparts $B_{\theta 1}(R_1)$ and $B_{\theta 3}(R_2)$ into (11) and (13). Note that one of the constants C_1, C_2'', C_3 is indetermined. It defines the amplitude of the fields of the nonlinear polaritons.

References

1. I. Zhelyazkov, O. Stoyanov and M. Y. Yu, Plasma Phys., 26 813 (1984)
2. Yu. M. Aliev and S. V. Kuznetsov, Soviet Physics-Lebedev Institute Reports, n 3 (1984); Yu. M. Aliev, S. V. Kuznetsov and A. P. Shivarova, Phys. Lett. A, submitted for publication (1985)
3. V. M. Agranovich, V. S. Babichenko and V. Ya. Chernyak, Pis'ma Zh. Eksp. Teor. Fiz., 32 532 (1980)
4. A. A. Maradudin, 2nd Int. School on Cond. Matter Phys., Varna, Bulgaria (1982)
5. V. K. Fedyanin and D. Michalache, Z. Phys. B - Cond. Matter 47 167 (1982)
6. M. Y. Yu, Phys. Rev. A, 28 1855 (1983)
7. A. Shivarova and N. Dimitrov, Int. Conf. on Plasma Phys., Lausanne, vol. 1, p. 43 (1984); Plasma Phys., 27 219 (1985)

[&]Permanent Address: Faculty of Physics, Sofia University, 1126 Sofia, Bulgaria

ELECTROMAGNETIC SURFACE WAVES AT THE INTERFACE BETWEEN TWO PLASMA MEDIA

Madgi M. Shoucri

Institut de recherche Hydro-Québec, IREQ, Varennes JOL 2PO,

P.Q., Canada

It is well known that surface waves can propagate at the interface between two plasma media. A great deal of attention is currently paid to study electromagnetic plasma modes which can propagate along a plasma interface, and to the possibility of their excitation by electron beams. A general dispersion relation for electromagnetic surface waves at the interface of beam-plasma systems, in the presence of an external magnetic field, has been previously reported[1]. However, in view of the complexity of this dispersion relation, only few particular cases have been treated in details, one of them is the well however quasistatic approximation[2]. In the present paper, we derive and study the dispersion relation for the electromagnetic plasma waves at the interface between two plasma media, in the absence of an external magnetic field. The model studied here consists of a plane interface (located at $x = 0$) separating a semi-infinite region ($x < 0$) designated as region I, from another semi-infinite region ($x > 0$) designated as region II. Region II contains a plasma of frequency ω_{p2}, while region I contains a plasma of frequency ω_{p1}. Perturbations of the form $\approx \exp i(k_z z - \omega t)$ are assumed to propagate along the interface in the z-direction, and no variation is assumed to exist in the ydirection. In the absence of an external magnetic field, the equations for the TE (B_z-mode) and the TM (E_z-mode) decouple. The equation of the electromagnetic E_z-mode is given by[3]:

$$\nabla_{\perp}^2 E_z + \frac{\omega^2}{c^2} (a_{1,2} - N_z^2) E_z = 0 \qquad (1)$$

where $N_z = k_z c/\omega$ and

$$a_{1,2} = 1 - \frac{\omega^2_{p1,2}}{\omega^2} \tag{2}$$

We set $K^2_{1,2} = \frac{\omega^2}{c^2}(N^2_z - a_{1,2})$ and look for solutions to Eq. (1) of the form:

$$\approx e^{-K_2 x} \qquad x > 0$$

$$\approx e^{+K_1 x} \qquad x < 0$$

The electromagnetic field components associated with this mode are given by[3]:

$$E_x = \frac{i}{k_z} \frac{N^2_z}{N^2_z - a_{1,2}} \frac{\partial E_z}{\partial x} \tag{3}$$

$$B_y = \frac{iN_z}{k_z} \frac{a_{1,2}}{N^2_z - a_{1,2}} \frac{\partial E_z}{\partial x} \tag{4}$$

Subscripts 1 or 2 are used for region I or II respectively. The boundary condition at the interface consists in the continuity of the electromagnetic field tangential components. We obtain the following dispersion relation[4]:

$$(\omega^2 - \omega^2_{p1})(1 + \frac{\omega^2_{p1}}{k^2_z c^2 - \omega^2})^{-\frac{1}{2}} = -(\omega^2 - \omega^2_{p2})(1 + \frac{\omega^2_{p2}}{k^2_z c^2 - \omega^2})^{-\frac{1}{2}} \tag{5}$$

which for $N_z \gg 1$ leads to the well-known quasistatic result for surface waves $\omega^2 = (\omega^2_{p1} + \omega^2_{p2})/2$.

The solution of Eq. (5) (for ${\omega^2}_{p1} > {\omega^2}_{p2}$) is shown in Fig. (1) for different values of the parameters $\alpha = \omega_{p2}/\omega_{p1}$, where ω/ω_{p1} is plotted against $k_z\ c/\omega_{p1}$. Solutions of Eq. (5) exist for $\omega_{p2} > \omega > (\omega^2_{p1} + \omega^2_{p2})^{\frac{1}{2}}\ /\sqrt{2}$, i.e. for $\alpha < \omega/\omega_{p1} < (1 + \alpha^2)^{\frac{1}{2}}/\sqrt{2}$. In Fig. (1) the curve for $\alpha = 0$ corresponds to the case where we have vacuum in region II ($\omega_{p2} = 0$). This curve starts from the origin and tends asymptotically to $\omega/\omega_{p1} = 1/\sqrt{2}$. For $\alpha \neq 0$ the curve starts at $\omega/\omega_{p1} = \alpha$ for $k_z\ c/\omega_{p1} = 0$, and tends towards the asymptotic value $(1 + \alpha^2)^{\frac{1}{2}}\ /\sqrt{2}$ for increasing values of $k_z\ c/\omega_{p1}$. For $\alpha = 0.8$ for instance, the whole frequency spectrum is concentrated in the range $0.8 < \omega/\omega_{p1} < 0.905$; similarly for $\alpha = 0.95$ the whole frequency spectrum is confined within the range $0.95 < \omega/\omega_{p1} < 0.975$. Hence for values of α close to unity (i.e. small density variation between region I and II), the whole frequency spectrum of the dispersion curve is confined within a narrow range close to $(\omega^2_{p1} + \omega^2_{p2})^{\frac{1}{2}}/\sqrt{2}$ when $k_z\ c/\omega_{p1}$ varies from zero to infinity.

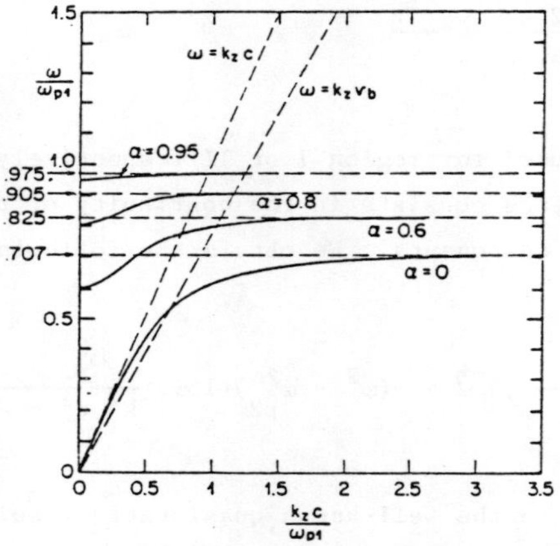

FIG. 1. Plot of ω/ω_{p1} against $k_z c/\omega_{p1}$ for the dispersion relation given by Eq. (5). The horizontal asymptotes are located at $\omega/\omega_{p1} = (1 + \alpha^2)^{\frac{1}{2}}/\sqrt{2}$.

1. M.M. Shoucri and A.B. Kitsenko, Plasma Phys. 10 633 (1968)
2. M.M. Shoucri and A.B. Kitsenko, Plasma Phys. 11 345 (1969)
3. M.M. Shoucri, IEEE Plasma Science, PS-9 109 (1981)
4. M.M. Shoucri, J. Apl. Phys. 50 702 (1979)

EXPERIMENTAL STUDY OF THE PHASE CONSTANT STABILIZATION TIMES FOR A PULSED PLASMA COLUMN CREATED BY A SURFACE WAVE.

A. Sola, A. Gamero, J. Cotrino & V. Colomer.
Departamento de Física de la Facultad de Ciencias. Córdoba (Spain).

Introduction The purpose of this paper is the study of the phase constant stabilization times of a surface wave propagating along a plasma column produced by the same wave, during the creation process of this plasma. The procedure to obtain long columns of plasma by coupling high power H.F. energy has been known since a few years ago. The structures permitting the obtainment of such a type of discharge are called surfatron and surfaguide, and allow the propagation of a travelling surface wave along the formed plasma column. The study of these stationary discharges is developed, fairly well, at this moment and its more important characteristics have been reported recently[1]. However, the study of these microwave discharges on impulsion regime, i.e., the H.F. energy producing the atom ionization is modulated by means of a square pulse signal, is not so developed as the stationary case, being the experimental data and the theoretical interpretation very limited[2,3,4]. In the present paper we study the effective phase constant $\bar{\beta}$ evolution of the travelling surface wave during the transitional creation and stabilization process of a plasma column produced by the same high power surface wave.

Experimental arrangement The arrangement used for the measurement of $\bar{\beta}$ is shown in Fig.1. The classical interferogram method utilized on the continuous wave case (stationary discharge) is used for the impulsion regime after an appropriate accommodation. Actually, we choose the time in which the measurement of $\bar{\beta}$ is made, from the moment when the H.F. energy is coupled to the plasma. This coupling has been made by means of a 2.45 GHz surfatron and the discharge tube was a Pyrex cilindrical one inside a cilindrical guide-wave (20 mm internal radius). The gas used in the discharge has been Argon in a laminar flowing regime at different mean pressure conditions. The H.F. energy has been modulated by a square pulse signal of a width 3 msec.& a period 10 msec., the incident power being maximun 200 w. inside the pulse.

The phase constant stabilization times measurement By means of the reflerred method, different plots of $\bar{\beta}$ versus the position z along the plasma column have been obtained at different times from the modulator pulse beginning, the position of the origin being placed at the coupling structure. Fig.2 shows how the curves move forward and make distance from the surfatron, as time proceeds. Fig.3 has been performed from Fig.2 and shows that more distanced positions along the plasma column is considered, more time is needed to achieve the stabilization value of $\bar{\beta}$. We estimate the stabilization time of $\bar{\beta}$ in a particular position of the plasma column as the time needed to go by so that the value of $\bar{\beta}$ decays down to the 10 % up of its final (stationary) value, this time being measured from the arrival of the ionization front at this particular position. In Fig.4, the positions of this ionization front (I.F.), at different instants from the modulator pulse beginning have, been plotted during the column creation process (curve I), the slopes of the tangents giving us the velocity v_I at each instant and position. In the same figure, the curve S has been performed by adding the estimated stabilization time, t_{Stb}, to the temporal coordinate in a particular position. The velocity of this estimated "stabilization front" (E.S.F.) can be obtained similary to the I.F. velocity. Both curves delimit three parts in the discharge zone: above curve I, where it does not yet exist ionization; under curve S, where $\bar{\beta}$ is estimated to be stabilized (at least 10 % up of its final value); and the intermadiate zone, where the plasma has been created but $\bar{\beta}$ is estimated to be unstabilized. The times t_{Stb} increase as z position are closer to the end of the column (segments AB), while the stabilization zones (segments BC) decrease with time and fall down asymtotically to zero.

Results and interpretations.

A. The $\bar{\beta}$ stabilization time along z, at different pressure conditions. These experiences were made in a 4.5 mm and 6 mm (4.5/6 mm) of internal and external radii and 150 cm length tube from 0.375 Torr to 18.5 Torr mean pressure, the power incident being 200 w. inside the pulse. On Fig.5, we show a plot of t_{Stb} versus z' (position from the end of the stationary column) for two different pressures. An exponencial dependence of t_{Stb} with z' is found in the majority of the column points. The same dependence has been for the other pressure conditions, at least along a part of the columns. There is a saturation of the stabilization time at the end of the plasma column, where the dependence is not expo-

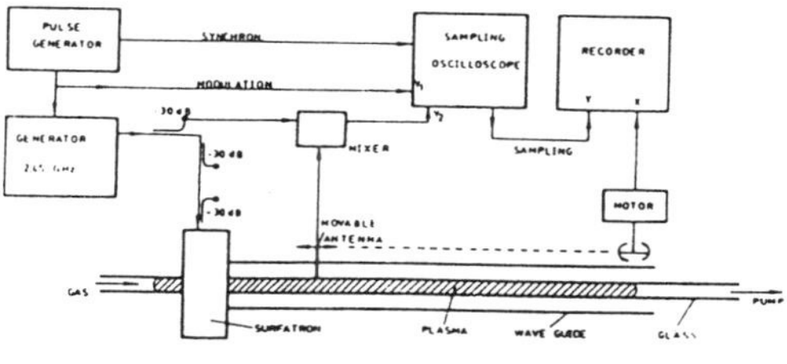

Fig. 1

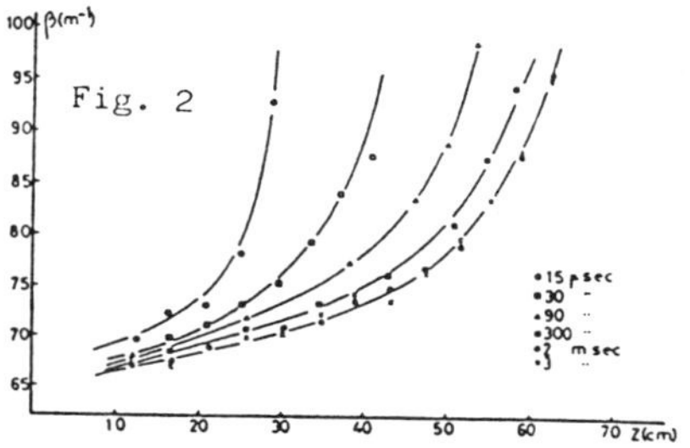

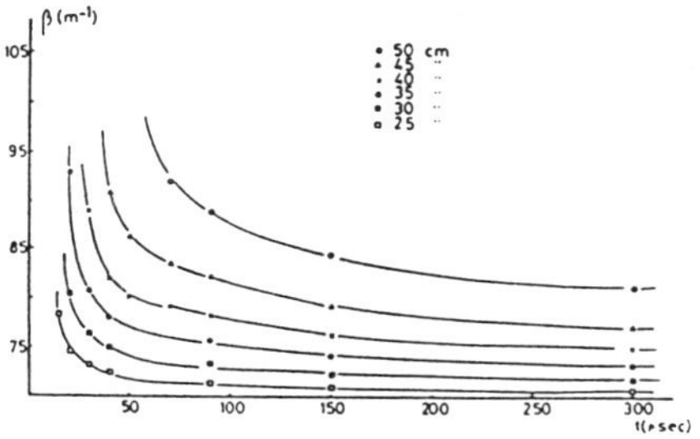

Fig. 3

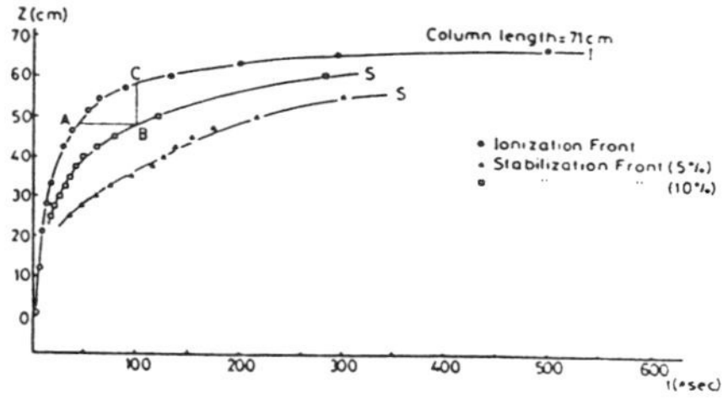

Fig. 4

nencial. The increase of t_{Stb} toward the column ends is interpreted as a consequence of the surface wave power diminution. That provokes the time necessary to get the stabilization value increase although the electronic density stationary value for such a type of discharges is smaller at the end of the columns. So, the increase of t_{Stb} from a small Δz (slat) of the column repecting the preceding-one depends on the ionizing available energy fall, this fall being function of the pressure for a given dicharge radius and a excitation frequency.

Analogous plots of t_{Stb} versus z' have been made by estimation of t_{Stb} as the necessary time to decay down to the 5 % up of the final value (curve S'). Similar dependece with z' has been found, the slopes m being closely the same in both estimated cases at the same pressure conditions (obviously t_{Stb} (5%) > t_{Stb}(10%)). Fig.6 shows the slopes dependence with the pressure. A minimun of m has been found at the same pressure at which a maximun stationary column length has been done. At this pressure, the surface wave power diminution toward the end of the column is the slowest, which agrees with the slowest increase of t_{Stb}.

B. The $\bar{\beta}$ stabilization time along z, at different incident power conditions. These experiences were made in both 150 cm length tube, 2.5/4 mm radii (tube A) and 4.5/6 mm radii (tube B). In the tube A, the in/out pressure were 6/3 Torr and in the tube B, 3/3 Torr. Three mean incident powers were used, the obtained results being plotted, in both cases, at the most different powers. These results showed that, in all the cases, the dependence of t_{Stb} wiht z' was exponential at least along the intermediate zone of the column. Also, there was a separation of the exponential behavior (saturation) at the column end. This separation happened independently on the incident power and the tube radius, the slopes of the right zones being closely the same at different incident powers, for each tube. In the tube A, the $\ln[t_{Stb}(10\%)]$ versus z' plots at different incident power, were superposed (Fig.7), while in the tube B there were different plots, the right zones remainning parallels, at different power (Fig.8). In this last case, a diminution of the incident power provoked a diminution of the saturation zone, while in the tube A provoked a diminution of the right zone. In general, one observed that less the used tube radii was, more rapidly the stabilization phenomena took place.

References

1.Moisan,Ferreira,Haglaoui,Henri,Hubert,Pantel,Ricard & Zakrzewski. Revue Phys.Appl.17,707 (1982).

2.Bloyet,Leprince & Moisan. Rev.Phys.Appl. 12,1719 (1977).

3.Bloyet,Leprince,Llamas Blasco & Marec. Phys.Lett.,82A, 391 (1981).

4.Llamas Blasco,Colomer & Rodríguez Vidal. Accepted in J. of Phys.D.(1985).

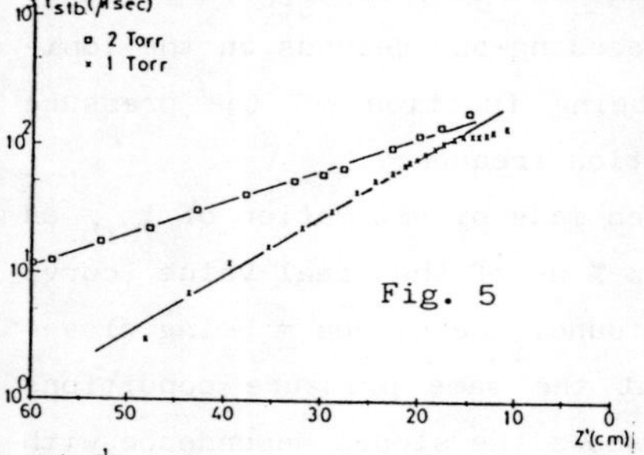

Fig. 5

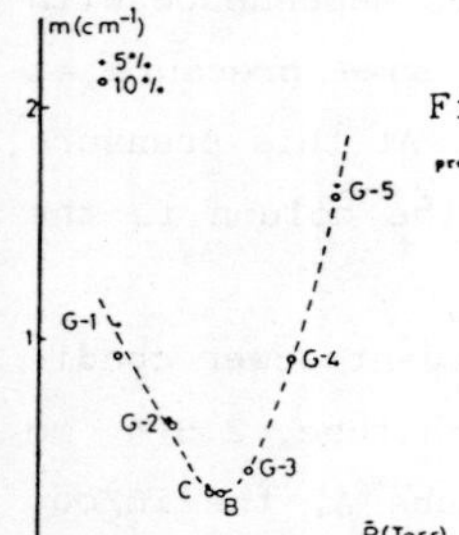

Fig. 6

pressure	in		out	
G-1	0.4	Torr	0.35	Torr
G-2	1.5	"	0.7	"
G-3	4	"	4	"
G-4	8.5	"	8	"
G-5	19	"	18	"
B	3.8	"	2	"
C	3	"	1	"

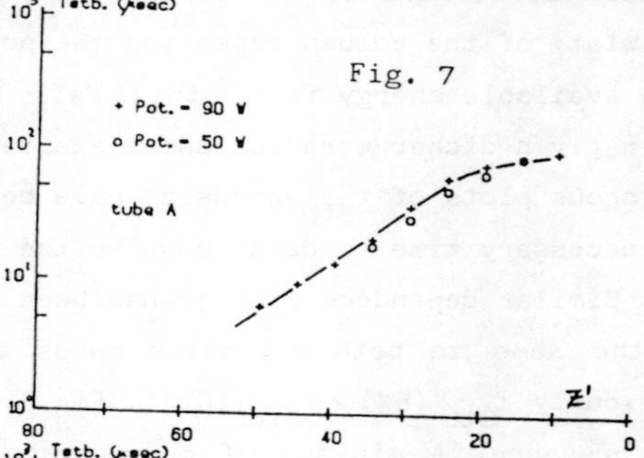

Fig. 7

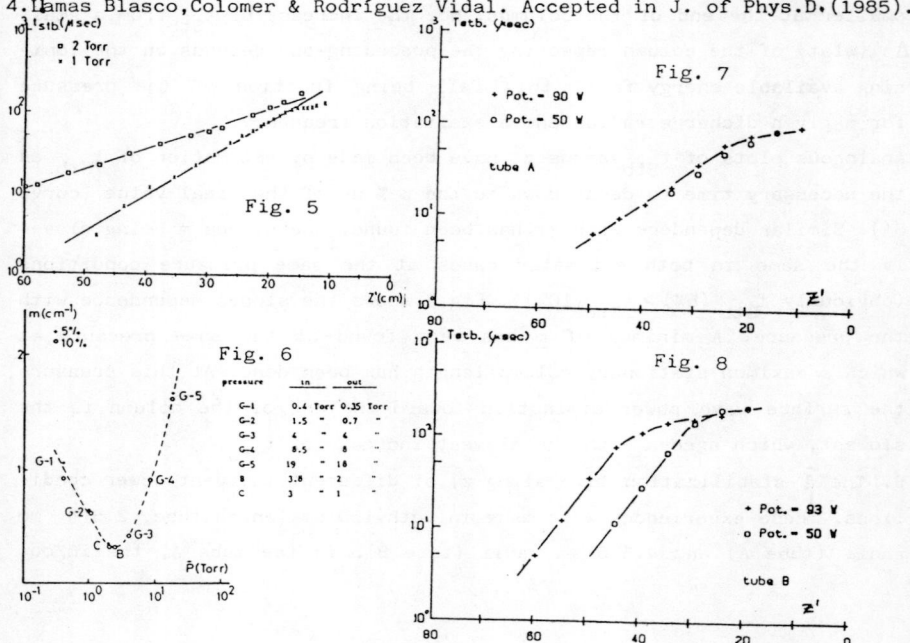

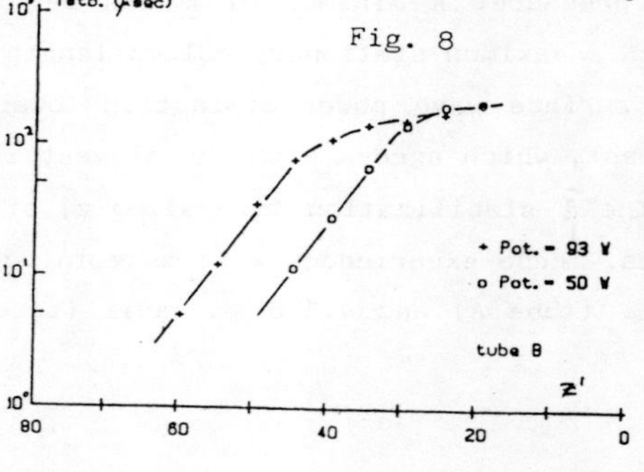

Fig. 8

NONLINEAR SURFACE WAVES ON A THIN PLASMA LAYER WITH ARBITRARY DENSITY VARIATION

L. Stenflo

Department of Plasma Physics, Umeå University, Umeå, Sweden

M. Y. Yu

Institut für Theoretische Physik, Ruhr-Universität Bochum, FRG

Abstract It is shown that simple analytical results can be obtained for the problem of solitary surface wave propagation in a thin plasma slab of non-uniform density.

Introduction Recently, the problem of surface waves existing on thin plasma layers has received a great deal of attention because of their importance in inertial fusion plasmas, fiber optics, as well as in condensed matter physics. Zhelyazkov et al.[1] considered nonlinear wave propagation on a thin plasma slab of constant density, and found that solitary surface waves can exist. In this paper, we generalize the problem to allow for an arbitrary density profile.

For the constant density thin layer case, Zhelyazkov et al.[1] obtained in the long wavelength limit ($k_z a \ll 1$, where k_z is the wave vector parallel to the slab surface, and $2a$ is the slab thickness), the dispersion relation

$$\frac{k_z^2 c^2}{\omega^2} = 1 + \frac{\omega^2 c^2}{a^2 \omega_{po}^2} , \tag{1}$$

where ω_{po} is the plasma frequency of the slab.

Formulation It is simple to generalize (1) to a slab with arbitrary density profile $n_o(x)$ by means of the method given in Ref. 2 (p. 116). One obtains

$$\frac{k_z^2 c^2}{\omega^2} = 1 + \frac{4\omega^2 c^2}{\left(\int_{-\infty}^{\infty} \omega_p^2(x)\,dx\right)^2} , \qquad (2)$$

for the long wavelength limit

$$k_z \left(\int_{-\infty}^{\infty} \omega_p^2(x)\,dx\right) \Big/ \omega_{p_{max}}^2 \ll 1 .$$

Solitons It is straightforward to obtain envelope soliton solutions. The nonlinear Schrödinger equation associated with the dispersion relation (2) is[3]

$$i\,\partial_\tau \tilde{E}_z + \frac{\partial\omega}{\partial k_z}\partial_\zeta \tilde{E}_z + \frac{1}{2}\frac{\partial^2\omega}{\partial k_z^2}\partial_\zeta^2 \tilde{E}_z - \frac{\partial\omega}{\partial\left(\int_{-\infty}^{\infty}\delta n\,dx\right)}\left(\int_{-\infty}^{\infty}\delta n\,dx\right)\tilde{E}_z = 0, \qquad (3)$$

where we have used the WKB ansatz $E_z = \tilde{E}_z(\zeta,\tau)\exp(i k_z z - i\omega t)$. Here, ζ and τ are the slow space and time variables ($\partial_\zeta \ll k_z$,

$\partial_\tau \ll \omega$). The expressions for $\partial\omega/\partial k_z$ and $\partial^2\omega/\partial k_z^2$, which are easily derived from (2), are similar to those given by (7) of Ref. 1. In fact, one needs only to redefine $\tilde{k}$ by

$$\tilde{k}^2 = 4\,k_z^2\,c^4 \Big/ \left(\int \omega_p^2\,dx\right)^2 .$$

Furthermore, from (2), we have

$$\frac{\partial\omega}{\partial\left(\int_{-\infty}^{\infty} \delta n\,dx\right)} = \frac{4\,\omega^3\,c^2}{\left(\int \omega_p^2\,dx\right)^2 (1 + 4\,\tilde{k}^2)^{1/2}} . \tag{4}$$

The low frequency density modulation of interest here is due to the ponderomotive-force enhanced ion acoustic waves. For the inhomogeneous slab, the latter are described by

$$(\partial_\tau^2 - c_s^2\,\partial_\zeta^2) \int_{-\infty}^{\infty} \delta n\,dx = \frac{c_s^2}{16\pi\,\omega^2\,T} \left(\int_{-\infty}^{\infty} \omega_p^2\,dx \right) \partial_\zeta^2 |\tilde{E}_z|^2 , \tag{5}$$

where we have noted that, for thin layers, $|\tilde{E}_z| \gg |\tilde{E}_x|$, and that $\tilde{E}_z$ is a slowly varying function of x in the plasma region.

We are interested in quasi-stationary localized solutions. For this purpose, we introduce the moving frame $\eta = \zeta - V\tau$. Equation (5) then yields

$$\int_{-\infty}^{\infty} \delta n\,dx = - \frac{|\tilde{E}_z|^2}{(1 - M^2)16\,\pi\,\omega^2\,T} \int_{-\infty}^{\infty} \omega_p^2\,dx , \tag{6}$$

where $M = V/c_s$ is the Mach number.

Equation (3) can now be solved with standard methods.[1] It can be shown that the usual sech solitons exist if $M^2 > 1$.

Summary To conclude, we have shown that one can easily find simple analytical results for solitary surface wave propagation in a narrow plasma layer with an arbitrary density profile. Our study thus generalizes that of the constant density slab model.

Acknowledgements This work was supported by the Sonderforschungsbereich 162 Plasmaphysik Bochum/Jülich. One of the authors (M.Y.Y.) would like to thank the Department of Physics, Umeå University, for hospitality during his stay there.

References

1. I. Zhelyazkov, O. Stoyanov, and M.Y. Yu, Plasma Phys. and Controlled Fusion 26, 813 (1984).

2. O.M. Gradov and L. Stenflo, Phys. Rep. 94, 111 (1983).

3. V.I. Karpman and E.M. Krushkal, Sov. Phys. JETP 28, 277 (1969).

SIMULATION STUDY OF ELECTRON BERNSTEIN WAVES IN BOUNDED PLASMA

D.Sultana, V.K. Decyk, & J.M. Dawson
University of California at Los Angeles
405 Hilgard Avenue, Los Angeles, California 90024

I. Introduction:

Propagation of surface waves in a magnetized bounded plasma has been studied for many years. In a finite plasma, the theoretical analysis becomes difficult because of the complex interaction of the particles with the boundary and the influence of the boundary on the fields.

Therefore, we have undertaken a numerical simulation study of electron waves in a magnetized plasma slab, bounded by vacuum. In the course of this study, a number of unexpected kinetic effects have been found.

II. The Simulation Model:

Our model is a 2-D electrostatic particle simulation code[1]. The plasma is confined between x = 0, L_x, with vacuum outside; the plasma is periodic in y. The model is shown in Fig. 1. Using periodicity in the y-direction, Poisson equation is solved with appropriate boundary conditions[1]. A constant magnetic field is applied in the z-direction, such that $\frac{\omega_{ce}}{\omega_{pe}} = .5$. Two cases were run; the first reflecting the particles from a rigid boundary; the second relied on the fixed ion charge and the magnetic field to confine the electrons. The system used for simulation contained 96 x 384 particles on a 32 x 128 grid. The normal mode eigenfunctions are found by using correlation techniques

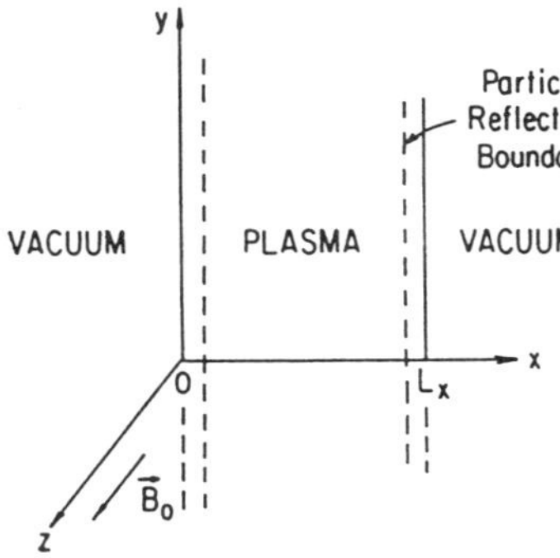

Fig. 1 The finite plasma model

while the normal mode frequencies are determined from power spectrum analysis.

3. Simulation results:

The dispersion relations obtained from simulation are shown in Fig. 2, together with fluid theory results[2]. As one can see, a new mode appears in the frequency range ω_{ce} to $2\omega_{ce}$ which is not predicted by fluid theory and thus must be a kinetic effect.

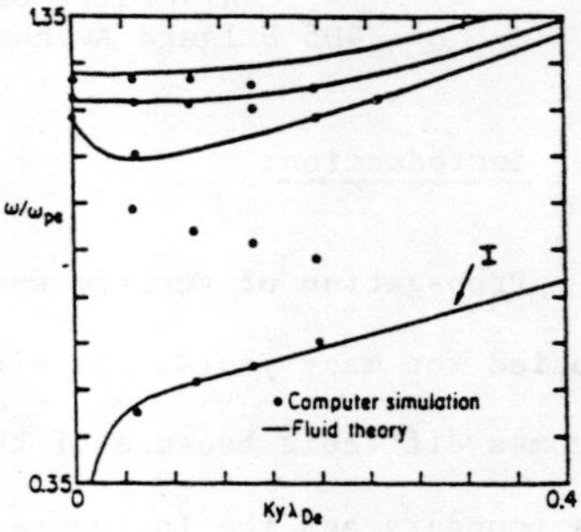

Fig. 2 Dispersion relations for Electron Waves.

The features of the normal modes can be illustrated in a space-frequency plot of $E_y{}^2(x,\omega)$ for fixed k_y (Fig. 3). The eigenfunctions corresponding to the spectral peaks of Fig. 3 are displayed in Fig. 4. It is clearly seen that there is a predominant tendency of these surface waves to be peaked at the left or right-hand boundaries. The new mode we have found in the frequency range ω_{ce} to $2\omega_{ce}$, is a right hand surface wave and the corresponding eigenfunction is displayed in Fig. 4(b). The wave-number corresponding to these waves is complex so that the eigenfunctions have both an evanscent and an oscillatory part. Unlike Bernstein waves, these waves show damping in time.

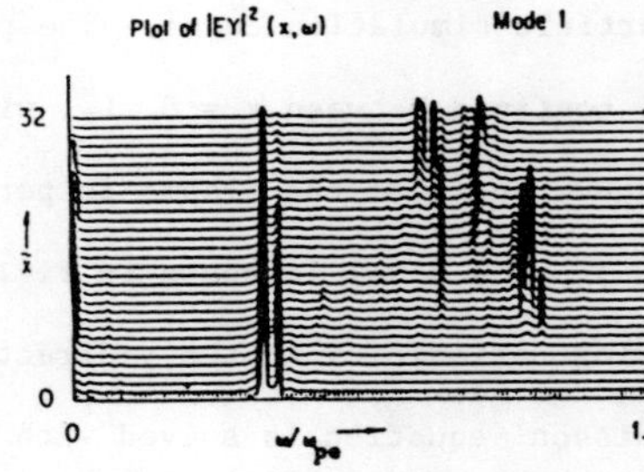

Fig. 3 Plot of $|E_y{}^2|$ versus position and frequency for $K_y\lambda_{De} = 0.049$ and $\tilde{V}_{th}=1.0$.

The damping of these surface waves increases almost linearly with the wavevector $k_y\lambda_{De}$. Similar effects have been predicted by Knodratenko[3] and Cheng, et al[4]. An estimate of the damping is obtained by measuring the decay time of the autocorrelation function.

Another kinetic effect observed was the splitting of the spectral lines corresponding to surface wave branches in Fig. 3. The eigenfunctions corresponding to these subpeaks have similar waveforms but are displaced relative to one another (Figs. 5a and 5b). In the case of the left-hand surface wave branch labelled curve 1 in Fig. 2, it is observed that the number of subpeaks increases with $k_y\lambda_{De}$. The frequency difference between two subpeaks or the total frequency width in case of several peaks increases linearly with k_yV_{th} (Fig. 6). The splitting is probably related to the guiding center drift of particles reflecting at the boundary.

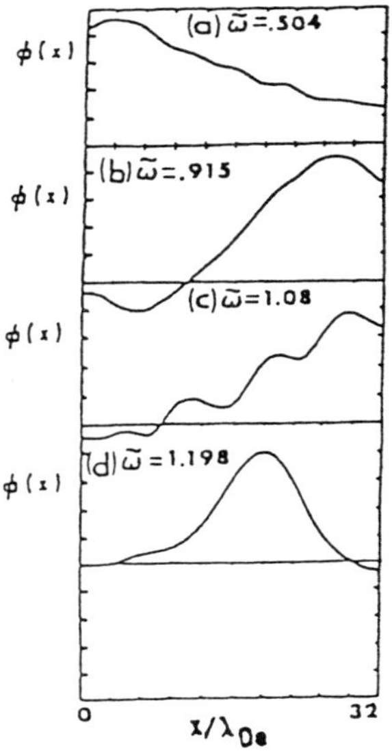

Fig. 4 Eigenfunctions corresponding to spectral peaks in Fig. 3.

To observe if the splitting is a sharp-boundary effect, we changed our boundary condition by taking the reflecting wall far away from the initial plasma boundary. All types waves mentioned in the earlier case and the splitting are still there. Some new effects were also observed: for the $K_y\lambda_{De} = 0$ mode, the space frequency plot of $|E_y(x,\omega)|^2$ shows a continuum of frequencies below $\omega \simeq 1.2\ \omega_{ce}$ (Fig. 7) and other spectral peaks also appear. A detailed study of these waves is in progress.

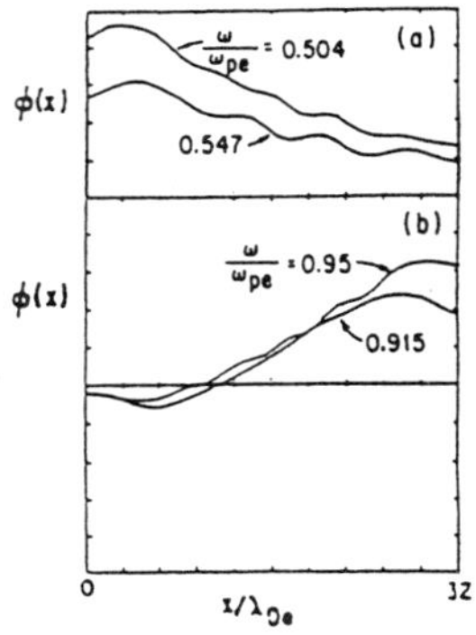

Figs. 5a,5b The left & right hand surface waves corresponding to subpeaks of the spectral lines in Fig. 3

CONCLUSION:

In this paper, we have studied the normal modes of a magnetized slab plasma with the magnetic field parallel to the vacuum plasma interface and the wave numbers perpendicular to the B-field. Several new effects have been found. First, a generalized right hand surface wave which is not predicted by fluid theory has been observed. Unlike the Bernstein wave, it is damped. A splitting of spectral lines is observed for the case of a sharp-boundary, as well as when the plasma forms its own boundary. Finally, a continuum spectrum has been observed in the case of a free boundary.

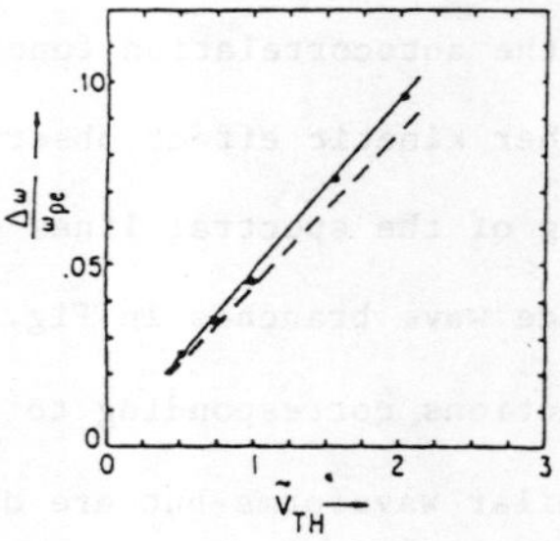

Fig. 6 Plot of the frequency difference between two subpeaks, $\Delta\omega$ against $K_y V_{th}$ for $K_y \lambda_{De}=0.049$ where the solid & dashed lines are obtained from simulation & the model where $\Delta\omega = \frac{2}{\pi} K_y V_\perp$ respectively.

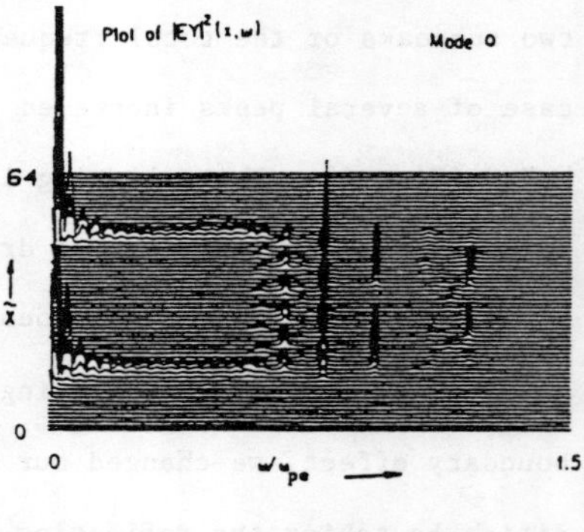

Fig. 7 Plot of $|E_y^2|$ vs. position and frequency for $K_y \lambda_{De}=0$ for the free-boundary case.

References:

1. V.K. Decyk and J. Dawson, J. Computational Phys., 30, 407 (1979)

2. V. Decyk, Ph.D. thesis, University of California, LA (1977)

3. A.N. Kondratenko, Nucl. Fusion 5, 267 (1965)

4. C.C. Cheng and E.G. Harris, Phys. Fluids 12, 1262 (1969)

Acknowledgments:

This work supported by USDOE.

PROPAGATION OF SURFACE POLARITONS IN ANISOTROPIC SEMICONDUCTORS

R.G.Tarkhanian

Institute of Radiophysics and Electronics, Academy of Sciences of the Armenian SSR, Ashtarak, 378410, USSR

1.The effect of anisotropy for the effective mass of free carriers and for crystal lattice is investigated on propagation of non-radiative surface electromagnetic waves, coupled to surface plasmons and optical phonons in uniaxial polar semiconductors. A number of new coupled surface modes are found that are absent in isotropic (cubic) materials. The number of branches for surface polaritons is originally shown to depend on the free carrier concentration.

Consider a uniaxial polar crystal fill in the half-space $z<0$; the half-space $z>0$ is taken to be vacuum. Crystal axis lies in the (x,z) plane and makes an angle θ with the z-axis. Denote by $\omega_{T\parallel}$ and $\omega_{T\perp}$ the frequencies of long-wavelength transverse optical phonons propagating along and across the crystal axis, $\frac{m}{\mu_\parallel}$ and $\frac{m}{\mu_\perp}$ are the corresponding components of the effective-mass tensor, m is the free electron mass. The crystal is described by a dielectric tensor [1]

$$\varepsilon_{ij}=\begin{pmatrix} \varepsilon_\perp\cos^2\theta+\varepsilon_\parallel\sin^2\theta & 0 & \sin\theta\cos\theta(\varepsilon_\parallel-\varepsilon_\perp) \\ 0 & \varepsilon_\perp & 0 \\ \sin\theta\cos\theta(\varepsilon_\parallel-\varepsilon_\perp) & 0 & \varepsilon_\perp\sin^2\theta+\varepsilon_\parallel\cos^2\theta \end{pmatrix}, \tag{1}$$

where

$$\varepsilon_\perp=\varepsilon_\perp^\infty\frac{\omega^2-\omega_{L\perp}^2}{\omega^2-\omega_{T\perp}^2}-\mu_\perp\frac{\omega_0^2}{\omega^2},\quad \varepsilon_\parallel=\varepsilon_\parallel^\infty\frac{\omega^2-\omega_{L\parallel}^2}{\omega^2-\omega_{T\parallel}^2}-\mu_\parallel\frac{\omega_0^2}{\omega^2}, \tag{2}$$

$\omega_{L\parallel}=\omega_{T\parallel}(\varepsilon_\parallel^0/\varepsilon_\parallel^\infty)^{\frac{1}{2}}$, $\omega_{L\perp}=\omega_{T\perp}(\varepsilon_\perp^0/\varepsilon_\perp^\infty)^{\frac{1}{2}}$, $\varepsilon_{\perp,\parallel}^0$ and $\varepsilon_{\perp,\parallel}^\infty$ are the static and high-frequency dielectric tensor components for the crystal lattice, $\omega_0^2=4\pi Ne^2/m$ and N is the free carrier concentration.

Surface waves are solutions of the Maxwell equations for which the electromagnetic fields are exponentially damped, as one leaves the surface. The general solution is a linear combination of ordinary (o) and extraordinary (e) waves. The vectors $\vec{E}^o$ and $\vec{H}^e$ are normal to the plane formed by the wave vector of corresponding wave and the crystal axis, while the vectors $\vec{E}^e$ and $\vec{H}^o$ lie in this plane. The surface wave is an ordinary in the case when the crystal axis is parallel to the surface and normal to the wave vector $\vec{K}$. The dispersion relation for this case can be written as

$$\frac{c^2K_\parallel^2}{\omega^2}=\frac{\varepsilon_\perp}{\varepsilon_\perp+1} \tag{3}$$

The nonradiative surface wave exists only if $K_\parallel>\frac{\omega}{c}$ and $\varepsilon_\perp<-1$. There are two surface modes, which exist, like in the isotropic medium, in regions $0<K_\parallel<\infty$ and $K_\parallel>\frac{\omega}{c}$. The surface wave is an extraordinary in the case when $\vec{K}_\parallel$ is parallel to the crystal axis $\vec{C}$ or when $\vec{C}$ is normal to the surface ($\theta=0$). When $\theta\neq 0$ and $\theta\neq\frac{\pi}{2}$, the nonradiative surface waves exist only if $\vec{K}_\parallel$ is parallel or normal to the plane (xz). The dispersion relation can be written as

$$\frac{c^2K_\parallel^2}{\omega^2}=\frac{\varepsilon_\parallel-\varepsilon_\perp\varepsilon_\parallel}{1-\varepsilon_\perp\varepsilon_\parallel} \tag{4}$$

for the case $\theta=0$. The surface waves exist only if $\varepsilon_\perp<0$, $K_\parallel>\frac{\omega}{c}$ and $\varepsilon_\parallel(\frac{K_\parallel^2c^2}{\omega^2}-\varepsilon_\parallel)<0$. Unlike the isotropic case, the number of modes, as well as the boundary of the existence region, depend on the carrier concentration. One of four solutions of the equation (4) exists in region $0<K_\parallel<\infty$. The rest three branches appear at $K_\parallel=\frac{\omega}{c}$ and exist over a finite range in $K_\parallel$. The number of branches for $K_\parallel\gg\frac{\omega}{c}$ increases with increasing N and can be 1,2 or 3.

2. Dispersion relations are obtained and analysed for surface modes of mixed phonon-plasmon vibrations in a thin slab of uniaxial polar semiconductor in the non-retardation limit $\frac{\omega}{K}\ll c$:

$$\varepsilon_{zz}=-K_\parallel d \tag{5}$$

for antisymmetrical modes, and

$$\varepsilon_{zz} = -K_{\parallel} d\, \varepsilon_{\perp} \left(\varepsilon_{\parallel} \cos^2\eta + \varepsilon_{zz} \sin^2\eta \right) \qquad (6)$$

for symmetrical modes; $tg\eta = \frac{K_y}{K_x}$, $K_{\parallel}^2 = K_x^2 + K_y^2$, L=2d is slab thickness. One can see that the frequencies of antisymmetrical waves are independent of the direction of propagation but depend on the orientation of the crystal axis relative to the surface. The frequency of the antisymmetrical modes decreases with increasing $K_{\parallel}d$, while the frequency of the symmetrical modes increases. As $K_{\parallel}d \to \infty$, the asymptotic values of the frequency approach the frequencies for unretarded surface waves at the interface between vacuum and a semiinfinite uniaxial crystal, and are defined by the equation [2]

$$\varepsilon_{\perp} \left(\varepsilon_{\parallel} \cos^2\eta + \varepsilon_{zz} \sin^2\eta \right) = 1. \qquad (7)$$

The number of solutions of (5) and (6) changes, depending on the values of θ, η, and N.

3. Crystal anisotropy makes possible the propagation of unretarded sinusoidal modes with fields falling off exponentially outside the slab. In an isotropic slab such modes exist only when retardation is included. In uniaxial semiconductor slab the non-radiative sinusoidal modes are described by the dispersion relations

$$\varepsilon_{zz} = -\frac{K_{\parallel}}{\nu} \tan(d\nu) \quad \text{and} \quad \varepsilon_{zz} = \frac{K_{\parallel}}{\nu} \cot(d\nu), \qquad (8)$$

where $\nu = \sqrt{-\varepsilon_{\perp}(\varepsilon_{\parallel} K_x^2 + \varepsilon_{zz} K_y^2)/|\varepsilon_{zz}|}$. These modes exist only if ν is real. For an n-CdS-type crystal in the case $\theta = 0$ there are three frequency regions in which sinusoidal modes can occur.

4. The effect of an external magnetic field $\vec{B}_0$ on surface electromagnetic waves in a semiconductor slab is investigated. The magnetic field is taken to lie in the plane parallel to the surface and normal to the wave vector $\vec{K}$. The slab thickness, as well as the wavelength, are large compared to the lattice spacing and the cyclotron radius. The dispersion relation is

$$\tanh(\alpha L) = -\frac{2\alpha\alpha_0}{\alpha^2 + \alpha_0^2 + K_{\parallel}^2 \varepsilon}, \qquad (9)$$

where $\mathcal{L}=\left(K_{\parallel}^2-\frac{\omega^2}{c^2}\varepsilon_v\right)^{\frac{1}{2}}, \mathcal{L}_0=\left(K_{\parallel}^2-\frac{\omega^2}{c^2}\right)^{\frac{1}{2}}, \ \varepsilon_v=\varepsilon_1-\frac{\varepsilon_2^2}{\varepsilon_1}, \ \varepsilon=\varepsilon_v-2+\frac{1}{\varepsilon_1}$,

$\varepsilon_1=\varepsilon_\infty\left[1-\frac{\omega_p^2}{\omega^2-\omega_c^2}\right]$, $\varepsilon_2=-\varepsilon_\infty\frac{\omega_c\omega_p^2}{\omega(\omega^2-\omega_c^2)}$, $\omega_c=\frac{eB_0}{m^*c}$,

$\omega_p^2=4\pi Ne^2/m^*\varepsilon_\infty$. Surface waves exist only if $\varepsilon<0$, $\mathcal{L}$ and $\mathcal{L}_0$ are real and positive. For the case of a semiinfinite medium the surface waves are described by the dispersion relation

$$\mathcal{L}+\mathcal{L}_0\varepsilon_v=\frac{\varepsilon_2}{\varepsilon_1}K_{\parallel}. \qquad (10)$$

One can see from (10), that positive and negative values of wave vector $\vec{K}_{\parallel}$ are not equivalent. Unlike to this case, $K_{\parallel}$ appears only at even powers in (9), so the reciprocity of $+\vec{K}_{\parallel}$ and $-\vec{K}_{\parallel}$ is restored, when the slab thickness is finite.

References.

1. L.E.Gurevich and R.G.Tarkhanian, Fiz.tverd.Tela.17,1944 (1975)

2.R.G.Tarkhanian, phys.stat.sol.(b) 72, 111 (1975)

SELF-EXCITATION OF ELECTROMAGNETIC SURFACE WAVES IN SEMICONDUCTORS IN STRONG ELECTRIC FIELD

R.G.Tarkhanian and K.M.Karapetian

Institute of Radiophysics and Electronics, of Sciences of the Armenian SSR, Ashtarak 378410,USSR

1.The instability of electromagnetic surface waves (ESW) propagating along the vacuum-semiconductor interface in presence of negative differential conductivity (NDC) is shown. this instability is observed in form of microwave radiation caused by radiative ESW generation.Radiation losses are being compensated by the constant current source energy.The generated surface waves, in contrast to bulk waves,are shown to propagate in the direction of external electric field $\vec{E}_0$, and may be excited only when the Columb interaction retardation is taken into account.

Consider a semiconductor in half-space $x<0$.Let a constant current of density $\vec{J}=\sigma(E_0^2)\vec{E}_0$,where σ is conductivity having squere field dependence,flow in the crystall in the z-axis direction. A small perturbation (fluctuation) of the electric $\vec{E}$ field results in a chunge of current density by a vector $\vec{J}$ with components ($\sigma E_x, \sigma E_y, \sigma E_z$),where

$$\sigma_d=\left(1+2\frac{d\ln\sigma}{d\ln E_0^2}\right)\sigma \qquad (1)$$

is differential conductivity.Consider the perturbations of form $\vec{E},\vec{H}\sim \exp[i(\vec{k}\vec{r}-\omega t)]$, whose frequencies satisfy the condition $\omega\tau << 1$(where τ is carrier energy relaxation time).Surface waves also correspond to perturbations in which the wave-vector tangential components K_y and K_z are real,and the normal components $K_x$are complex, where$\mathrm{Im}K_x > 0$ in vacuum ($X>0$) and $\mathrm{Im}K_x < 0$ in crystal($X<0$).Linearizing Maxwell's equations over small fluctuations of $\vec{E}$,$\vec{H}$ and $\vec{J}$ and solving the corresponding dispersion equation for TH-type waves, propagating along the constant electric field $\vec{E}_0$ we obtain [1]

$$K_x^2=\left(\frac{\omega^2}{c^2}-\frac{K_z^2}{\varepsilon(\omega)}\right)\varepsilon_d(\omega)\,, \qquad (2)$$

where $\varepsilon(w)=\varepsilon+\frac{4\pi i}{w}\sigma$, $\varepsilon_d(w)=\varepsilon+\frac{4\pi i}{w}\sigma_d$, ε is the static dielectric constant of crystall lattice. Using the fields continuity conditions at the interface X=0, we obtain the dispersion relation for ESW in form

$$\frac{c^2K_z^2}{w^2}=\frac{\varepsilon_d(w)-1}{\varepsilon_d(w)-\varepsilon^{-1}(w)}\,. \qquad (3)$$

This is the 4-th order algebraic equation with respect to w. In the frequency range

$$\frac{1}{\tau}\gg \mathrm{Re}\,w \gg 4\pi\left(\frac{\sigma}{\varepsilon},\frac{|\sigma_d|}{\varepsilon}\right) \qquad (4)$$

this equation has an approximate solution

$$w=\Omega+i\gamma=cK_z\sqrt{\frac{\varepsilon+1}{\varepsilon}}+i\frac{2\pi(\sigma-\varepsilon\sigma_d')}{\varepsilon(\varepsilon^2-1)}\,, \qquad (5)$$

where Rew>>Imw. In this case, with accuracy up to the terms of order $\frac{\gamma}{\Omega}$ we will have

$$K_x=K_z\sqrt{\varepsilon}+i\frac{\sqrt{\varepsilon+1}}{c}\left(\varepsilon\gamma+\frac{4\pi\sigma_d}{\varepsilon+1}\right),\; K_{ox}=\frac{K_z}{\sqrt{\varepsilon}}+i\sqrt{\varepsilon+1}\,\frac{\gamma}{c} \qquad (6)$$

Using now the expressions (5) and (6) and conditions $\mathrm{Im}K_{ox}>0$, $\mathrm{Im}K_x<0$, we obtain the conditions of instability ($\gamma>0$):

$$\sigma_d<0,\quad |\sigma_d|>\frac{\sigma}{\varepsilon-2} \qquad (7)$$

Nonlinear effects limit the increase of amplitude ($E\sim e^{\gamma t}$) and result in a constant field amplitude in stationary conditions. In this case the increment factor becomes equal to zero, and K_{ox} becomes real: $K_{ox}=K_z\varepsilon^{-\frac{1}{2}}$. Hence follows that the surface wave has a radiational character, i.e. it is radiated into vacuum. The output angle θ, at which the wave leaves the surface, is determined by the relation

$$\mathrm{tg}\,\theta=\frac{K_{ox}}{K_z}=\frac{1}{\sqrt{\varepsilon}}\,. \qquad (8)$$

For GaAs this angle $\theta\sim15°$. Taking the sample length d along $\vec{E}_o$ equal to d=1 cm and $K_z=\frac{2\pi m}{d}$, m=1,2,... we obtain from(5) the generated waves frequencies $\Omega \cong 2\cdot 10^{11}$ sec^{-1}.

2.It was shown that in a thin semiconductor slab $|x|<\ell$ in the presence of NDC, self-excitation of ESW is possible, which propagate in a direction normal to constant electric field $\vec{E}_0$. No potential waves (bulk or surface) may be excited in that direction. Wave instability is possible only for symmetric TE-type waveguide modes, for which the increment factor is

$$\gamma \sim |\sigma_d| (\ell K_y)^2 \qquad (9)$$

while the field localization depth outside the crystall is sufficiently larger than the slab thickness [2].

3.SEW excitation conditions have been analyzed in heterostructures in the presence of heating electric field parallel to heterojunction plane. Unlike to the case of semiconductor-vacuum interface, in heterostructure the self-excitation of TH-type non-radiative ESW is shown to be possible, which propagate along the constant electric field $\vec{E}_0$. Two cases of NDC appearence are considered: 1) high-mobility electrons transition into the upper lying valley under heating by constant field; 2) high-mobility electrons transition from the n-GaAs region close to surface into the N-$Al_xGa_{1-x}As$ conductive band, whose bottom is below the L-minima energy for GaAs. Besides, in the second case the instability threshold is much lower than in the first case. Besides, in the second case one should also take into account the wave magnetic field tangential component discontinuity at the heterojunction, due to the presence of surface current $J_s = \sigma_d^s E$. In this case the dispersion equation for ESW has the following form:

$$\frac{\varepsilon_1(\omega)}{K_{x1}} - \frac{\varepsilon_2(\omega)}{K_{x2}} = \frac{4\pi\sigma_d^s}{\omega}, \qquad (10)$$

where subscripts 1 and 2 respectively denote the parameters of broadband and narrowband semiconductors, and

$$K_{x\alpha}^2 = \varepsilon_\alpha(\omega)\frac{\omega^2}{c^2} - K_z^2; \quad \alpha = 1, 2 \qquad (11)$$

Taking the broadband semiconductors conductivity equal to zero, for simplicity, we obtain the following expressions for ESW frequency and increment factor:

$$\Omega = \frac{2\pi|\sigma_d|}{\sqrt{\varepsilon(4\pi^2\sigma_d^2 - \varepsilon c^2)}} c K_z ; \quad \gamma = \frac{2\pi\sigma}{\varepsilon} \frac{\varepsilon c^2 - 2\pi^2\sigma_d^2}{4\pi^2\sigma_d^2 - \varepsilon c^2} . \tag{12}$$

Hence follows that the instability criterion has the form:

$$\sqrt{\varepsilon} < \frac{2\pi|\sigma_d|}{c} < \sqrt{2\varepsilon} . \tag{13}$$

In the case 1) the corresponding expressions are

$$\Omega = \sqrt{\frac{\varepsilon_1 + \varepsilon_2}{\varepsilon_1 \varepsilon_2}} c K_z ; \quad \gamma = \frac{2\pi\varepsilon_1}{(\varepsilon_2^2 - \varepsilon_1^2)} \left(\frac{\varepsilon_1}{\varepsilon_2} \sigma - \sigma^d \right), \tag{14}$$

$$\varepsilon_2 > 2\varepsilon_1 ; \quad \sigma^d < 0 ; \quad |\sigma^d| > \frac{\varepsilon_1 \sigma}{\varepsilon_2 - 2\varepsilon_1}, \tag{15}$$

where ε_1 and ε_2 are static dielectric constants(we have taken $\varepsilon_1 = \varepsilon_2$ in (10) and (11). One may observe from these expressions that the account of surface current essentially changes the ESW instability criterion and frequencies.

References

1. R.G.Tarkhanian. Fiz.tverd.Tela,22,1467,(1980)
2. R.G.Tarkhanian. Fiz.tverd.Tela.22,3186,(1980)
3. R.G.Tarkhanian. K.M.Karapetian Fiz.Tekh.Poluprovodnikov.17,742(1983)

SUPPRESSION OF IONIZATION WAVES BY HIGH-FREQUENCY SURFACE WAVE

E. Tatarova and T. Stoychev
Faculty of Physics, Sofia University, 1126 Sofia, Bulgaria

Introduction

The appearance and the propagation of ionization waves (moving striations) in the positive column of gas discharge is the most common instability in low-temperature plasma which can be always (and easily) observed.[1] The spontaneous excitation of striations is undesirable effect for the application of low-temperature plasma in all types of discharge devices, and their stabilization and suppression is an important problem.[1-3] Since the application of alternating electric field to plasma can strongly effect low-frequency instabilities[4] and suppress them, the study of the mechanisms of the nonlinear interactions of the ionization waves with other plasma eigen modes is of interest. It is suitable to include the high-frequency (HF) surface waves (SWs) in such nonlinear interactions because of their easy excitation and pronounced propagation.[5] The experiments[6,7] with HF amplitude-modulated SWs show effect of suppression of the LF fluctuations due to ionization waves. But in this case different types of nonlinear interactions lead to appearance of forced waves in the LF range, so the main effect in these experiments is synchronization of the fluctuations.

The experiments reported here are directed to studying the possibilities for suppression of ionization waves by the HF field of SWs. The experimental data for the quenching of the amplitudes of the ionization waves are in accordance with the measured changes of the electron energy distribution function when a SW is launched.

Experimental results

The experiments are performed in the positive column of hot cathode gas discharge in neon. The presented data are obtained at pressure $p = 2 \times 10^{-2}$ Torr and discharge current $I_d = 40$ mA. The plasma parameters are as follows: plasma density $n_e = 4.7 \times 10^{8}$ cm^{-3} and electron temperature $T_e =$

5.2 eV.

Under the investigated gas discharge conditions spontaneous excitation of ionization waves is observed. The spectrum of the LF plasma fluctuations is continuous in a frequency range up to 200 kHz with a high peak of the noise intensity at frequency f_{m1} (Fig. 1a). A second (lower) maximum noise intensity at frequency f_{m2} close to the double frequency ($f_{m2} \gtrsim 2f_{m1}$) is also evident whereas the third one (at $f_{m3} \gtrsim 3f_{m1}$) is not pronounced. (The zero peak is due to the operation of the receiver of the LF fluctuations.) The spectral

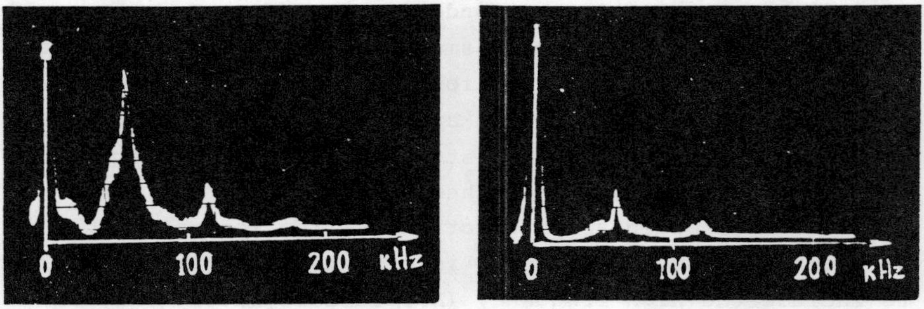

Fig. 1. LH fluctuation spectra without (a) and with (b) propagating HF SW (53 cm away from the cathode; 20 cm away from the SW launcher (b); ordinate scale 5 dB/div.; f_{m1} = 55 kHz, f_{m2} = 115 kHz; f_{m3} = 165 kHz (a)).

component of frequency f_{m1} is almost undamped along the plasma column (the measured space damping rate is $\gamma_i = 3 \times 10^{-3}$ cm^{-3}). The results for the phase changes of the LF oscillations show that their dispersion law (fig. 2) corresponds to that of the moving striations at comparatively higher gas pressure (backward ionization waves). In the frequency range between 50 and 80 kHz the measured phase velocity vary between 7×10^5 and 1.4×10^6 cm/s.

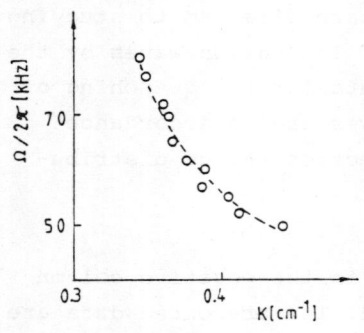

Fig. 2. Dispersion of the self-excited ionization waves (frequency Ω vrs. wavenumber k).

In order to have more complete information about the nature of the LF fluctuations we have measured the electron energy distribution function. According to the qualitative explanation for the existence of the striations[8], there is a group of fast electrons in the striation whose energy at first increases (with the change of the wave phase). This results in the appearance of a distinct bump in the high energy tail of the electron energy distribution function. The bump moves to the lower energy values[9] since the fast electrons lose energy at inelastic collisions (excitation and ionization). The obtained data for the electron energy distribution function (Fig. 3) show that in the

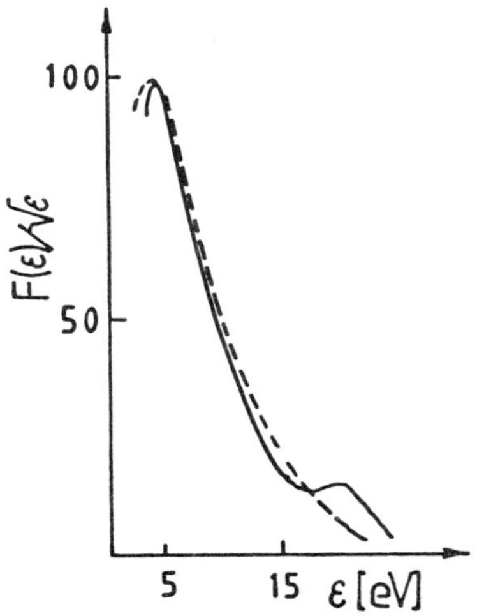

Fig. 3. Electron energy distribution function without (——) and with (– –) launched HF SW.

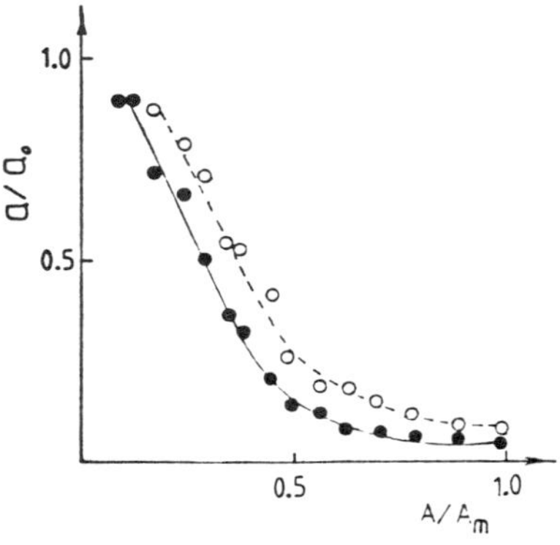

Fig. 4. Relation between the amplitude a (normalized to its value a_0 without SW) of the self-excited ionization waves at frequency f_{m1} = 55 kHz and the amplitude A of the launched HF SW (normalized to its maximum value A_m); 20 (•) and 40 (o) cm away from the SW launcher.

region of higher energy values the distribution is non-maxwellian with a marked bump at about 20 eV. In the region of comparatively lower electron energy values it is close to maxwellian with electron temperature of 5.2 eV. The results obtained for the electron energy distribution also confirm[9] the existence and the propagation of ionization waves in the positive column of the discharge.

When a HF SW is excited in the plasma strong changes in the behaviour of the ionization waves are observed. (The characteristics of the propagating SW launched at frequency 130

MHz are: wavelength λ = 7 cm and space damping rate γ = 2.2 x 10^{-1} cm^{-1}.) First of all, the measured LF spectrum shows that the spontaneous excited ionization waves are quenched. The amplitude of the fluctuations with maximum intensity (at frequency f_{m1}) is strongly suppressed (about 17 times (Fig. 1b)) in respect to its value without SW. The suppression depends on the amplitude of the HF SW (Fig. 4). At higher amplitudes of the applied HF signal the amplitudes of the self-excited ionization waves reach saturation at a value which is about 15 times (in average) less than that without HF field. The efficiency of the suppression slightly decreases with the distance from the launcher (probably because of the damping of the SW). The results for the electron energy distribution function in the presence of propagating SW confirm the data for the suppression of the amplitudes of the ionization waves. The local bump in the tail of the distribution without HF signal vanishes and the energy distribution tends to maxwellian in the whole energy region (Fig. 3).

Conclusion

A suppression of fluctuation noise due to spontaneous ionization waves in gas discharge plasma is observed when a HF SW is launched in the plasma. Concerning the mechanism of the suppression we suppose that the nonlinear interaction of the HF SW and its side-bands arising in result of a modulation of the HF wave by the LF fluctuations leads to an appearance of a ponderomotive force which influences strongly the damping of the ionization waves.

References

1. P. S. Landa, N. A. Miskinova and Yu. V. Ponomarev, Usp. Fiz. Nauk, 132 601 (1980)
2. G. Bekefi, 'Principles of Laser Plasma', Wiley (1976)
3. H. Amemiya, Plasma Phys., 25 735 (1983)
4. É. M. Barkhudarov, A. Sh. Dzagnidze, V. F. Lygin and D. D. Tskhakaya, Fiz. Plazmy, 10 189 (1984)
5. M. Moisan, A. Shivarova and A. W. Trivelpiece, Plasma Phys., 24 1331 (1982)
6. E. Tatarova, T. Stoychev and A. Shivarova, Phys. Lett. A, in press (1985)
7. E. Tatarova, T. Stoychev and A. Shivarova, J. Phys. D: Appl. Phys., submitted for publication (1985)
8. B. N. Klyarfel'd, JhETF, 22 66 (1952)
9. S. W. Rayment, J. Phys. D: Appl. Phys., 7 871 (1974)

FORCED EXCITATION OF LOW-FREQUENCY WAVES BY AMPLITUDE-MODULATED SURFACE WAVE AND THEIR INTERACTION WITH SELF-EXCITED PLASMA MODES

E. Tatarova, T. Stoychev and A. Shivarova*&
Faculty of Physics, Sofia University, 1126 Sofia, Bulgaria
*Ruhr-Universität Bochum, Institut für Experimentalphysik II, 4630 Bochum, F. R. Germany

Introduction

Although the problem of suppression/synchronization of ionization waves is important for many applications of gas discharge plasma, by using the method of external excitation of low-frequency (LF) waves it has been studied only in few works.[1,2] It was also observed that fluctuation noise due to the ionization wave perturbations can be suppressed if the 'artificially' launched wave appears in the plasma in result of nonlinear interaction of high-frequency (HF) surface waves (SWs).[3]

Here we investigate the excitation of forced oscillations and the splitting of the continuous fluctuation spectrum of the self-excited ionization waves in gas discharge plasma to separate spectral components when a HF amplitude-modulated (AM) SW is launched. The nature of these spectral components appearing in result of nonlinear wave interactions is discussed.

Experimental results

The experiments are performed in the positive column of hot-cathode gas discharge (in glass tube with radius R = 4 cm) in neon at gas pressure $p = (1 - 2.5) \times 10^{-2}$ Torr and discharge current $I_d = (30 - 100)$ mA (plasma density $n \sim 10^9$ cm^{-3} and electron temperature $T_e \sim 10^5$ K).

An azimuthally-symmetric SW of frequency $f_o = 130$ MHz amplitude-modulated at a frequency $f_m = (20 - 100)$ kHz is launched into the plasma. The LF spectra are observed by using a receiver of LF fluctuations.[4] The phase velocity of the LF forced and eigen waves is obtained by phase shift measurements (the signals are detected by photomultipliers).

The spectrum of the LF self - excited fluctuations is

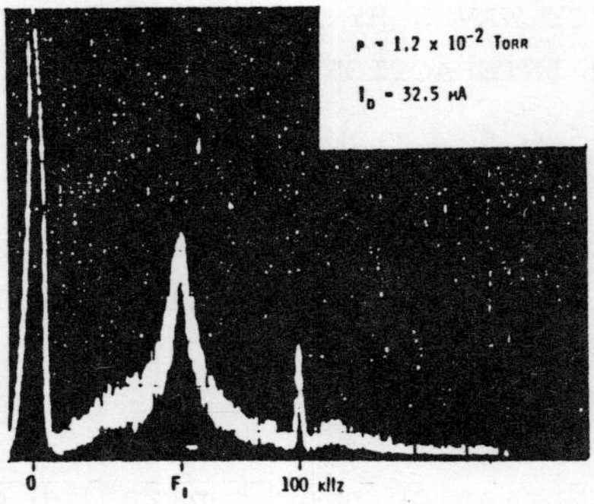

(a)

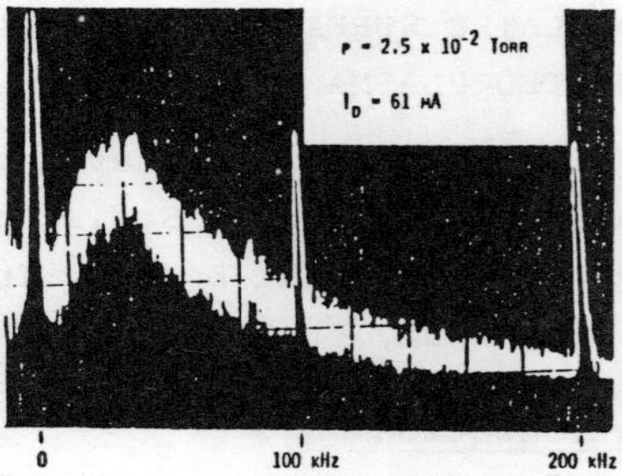

(b)

Fig. 1. Spectra of self-excited LF fluctuations continuous in a frequency range up to 100 - 150 kHz having maximum intensity value at frequency f_i = (20 - 60) kHz (Fig. 1a) and it broadens (Fig. 1b) with increasing the discharge current. The self-excited oscillations propagate according to the dispersion of backward ionization waves (Fig. 2).

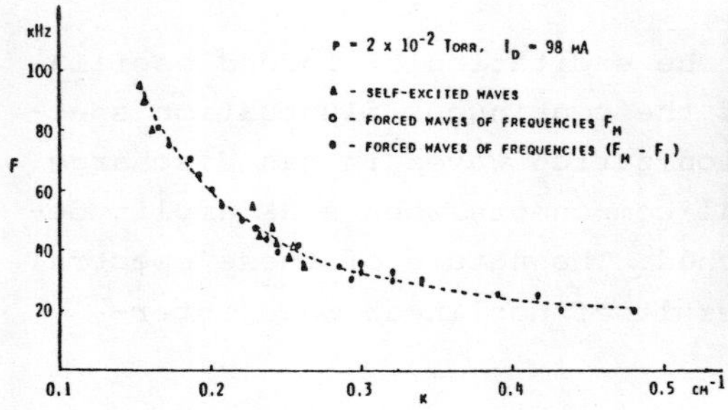

Fig. 2. Dispersion curves

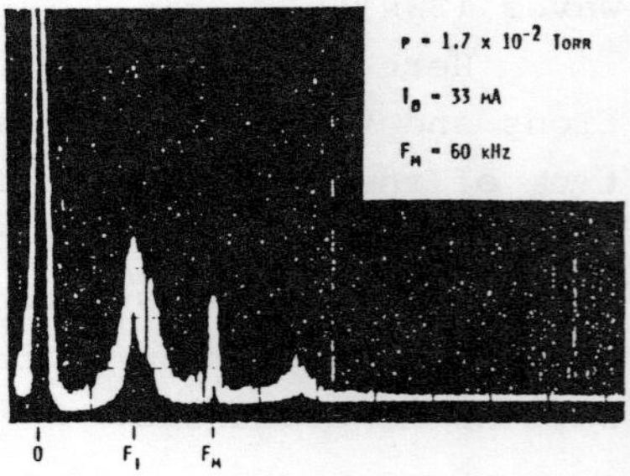

Fig. 3. LF spectrum with launched HF AM SW

When a HF AM SW is launched, excitation of forced oscillations at the modulation frequency is observed (Fig. 3). At comparatively low value of the modulation frequency ($2f_m \lesssim f_i$) forced oscillations at its second harmonic also appear[3]. The amplitudes of the forced oscillations of frequency f_m depend on the degree of coincidence of the modulation frequency with the frequency of the LF fluctuations with maximum intensity. They have maximum value at $f_m \simeq f_i$ (Fig. 4). The excited forced oscillations follow the dispersion of the ionization waves (Fig. 2).

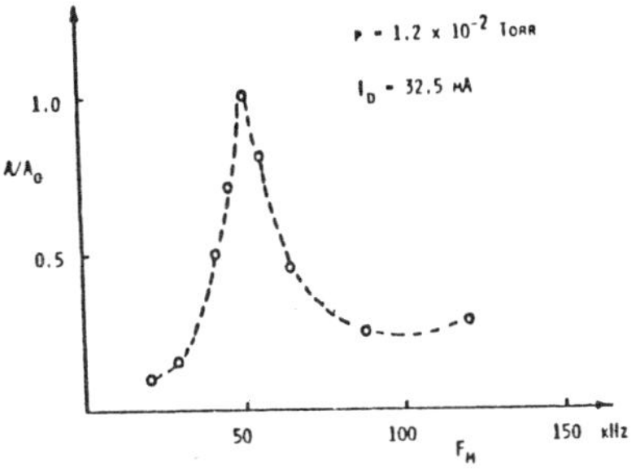

Fig. 4. Dependence of the amplitudes of the forced oscillations at the modulation frequency on its value

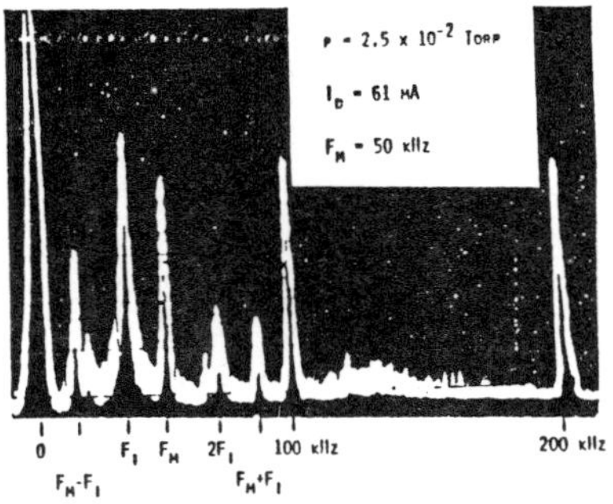

Fig. 5. LF spectrum with launched HF AM SW

By increasing the discharge current and the modulation index, the amplitudes of the self-excited and forced perturbations increase respectively, and an effect of their nonlinear interaction is observed. When a HF AM SW is launched the broad self-excited fluctuation spectrum (Fig. 1b) splits into separate spectral components (Fig. 5) with different nature : i) from the self-excited spectrum, the spectral component at frequency f_i and its second harmonic $2f_i$ (which is not decleared in the LF spectrum without HF signal) are the only ones which manifest their existence; ii) the forced oscillations at frequency f_m which are result of nonlinear nonresonant interactions of the three spectral components of the HF AM SW are also observed; iii) in result of nonlinear nonresonant interactions of the self-excited and forced perturbations, LF signals appear at their combination frequencies $(f_m \pm f_i)$; $f_m > f_i$. Except the spectral component of frequency f_i (and partially that of frequency $2f_i$) all others existing in the broad LF spectrum are suppressed (up to 80% in the LF part of the spectrum $(f < f_i)$). The forced oscillations at the combination frequencies also appear as waves of the type of backward ionization waves (Fig. 2).

Theoretical model

The launched HF AM SW consisting of three spectral components of frequencies $\omega_1 = \omega_o$, $\omega_{2,3} = \omega_o \mp \Omega_m$ $(\omega_o = 2\pi f_o$, $\Omega_m = 2\pi f_m)$ and wavenumbers k_n $(n = 1,2,3)$ causes an excitation of forced perturbations of T_e and of the plasma density n at the difference frequencies (Ω_m, $2\Omega_m$) of each two of the HF waves.[5] The behaviour of the ionization instability in the presence of the forced oscillations at frequency f_m can be approximately considered in a manner similar to that of a classical Van der Pol oscillator[2]:

$$\frac{d^2n}{dt^2} - (\alpha_o - 2\beta n - 3\gamma n^2)\frac{dn}{dt} + \Omega_i^2 n = B\Omega_m^2 \sin(\Omega_m t + \xi) \qquad (1)$$

where $\Omega_i = 2\pi f_i$, $\alpha_o = \alpha - \nu$, α, β and γ are the constants in the series expansion of the ionization source, ν is the ion-neutral collision frequency, B is the density disturbance[5] at frequency f_m and $\xi = \kappa z + \xi_o$ ($\kappa = k_1 - k_2 \simeq k_3 - k_1$, ξ_o = const.). Eq. (1) solved according to the method of asymptotic expansion in the nonresonant case[8] gives:

$$n = \frac{a_o}{2}\{a\cos(\Omega_i t + \psi_o) + u\sin(\Omega_m t + \xi) + \varepsilon\frac{\beta a_o}{2\alpha_o}au[\frac{q+1}{q(q+2)} \qquad (2)$$
$$\cos((\Omega_m + \Omega_i)t + \xi + \psi_o) + \frac{q-1}{q(q-2)}\cos((\Omega_m - \Omega_i)t + \xi - \psi_o)]\}$$

where a is the amplitude of the self-excited waves, ψ_o = const., $a_o = (4\alpha_o/3\gamma)^{1/2}$, $u = 2Bq^2/a_o(1 - q^2)$, $\varepsilon = \alpha_o/\Omega_i$ and $q = \Omega_m/\Omega_i$. The perturbations at frequencies Ω_i, Ω_m as well as at ($\Omega_m \pm \Omega_i$) obtained in the experiment appear also in (2). The dependence of the amplitude of the forced oscillations at the modulation frequency of its value shows a resonance responce (as in Fig. 4). The ratio of the amplitudes of the signal generated at frequencies ($f_m - f_i$) and ($f_m + f_i$) estimated from (2) is 1.5 whereas the corresponding experimental value (Fig. 5) is 2.3.

References

1. K. Ohe and S. Takeda, Beit. Plasmaphys., 14 55 (1974)
2. H. Amemiya, Plasma Phys., 25 735 (1983)
3. E. Tatarova, T. Stoychev and A. Shivarova, Phys. Lett. A, in press (1985)
4. T. Stoychev and V. Dimitrova, Ann. Univ. Sofia, Fac. Physique, 69 92 (1978-9)
5. A. Shivarova, E. Tatarova, T. Stoychev and K. Ivanova, Proc. Int. Conf. Plasma Phys., Lausanne, vol. 1, p. 31
6. N. N. Bogoliubov and Y. A. Mitropolsky 'Asymptotic Methods in the Theory of Non-Linear Oscillations', Gordon&Breach Sci. Publ., N. Y. (1961)

[&] Permanent Address: Faculty of Physics, Sofia University, 1126 Sofia, Bulgaria

COUPLED NONLINEAR ELECTRON–PLASMA AND ION–ACOUSTIC WAVES IN A THIN PLASMA LAYER

T. Vodenicharova
Department of Physics, Institute for Foreign Students, BG-1111 Sofia, Bulgaria

I. Zhelyazkov
Faculty of Physics, Sofia University
BG-1126 Sofia, Bulgaria

M. Y. Yu
Institut für Theoretische Physik, Ruhr-Universität Bochum, D-4630 Bochum 1, F. R. Germany

In a recent paper[1] we have shown that the propagation of finite-amplitude high frequency antisymmetric TM surface waves on a thin unmagnetized plasma layer of thickness 2a is governed by the following evolution equation

$$i\left(\partial_\tau \tilde{E}_x + \frac{\partial\omega}{\partial k_z}\partial_\zeta \tilde{E}_x\right) + \frac{1}{2}\frac{\partial^2\omega}{\partial k_z^2}\partial_{\zeta\zeta}\tilde{E}_x - \frac{\partial\omega}{\partial\tilde{n}}\tilde{n}\tilde{E}_x = 0 \tag{1}$$

where $\tilde{E}_x(\zeta,\tau)$ is the slowly varying wave amplitude $\{E_x(z,t) = \tilde{E}_x(\zeta,\tau)\exp i(k_{zo}z - \omega_o t)\}$ and $\tilde{n} = \delta n_e/n_o$. The expressions for the nonlinear frequency shift, the wave group velocity, and the group velocity dispersion calculated from the wave dispersion relation

$$k_z a = (-\varepsilon_p/\sqrt{2})(1+X)^{1/2} \tag{2}$$

[with $X = (1 - 4\sigma^2/\varepsilon_p^2)^{1/2}$, $\varepsilon_p = 1 - \omega_p^2/\omega^2$] are

$$\frac{\partial\omega}{\partial\tilde{n}} = \omega_o/2, \quad \frac{\partial\omega}{\partial k_z} \equiv v_g = -\sqrt{2}\,a\,\frac{\omega_o^3}{\omega_{po}^2}X/(1+X)^{3/2}$$
$$\frac{\partial^2\omega}{\partial k_z^2} = -\frac{2a^2\omega_o^3}{\varepsilon_p\omega_{po}^2}\frac{X}{(1+X)^3}\left\{3\left(1-\frac{\omega_o^2}{\omega_{po}^2}X\right) - 2\left(\frac{1}{X} - X\right)\right\} \tag{3}$$

where $\omega_p^2 = \omega_{po}^2(1+\tilde{n})$. We note that in the range of backward wave propagation[1] $\partial^2\omega/\partial k_z^2 \equiv \partial v_g/\partial k_z$ is positive.

The wave electric field envelope is coupled to the nonlinear plasma response via the ponderomotive force and by using a standard procedure[2] we obtain

$$\delta n_e = n_o\left\{\exp\left(\frac{e\phi}{T_e} - \frac{\omega_{po}^2}{\omega_0^2}\frac{|\vec{E}|^2}{16\pi n_o T_e}\right) - 1\right\} \tag{4}$$

where ϕ is the ambipolar potential. We remember that because of their large mobility electrons react to the forces via a Boltzmann distribution. After normalizing $e\phi$ with respect to T_e and $|\vec{E}|$ to $(4\pi n_o T_e)^{1/2}$ we finally have

$$\tilde{n} = \exp\left(\Phi - \frac{1}{4}\frac{\omega_{po}^2}{\omega_0^2}\bar{E}^2\right) - 1 \tag{4a}$$

where $\Phi = e\phi/T_e$, $\bar{E}^2 = |\vec{E}|^2/4\pi n_o T_e$. By introducing dimenssionless time $\bar{\tau} = \tau\omega_{pi}$ and coordinate $\bar{\zeta} = \zeta/\lambda_D$ (where ω_{pi} is the ion plasma frequency and λ_D the Debye length) and by letting $\bar{v}_g = v_g/c_s$, $\bar{k}_z = k_z\lambda_D$ Eq. (1) becomes

$$i\epsilon\left(\partial_{\bar{\tau}}\tilde{\bar{E}}_x + \bar{v}_g\partial_{\bar{\zeta}}\tilde{\bar{E}}_x\right) + \frac{1}{2}B\partial_{\bar{\zeta}\bar{\zeta}}\tilde{\bar{E}}_x - \frac{\bar{\omega}_o}{2}\tilde{n}\tilde{\bar{E}}_x = 0 \tag{1a}$$

where

$$B = -\frac{2\bar{a}^2\bar{\omega}_o^3}{\varepsilon_p}\frac{X}{(1+X)^3}\left\{3\left(1-\bar{\omega}_o^{2}X\right) - 2\left(\frac{1}{X} - X\right)\right\}$$

$\bar{a} = a/\lambda_D$, $\bar{\omega}_o = \omega_o/\omega_{po}$, $\epsilon = (m_e/m_i)^{1/2}$. We look for a stationary solution for $\tilde{\bar{E}}_x(\zeta,\tau)$ in the form

$$\tilde{\bar{E}}_x(\bar{\zeta},\bar{\tau}) = E(\bar{\zeta} - M\bar{\tau})\exp\{i[Z(\bar{\zeta}) + T(\bar{\tau})]\} \tag{5}$$

where M is the Mach number which gives the fraction of the soliton velocity in terms of ion sound speed c_s. If one substitutes expression (5) in Eq. (1a) one obtains (with $\bar{\zeta} - M\bar{\tau} \equiv \xi$)

$$A\frac{d^2E}{d\xi^2} = (\lambda - d)E + Ed\exp\left(\Phi - \frac{1}{4\bar{\omega}_o^2}E^2\right) \tag{6}$$

with $A = B/2$, $d = \bar{\omega}_o/2$, $\lambda = \Delta + (\epsilon^2/B)\left[(M - \bar{v}_g)\bar{v}_g + \frac{1}{2}(M - \bar{v}_g)^2\right]$. Here $\Delta = \epsilon dT/d\bar{\tau}$ and $dT/d\bar{\tau}$ is the normilized frequency shift. Following Rao & Varma[3] we find for the ambipolar potential Φ the equation

$$\frac{d^2\Phi}{d\xi^2} = -M(M^2 - 2\Phi)^{-1/2} + \exp\left(\Phi - \frac{1}{4\bar{\omega}_o^2}E^2\right) \tag{7}$$

Eqs. (6) and (7) thus constitute a coupled set of nonlinear equations for the propagation of stationary solutions for E and Φ in the entire range of Mach numbers, $0 < M < 1$.

For solving Eqs. (6) and (7) we use the method developed in Ref. 3, namely under the boundary conditions

$$E,\ \Phi,\ \frac{dE}{d\xi},\ \frac{d\Phi}{d\xi} \longrightarrow 0 \quad \text{as } |\xi| \longrightarrow \infty \tag{9}$$

we seek the solution for $\Psi \equiv (1/4\bar{\omega}_0^2)E^2$ in the form

$$\Psi = \sum_{n=1}^{\infty} b_n \theta^n \tag{10}$$

where $\theta = \Phi/M^2$. Finally we get

$$\theta(\xi) = \frac{\beta_1 \beta_2 \operatorname{sech}^2 k\xi}{\beta_1 - \beta_2 \tanh^2 k\xi} \tag{11}$$

where

$$\beta_{1,2} = [-2\alpha_2 \mp (4\alpha^2 - 18\alpha_1\alpha_3)^{1/2}]/3\alpha_3, \qquad k = \sqrt{\beta_1\beta_2}/2M$$

$$\alpha_1 = -1-\alpha, \quad \alpha_2 = -\frac{3}{2} - b_2 + \frac{\alpha^2}{2}, \quad \alpha_3 = -\frac{5}{2} - b_3 + \alpha b_2 - \frac{\alpha^3}{6}$$

The corresponding solution for E^2 is

$$E^2 = 4\bar{\omega}_0^2 (b_1\theta + b_2\theta^2 + b_3\theta^3) \tag{12}$$

Here

$$b_1 = M^2 - 1 - 2M\left(\frac{P}{V} + \frac{\lambda}{A}\right), \quad \text{where } P = V(\lambda - d) + 1,\ V = 1/d$$

$$b_2 = F_2/F_1, \quad b_3 = F_4/F_3, \quad \alpha = b_1 - M^2$$

$$F_1 = VA\left[b_1 + 2M^2\left(\frac{P}{V} + \frac{\lambda}{A}\right) + 8M^2\lambda\frac{V}{A} + 6(1 - M^2 + b_1)\right]$$

$$F_2 = -\frac{1}{2}VAb_1(3-\alpha) + AM^2(\alpha^2 - M^2) + 2dV\alpha b_1 M^2 + 2AM^4(1+\alpha) + 4\lambda M^6$$

$$F_3 = -10M^2Ab_1P + 9b_1VA(1+\alpha) + VAb_1^2 + 14M^2Vb_1\lambda$$

$$F_4 = -8AM^2b_2\left(-\frac{M^2}{2} + \frac{\alpha^2}{2} - 2P\right) - VA\left[b_1\left(\frac{5}{2} - b_2\alpha + \frac{\alpha^3}{6}\right) + 6b_1b_2\left(\frac{3}{2} - \frac{\alpha^2}{2}\right.\right.$$
$$\left.\left.+ b_2\right) + 12b_2^2(1+\alpha)\right] + 2AM^2\left[b_1\left(-\frac{M^2}{2} + b_2\alpha - \frac{\alpha^3}{6}\right) + 4b_2\left(-\frac{M^2}{2} - Vb_2\lambda + \frac{\alpha^2}{2}\right)\right.$$
$$\left. - 4Vb_2^2\lambda/A\right] - 2AM^2\left\{b_1\left[\frac{V}{A}b_1 d\left(\frac{\alpha^2}{2} - b_2\right) - b_2\alpha\right] + 4b_2\frac{V}{A}(-b_1 d\alpha + b_2\lambda)\right.$$
$$\left. + 4b_2\lambda\frac{V}{A}\right\} + 2AM^4\left[b_1\left(\frac{3}{2} + b_2 - \frac{\alpha}{2}\right) + 3b_2(1+\alpha)\right] + 4M^6(2b_2\lambda - b_1\alpha).$$

The numerical investigation of expression (12) with $\sigma = 0.1$ and $k_z a = 0.48$ ($\bar{\omega}_0 = 0.82$) -- in the linear-ion-response limit at the same conditions single-hump solitons exist -- leads to the following conclusions: (i) For $\Delta \leqslant 0.014$ there occur three

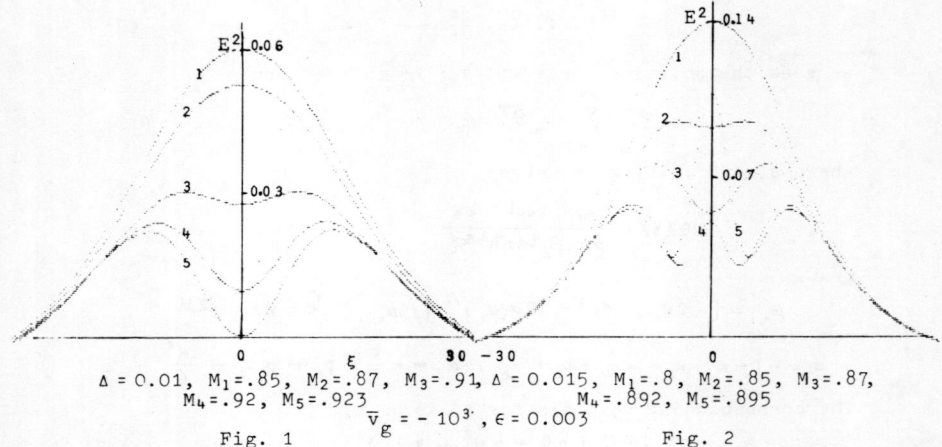

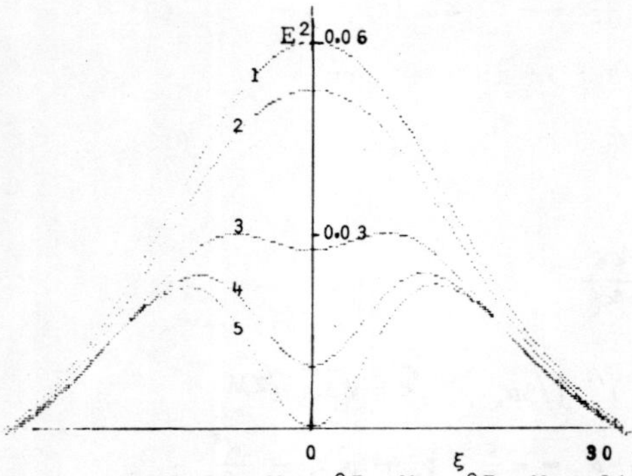

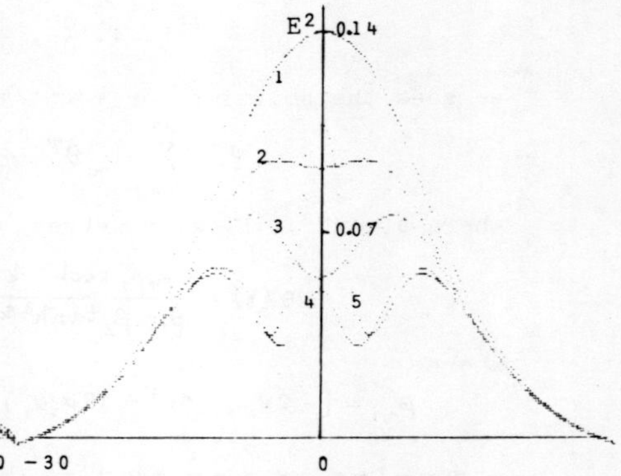

$\Delta = 0.01$, $M_1 = .85$, $M_2 = .87$, $M_3 = .91$, $M_4 = .92$, $M_5 = .923$

Fig. 1

$\Delta = 0.015$, $M_1 = .8$, $M_2 = .85$, $M_3 = .87$, $M_4 = .892$, $M_5 = .895$

Fig. 2

$\overline{v}_g = -10^3$, $\epsilon = 0.003$

types of solitons, namely single-hump, double-hump and Nishikawa solitons[4] (Fig. 1). (ii) For $\Delta > 0.014$ there exists a very narrow region of Δ, where in the place of Nishikawa solitons at large values of M triple-hump solitons appear (Fig. 2). (iii) The main difference between our solutions (11) and (12) and those of Rao & Varma[4] (expressions (49) and (48) in their paper) is that we retain both α_3 and b_3. Nevertheless $b_3\theta^3$ in (12) is negligibly small, the 'role' of b_3 in the expression for α_3 is important. The question for the contributions of the other terms, respectively $b_4, b_5, \ldots$, in expansion (10) is still open although one may expect that they might generate multi-hump solitons. A clear picture for the transition from standing ($M \to 0$) to sonic ($M \to 1$) surface-wave solitons can be obtained via a numerical solving of Eqs. (6) and (7) which is in progress.

Acknowledgment: This research was partially supported by the Sonderforschungsbereich 162 Plasmaphysik Bochum/Jülich.

References

1. I. Zhelyazkov, T. Vodenicharova, and M.Y. Yu, in Proc. of the 1984 ICCP (Lausanne) p. 46. 2. V.E. Zakharov, ZhETF 62, 1745 (1972). 3. N.N. Rao and R.K. Varma, J. Plasma Phys. 27, 95 (1982). 4. K. Nishikawa, H. Hojo, K. Mima, and H. Ikezi, Phys. Rev. Lett. 33, 148 (1974).

INTERACTION OF GAS WITH METAL CLUSTER SURFACE IN THE HIGH-TEMPERATURE PHASE OF METAL VAPORS

by

S.J. WANG
Oberflachentechnologie, Portnerskamp 2,
Postfach 1160, D-4270 Dorsten

1. In order to understand catalytic processes during the vapor deposition fundamental investigations of gas metal clusters reaction seem to be important. Especially investigations of small clusters in the high-temperature phase of metal vapors are essential for the future development of PVD and CVD processes.

2. These phenomena may also occur in solid materials subjected to intense radiation.

3. Results from theoretical wavemechanical calcu ations will be demonstrated. For example: Hadrogen on Li-clusters,

   Co, O_2 on Ni, W and Cu clusters.

4. Results from experimental gas metal reaction kinetics will be presented.

5. Industrial needs and application of theoretical and experimental results in the industry.

Surface Wave Effects in RF Pumped Lasers

R. Waynant, W.M. Bollen [a] and C.P. Christensen [b]
Naval Research Laboratory, Washington, DC 20375

Radio frequency excitation was used for the first gas laser (He-Ne) by Javan, Bennett and Herriot[1], but it fell from favor with laser designers due to the lower cost and simplicity of dc circuitry. Recently rf pumping has made a comeback in the infrared. Solid state power supplies and waveguide laser construction have simplified rf design to the point where rf excitation - and its lack of contaminating electrodes in contact with the gas - can compete with dc excitation for long operating lifetime. Long life is important in the ultraviolet where rare gas halide (RGH) lasers have made a large impact. RF pumping of low power RGH oscillators is attractive since the halogen gas mixtures used in RGH lasers are extremely corrosive. Some success has been obtained[2,3] by pulsed rf excitation of xenon fluoride (XeF) and xenon chloride (XeCl). Already these rf discharge systems compare favorably in terms of volume efficiency to the dc pulsed systems. In addition, the pulse widths obtained by rf pumping can be much longer than those typically obtained, in a dc avalanche-type discharge - 320 ns pulses have been obtained, for example, from XeCl.[4] It is thought, however, that the performance of these rf pumped lasers is severely degraded by the inadvertent generation of surface waves.

In rf laser excitation systems, the rf fields must be closely coupled to the laser gas mixture. This is done by placing a dielectric tube (sapphire or quartz) into a waveguide section or a resonant cavity into which the rf energy flows. Observations of the emission from a number of experimental systems has uncovered a fundamental difficulty. The observation was the collapse of the initially uniform discharge leaving emission only from the region near the walls of the dielectric containing the gas. This collapse can be seen in fast photography (Fig 1) and from a radially scanning apparatus which shows that the collapse occurred regardless of the gas (See Fig 2 and 3).

a) Mission Research Corporation, Alexandria, VA
b) Potomac Photonics, Inc., Alexandria, VA

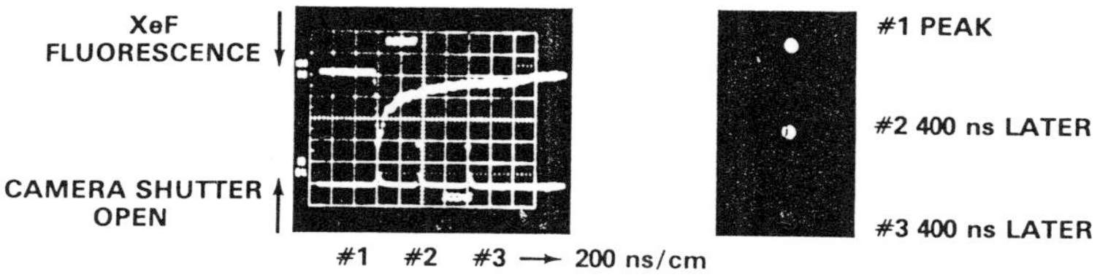

Fig 1. Three fast-frame photos looking down the discharge axis.

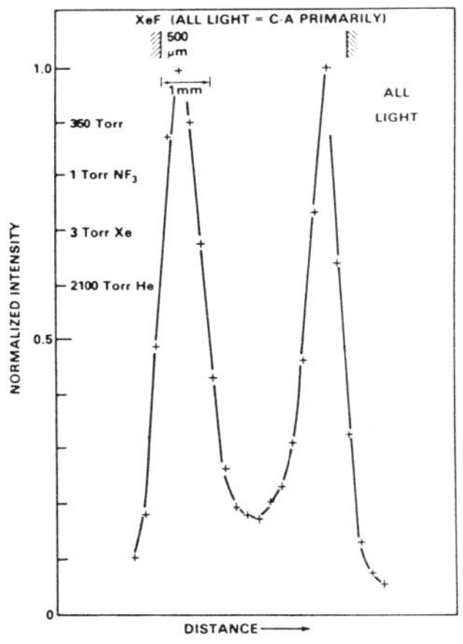

Fig 2. Radial profile of emission from rare-gas halide mixture.

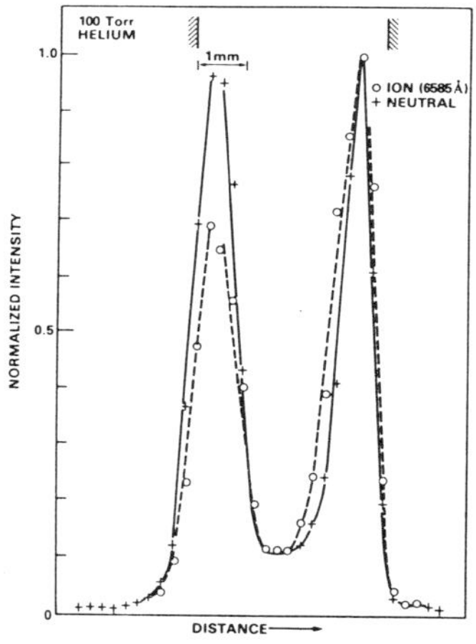

Fig 3. Radial profile of emission from ion and neutral helium lines

This collapse to the wall is attributed to the cut-off of normal waveguide modes by the high electron density of the plasma and to the subsequent establishment of surface-wave propagation at the dielectric/plasma interface. The existence of surface waves was confirmed by the detection of radiation propagating from the waveguide structure using a microwave receiver horn. Laser emission terminated with the onset of the collapse since resonant cavity modes are fed from the center of the discharge. Since the establishment of high electron density conditions for cut-off takes place at a time dependent on excitation power, it makes sense that longer laser pulse widths were observed at near-threshold power densities.

Since the generation of surface waves has been established, it remains to decide how best to either eliminate them or use them to advantage. The question of eliminating them is difficult since intense pumping is necessary for uv lasers. Use of tubes with multiple capillary tubes increases the surface area and seems to improve the surface wave coupling to the plasma. A second approach might be to change the electron density distribution by using an axial magnetic field. This has been shown to work at low pressures by Moisan.[5]

On the other hand if a way is found to make a uniform plasma that is fed by a surface wave, then surface waves can be a versatile way to excite lasers. Moisan[6] has demonstrated a He-Ne laser excited by surface waves. At low pressures the axial distribution shows maximum excitation in the center of the tube, but at pressures as high as a few Torr some tendency to peak near the walls has been seen. At NRL surface wave plasmas have been excited using a pulsed "Surfatron" launcher (Fig 4) at 10 GHz using pressures comparable to those in rare gas halide lasers. Under these conditions it would appear that there is a strong tendency for the plasma to peak near the walls just as in the waveguide method of excitation. Surfatron excitation is inherently simpler and more convenient to use than our previous methods of placing a tube in a waveguide or cavity. It also will allow the freedom of easily placing an axial magnetic field around the tube to try to change the electron distribution.

Fig 4. Surfatron launcher for the investigation of surface wave laser excitation.

In summary, the excitation of surface waves causes a problem during rf excitation of atmospheric pressure rare-gas halide lasers. A better understanding of surface waves may allow techniques to be developed that will overcome the non-uniform electron distribution. On the other hand it appears that at low pressures the non-uniform electron distributions are not detrimental and that rf generated surface waves can be conveniently used to excite low pressure neutral gas and ionic lasers.

References

1) A. Javan, W.R. Bennett Jr., and D.R. Herriott, Phys. Rev. Lett. 6, 106 (1961).

2) P.J.K. Wisoff, A.J. Mendelsohn, S.E. Harris and J.F. Young, IEEE J. Quantum Electron. QE-18, 1839 (1982).

3) C.P. Christensen and R.W. Waynant, Appl. Phys. Lett. 41, 794 (1982).

4) C.P. Christensen, R.W. Waynant and B.J. Feldman, Appl. PHys. Lett. 46, 321 (1985).

5) M. Moisan, C.M. Ferreira, Y. Hajaoui, K. Henry, J. Hubert, R. Pantel, A. Ricard and Z. Zakrzewski, Rev. Phys. Appl. 17, 707 (1982).

6) C. Moutoulas, M. Moisan, L. Bertrand, J. Hubert, J.L. Lachambre, and A. Richard, Appl. PHys. Lett. 46, 323 (1985). Also see C. Moutoulas, M. Moisan and Z. Zakrzewski, Proc. Conf. on Lasers and Electro-optics, p. 114, 1985.

Effect on the antenna admittance of an ion sheath around a planar or cylindrical antenna immersed in a cold plasma.

M.P.H. Weenink
Department of Electrical Engineering
Eindhoven University of Technology
5600 MB Eindhoven, The Netherlands

Abstract: The admittance of a plane or cylindrical antenna, immersed in a collisionless unmagnetized plasma, is calculated as a function of frequency. The effect of an ion sheath is accounted for. Various sheath models are considered with respect to their influence on the antenna performance. The calculations are based on full wave solutions; resonant energy absorption occurs for frequencies less than the plasma frequency for smooth sheath profiles.

Introduction: We consider first a plane plate antenna with an infinitesimally thin gap (along the z-axis of a cartesian coordinate system), across which an alternating voltage $V\exp i\omega t$ is applied. The y-axis lies in the plane of the antenna. In the absence of a sheath no surface waves are excited and the radiation into the plasma is either absent, for $\omega < \omega_p$, or isotropic in all directions, for $\omega > \omega_p$. The antenna admittance Y for $\omega < \omega_p$ is zero for this case. Y is defined as the radiated power per meter gaplength divided by V^2. The simplest sheath model is that of a vacuum layer (thickness d). In ref. [2] radiation patterns, including surface wave excitations are given for various values of ω, ω_p and d. In this model no resonant energy absorption takes place.

The next, somewhat more realistic, sheath model consists of a linearly increasing electron density profile up to d and a homogeneous plasma for $x > d$. This model has been considered in ref. [1]. There resonant energy absorption takes place if $\omega < \omega_p(\infty)$. The admittance has been calculated for $\omega < \omega_p(\infty)$ based on a full wave solution of Maxwell's equations for $\omega d/c << 1$, i.e. for sheath thicknesses that are relevant for ionospheric conditions. In this model there is a discontinuity in the slope of the density profile at x = d. It is well known that a discontinuity leads to reflections. In order to reduce the effect of such an unrealistic discontinuity a smooth profile was chosen with two free parameters. $\omega_p^2(x) = \omega_p^2(\infty)(1 - x_o/x)$. This profile was only used for a cylindrical antenna ($x \to r$), ref. [3]. For this case the admittance has been calculated for all frequencies, with the antenna radius r_o and $\omega_p(\infty)$as free parameters.

The shortcoming of the theories mentioned above, for $\omega < \omega_p$ are the blowing up of the electric fields at the resonance. This necessitates a nonlinear treatment if one prefers to adhere to a cold plasma model. It is however more realistic to invoke either collisions to smear out the resonance, or to take into account temperature effects, such as partial mode conversion. Both effects are treated in ref. [4], [5].

<u>Antenne admittance</u>. In absence of a sheath (plasma homogeneous up to the antenna the admittance Y is given by

$$Y_{hom} = \frac{\omega}{c} \frac{\varepsilon_o}{\mu_o} \varepsilon_r \text{ for } \omega > \omega_p, \quad Y_{hom} = 0 \text{ for } \omega < \omega_p, \tag{1}$$

where ε_r is the relative permittivity ($= 1 - \omega_p^2/\omega^2$).
The admittance of an antenna surrounded by a vacuum sheath ref. [2] is, in the absence of surface waves, given by

$$Y = Y_{hom} \cdot \frac{2}{\pi} \int_o^{\pi/2} \frac{d\alpha}{|\Delta(\alpha)|^2}, \tag{2}$$

$$\Delta(\alpha) = \cos\left(\frac{\omega d}{c} \sqrt{1 - \varepsilon_r \sin^2\alpha}\right) - i\varepsilon_r \frac{\sqrt{1-\varepsilon_r \sin^2\alpha}}{\cos\alpha} \sin\left(\frac{\omega d}{c} \sqrt{1 - \varepsilon_r \sin^2\alpha}\right). \tag{3}$$

The zero's of $\Delta(\alpha)$ represent surface waves. If $d \to 0$ $\Delta(\alpha) \to 1$, i.e. isotropic radiation, i.e. energy flux the same in all directions. As an example we give in fig. 1 the radiation pattern for $\omega c/d = 40$ and $\omega = 10\,\omega_p$.

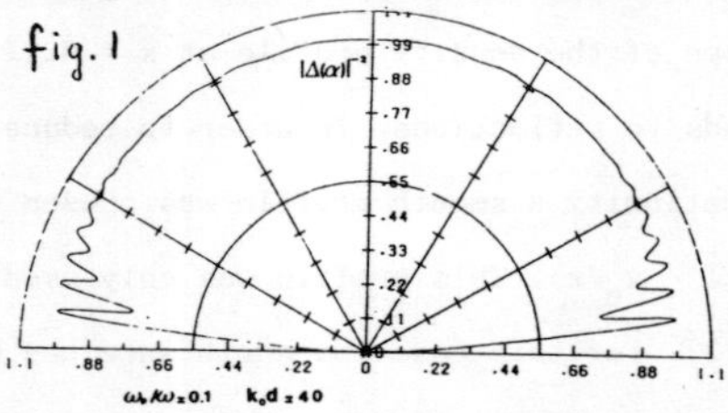

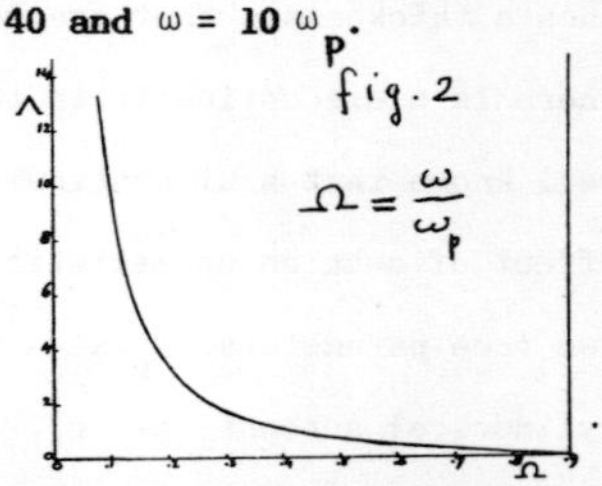

For $\omega < \omega_p$ only surface waves are excited. The contribution of surface waves is given in ref. [2], and derived in ref. [6]. In ref. [1] the admittance has been calculated for $\omega < \omega_p(\infty)$ and the linear sheath profile, by means of a full wave solution (see fig. 2).

In figure 3 the admittance for a smooth density profile is shown as a function of frequency for five different values of $\omega_p(\infty)$.

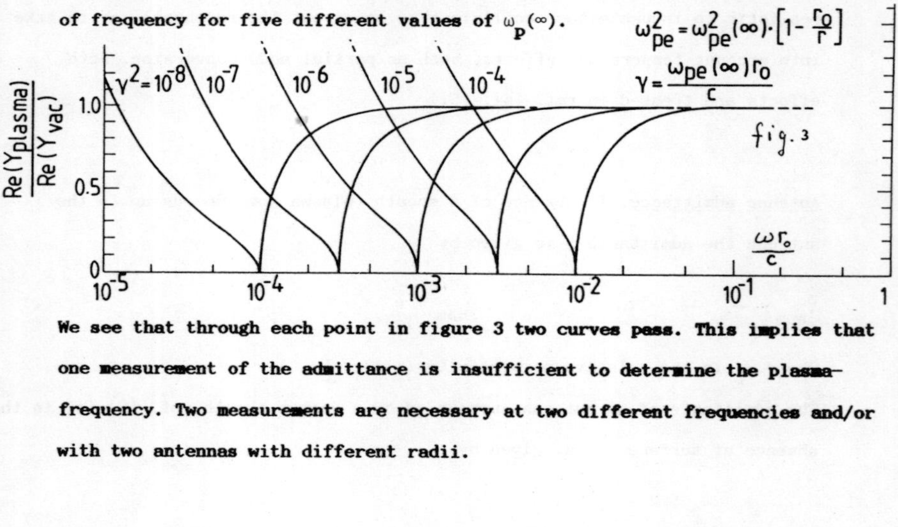

We see that through each point in figure 3 two curves pass. This implies that one measurement of the admittance is insufficient to determine the plasma-frequency. Two measurements are necessary at two different frequencies and/or with two antennas with different radii.

References:

[1] M.P.H. Weenink, Resonant energy absorption in the ion sheath of a plate antenna, Radio Sci., 17, 411-419, 1982.

[2] P.A. Beeckman, M.P.H. Weenink, Radiation properties of a vacuum insulated, infinite, flat, slotted plate antenna in a cold, isotropic, homogeneous plasma, Radio Sci., 18, 651-663, 1983.

[3] L.P.J. Kamp, M.P.H. Weenink, H.J.L. Hagebeuk, Admittance of a cylindrical antenna coated with an inhomogeneous cold plasma sheath with a resonance, Accepted for publication in Radio Sci. 1985.

[4] L.P.J. Kamp, M.P.H. Weenink, Propagation of electromagnetic waves in a planarly stratified, isotropic warm plasma with a resonance layer, Physica 122 B+C, 89-112, 1983.

[5] L.P.J. Kamp, Linear and nonlinear wave propagation and mode conversion in an inhomogeneous unmagnetized plasma with a resonance, Thesis Eindhoven University of Technology, Eindhoven, the Netherlands, 271 p., 1984.

[6] P.A. Beeckman, M.Sc. thesis, ET-14-81, Eindhoven University of Technology, Eindhoven, the Netherlands, 1981.

RAYLEIGH WAVES IN THERMOELASTICITY WITH RELAXATION TIMES

R. Wojnar, IPPT of Polish Academy of Sciences, ZTK
ul. Świętokrzyska 21, 00-049 Warszawa, Poland

1. Introduction

To remove the known paradox of classical thermoelasticity that a thermoelastic disturbance propagates with an infinite speed, new thermoelasticity theories have been proposed in the literature. One of them, called the thermoelasticity with one relaxation time (i), was proposed by Kaliski[1] and Lord and Shulman[2]. Another, called the thermoelasticity with two relaxation times, (ii), was given by Green and Lindsay[3]. Rayleigh waves in a thermoelastic semi-space with one relaxation time were investigated by Nayfeh and Nemat-Nasser[4]. In the present article we study the surface waves in a thermoelastic semi-space made of Green-Lindsay material. We assume that the body is homogeneous and isotropic and obtain a dispersion relation involving frequency and length of the wave vector, the thermoelastic coupling constant ε and two relaxation times τ_o and τ_1. Particular cases are investigated in some detail.

2. Basic relations

For a linear isotropic homogeneous thermoelastic body and in the absence of body force and heat supply fields, the basic equations of both mentioned theories are[5] :

$$\begin{aligned} &E_{ij} = \tfrac{1}{2}\left(u_{i,j} + u_{j,i}\right); \qquad S_{ij,j} = \rho\, \ddot{u}_i \\ &S_{ij} = 2\,\mu\, E_{ij} + \lambda\, E_{kk}\, \delta_{ij} - \gamma\left(1 + t_1\, \partial/\partial t\right) \Theta\, \delta_{ij} \\ &-q_{i,i} = C_E\left(1 + t_o\, \partial/\partial t\right) \dot{\Theta} + \gamma\, \Theta_o\, \dot{E}_{kk} \\ &\left(1 + t^{*}\, \partial/\partial t\right) q_i = -\,K\, \Theta_{,i} \end{aligned} \tag{2.1}$$

Here u_i, E_{ij}, S_{ij}, q_i and Θ, in this order, represent the components of the displacement, deformation, stress, heat flux and temperature above a reference state Θ_o; ρ, λ and μ, C_E and K are the density of medium, Lamé moduli, specific heat for zero deformation and coefficient of heat conductivity, respectively; $\gamma = (3\lambda + 2\mu)\alpha_o$, where α_o is the coefficient of thermal expansion; moreover, t^*, t_o and t_1 are the relaxation times. If

$$(2.2) \qquad t_o = t_1 = 0 \qquad \text{and} \qquad t^* > 0$$

the system (2.1) represent (i) theory; if

$$(2.3) \qquad 0 < t_o \leq t_1 \qquad \text{and} \qquad t^* = 0$$

we deal with (ii) theory.

By eliminating E_{ij}, S_{ij} and q_i from (2.1) we arrive at the displacement-temperature equations of thermoelasticity with two relaxation times:

$$(2.4) \qquad \begin{aligned} &\mu\, u_{i,jj} + (\lambda + \mu)\, u_{j,ji} - \gamma(\Theta + t_1 \dot{\Theta})_{,i} = \rho\, \ddot{u}_i \\ &K\, \Theta_{,ii} = C_E(\dot{\Theta} + t_o \ddot{\Theta}) + \gamma\, \Theta_o\, \dot{u}_{j,j} \end{aligned}$$

3. Travelling wave

Let in x_1-direction propagate a plane thermoelastic wave independent of x_3, $u_i = u_i(x_\alpha, t)$, $\Theta = \Theta(x_\alpha, t)$, $i=1,2,3$, $\alpha=1,2$. We search a solution of (2.4) in the form of a travelling wave (with transversaly damped amplitude)

$$(3.1) \qquad (u_1, u_2, \Theta) = (U, V, \vartheta) \exp\left[-A\, x_2 + i(k\, x_1 - \Omega t)\right]$$

The following set of equations for unknown constants U, V and ϑ is obtained from (2.4)

$$(3.2) \qquad \begin{aligned} &(\omega^2 + \beta^{-2}\alpha^2 - q^2)U - (1-\beta^{-2})\alpha i q\, V - m\, c_1^{-1} \varkappa i q(1 - i\omega\tau_1)\,\vartheta = 0 \\ &-(1-\beta^{-2})\alpha\, i q U + (\omega^2 + \alpha^2 - \beta^{-2} q^2)V + m\, c_1^{-1} \varkappa\, \alpha(1 - i\omega\tau_1)\,\vartheta = 0 \\ &-\eta\, \omega q\, U - \eta\, i\, \omega\, \alpha V + c_1^{-1}\left[\alpha^2 - q^2 + \omega(i + \tau_o \omega)\right]\vartheta = 0 \end{aligned}$$

Here

(3.3) $\omega = \Omega/\Omega^{*}$, $\alpha = Ac_1/\Omega^{*}$, $q = kc_1/\Omega^{*}$, $\varepsilon = m\eta\varkappa$, $\beta^2 = c_1^2 c_2^{-2}$

(3.4) $\tau_o = \Omega^{*} t_o$, $\tau_1 = \Omega^{*} t_1$

and $\Omega^{*} = c_1^2/\varkappa$ is Chadwick-Sneddon unit of frequency. Moreover

(3.5) $c_1^2 = (\lambda+2\mu)/\rho$, $c_2^2 = \mu/\rho$, $m = \gamma/(\lambda+2\mu)$, $\eta = \gamma\Theta_o/K$, $\varkappa = K/C_E$

The condition of existence of non-zero solution of (3.2) gives the following relations for eigenvalues α :

(3.6) $\alpha_1^2 = q^2 - \beta^2\omega^2$, $\alpha_2^2 + \alpha_3^2 = -P-Q-R$, $\alpha_2^2\alpha_3^2 = PQ - q^2 R$

where

(3.7) $P = \omega^2 - q^2$, $Q = -q^2 + \omega^2\tau_o + i\omega$, $R = \varepsilon\omega(1+\omega\tau_1)$

The system (3.2) yields the following equations for U, V, ϑ :

$$V^{(1)} = -\frac{q}{i\alpha_1}U^{(1)}, \qquad \vartheta^{(1)} = 0 \tag{3.8}$$

$$V^{(j)} = -\frac{\alpha_j}{iq}U^{(j)}, \quad \vartheta^{(j)} = \frac{\omega^2 + \alpha_j^2 - q^2}{m\varkappa iq(1-i\omega\tau_1)}c_1 U^{(j)}, \quad j=2,3$$

4.Surface waves

We assume that the semi-space $x_2 \geqslant 0$ is unloaded and a free exchange of heat takes place on its surface. Thus for x_3-independent fields, at $x_2=0$ the following boundary conditions hold true:

(4.1) $u_{1,2} + u_{2,1} = 0$, $\lambda u_{1,1} + (2\mu+\lambda)u_{2,2} - \gamma(\Theta + t_1\dot{\Theta}) = 0$, $u_{3,2} = 0$

(4.2) $\eta_1\Theta + \eta_2\,\partial\Theta/\partial x_2 = 0$ for constants $1 \geqslant \eta_1 \geqslant 0$, $\eta_2 \geqslant 0$

To obtain (4.1) the eq.$(2.1)_3$ was used. We substitute the general form of our travelling wave

$$(u_1, u_2, \Theta) = \sum_{j=1}^{3}(U^{(j)}, V^{(j)}, \vartheta^{(j)})\exp\left[-A_j x_2 + i(kx_1 - \Omega t)\right] \tag{4.3}$$

where $A_j = \alpha_j c_1/\varkappa$, into $(4.1)_{1,2}$ and (4.2) to obtain :

$$(\alpha_1^2+q^2)U^{(1)} + 2\alpha_1\alpha_2 U^{(2)} + 2\alpha_1\alpha_3 U^{(3)} = 0$$

$$2q^2 U^{(1)} + (2q^2 - \beta^2\omega^2)U^{(2)} + (2q^2 - \beta^2\omega^2)U^{(3)} = 0 \tag{4.4}$$

$$(\eta_o - \eta_2\alpha_2)(\omega^2 + \alpha_2^2 - q^2)U^{(2)} + (\eta_o - \eta_2\alpha_3)(\omega^2 + \alpha_3^2 - q^2)U^{(3)} = 0$$

Here $\eta_0 = \eta_1 \varkappa / c_1$ and eqs. (3.8) were used. The condition of existence of non-zero solution of (4.4) leads after some transformations to the desired dispersion relation

$$(4.5)\quad \left[q\, G\, \eta_2(\alpha_2^2+\alpha_3^2+\omega^2-q^2+\alpha_2\alpha_3) + \eta_0(q^2-\omega^2+\alpha_2\alpha_3)\right]^2 = (q\, G\, \eta_0 + \eta_2\alpha_2\alpha_3)^2(\alpha_2^2+\alpha_3^2+2\alpha_2\alpha_3)$$

where we have defined

$$(4.6)\qquad G = \frac{q}{\alpha_1}\left(\frac{\alpha_1^2+q^2}{2q^2}\right)^2$$

5. Discussion

For a limiting case $\varepsilon=0$, eq. (4.5) admits the solution

$$(5.1)\qquad G^2 = 1-\omega^2/q^2, \qquad q^2 = \omega^2\tau_0 + i\omega$$

Eq. $(5.1)_1$ is classical equation for Rayleigh wave speed, while $(5.1)_2$ is a dispersion relation for a thermal wave.

For the special case: $\eta_1=0$, $\eta_2=1$, eq.(4.5) reduces to

$$(5.2)\quad G^2\left[q^2-\omega^2(\tau_0+\varepsilon\tau_1) - i\omega(1+\varepsilon) + \alpha_2\alpha_3\right]^2 = (\alpha_2^2\alpha_3^2/q^2)\left[2q^2-\omega^2(1+\tau_0+\varepsilon\tau_1) - i\omega(1+\varepsilon) + 2\alpha_2\alpha_3\right]$$

If $\omega\to 0$, and $q\to 0$ in such a way that $\omega/q \to$ constant, then for every $\tau_1 \geqslant \tau_0 > 0$ from (5.2) we recover Lockett's[6] result

$$(5.3)\qquad G^2 = (1+\varepsilon-\omega^2/q^2)/(1+\varepsilon)$$

If $\tau_1=\tau_0=\tau$, eq.(4.9a) of ref.[4] is recovered from our (5.2). However the amplitudes (3.8) are not the same as in[4].

References

1. S. Kaliski, Bull. Acad. Pol. Sci. techn. 13 253 (1965)
2. H.W. Lord, Y.Shulman, J.Mech. Phys. Solids 15 299 (1967)
3. A.E. Green, K.A. Lindsay, J. Elasticity 2 1 (1972)
4. A. Nayfeh. S. Nemat-Nasser, Acta Mech. 12 53 (1971)
5. J. Ignaczak, The Shock and Vibration Digest 13 3 (1981)
6. F.J. Lockett, J. Mech. Phys. Solids 7 71 (1958)

LOCAL DENSITY OF ELECTRONIC STATES AT THE SURFACE

by

L. WOJTCZAK, S. ROMANOWSKI, W. STASIAK
Institute of Physics, University of Łodź, Łodź, Poland.

S. W. TEMKO
Moscow Geological Prospecting Institute, Moscow K-9, USSR

The local density of electronic states at the surface is considered and applied to investigate the photoemission and the reflectivity of light at the surface. As an example the calculations are given for the copper films with respect to various orientations of the crystallographic axes.

Another applications are connected with the description of the interfaces and their properties. The detailed calculations are performed for films of copper (100) covered by a thin layer of copper (110) and copper (111), respectively. The results are reported for different boundary conditions described by an effective parameter introduced at the surface and responsible for the surface deformations.

The method allows us to discuss the chemisorption of atoms, molecules and ions on the surface and also to consider the nature of chemisorption phenomena.

DIFFRACTION OF LIGHT AND LOCALIZED SURFACE MODES AT A DIELECTRIC RIDGE OR GROOVE

W. Zierau[a] and A. A. Maradudin[b]

[a]Inst. f. Theor. Physik II, Universität Münster, Domagkstr. 75, D4400 Münster, W. Germany

[b]Dept. of Physics, University of California, Irvine, Calif. 92717, USA

The influence of surface roughness on the excitation and propagation of surface waves has been studied from various points of view in the past (for a review see e.g.[1]). In this paper we are interested in the effect of a single structure on an otherwise flat surface. The profile to be considered will be either a ridge or a groove. Taking a dielectric material the question arises to what extent there will be a resonant excitation of surface polaritons by the scattering of light at such surface profile. A further aspect is the occurence of localized modes in the vicinity of such isolated surface structure.

In the following we restrict ourselves to a one-dimensional Gaussian profile and assume p-polarized light incident in the x_1x_3 -plane on a medium characterized by the dielectric constant $\varepsilon(\omega)$.

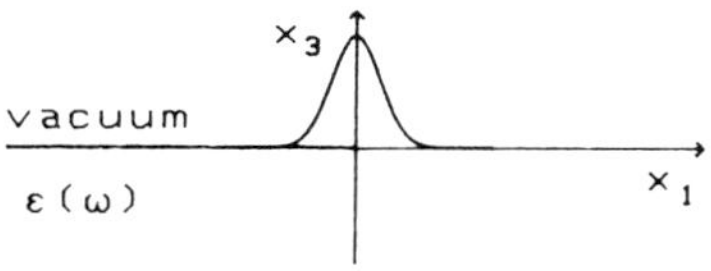

$$\zeta(x_1) = A \exp(-x_1^2/R^2)$$

Fig. 1 Surface profile. Ridge: $A>0$, Groove: $A<0$

Taking Maxwell's equations the solution in terms of the

magnetic field vector above the surface is given by

$$H_2^>(x_1 x_3 \omega) = e^{ikx_1 - i\alpha_o(k\omega)x_3} + \int \frac{dq}{2\pi} A(q|k)\, e^{iqx_1 + i\alpha_o(q\omega)x_3} \quad (1)$$

with

$$\alpha_o(q\omega) = (\omega^2/c^2 - q^2)^{1/2} \qquad q^2 < \omega^2/c^2$$
$$= i\,(q^2 - \omega^2/c^2)^{1/2} \qquad q^2 > \omega^2/c^2 \quad (2)$$

The coefficients $A(q|k)$ are determined from the boundary conditions at the surface. This is achieved by applying Rayleigh's hypothesis together with Green's theorem and the extinction theorem in a manner recently described in [2] . These calculations yield an integral equation:

$$\frac{J(\alpha(p\omega)+\alpha_o(k\omega)\,|\,p-k)}{\alpha(p\omega)+\alpha_o(k\omega)}\,(\alpha(p\omega)\alpha_o(p\omega) - pk)$$
$$-\int \frac{dq}{2\pi}\,\frac{J(\alpha(p\omega)-\alpha_o(q\omega)\,|\,p-q)}{\alpha(p\omega)-\alpha_o(q\omega)}\,(\alpha(p\omega)\alpha_o(q\omega)+qp)\,A(q|k) = 0 \quad (3)$$

with $\alpha(p\omega) = (\varepsilon(\omega)\omega^2/c^2 - p^2)^{1/2} \qquad \mathrm{Re}(\alpha)>0,\ \mathrm{Im}(\alpha)>0 \quad (4)$

and $J(\alpha|Q) = 2\pi\delta(Q) - i\,\alpha\, I(\alpha|Q)\ . \quad (5)$

The surface profile is contained in $I(\alpha|Q)$:

$$I(\alpha|Q) = \int dx_1\, e^{-iQx_1}\, \frac{1 - e^{-i\alpha\zeta(x_1)}}{i\alpha} \quad (6)$$

The inhomogeneous integral equation is solved using the ansatz:

$$A(p|k) = 2\pi\delta(p-k)\,R_o(k\omega) - 2i\,B(p|k) \quad (7)$$

with $R_o(k\omega)$ being the Fresnelcoefficient for the reflection at a planar surface. The efficiency of the diffuse scattering can be given in terms of the coefficients $B(p|k)$:

$$I(p|k) = \frac{4}{L_1}\frac{\alpha_o(p\omega)}{\alpha(k\omega)}\left|B(p|k)\right|^2 \quad (8)$$

L_1 is the length of the surface $x_3=0$ in x_1 direction.

Calculations have now been carried out for light of wavelength $\lambda = 4579$ Å incident on a Ag surface with $\varepsilon(\omega)=-7.5+0.24i$ [3]. A numerical solution of the integral equation (3) has to take into account the resonance from the surface polariton in the integrand. We therefore applied a Gauss-Legendre integration scheme with a variable number of nodes (600 points) and a finite cut-off. The integral in eq. (6) was solved by means of a series expansion where according to the ratio A/R up to 25 terms have to be taken. Results for the scattering coefficient are shown in Fig. 2.

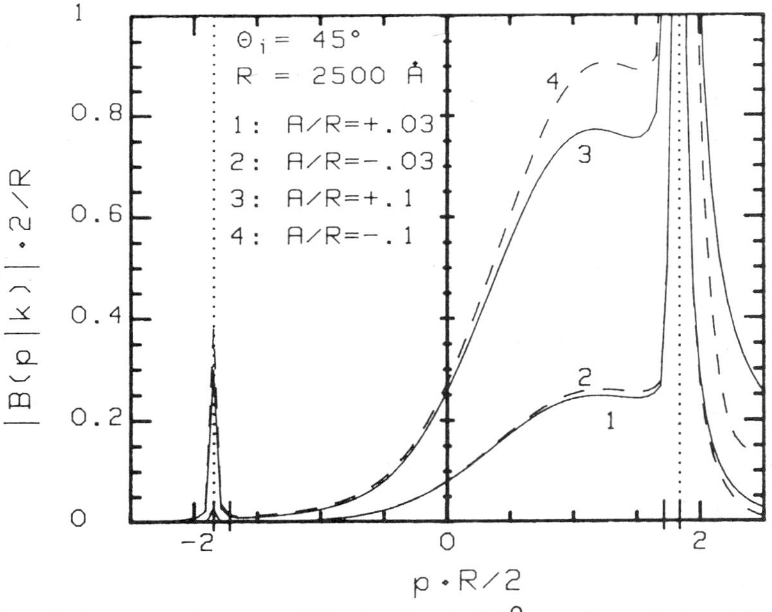

Fig. 2 Scattering coefficient at 45° angle of incidence.

The difference between a ridge (A>0) and a groove (A<0) becomes significant only for larger values of A/R. In either case the resonance from the excitation of the surface polariton is clearly displayed at $p=k_{sp}$.

We have also studied the corresponding homogeneous

problem (eq. (3)) for the existence of localized modes. As its solution requires the finding of zeros of complex determinants we solved the problem of a perfect conductor ($c \to \infty$) first. Taking the dielectric constant $\varepsilon(\omega)=1-\omega_p^2/\omega^2$ one obtains a series of modes converging against $\omega_p/\sqrt{2}$. In Fig. 3 the dependence of the first 3 modes on the magnitude of the surface profile is shown.

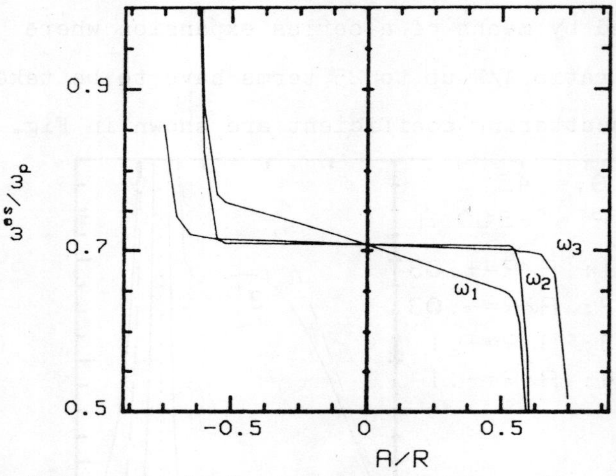

Fig. 3 Localized electrostatic modes ω^{es}.

The electromagnetic mode for the lowest frequency at A/R=0.03 was found to be ω/ω_p=0.637 with a vanishing imaginary part.

References

1. A. A. Maradudin, in Surface Polaritons, eds. V. M. Agranovich and D. L. Mills (North Holland, Amsterdam, 1982), p. 405.
2. G. Brown, V. Celli, M. Haller, A. A. Maradudin and A. Marvin, Phys.Rev. B31 4993 (1985) and Refs. therein.
3. P.W. Johnson and R.W. Christy, Phys.Rev. B6, 4370 (1972).

AUTHOR INDEX